THE PHYSICS OF MULTIPLY AND HIGHLY CHARGED IONS

The Physics of Multiply and Highly Charged Ions

Volume 2
Interactions with Matter

Edited by

FRED J. CURRELL
Department of Physics,
The Queen's University,
Belfast, Northern Ireland

SPRINGER-SCIENCE+BUSINESS MEDIA, B.V.

A C.I.P. Catalogue record for this book is available from the Library of Congress.

ISBN 978-90-481-6409-7 ISBN 978-94-017-0544-8 (eBook)
DOI 10.1007/978-94-017-0544-8

Printed on acid-free paper

Contents

Part II Interactions with Gaseous Targets

List of Figures

List of Tables

Preface

A fairly cavalier (but arguable) claim is that all of chemistry and a large portion of atomic physics is concerned with the behaviour of the 92 naturally occurring elements in each of 3 charge states (+1, 0, -1); 276 distinct species. The world of multiply and highly charged ions provides a further 4186 species for us to study. Over 15 times as many!

It is the nature of mankind in general to explore the unknown. This nature is particularly strong in physicists although this may not be readily apparent because the explorations undertaken by physicists occur in somewhat abstract 'spaces'. It is, then, no surprise that we have begun to explore the realm of multiply and highly charged ions. Over the past few decades, a consistently high quality body of work has emerged as the fruits of this exploration. This internationally based subject, pursued in universities and research laboratories worldwide, has expanded beyond its roots in atomic physics. We now see it embracing elements of surface science, nuclear physics and plasma physics as well as drawing on a wide range of technologies. It offers new tests of some of our most fundamental ideas in physics and simultaneously new medical cures, new ways of fabricating electronic gadgets, a major hope for clean sustainable energy and explanations for astrophysical phenomena. It is deeply fundamental and widely applicable.

The aim of this textbook and its companion volume is to present both a comprehensive introduction and an up to date review of this emerging body of work. To this end, some thirty scientists working at the forefront of this subject have taken some time away from the cliff-face to compose a set of self-contained but interlinked chapters.

The resultant volumes are intended for a wide range of interested readers with a consistent effort having been made to be instructive. To this end the volumes are suitable for anyone who has completed an undergraduate degree in physics or is in the final stage thereof, with supporting reading of the cited texts. To become a fully equipped researcher one has to make two transitions. Firstly one must be able to understand the forefront of knowledge in a particular subject area. Secondly one must become able to push that forefront back, revealing the unknown. The material in these volumes seeks to take the reader to that cutting edge and provide a glimpse what might lay ahead. Much of the material is then at this edge to help stimulate the first and hopefully the second transition.

So how should a new postgraduate working in this subject area approach the material presented herein? Two approaches recommend themselves, depending of the reader's personal preference. A 'top-down' approach would be to skim the chapters in the sources and applications sections of volume 1 before going to material of more specific interest. A 'bottom-up' approach would be to start with the chapter most closely associated with the reader's particular field of study and work outwards from there. The point here is that the textbook (like the subject matter itself) isn't linear. Comprehensive lists of contents figures and tables have been provided in addition to an index to help the reader to explore the volumes as best suits. The introduction also draws out some of the common threads running though this volume.

The more experienced practitioner, already established in the field, should also find this material of great interest. For this reader no advice is necessary as to how to approach the material, as a conceptual map will already be in place. This material will augment and reshape that conceptual map.

It only remains to thank the authors of the chapters for taking the time away from the cliff-face of discovery to write the following chapters. Special thanks also to Wendy Rutherford for helping typeset the volume and doing numerous other 'little jobs' that all add up to a great deal of help and finally to Liz, the brightest star in my cosmos and a truly special person.

This book is dedicated to the reader, especially if he will use the material presented in the following pages as stimulus to take our subject further.

Introduction

As the subtitle suggests, this volume deals with the processes which occur when multiply and highly charged ions interact with matter. In this context, the distinction between multiply charged and highly charged is purely a matter of degree with members of the research community tending to use the terms interchangeably. Matter is broadly interpreted as relatively complex systems containing bound electrons. Interactions with more fundamental particles are discussed in the accompanying volume, "Sources, Applications and Fundamental Processes".

As an atom is successively stripped of its electron (i.e. it becomes more highly charged), it becomes a system further out of equilibrium; it is more 'electron-starved' and hence acts as an ***electron scavenger***. For more highly charged ions, this electron scavenging nature becomes more apparent; the ion is generally able to transfer more electrons from nearby matter and do so more quickly. Indeed such highly charged ions are the most chemically reactive species known. This reactivity is most apparent when highly charged ions interact with electron–rich forms of matter. Once electrons are transferred, the system can stabilize in a wide variety of ways, as is discussed in this volume.

Perhaps the most electron-rich system is solid matter. The interface between the solid and the vacuum is a region particularly amenable to study. This is one of the reasons for the ongoing interest in interactions between highly charged ions and surfaces. In particular, when a highly charged ion approaches a surface slowly (i.e. it has low kinetic energy), the effects of its large potential energy become most apparent. This subject is dealt with in a tutorial manner in chapter 1, "Inelastic Interaction of Slow Ions with Clean Solid Surfaces". Not only does this chapter provide a grounding in the subject, it also contains a concise introduction to some of the relevant experimental techniques and tools. In particular a brief review of the ion sources and beam preparation systems used serves to make this volume self contained. Accordingly, consultation of this chapter is also recommended to readers interested in interactions between multiply charged ions and other forms of matter. Interested readers are also referred to the companion volume, "Sources, Applications and Fundamental Processes" for more complete discussions of the types of ion source available and details of their use.

Not only does chapter 1 discuss the apparatus and preparation of clean surfaces, it also provides an introduction to the underlying basic mechanisms involved in the interaction between multiply charged ions and surfaces. From the understanding of basic mechanisms, a number of current research topics are discussed, with particular reference to the various ***electron emission*** and ***sputtering*** mechanisms which can occur, i.e. the escape of electrons and atomic species respectively into the vacuum. This chapter also serves to set the scene for the next two chapters, "Interaction of Slow Highly Charged Ions with Surfaces" and "Interaction of Hollow Atoms with Surfaces". Both of these chapters deal with a particularly fascinating atomic physics species, the ***hollow atom***. Hollow atoms, as the name implies, have several electrons in higher lying orbitals, while vacancies exist in lower lying orbitals. Such systems can be formed when the rate of electron capture sufficiently exceeds the rate of decay to lower lying orbitals. Why then are the electrons captured into higher lying orbitals, rather than being predominantly captured into the lowest lying empty orbitals? The answer is given through a simple and yet powerful concept of the ***classical over barrier mechanism*** discussed in chapter 2. Essentially the potential formed as the incoming ion approaches the surface has a saddle point. Electrons cross this saddle point when they are energetically allowed to (i.e. quantum tunneling is neglected), to find themselves in high-lying Rydberg orbitals of the approaching ion. Much of the discussion of chapter 3 deals with how the 'flow of charge' can be modeled in such systems, particularly through consideration of the electron density. These model calculations of the cascade-like filling of hollow atoms are also used to explain potential energy transfer to produce Auger electrons and ***plasmons***, i.e. collective lattice excitations.

Chapter 4, "Interactions of Highly Charged Ions with C_{60} and Surfaces" forms a bridge between the most electron-rich targets considered (i.e. surfaces, as discussed above) and somewhat less electron-rich clusters, in particular, C_{60}. The interactions between this system and highly charged ions also serves to inform our understanding of ion-surface interactions since it is possible to distinguish between distant and close interactions through coincidence techniques. Although both type of interaction occur in an ion-surface interaction, it is not possible to make a clear distinction because both type of interaction occur for even the most glancing of collisions, since every ion trajectory eventually results in a close encounter. Again the formation of hollow atomic systems is discussed in this chapter, with comparisons being made between formation through interaction with surfaces and clusters.

Part II of this volume deals with the interaction between multiply charged ions and predominantly atomic targets although molecular and ionic targets are also discussed. A number of different experimental and theoretical techniques are described as collisions in various regimes are considered.

In chapters (5 – 10) collisions between slow multiply charged ions and quite simple, one or two electron targets (H, D, He, H_2) form the main focus of the discussion. These collision systems are highly relevant to the fusion and astrophysics communities (see chapter 4 of volume 1: "Highly Charged Ion Collision Processes in High Temperature Fusion Plasmas" for further details). However, this is not the only reason for considering such collision systems. Since relatively few electrons are involved, the natures of the component processes involved is most apparent. Hence, these chapters form a closely linked set through which the reader can gain a basic understanding of the processes involved these low energy collisions.

Chapter 5, "Photon Emission Spectroscopy of Electron Capture and Excitation by Multiply Charged Ions" introduces the capture process, again in terms of the classical over barrier model. The results of this model are related to experimental findings deduced through the measurement of visible and ultra-violet photons produced as the captured or excited electrons decay.

Chapters 6 and 7, "Low Energy Electron Capture Measurements using Merged Beams" and "Application of The Beam Guide Technique to Low Energy Collision Experiments" respectively, describe experimental techniques whereby particularly slow collisions can be studied. Two complimentary techniques are described (***merged*** and ***guided beams***) which allow one to make difficult measurements at such low collision energies. These chapters describe the techniques and findings in detail, including descriptions of some of the precautions which must be taken to ensure the measurements' validity. Interactions between targets such as H, D, He and H_2 and very slow multiply charged ions are used to illustrate the techniques discussed.

Once such low energy collisions are considered, one has to abandon simple classical approaches to a theoretical description, using either ***semi-classical*** or ***fully quantal treatments***. Chapter 8, "Theoretical Description of Low Energy Collisions: Close-coupling Semiclassical Treatments" describes a close-coupling semi-classical theoretical treatment applicable to atomic or molecular systems. In particular, collisions involving various multiply-charged ions and either H or H_2 are considered. Chapter 9, "Quantum Dynamics of Ion-Atom Collisions" outlines a fully quantum mechanical model capable of treating the dynamics of inelastic, rearrangement or ionization processes taking place in collisions involving multiply charged ions with atomic targets. In particular collisions with H or He are used to illustrate the methods discussed.

One issue which has often made clear comparison between theoretical and experimental results difficult has been the existence of an unknown fraction of metastable ions mixed in with ground-state ions extracted from ion sources used to make experimental measurements. The metastable ions of course have a different electron wavefunction and internal energy so the collision dynamics is different. 'Double Translational Energy Spectrometry' is an ingenious experimental technique able to facilitate clean experiments whereby either pure ground-state or metastable beams are used to initiate the collision. This technique is described in chapter 10, "State-selective Electron Capture by Translational Energy Spectrometry", in terms of the wider framework of translational energy spectrometry. Once again collisions involving various multiply-charged ions (ground and metastable states) and either H or H_2 are considered.

In chapter 11, "Ionization and Excitation of Atomic Li by Fast Ions", the interactions between much faster (nearly relativistic) Ar^{18+} ions and atomic Li are described. This target is a unique system in which to study ionization and excitation because of its tightly bound inner shell and weakly bound outer shell. Processes such as single-electron ejection and production of doubly vacant K-shell states are described with an informative analogy being made with interactions between photons and atomic targets.

The final chapter of this volume, "Ion–Ion Collisions", describes the types of interaction which can occur when two different ionic species interact. In particular, the discussion focuses on crossed-beam measurements with charge-exchange and ionization processes being diagnosed through the use of coincidence techniques. In addition to total cross-sections, angular differential cross-sections are considered since they provide stringent tests of existing theories on a fundamental level and more detailed insight into the collision process involved.

Collectively the chapters in this volume provide both a tutorial introduction and a comprehensive review of current theoretical and experimental techniques involved in in studying interactions between the most chemically reactive species known and various forms of matter and reviews results which illuminate our understanding of the kinds of interactions which occur.

I

INTERACTIONS WITH SURFACES AND CLUSTERS

Chapter 1

INELASTIC INTERACTION OF SLOW IONS WITH CLEAN SOLID SURFACES

Potential and kinetic electron emission, plasmon excitation, potential sputtering

HP. Winter and F. Aumayr
Institut für Allgemeine Physik
Vienna University of Technology, Austria
winter@iap.tuwein.ac.at

Abstract We present a tutorial treatment of the interaction of slow ions with clean solid surfaces, with special emphasis on experimental aspects. We start by describing the underlying basic mechanisms and some relevant applications, and then present the established experimental techniques for investigating different reactions, including production of slow singly and multiply charged ion beams and preparation of clean solid surfaces for pertinent studies. Following this outline, we discuss the current status of research in four selected areas of the low ion impact energy regime to which we have contributed and where in recent years important progress could be achieved: Electron emission due to ion potential energy (potential emission - PE), kinetic electron emission (KE) near the ion impact energy threshold, excitation of plasmons by the ion potential energy (potential excitation of plasmons - PEP), and sputtering of insulator surfaces related to the ion potential energy (potential sputtering - PS).

Keywords: ion beam impact, interactions with surfaces, potential electron emission, kinetic electron emission, plasmon excitation, potential sputtering

1. Introduction: Inelastic Slow Ion-surface Interaction - Basic Mechanisms, Characteristic Features and Typical Applications

If an energetic particle hits the surface of a solid, it can initiate a plethora of different processes (cf. fig. 1.1), depending on its kinetic and internal energy (excitation, ionisation, multiple-ionisation), the species, structure and status of

F.J. Currell (ed.), The Physics of Multiply and Highly Charged Ions, Vol. 2, 3-45.

the target surface, and the given scattering geometry. Most of these processes have some practical applications which either need to be taken into account in particular situations (e.g., for plasma-wall interaction or the energy- and charge balance for gaseous electronics), or can be utilized for analytical purposes (particle counting, investigation of the structure and electronic state of solid surfaces). In order to judge the relevance of particular processes and also to make good use of them, it is essential to understand their underlying physical mechanisms in sufficient detail, to which purpose this field has now been studied already for more than a century. However, only with the wider availability of well-characterised surfaces under ultra-high vacuum conditions we have acquired the experimental tools which are indispensable for meaningful research in this complex field.

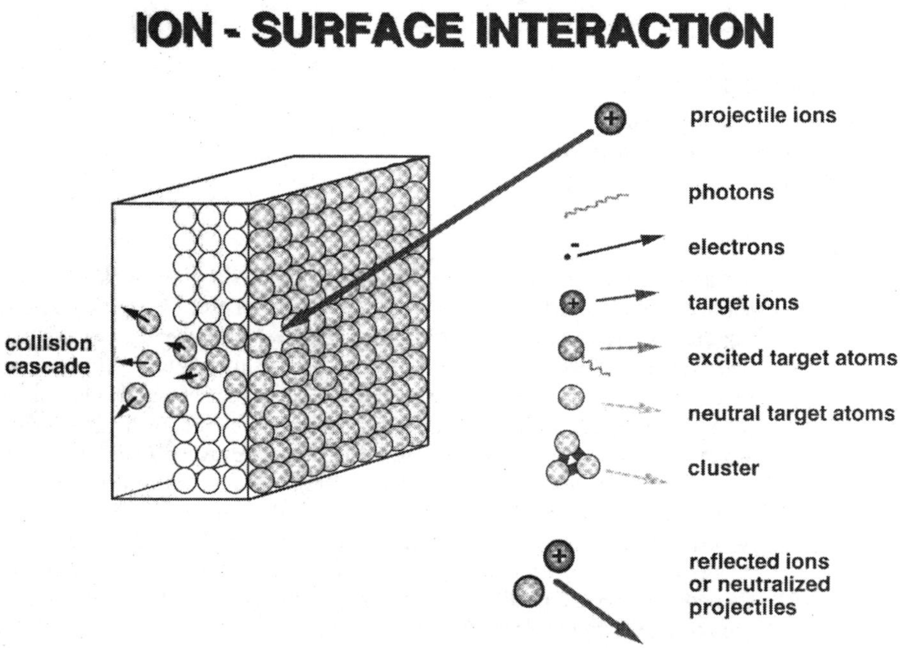

Figure 1.1. Typical processes taking place during slow ion - surface interactions: Particle penetration (electronic and nuclear stopping, production of collision cascades and defects), projectile reflection and emission of secondary particles (neutral and ionized atoms, molecules and clusters, and electrons and photons).

At low impact velocity the potential energy of the projectile becomes especially relevant, as has first been clearly demonstrated in the classical measurements of electron emission, potential (PE) yields for impact of slow singly and multiply charged ions on clean metal surfaces [1–3]. Such PE yields become rather large for impact of multiply charged ions (MCI) where the transient formation of "hollow atoms" in the course of ion neutralisation has achieved special attention [4–9]. In the same context also the properties of highly charged ions and their spectroscopical applications are of interest [10], as is the neutralisation and relaxation of MCI just at the surface and inside the bulk of a solid [11].

With regard to ion-induced electron emission one has to distinguish between the influences of the potential and the kinetic projectile energy [9] as indicated in fig. 1.2, and consequently one should use the term "highly charged ion" only in cases where PE is dominant.

PE is the result of various Auger processes [1–3] which for impact of MCI are always initiated by resonant electron capture from the surface into an approaching ion. A "hollow atom" is created because actually a number of electrons can be resonantly captured into highly excited states of the projectile before the electron emission sequence starts by way of autoionisation. Further explanations and results for PE will be given in section 1.3.

Electron emission due to the kinetic projectile energy ("kinetic emission - KE") is a different process where the projectile has to penetrate the surface in order to produce electrons in the target bulk which then can diffuse to the surface and into vacuum [12–14]. In section 1.4 we will concentrate on KE mechanisms which are dominant at and near the respective threshold, as in contrast to PE a certain minimum kinetic energy of the projectile is required. However, this threshold depends strongly on the KE processes which are involved for given combinations of ion and target surface (metal, semiconductor, insulator; [15, 16]). This is clearly apparent from fig. 1.3 for the impact of singly charged ions on atomically clean gold; measured KE yields decrease with falling impact velocity much less rapidly for heavier projectile ions where actually no clear threshold is detectable if sufficiently sensitive measurements are conducted.

Impact of fast charged particles (electrons, ions) on free-electron metals can initiate collective oscillations of the target electron gas (so-called plasmons; [17]). Recently it has been found, however, that such plasmons can also be produced by slow ions [18–20, 11], if the projectile has a sufficiently high potential energy ("potential excitation of plasmons- PEP"). The signature of plasmon excitation is their subsequent one-electron decay which leaves a characteristic feature in the corresponding ion-induced electron energy distribution (see fig. 1.4 and further information in section 1.5).

For some insulator surfaces a rather interesting process can be initiated by impact of MCI. In contrast to the common sputtering, kinetic which is observed for all kinds of target materials, the rapid capture of electrons from a small surface region can cause defects which give rise to ejection of neutral and to some lesser extent also ionized target particles. Such so-called "potential sputtering - PS" has first been explained by some type of Coulomb explosion [21–23], considering the mutual repulsion of ionized target particles in the electron-

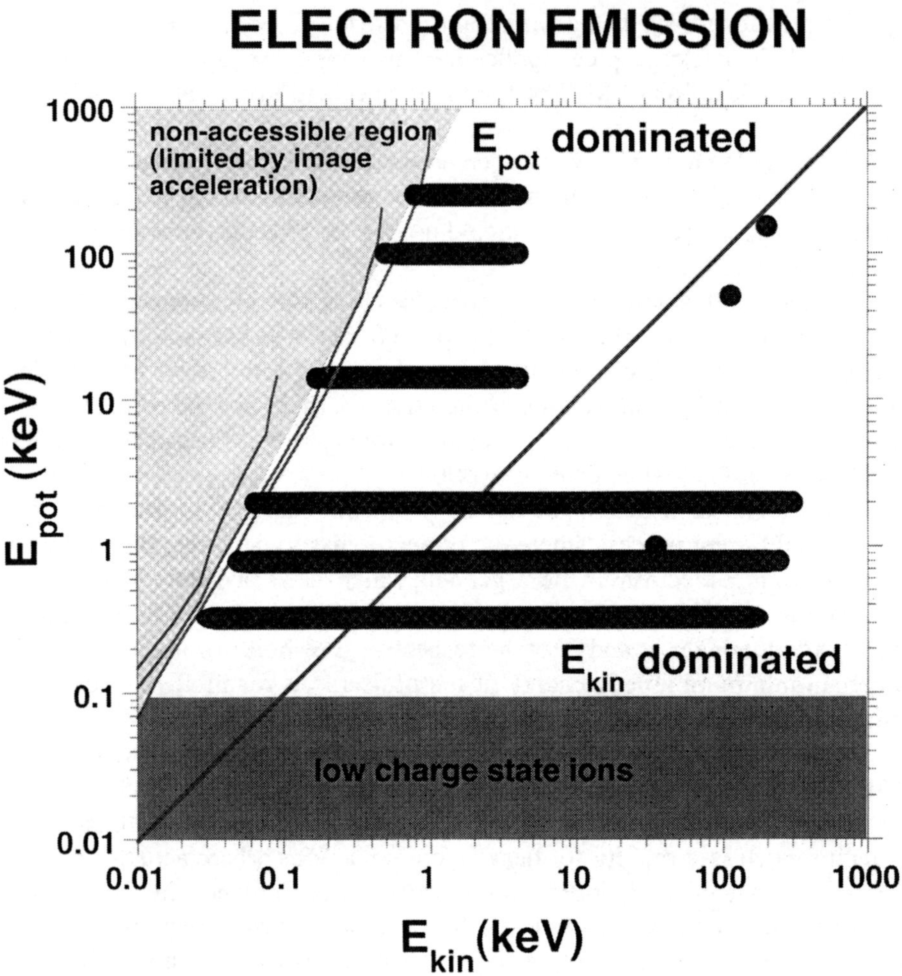

Figure 1.2. Qualitative distinction for experimental studies on ion-induced electron emission conducted for different potential- and kinetic projectile energies. Only data on the upper left side of the separation line belong to "highly charged ions". The curves which limit the area characterised as the "non-accessible region" indicate image charge acceleration (see section 1.3.2) for Ar^{q+}, Xe^{q+} and Th^{q+} and therefore the respectively lowest possible HCI impact energy [9].

depleted zone. Careful investigations of MCI-induced sputtering of a clean Si surface have shown that only the yield of the ionized target particles increases with the projectile charge [24]. At considerably higher impact energies a clear effect of the projectile charge on the secondary ion emission yield has been found and explained by Coulomb explosion [25]. Later on the same mechanism was invoked in order to explain the increase of neutral sputtering yields

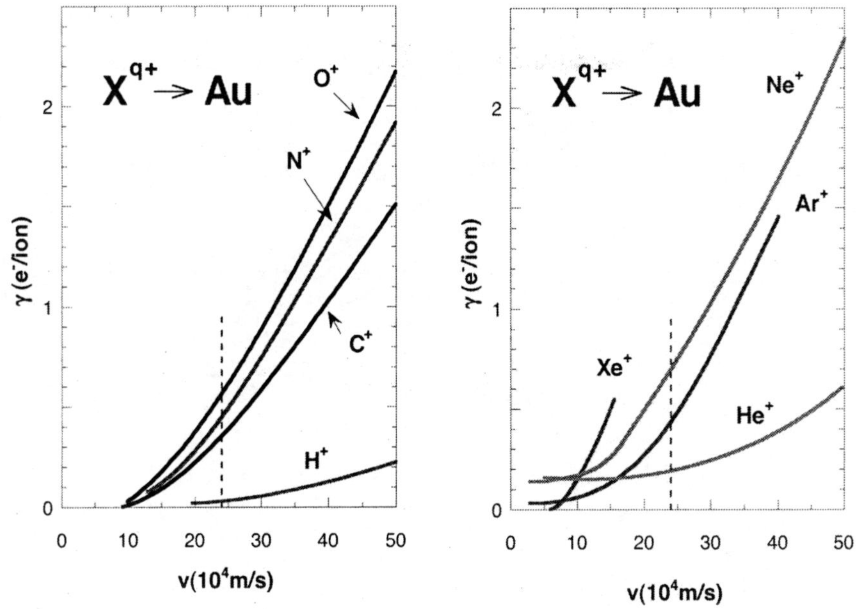

Figure 1.3. Total electron yields γ vs. ion velocity v measured for impact of different singly charged ions on atomically clean polycrystalline gold [15]. The vertical dashed line indicates the eKE threshold velocity.

beyond a certain minimum projectile charge [8]. However, in the course of experiments conducted at rather low MCI impact energy with alkali halide and some other insulator targets, the strong PS could be explained from the defect production, initiated by MCI-induced electron capture from the target surface [26–29] ("defect-mediated desorption", see section 1.6). Similar to electron emission, recently also for sputtering a clear distinction between kinetic- and potential-energy related origins has been proposed (see fig. 1.5 from ref. [9]). PS and related MCI-surface interactions might become of practical interest for new surface-analytical and nano-technological applications.

In summing up, the interaction of slow ions with solid surfaces can initiate processes which strongly depend on the kinetic and potential energy of the projectile ion as well as on properties of the target surface. Fig. 1.6 shows the multitude of processes which may be caused by impact of a slow MCI on a clean metal surface. We can distinguish an "above-surface" phase which is dominated by hollow-atom formation and decay, from a subsequent "at-" and "below-surface" phase where always a close cooperation of potential and kinetic energy-related effects will take place. Recent experiments described in sections 1.3 and 1.4 have convincingly established that PE created in the above-

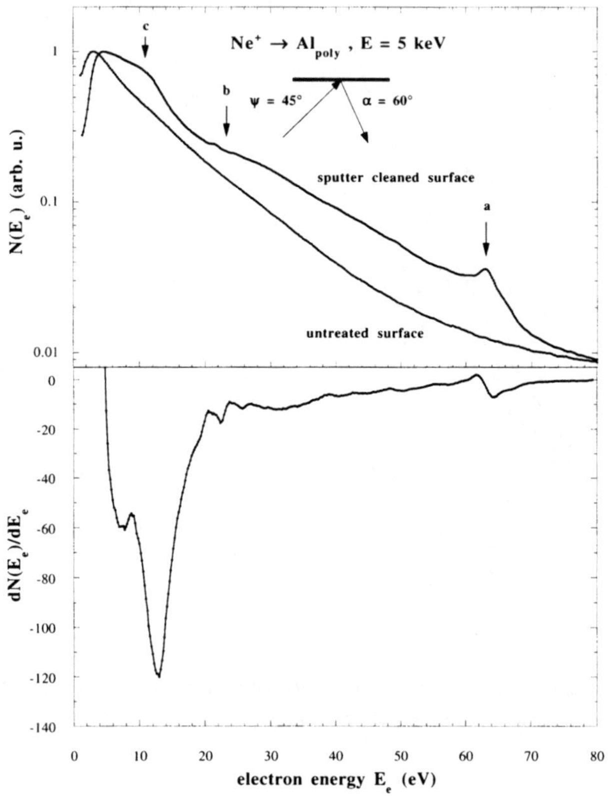

Figure 1.4. Top: electron energy spectra from clean and adsorbate-covered polycrystalline Al for bombardment with 5 keV Ne^+ ions under 45° incidence- and 60° electron emission angle (from ref. [20]). Structure labelled (a) is from Al-LMM Auger de-excitation from sputtered Al atoms, the small peaks labelled (b) result from Ne^{**} autoionisation lines which Doppler-shift for different emission angle, and structure labelled (c) is due to bulk-plasmon decay.
Bottom: Differentiated electron spectrum for a clean polycrystalline Al target for better visualization of the various structures.

surface phase can be clearly distinguished from subsequent electron production which strongly depends on the kinetic projectile energy involved.

Our present-day understanding of the above-surface phase is thus already satisfactory as far as metal targets are concerned, and it explains how "free" hollow atoms and ions can be produced by sending MCIs along the 'near-surface' region. On the other hand, neutralization- and relaxation mechanisms at and below the surface are still less well understood because they depend on the ion's potential and kinetic energies in a non-additive and usually rather complex way as, such as for potential sputtering (PS) and excitation of target plasmons due to (or supported by) the potential MCI energy. The fast Auger

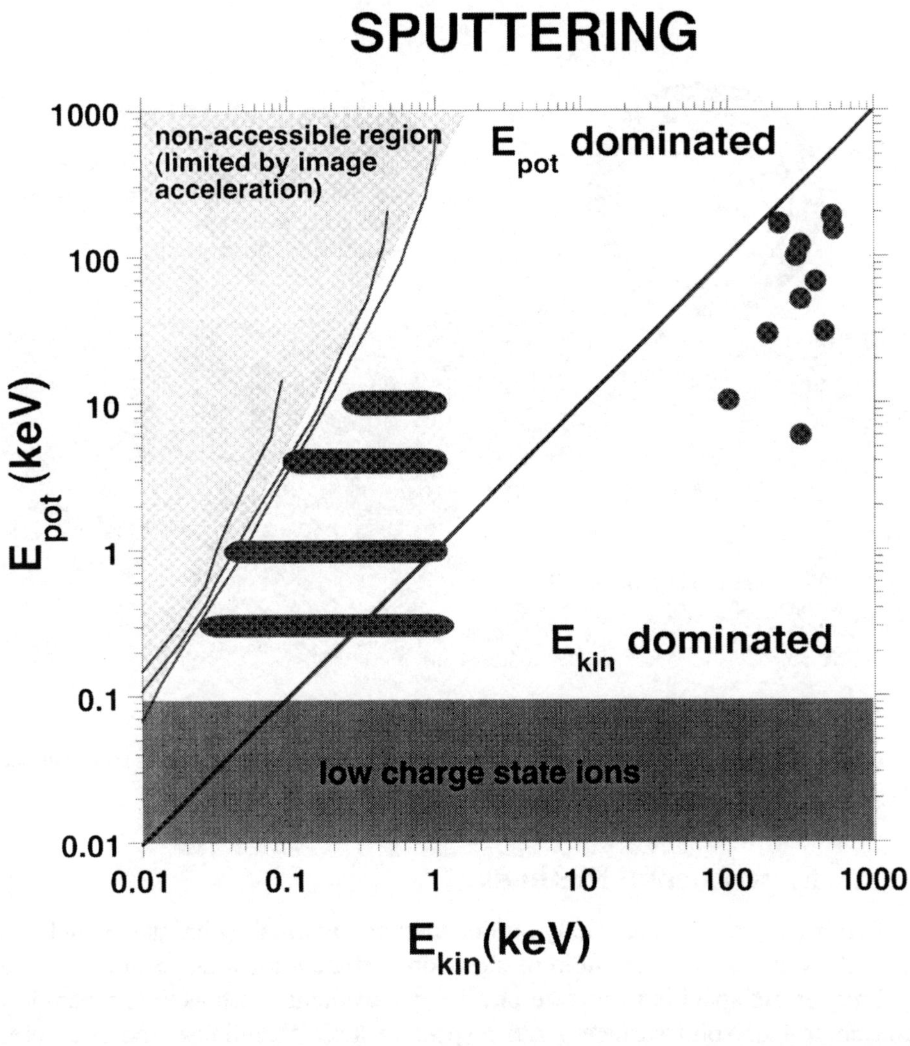

Figure 1.5. Qualitative distinction for experimental data on potential sputtering (PS) and related processes obtained for different potential- and kinetic projectile energies (from ref. [9]). Only data on the upper left side of the separating line belong to "highly charged ions". The curves which limit the area characterised as the "non-accessible region" indicate image charge acceleration for Ar^{q+}, Xe^{q+} and Th^{q+} and therefore the respectively lowest possible HCI impact energy.

electron emission during the collapse and relaxation of hollow atoms below the surface is an especially interesting signature for learning more about the late history of MCI-surface interaction [11].

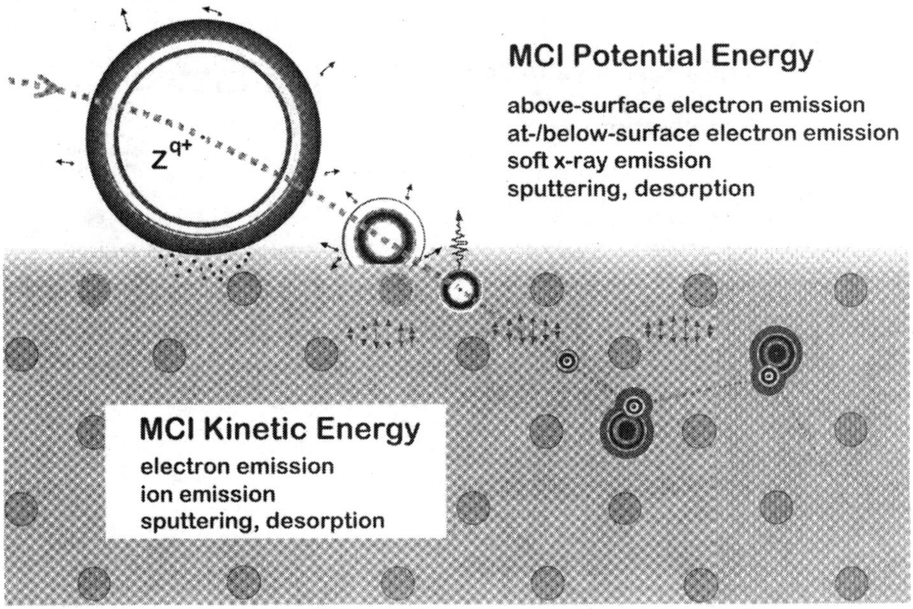

Figure 1.6. Hollow atom development and decay during slow MCI impact on a metal surface (cf. text).

2. Experimental Methods

In this chapter we will briefly outline the experimental techniques which are nowadays available in the field of slow ion-surface collisions. Some of these techniques are specific to surface physics experiments, such as the preparation and control of solid surfaces with respect to their cleanliness and structure, while others like the application of ultrahigh vacuum (UHV) apparatus and the transport and manipulation of singly and multiply charged ion beams are used in many other fields as well. We then describe some methods for the detection and energy-analysis of ion-induced slow electrons and for the investigation of scattered and sputtered particles which occur in neutral and ionized states as a result of slow ion-surface interaction.

2.1 Production and characterisation of slow singly and multiply charged ion beams

The production of ions, their selection with respect to charge, mass and momentum and their transport to the experimental area of interest are well-established experimental techniques [30]. Of special interest are multicharged ion (MCI) sources which nowadays can deliver beams of Z^{q+} ions (atomic number Z, charge state q) for virtually any chemical species up to the fully stripped ("naked") ions ($q = Z$). In the following we describe some MCI sources which naturally can also deliver slow singly charged ions.

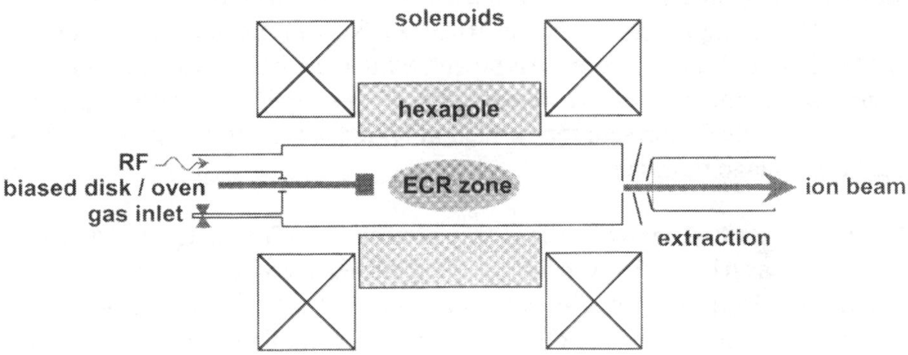

Figure 1.7. Main components of an ECRIS for MCI production (cf. text).

2.1.1 ECRIS - Electron Cyclotron Resonance Ion Source. An ECRIS consists of main components as shown in fig. 1.7. A discharge chamber filled with the working gas (pressure typically $\leq 10^{-5}$ mbar; ion production from non-volatile substances see below) is immersed in a "min-B" magnetic field geometry (i.e., the magnetic field strength increases from the plasma center outward), providing the necessary ion confinement. Such magnetic field configurations are generated by solenoids (which for larger set-ups may be superconducting) and multipole magnetic fields produced by permanent magnets (Sm - Co,Fe - Nd - B). Microwave radiation with a frequency up to 28 GHz is fed into the discharge chamber to produce a plasma wherein the electrons are heated by electron cyclotron resonance (ECR) when drifting through the so-called ECR zone (cf. fig. 1.7), where the electron cyclotron frequency $v_{ec} = e.B/2\pi.m_e$ matches the applied microwave frequency (e and m_e are electron charge and mass, respectively, and B is the local magnetic field). The electrons can acquire hundreds of keV and thus ionise the magnetically confined ions which stay rather cold. This permits efficient step-by-step electron impact ionization up to high q values and extraction of high-quality (low-emittance) MCI beams. The energy spread of extracted ions is typically between 5 and 10 eV times charge state q.

Production of ions from solid compounds can be achieved via sputtering or melting from suitable electrodes brought near the ECR zone, or by evaporation from a crucible inside the discharge chamber. Stable operation of the ECRIS plasma requires additional electrons which can be delivered from a hot filament or by ion-induced electron emission from oxide-coated inner walls of the discharge vessel, or probes biased negative with respect to the ECR plasma. Since its invention in the late sixties [31] ECRIS have undergone rapid development. Of special interest for the experimenter is the actual trend to rather compact all-permanent magnet ECRIS which consume astonishingly low microwave power. For example, the "Nanogan 10 GHz" [32] can produce 10 eμA of Ar^{8+} with only one Watt of microwave power, which can be furnished by low-cost, rugged solid-state microwave transmitters. In this version the vacuum pumping is done exclusively through the extraction region, superseding any pumping on the high voltage terminal.

2.1.2 EBIS - Electron Beam Ion Source. The EBIS utilises a high-density electron beam for both ion confinement and step-by-step ionization [33]. An intense electron beam with typical length of about one meter, a diameter of about 0.1 mm and a current density of up to several thousands A/cm^2 inside a series of drift tubes which can be connected to different potentials is focused by a strong magnetic field of typically one Tesla produced by a long (usually superconducting) solenoid (see fig. 1.8).

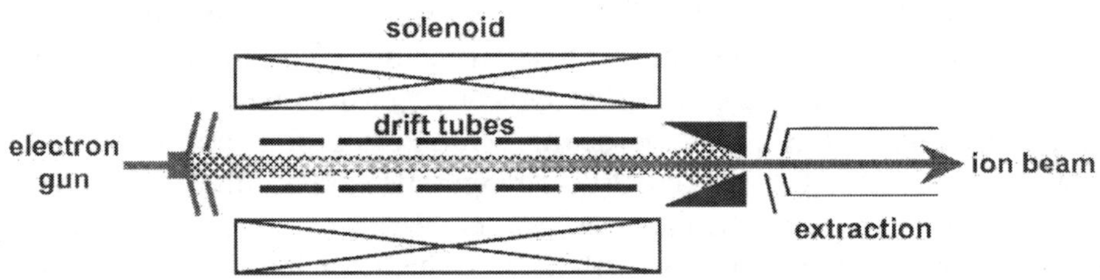

Figure 1.8. Main components of an EBIS for MCI production (cf. text).

The electron beam, by virtue of its strong negative space charge, provides strong confinement for positive ions as long as appropriate potential barriers are applied on both ends (see fig. 1.8). The ion confinement can last from a few milliseconds up to many seconds until the full space-charge compensation of the electron beam is reached.

The ions are brought into the electron beam either by gas injection or from external ion sources and will be ionised in a step-by-step manner. The lower the surrounding background gas pressure, the more efficiently this ionization proceeds. The background pressure can be kept well below 10^{-10} mbar by cryogenic pumping on the inner walls of the superconducting solenoid. At any time during the ion confinement only a narrow group of MCI charge states is present. In the short-pulsed mode, ions are extracted by rapidly lowering one of the axial potential barriers (extraction times typically 5 - 50 μs). However, the EBIS may also be operated c.w. in the so-called "leaky mode" which delivers less highly charged ions than in the (short) pulsed mode. EBIS are capable of producing up to fully stripped Xe^{54+} ions (about 10^4 per pulse) or fully stripped Ar^{18+} (about 10^8 per second; [34]). Typically, the emittance is $\leq 10\pi$.mm.rad and thus smaller than for the ECRIS.

2.1.3 EBIT - Electron Beam Ion Trap. EBIS and EBIT have as common working principle the step-by-step ionization of ions which are trapped in the space charge of a dense electron beam. The EBIT was developed by Marrs et al. [35]. It involves a much shorter trap (only a few cm) than the EBIS, which is easier to realise and probably gives more stability. Originally, the EBIT was devised for studying soft X-ray radiation from trapped MCI subjected to electron impact excitation and ionization for up to H-like U^{91+} [36]. However, the MCI can also be extracted along the electron beam direction, in which case similar ion charge spectra and MCI yields are obtained as with EBIS. For production of MCI from non-gaseous compounds, singly charged ion species can be injected in a similar way as for the EBIS, e.g., from a metal vapour vacuum arc source ("MEVVA"; Brown [37]). EBITs are relatively small devices comparable in size, technology and maintenance costs to medium-performance electron microscopes. Recently, a very compact EBIT has been developed with the magnetic field for electron beam confinement produced from permanent magnets [38], which therefore needs no cryogenic components.

2.1.4 Characterisation and transport of slow ion beams. Ion beams are well characterised by their emittance and brightness (see below) and can only be efficiently transported and decelerated to low impact energy with a sufficiently small energy spread. If we regard an ion beam drifting along the z-direction and neglect the ion beam space charge (see below), we may substitute [39] the six-dimensional phase space $\{x, y, z; p_x, p_y, p_z\}$ by its two-dimensional sub-spaces $\{x, p_x\}$ and $\{y, p_y\}$.

By further assuming that $p_x, p_y \ll p_z$, we define $x' \equiv p_x/p_z = \tan\alpha_x \approx \alpha_x$; $y' \equiv p_y/p_z = \tan\alpha_y \approx \alpha_y$, where α_x and α_y are the angles of a particular ion trajectory with the z-axis for a given z in the x-z- and y-z-planes, respectively. These angles can be measured by directing the ion beam onto a slotted plate followed by a movable wire detector. Liouville's theorem simply states that the "areas" $\varepsilon_x\left(x, x'\right)$ and $\varepsilon_y\left(y, y'\right)$, which result from such measurements and are called (two-dimensional) ion beam <u>emittances</u>, will stay constant along a drifting ion beam. For cylindrically symmetric ion beams the emittance can be given as the area of a two-dimensional phase-space measured in units of π.millimeter.milliradian. For an accelerated or decelerated ion beam this formally remains only valid for the relativistic ion momenta, leading to the "normalised emittance" [39] $\varepsilon_n \equiv \beta.y.\varepsilon$, where β and γ have their usual meaning in relativity. Based on the concept of emittance, the most relevant figure of merit of an ion beam is its ion current density in phase space (so-called brightness or normalised brightness, respectively): $B \equiv I/\varepsilon_x.\varepsilon_y$; $B_n \equiv I/\varepsilon_{n,x}.\varepsilon_{n,y}$. The brightness is measured in units of ampere/$(\pi.\mathrm{mm.mrad})^2$. We can simply state that ion beams possess higher quality the lower their emittance and the higher their brightness. If ion beams are subject to non-negligible space charge and, consequently, strong expansion by the Coulomb interaction, the above given pre-conditions for the validity of Liouville's theorem break down. Since c.w. or slowly pulsed positive ion beams drifting in good vacua (typically above 10^{-7} mbar) remain space-charge compensated by the electrons which they produce in collisions with the background gas molecules, the concepts of emittance and brightness are still applicable. However, influence of space charge along acceleration- or deceleration regions as well as imperfect ion-optical systems will cause a considerably larger "effective" emittance. This concerns ion beam transport in ion-optical systems through experimental environments which can be characterised by their "acceptance" according to the same criteria as for the ion beam emittance.

2.2 Preparation of target surfaces

Well defined target surfaces are a basic requirement for reproducible experimental results to be compared meaningfully to theory. Here we distinguish polycrystalline and single crystalline surfaces. In case of polycrystalline materials the sample cleanliness is important but structural aspects cannot be neglected. Especially for low ion energies possible preferential orientation of the grains of a polycrystal may cause channelling leading to structural effects in particle penetration, scattering and sputtering. In the experiments discussed later on, polycrystalline samples were used for studying ion-induced electron emission and sputtering. In some of these cases targets have been prepared and then transported into the collision area by a manipulator. For more advanced electron emission- and especially ion scattering experiments, single crystal surfaces are necessary. In fig. 1.9 we show a sketch of the typical scattering geometry.

In general, the target preparation follows conventional recipes [40–42]. Metal single crystals are polished and oriented before being transferred into the UHV chamber. Stress-free mounting on the usual UHV goniometers is achieved by cutting a groove into the side of the sample and holding it with clamps there. Target preparation consists of sputtering and annealing cycles. The target cleanliness is usually checked by Auger-Electron-Spectroscopy (AES) or ion scattering spectrometry (ISS) [41, 43].

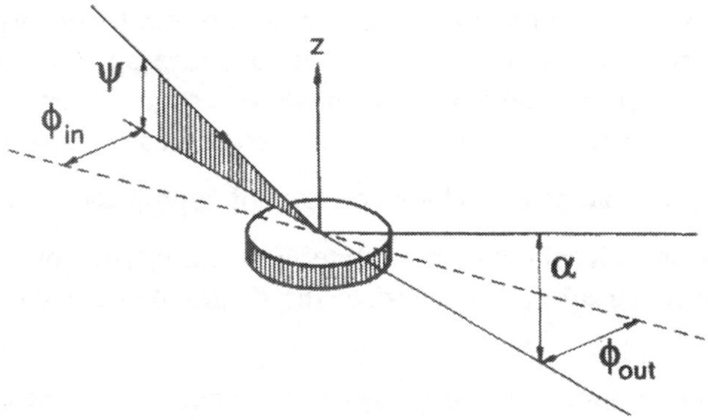

Figure 1.9. Sketch of scattering geometry for collisions of ions and atoms from surfaces for grazing angle of incidence Ψ_{in} (typically less than a few degrees). By varying the azimuthal angle ϕ_{in} one can choose between low-index (channelling conditions) or high-index to random scattering conditions. α denotes the exit angle for the scattered projectile or the angle of ejection of secondary particles which have been produced in the course of the collision.

Annealing temperatures for single-crystalline surfaces can be kept below 0.5 T_m where T_m is the bulk melting temperature. Higher temperatures may cause severe segregation of impurities. In cases of obnoxious impurities suitable gas-surface reactions, e.g. with oxygen to remove carbon or with hydrogen to remove oxygen, may be necessary. The structural quality of the surface is controlled in many cases by low electron energy diffraction (LEED) or by surface channelling effects [41, 43]. For low index directions of a single crystal surface characteristically shaped spatial ion distributions are observed [44, 45], whereas for high index- or "random" directions the angular peak width of reflected ions is a good indicator of the "flatness" of a surface [42].

For surfaces, semiconductor the sputtering and annealing procedure is less effective or simply destructive. Due to the covalent bonding, annealing of a sputtered semiconductor surface is poor. Semiconductor surfaces are prepared by following chemical recipes in case of Si and a heating process in vacuum [40, 46]. In some cases, e.g. for III-V semiconductors, cleaning in vacuum is possible. In the case of oxides the extreme sensitivity to ion bombardment causes problems. Beside structural damage by preferential sputtering, i.e. the depletion of oxygen in the near surface region, defect creation causes severe changes of surface properties. Thin oxide layers can be grown in situ on the proper metal substrate [47, 48].

For surfaces, insulator a severe difficulty in ion beam experiments is caused by the generation of electrical charges at the surface. Both the impact of positively charged ions and the consequent electron emission contribute to a positively charged surface layer. This will not only change the impact energy and beam geometry, but also the energy distribution of emitted charged particles. Since such energies are usually very low, even a charge-up by less than one Volt can severely influence measured total yields of charged secondary particles. Charge building-up at the surface can be overcome in several ways.

- Flooding the target with charge carriers of appropriate polarity.
- Deposition of insulating target materials as thin (μm) or ultra thin films (nm) on metal substrates, to reduce the electrical resistance of the surface layer.
- Heating of a sample up to temperatures where it becomes a good ionic conductor (e.g. for alkali halide targets).

Experimental set-ups for studying low-energy electron emission need magnetic shielding from disturbing fields, in particular for the earth magnetic field (cf. Eder et al. [20]).

2.3 Techniques for studying inelastic processes initiated by slow ion - surface impact

2.3.1 Total yield and number statistics of slow electrons. The total electron yield γ (mean number of electrons emitted per single projectile impact) can be determined from the fluxes of the incoming projectiles I_p and the emitted electrons $I_e = \gamma . I_p / q$ (so-called current measurement). With charged projectiles this can be accomplished in a straightforward manner by measuring target currents with and without permitting the electrons to leave the target, which is done by appropriate target biasing. Precautions have to be taken against possible disturbances from charged particle reflection, secondary ion emission and, especially, spurious electron production due to impact of reflected or scattered projectiles or electrons, all such effects possibly causing additional electron emission from the target region. In general, for current measurements ion currents of typically more than one nA are necessary.

A superior technique for determination of total electron yields γ involves the electron emission statistics (ES), i.e. the probabilities W_n for ejection of 1,2,...,n electrons per incident projectile, from which the total yield is obtained as the first moment of the W_n [49–51].

Fig. 1.10 shows an appropriate set-up where the incoming ions can be accelerated or decelerated by a four-cylinder lens assembly to any desired nominal impact energy, before hitting a target surface under normal incidence. In some experiments the ion impact energy is only limited by the projectile ion image charge interaction with the surface [52]. Practically all electrons ejected from the target with energies smaller than 60 eV into the full 2π solid angle are deflected by a highly transparent (96 %) conical electrode and then, after extraction from the target region, accelerated and focused onto a surface barrier

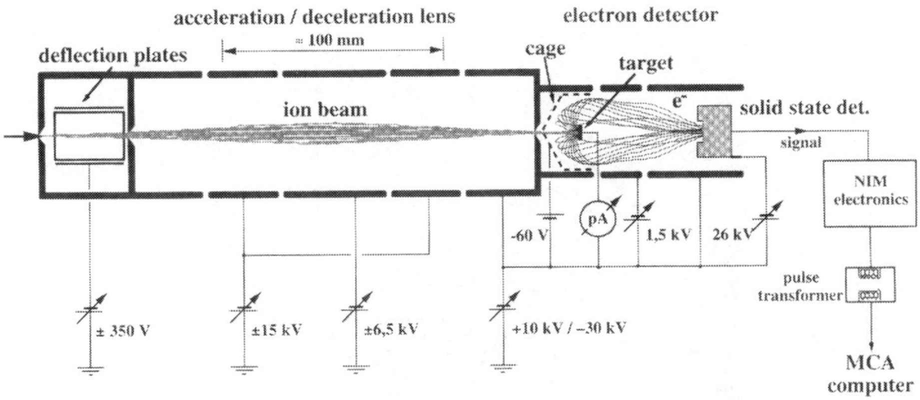

Figure 1.10. Experimental setup for measuring the electron emission statistics [15].

detector connected to $U_e \geq +20kV$ with respect to the target. The resulting ejected electron trajectories have been indicated in fig. 1.10. The probability W_0 that no electron is emitted cannot be determined directly, but may practically be neglected for yields $\gamma \geq 3$. This ES technique requires ion fluxes at the target surface of less than 10^4 projectiles/s and is therefore ideally suited for comparably tiny MCI beams from EBIS and EBIT (see above). Additionally, charging-up of insulator surfaces under HCI bombardment will completely be avoided [53, 54]. Furthermore, since MCI-induced potential electron emission depends strongly on the MCI charge state, their respective ES spectra can be utilized to distinguish between different MCI species with equal or nearly equal charge-to-mass ratios present in "mixed" ion beams [55].

Apart from giving access to precise total electron yields, the emission number statistics is also of genuine interest by itself since it can provide further useful information on the total number of electrons involved in particular emission processes and the mean "single electron emission probability": Only a fraction of all electrons involved in an ion-induced event can actually escape into vacuum [56].

2.3.2 Ejected electron energy distribution. For electron energy analysis various types of spectrometers can be applied. Electron spectrometers consist essentially of electrodes producing a well defined electric field. The notation of such spectrometers refers primarily to the shape of these electrodes (parallel plate-, cylindrical mirror-, cylindrical- or spherical spectrometer). In fig. 1.11 we show a setup with parallel plate electron spectrometer.

Electrons from the target surface enter the electric field through an entrance slit and, after appropriate deflection, leave the field by passing the exit slit. An

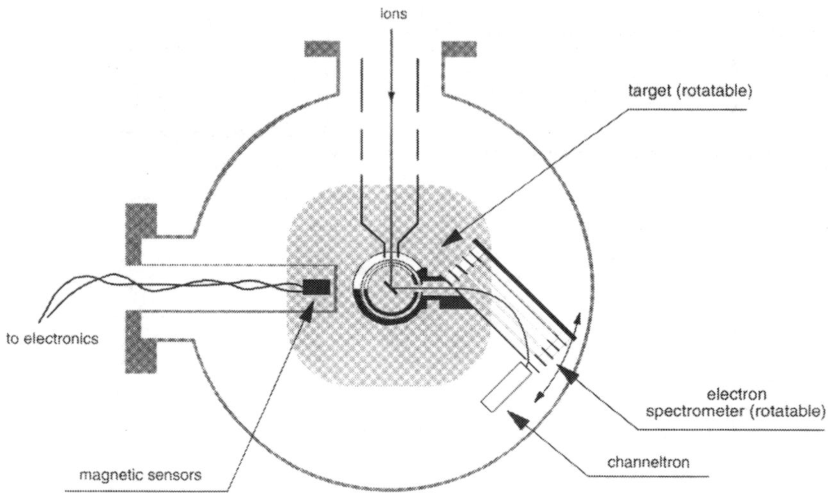

Figure 1.11. Experimental setup for measuring electron energy distributions [20].

important quantity of the spectrometer is the solid angle defined as the product of the polar and azimuthal acceptance angles. Another important property of the spectrometer is the (relative) energy resolution $\Delta E/E$ defined as the accepted energy interval ΔE divided by the electron energy E. The quality of a spectrometer is also determined by its focusing power since electron spectrometers can focus in different orders. Various aspects have to be considered if a choice is to be made between different types of spectrometers. Favourable features of the spectrometers are high resolution, efficiency and simplicity of design. High resolution is required when individual Auger lines are to be measured. Natural line width and separation of adjacent lines are of the order of 1 eV. Accordingly, resolutions $\Delta E/E$ of 10^{-4} to 10^{-3} are required using, e.g. 1000 eV electrons. Such a resolution may in principle be realised by reduction of the spectrometer slits or by deceleration of the electrons. The latter method takes advantage of the fact that $\Delta E/E$ is usually constant so that a reduction of E will also decrease ΔE. The deceleration method is advantageous, since the loss in the spectrometer efficiency is relatively low. Furthermore, the deceleration method allows for varying ΔE during the measurements. For high resolution measurements one has to reduce the earth magnetic field by a factor of about 100 which can be achieved by mumetal shielding inside the scattering chamber.

Spurious electric fields must also be avoided which are generally produced by electrons collected at insulating surfaces.

2.3.3 Projectile scattering and energy-loss, secondary ion emission.

Electrostatic analysers are also used for the investigation of inelastically scattered ions according to their charge state distribution and energy loss spectra, and for investigation of secondary ion emission. Angular distributions of the scattered projectiles can be measured with a moveable diaphragm behind the target region and a channeltron for scattered particle detection [42]. When combining movable slits with deflection plates and a position-sensitive detector, scattered particle charge state distributions and -angular distribution can be measured in conjunction (see fig. 1.12).

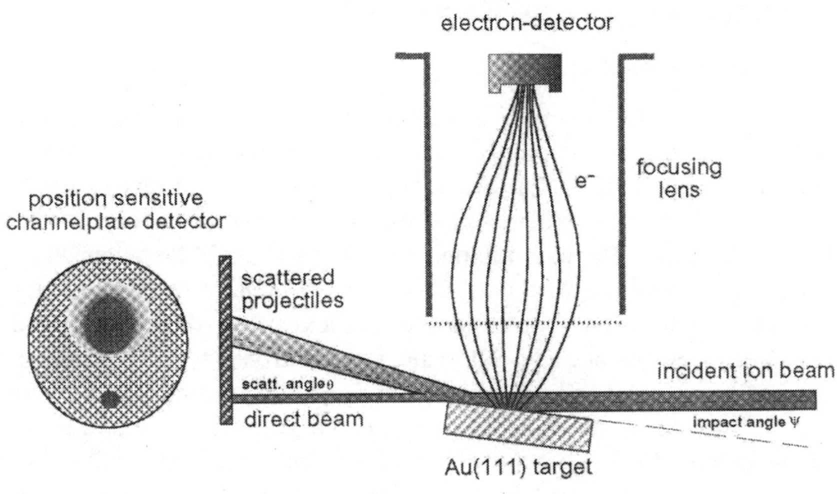

Figure 1.12. Experimental setup (schematically) for measuring electron emission from grazing incidence ion-surface interaction in coincidence with scattered projectiles [57, 58].

One advantage of this scheme is that scattered fast neutrals can also be taken into account, which in some cases are more important than the scattered ions. Another interesting aspect concerns the detection of negative ions after scattering. Electrostatic analyzers can also be applied for investigating secondary ion emission. For such measurements it is rather important to avoid surface charge-up because the energy of secondary ions is generally quite low. The kinetic energy (energy loss/gain) of both charged and uncharged secondary particles may also be determined by time-of-flight (TOF) methods, for which technique one has to apply pulsed primary ion beams [59].

2.3.4 Measurement of potential sputtering yield. Bombardment of certain classes of insulator surfaces by slow ions may cause sputtering and secondary ion emission, which cannot be explained by the elastic transfer of kinetic projectile energy to the target particles, but rather by deposition of some part of the projectile potential energy. Investigation of "potential sputtering" effects has so far mainly been carried out by means of the quartz microbalance technique which determines the mass loss from thin polycrystalline insulator films having been deposited onto the face of a quartz oscillator crystal (cf. fig. 1.13). If such films are then bombarded by slow ions, the change in the quartz oscillator frequency gives access to the mass loss from the film. By taking into account the ion current the respective sputtering yield can be derived (see, e.g., [60] and refs. therein).

3. Potential Electron Emission

3.1 Introduction

The influence of the potential projectile energy on electron emission is naturally larger the higher the charge state of the impinging ions (cf. fig. 1.14). A slow multicharged ion (MCI) which approaches a solid surface catches electrons from the latter into highly excited states and transiently becomes a neutral "hollow atom" (HA). Rapid autoionisation of this HA will be balanced by further ongoing electron capture until close surface contact when shielding sets in, and inner-shell vacancy recombination at and below the surface produces further slow electrons and characteristic fast electrons (see [11]) and soft X-rays. Until its complete neutralization the projectile continues to be attracted toward the surface by its decreasing image charge which causes a minimum impact velocity. Many aspects of this HA scenario have been studied by various groups for different ion species and charge states at metal, semiconductor and insulator surfaces with a number of experimental methods. The most detailed comparison with available theory is possible if electron emission and projectile energy losses and charge changes are studied in mutual coincidence for scattering along well-defined trajectories at smooth mono-crystalline target surfaces. For insulator surfaces so called "potential sputtering" (PS) and strong secondary ion emission (cf. section 1.6) can be induced by the multi-electron capturing MCI.

Our current understanding of the above-surface part of MCI-surface interactions is satisfactory and can explain why "free" hollow atoms and ions are produced by funneling MCI through thin capillaries or colliding them on clusters or fullerenes. On the other hand, processes occurring at and below the surface are less well understood since they depend on the potential and kinetic ion energies in a clearly non-additive way. This is particularly true for the ex-

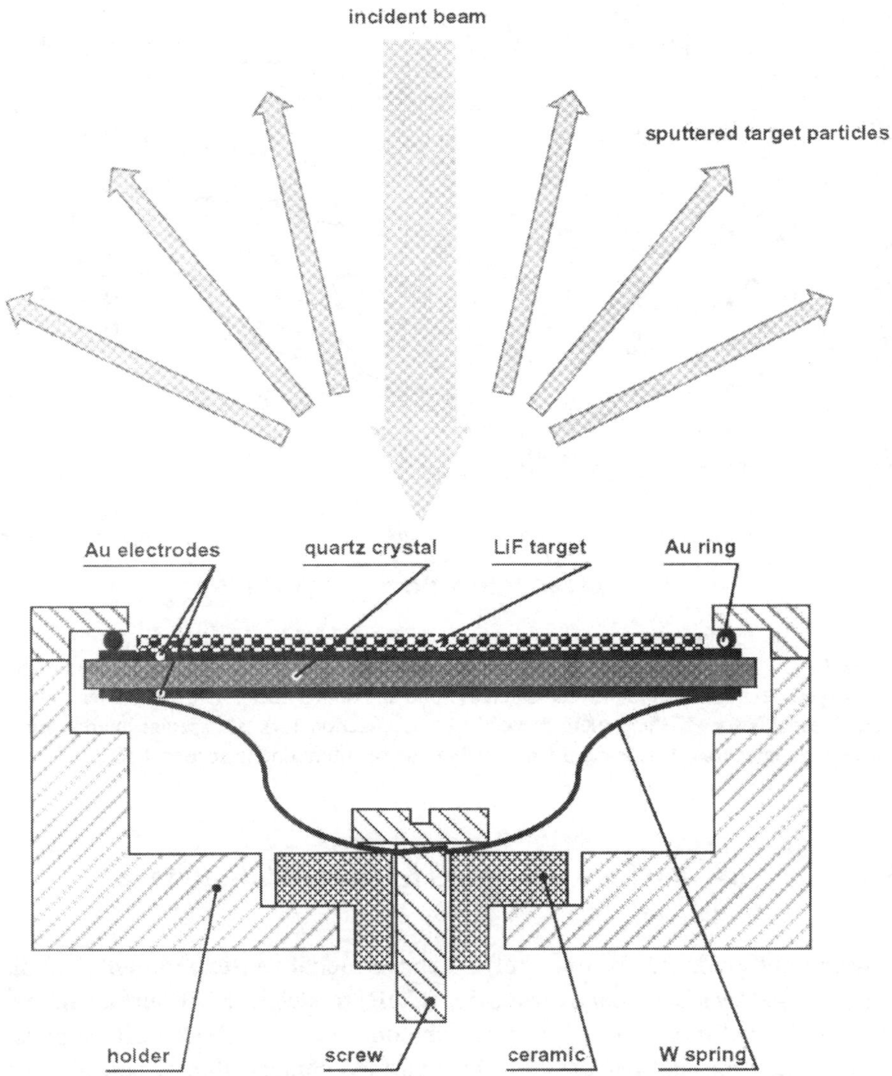

Figure 1.13. Quartz crystal microbalance for measuring total sputtering yields [60]. The relative mass loss of the thin target film due to sputtering is proportional to the relative change of the quartz resonance frequency.

citation of target plasmons due to the potential MCI energy (cf. section 1.5), and also potential sputtering (section 1.6).

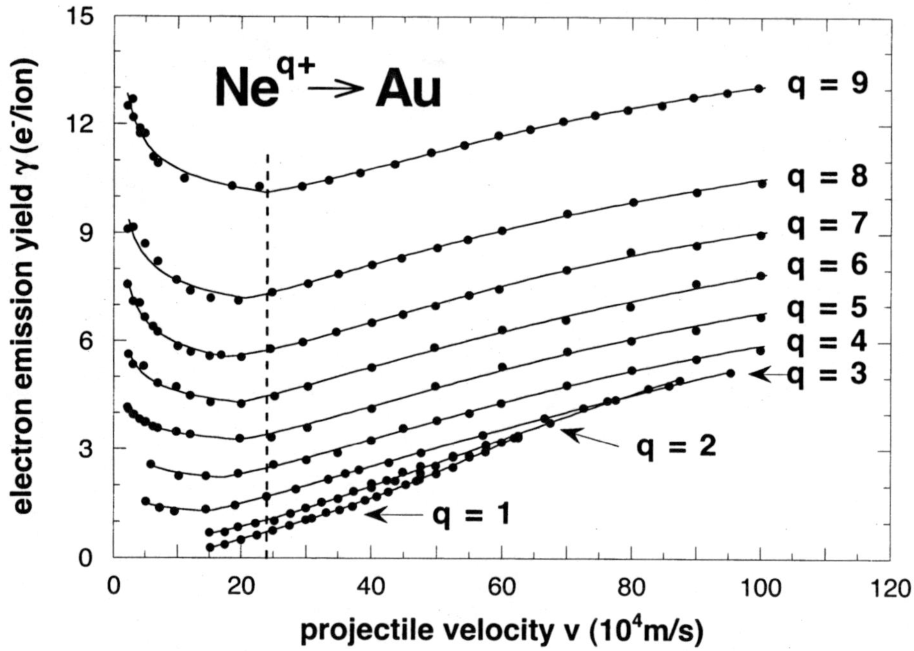

Figure 1.14. Total electron yields vs. ion velocity for impact of Ne^{q+} ($q = 1 - 9$) on clean gold (experimental data from [15] indicated by solid and dotted lines). The dashed vertical lines mark thresholds for eKE and pKE, respectively (cf. section 1.4). Of special interest are the converging yields for q = 1, 2 and 3 which will again be referred to in section 1.4.1

3.2 Above-surface processes in slow MCI-surface interaction

Important guidance for many of the above mentioned experimental studies was provided by a classical over-barrier (COB) model for MCI-surface interaction of J. Burgdörfer et al. [4, 61], which considers the "above-surface phase" of HA formation and decay until close surface contact, that is before strong shielding by the target electrons sets in. According to this COB model, inside a critical distance $d_c \approx (2q)^{1/2} / W_\phi$ (atomic units, q .. MCI charge state, W_ϕ .. surface work function) the potential barrier between a metal surface and the approaching MCI drops below the Fermi level and electrons from near the Fermi edge can rapidly pass over to the MCI. Since resonant classical over-barrier transitions are favored, comparably highly excited projectile states will be populated. With further approach these states will be shifted upwards in energy due to image interaction and screening of the projectile charge by the already captured electrons. In this way successively lower projectile n-shells will be populated by cascading from autoionization transitions and because the now up-shifted levels fall into resonance with occupied target states.

At the same time previously populated projectile states are emptied by resonance ionization into unoccupied conduction band states, autoionization and promotion above the vacuum level. This complex interplay of electronic transitions goes on during the projectile's approach towards the surface and gradually shifts its electronic population to lower n-shells. Electrons lost from the projectile are rapidly replaced by resonance neutralization and the projectile stays thus practically neutral while still above the surface. This highly transient situation characterizes the above-surface phase of the so-called "hollow atom" (HA). Projectiles yet to be fully neutralised are accelerated toward the surface by their image charge and the resulting kinetic energy gain ΔE_{im} is correlated to the distance d_c as given by the COB model. Assuming stepwise neutralization ("staircase-approximation"), we obtain $\Delta E_{im} \approx 0.25 W_\phi . q^{3/2}$, of which about three quarters (the so called "classical lower limit") are gained by the projectile until it reaches the distance d_c for first resonance capture. ΔE_{im} has been measured in two ways. For grazing incidence on a smooth monocrystalline surface specular reflection of neutralized ions takes place for which the outgoing trajectory is more steeply inclined toward the surface than the incoming trajectory where the image charge acceleration takes place [62, 63]. Alternatively, because of the lower projectile impact energy limit due to ΔE_{im}, saturation of the total electron yield will take place toward zero nominal impact energy [52].

For insulators (e.g., ionic crystals as LiF) one has also to take into account a frequency- dependent dielectric surface response and appropriate potentials for modelling the electronic interaction between projectile and surface [64]. ΔE_{im} measurements for grazing incidence of MCI on LiF [65] are as well reproduced by the such extended COB model as the measurable weak dependence of ΔE_{im} on the surface-parallel MCI velocity component [66]. A somewhat different COB model for metals and insulators has been presented by Thumm et al. ([67], see also U. Thumm in this book). The above described neutralization dynamics leads to the emission of a large number of electrons (cf. fig. 1.14 and 1.15). At low projectile impact velocities the total electron yield first increases linearly with the projectile potential energy. This increase becomes less pronounced as soon as transitions into inner shell vacancies become possible and the energy is carried away by faster Auger electrons [5, 68].

The above-surface HA dynamics described here is also relevant for the survival of "free HA" and "free hollow ions" [69] far away from an electron-donor surface. Funnelling slow MCI through capillary arrays (typically 100 nm diam. in a 1 μm thick Ni sheet) produces downstream ion charge state distributions which can be satisfactorily explained by the COB model [70]. Recently, a similar technique has been applied by transmitting 3 keV Ne^{7+} through 80 nm sized tracks which have been etched into thin Mylar foils [71]. "Free hollow ions" can

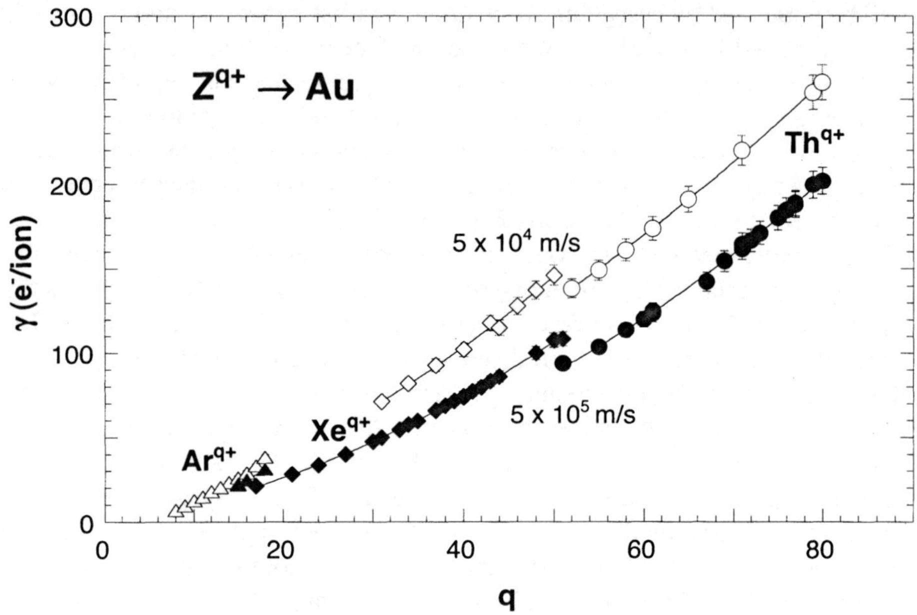

Figure 1.15. Charge-state dependent electron emission yield for MCI hitting a clean Au surface at two different impact velocities (data from [52]).

also be produced by passage of MCI through a very thin foil, as demonstrated with 6.9 keV/amu Au^{68+} in 10 nm thick diamond-like carbon foils [8]. Still another access to production of "free hollow ions" could be the back-bouncing of HA from insulator surfaces, as suggested by soft-X ray spectra measured for impact on hydrogen-terminated silicon [72].

3.3 Hollow Atom dynamics at and below the target surface

If a MCI which has reached its HA phase finally touches the surface, it will first become partially re-ionized by "peeling-off" of its still loosely bound electrons which are screened by the target electron gas [5]. Faster projectiles can penetrate into the target bulk and will then become screened by a dynamic electron cloud [73, 74]. Consequently, below the surface only a much smaller HA than its previous version can be formed, with its empty inner shells being rapidly filled by Auger and/or radiative transitions. This at and below surface HA relaxation has been studied to considerable detail by means of Auger electron and soft X-ray spectroscopy. It can be modelled with atomic structure codes for the screened projectile by considering its transport inside the target bulk [6, 11, 75].

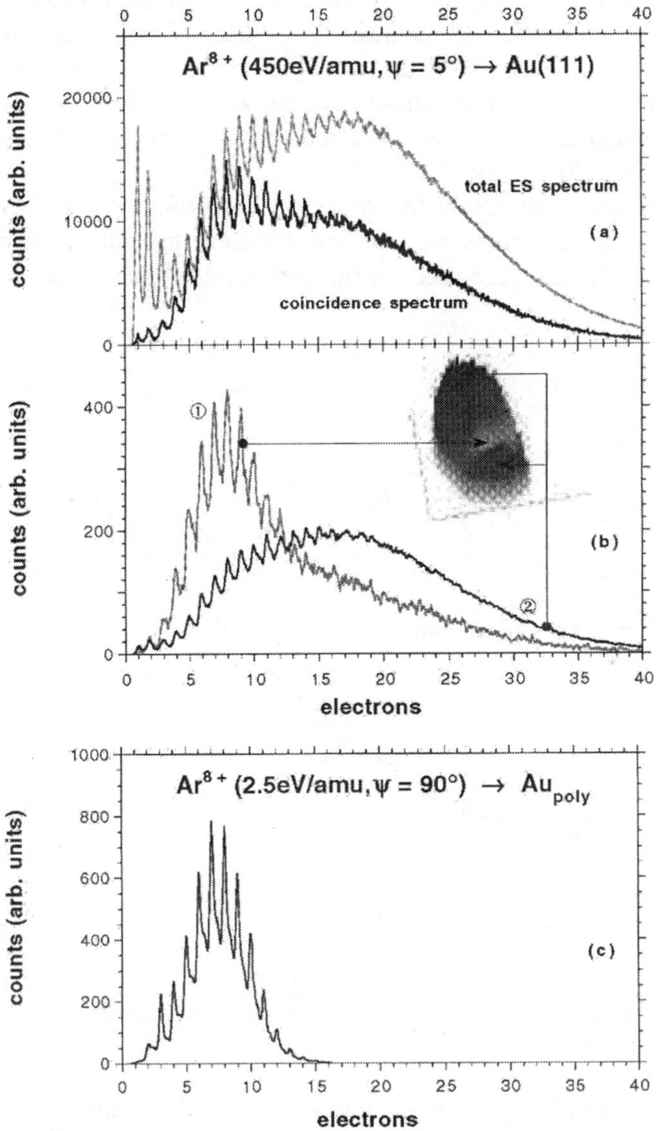

Figure 1.16. a) Total (non-coincidence) and coincidence ES (electron emission statistic) spectra for the case of 0.45 keV/amu Ar^{8+} ions impinging under a grazing angle of 5° onto a Au(111) single crystal surface (from ref. [57]).
b) ES spectra in coincidence with the central part (1) and with the wings of the scattering distribution (2); see inset.
c) For comparison: ES spectrum obtained for 2.5 eV/amu normal incidence Ar^{8+} projectiles on polycrystalline Au (from ref. [78])

Plasmon losses of K-Auger electrons which are ejected from a HA just along the surface permit further refined modelling calculations for the respective HA dynamics [76]. Separation of contributions to slow electron emission from the above and below surface phases can be achieved with coincidence measurements which distinguish between the correspondingly different projectile trajectories [57, 77] (cf. fig. 1.16).

A peculiar case was found for impact of slow multicharged fullerenes on clean metal surfaces, where the potential MCI energy will not be released as PE (cf. fig. 1.17) but rather used for fast projectile fragmentation [79].

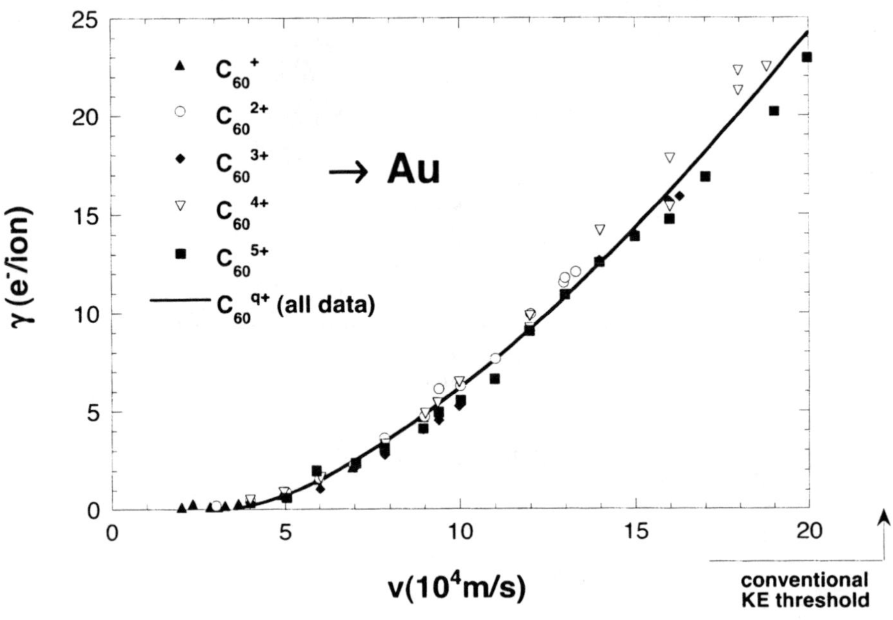

Figure 1.17. Electron emission yield vs. impact velocity for multicharged fullerene ions hitting a clean Au surface (data from [79]).

As a final remark in this short review on PE we state that the potential ion energy can cause both neutral and ionized target particles to be ejected from an insulator surface ("potential sputtering" – PS, cf. section 1.6).

4. Kinetic Electron Emission near the Impact Energy Threshold

4.1 Introduction

For impact of slow ions on a clean metal surface, the potential emission (PE) yield generally decreases toward higher impact velocity v (see section 1.3), whereas the yield for kinetic emission (KE), starting above some threshold, increases with v. KE is usually ascribed to screened-projectile stopping in the quasi-free electron gas (eKE) of the metal, and to electron promotion into continuum in binary ion collisions with target atoms (pKE). The eKE threshold depends on work function W_ϕ [80] and density-of-states (S-DOS) of the target surface, and the pKE contribution can be estimated from the corresponding quasimolecular orbital-correlation diagram [16]. However, it has been shown [16] that the expected sharp eKE threshold is "softened" because of partial localization of quasi-free electrons at the surface, which below the eKE threshold may cause a stronger KE contribution ("sKE") than from pKE. Moreover, it has been observed at very low impact velocity that for some projectile-target combinations the KE yield levels off rather slowly with decreasing v. The latter effect is clearly more pronounced for heavier projectiles and can be explained by many-electron excitation in the impact zone due to spatial and temporal projectile energy localization ("mKE") [16]. If an ion collides with an insulator surface, neither eKE nor sKE are possible. Then, at a given v, the total electron yield may become smaller for higher ion charge despite the larger PE contribution (cf. fig. 1.18), because the pKE contribution decreases with higher projectile charge [53, 54]. A similar trend was recently also observed for slow MCI impact on clean gold [15] (see fig. 1.14 in section 1.3).

4.2 Mechanisms for ion-induced kinetic emission (KE) below the "classical threshold"

There are four possible mechanisms of KE, which can be distinguished by the respective primary excitation process. In the first case (A) the valence electrons of the solid become excited in binary collisions with the projectile moving in an idealized Fermi electron gas and are thereby ejected into the vacuum [80]. In this process, which we will call "electronic KE (eKE)", the electron excitation spectrum is very narrow and has a sharp cutoff at $2k_F v$ (so-called "classical KE threshold" with k_F being the Fermi vector). If the excitation above the Fermi level is not sufficient, i.e., if $v \leq W_\phi/2k_F$ (approximate definition for the "classical threshold velocity" $v_{th,e}$), no electron emission should be possible within this eKE mechanism.

The second process (B) involves electron promotion (pKE) in close binary collisions between projectiles and target atoms (or mutual target-atom collisions) at or below the surface. These collisions temporarily create quasi-molecules in which some electronic levels may be sufficiently strongly promoted to higher orbital energies and thus give rise to electron emission in subsequent de-excitation steps (so-called Fano-Lichten process [81, 82]). Compared to the perturbation of valence orbitals in eKE, in pKE the perturbation of localized (core) or semilocalized ("semicore") orbitals is involved. The pKE process

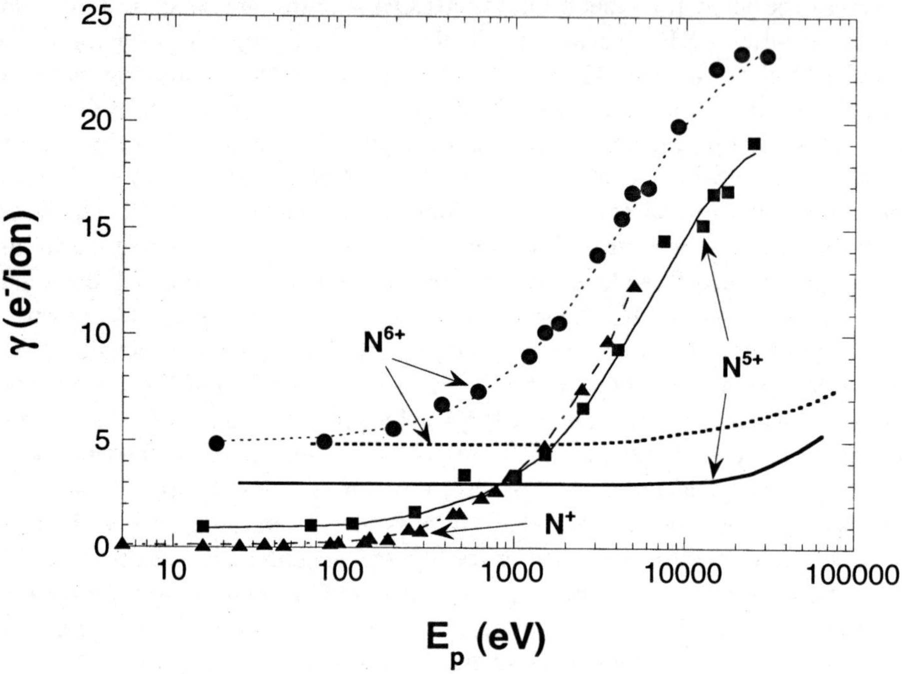

Figure 1.18. Electron emission yield vs. impact velocity for singly and multiply charged nitrogen ions hitting a LiF surface (data points). Results for a clean Au surface (lines) are shown for comparison (from [53]).

is characterized by the existence of a "critical" internuclear collisional distance at which the probability of electron excitation into the continuum abruptly increases. This excitation strongly depends on the velocity at and the "orbital dynamics" of the quasimolecule near the critical distance, which is specific for the species of the two collision partners. Because of a rather abrupt change of the probability for pKE at this critical distance it is common to speak about a threshold kinetic energy $E_{th,p}$ at the impact velocity $v_{th,p}$, although this "threshold" is not as sharp as in the case of eKE. This behaviour is common to one-electron non-adiabatic processes.

The pKE mechanism can be further subdivided by considering particular de-excitation processes during separation of the collision partners: for sufficiently high excitation above the Fermi level excited electrons can be directly emitted into vacuum [14] ("electron promotion followed by molecular autoionization"), or Auger de-excitation can take place for a core-level vacancy which is created

during the collision. A signature of the latter process is the appearance of sharp peaks in the respective electron emission spectra.
eKE and pKE are well established processes and until recently have been regarded as KE processes in general. However, it turned out that the respective models have difficulties in explaining the subthreshold impact velocity region below $v_{th,e}$ [16], where the remaining KE yield becomes gradually more significant for heavier projectiles, e.g., for impact of Xe^+ or Au^+ on Au [15]. In order to resolve this problem, the idealized concept of plane-wave like free electrons and of a projectile moving with constant velocity has been re-examined. If the electrons are assumed to be partially localized or the particle does not move along a straight trajectory with constant speed, smearing-out of the threshold takes place and electrons can be emitted at impact velocities below $v_{th,e}$. Localization of valence-band electrons can be due to:

- (C.1) a d-like character of the valence band [83],
- (C.2) a spatial confinement of valence electrons by the presence of the solid surface [16], and
- (C.3) the intrinsic partially localized character of the valence wave function caused by the quantum-mechanical requirement of its orthogonality to the inner-shell wave functions.

Model (C.1) was invoked in [83] in order to identify the mechanism for KE from grazing collisions of keV protons with a single-crystalline Cu surface. The electrons were assumed to be excited directly from the Cu d-band by several distant collisions of the neutralized hydrogen projectile. In [16] the partial localization of the valence-band electrons was related to their confinement by the surface of the solid (C.2).

In [84] the excitation is ascribed to the localized charge transfer process from the ionization level of the projectile into the continuum of the solid. In [16] the concept of surface excitation is generalized by including a direct perturbation of the substrate by the projectile potential. This KE mechanism was labelled "surface-assisted electronic KE (sKE)". Collision of the projectile with semilocalised electrons is a process which has no classical analogue as compared to the collisions with free (plane-wave like) electrons in eKE. The main consequence is that sKE involves no cut-off impact velocity but rather a smooth exponential decay toward lower impact velocity.

(D) Although model (C) is applicable to KE for a wide range of combinations of projectiles and targets (cf. fig. 1.19), difficulties remained for the interpretation of experiments with heavier projectiles like Xe^+ or Au^+. It was proposed in [16] that the distribution of excited electrons (or at least part of it) is modified (broadened) by electron-electron (e-e) interaction. The key assumption for such e-e interaction to be relevant is a localization of the electronic excitation in the impact zone. It is conjectured that a sufficiently long (a few femtoseconds) localization is caused by the slow passage of the projectile through the surface. This process seems to be more pronounced if the projectile is heavier and has more electrons, and will be labeled as "mKE" (m for many-electron).

In fig. 1.19 we show a comparison of measured total KE yields for impact of various singly charged ions on polycrystalline gold, with calculated KE yields taking into account the above explained sKE (solid lines) and mKE processes (dashed lines).

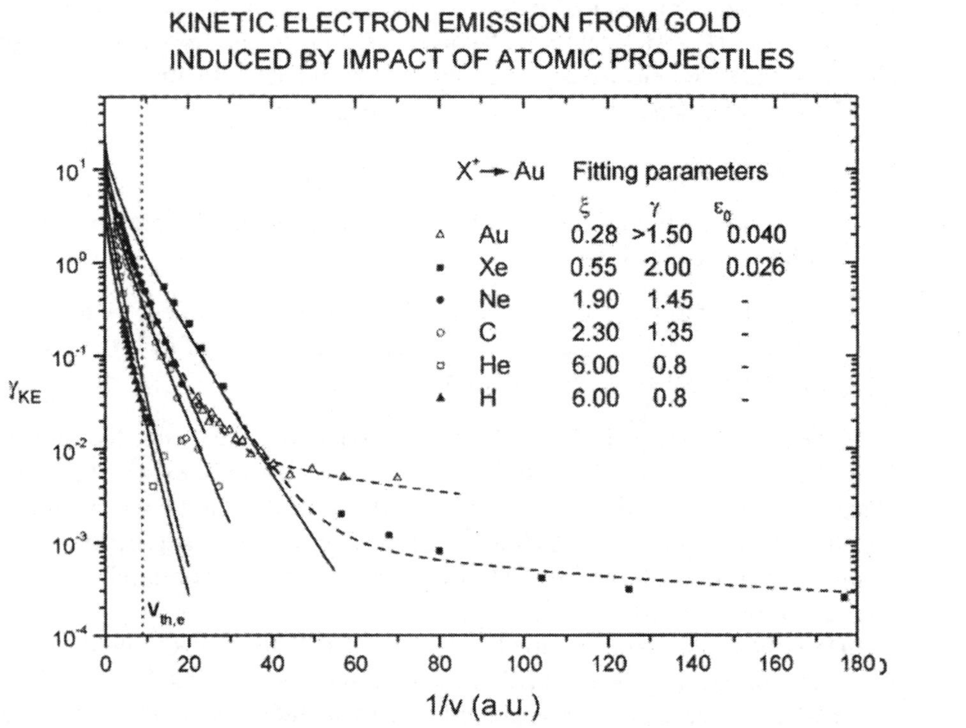

Figure 1.19. Experimentally determined total KE yields γ_{KE} for different singly charged ions colliding with a clean gold surface, in dependence on the inverse ion impact velocity. The solid and dashed lines are results of calculations based on the sKE and mKE mechanism, respectively [16]. The vertical dotted line indicates the KE subthreshold region.

For kinetic electron emission from insulators eKE and sKE can practically be neglected. For an educative model system, electron emission has recently been studied for impact of H^0 projectiles (impact energies 1 - 20 keV) grazingly scattered off a LiF(001) single-crystal surface (impact angle 0.5^0 to 1.8^0 with respect to the surface plane) [85]. By using neutral hydrogen atoms as projectiles, image charge effects on the incoming trajectory and potential electron emission as an electron source can be excluded. Under grazing incidence conditions, the projectiles are scattered from the LiF(001) surface plane along well defined trajectories ("surface channelling"). Impact events, where projectiles penetrate the surface plane at surface imperfections can be discriminated by measuring emitted electrons and surface channelled projectiles in coincidence. Electron emission induced by those projectiles will be less affected by transport phenomena in the bulk and subsurface region than for impact under larger angles. By variation of impact angle and energy, the distance of closest approach of the projectiles to the surface plane can be tuned from one to several a.u. (cf. fig. 1.20). In this way long-range interactions of hydrogen atoms with the ionic crystal can be probed and separated from interactions that require close encounters. From this type of measurements position- and velocity-dependent electron production rates can be derived, which reveal an electron promotion mechanism for the observed (kinetic) emission of electrons from LiF.

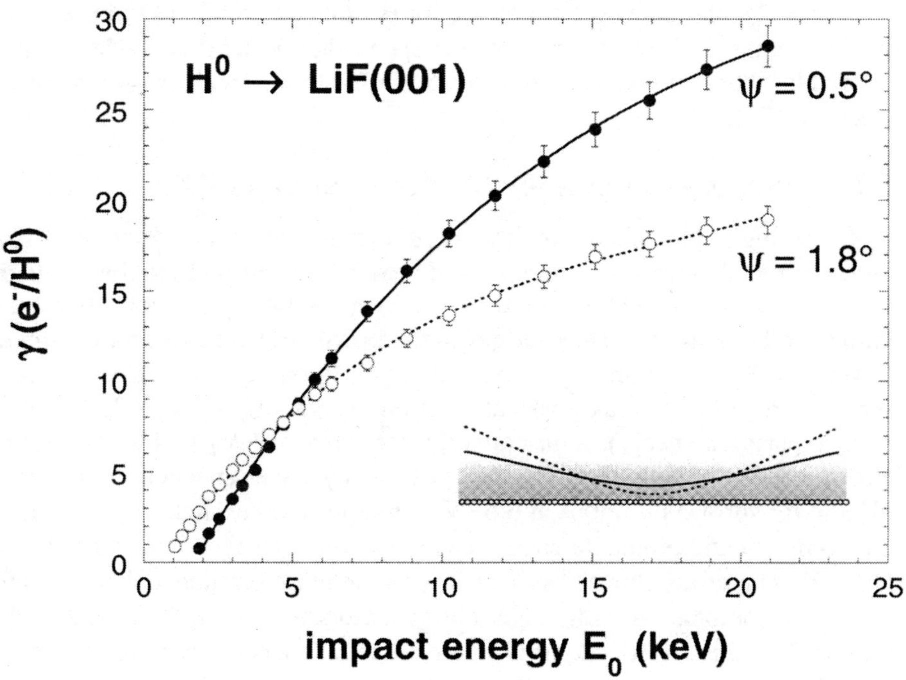

Figure 1.20. Total electron yield as a function of impact energy E for grazing incidence of H^0 on LiF(001) under 0.5^0 (full circles) and 1.8^0 impact angle (open circles), respectively (from [85]).

5. Potential Excitation of Plasmons

5.1 Introduction

Recently it has been found that electron spectra for impact of slow singly [18, 19] and multiply charged ions [19] on quasi-free electron-metals may contain some contributions from one-electron decay of plasmons. The respective signature appears at electron energies $E_p - W_\phi$, (E_p.. target bulk- or -surface plasmon energy). So-called "potential excitation of plasmons" ("PEP") arises if the ion potential energy exceeds $E_p + W_\phi$, but plasmons can also "indirectly" be excited by sufficiently fast electrons (≥ 35eV) from KE. In our own experimental studies on PEP ([20], see below), we found only small contributions to the respective total electron yield, and surface plasmons were generally less probably excited than bulk plasmons. Furthermore, the PEP process appears to be quasi-resonant with respect to the required plasmon energy which suggests that for MCI no significantly stronger PEP contribution should be expected than for singly charged ions.

A special situation was encountered for H^+ impact on Al(111) where a quite strong structure in the respective electron spectra has been explained by plasmon decay [86], but is probably caused by the diffraction of multiply-scattered KE electrons [20].

5.2 Some examples for the observation of PEP

According to theoretical predictions [87] for impact of slow ions ($v < 1au$) on clean metal surfaces an important de-excitation channel for the projectile potential energy should be offered by plasmon excitation (i.e. collective oscillations of the quasi-free electron gas in the solid). Electron spectra for impact of singly [18, 19] and multiply charged ions [19] on quasi-free electron metal surfaces show fairly weak peaks at electron energies $E_p - W_\phi$ (E_p.. bulk- or surface plasmon energy). Consequently, for clean Al ($W_\phi \approx 4.3$eV) the one-electron decay of bulk plasmons ($E_p \approx 15.3$eV) leaves an electron peak at 11 eV and for surface plasmons at 6.6 eV. Bulk plasmon excitation in aluminum due to the kinetic projectile energy can only proceed beyond ca. 40 keV/amu [17–19]. Therefore, slower ions can only excite plasmons due to their potential energy, in competition with Auger electron emission. Such "potential excitation of plasmons – PEP" requires an ion potential energy exceeding the sum of $E_p + W_\phi$.

However, "indirect" excitation of plasmons can take place by sufficiently fast electrons (≥ 35eV) which are produced from KE in the same collision process [88]. Guided by recent studies on PEP [18, 19], we have measured electron spectra for impact of ≤ 10keV H^+, H_2^+, He^+, Ne^{q+} and Ar^{q+} (q = 1, 2) on

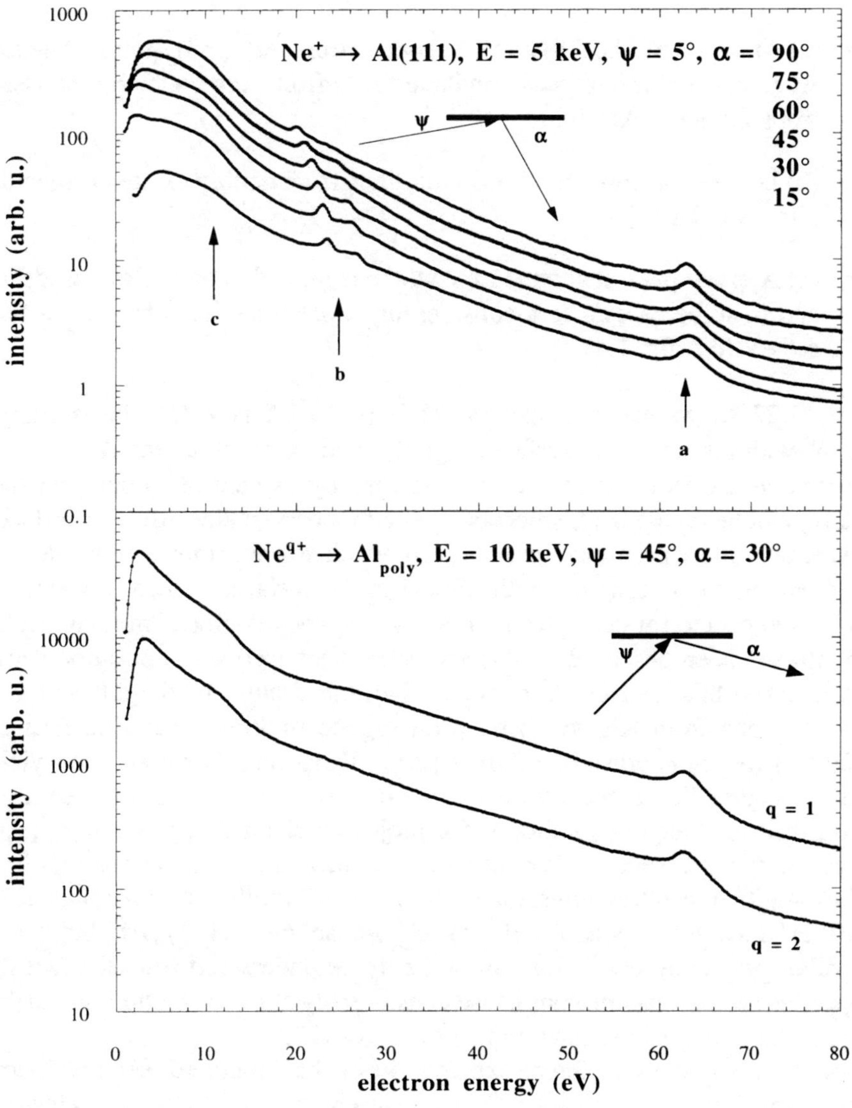

Figure 1.21. (a) Electron spectra (in arbitrary units) for impact of 5 keV Ne^+ on clean Al(111) for 5^0 ion incidence- and various electron emission angles. The different spectra have been off-set with respect to each other to afford a better comparison.
(b) Electron spectra (in arbitrary units) for impact of 10 keV Ne^+ and Ne^{2+} on clean polycrystalline Al for 45^0 ion incidence- and 30^0 electron emission angles. The two spectra have been set-off with respect to each other for better comparability (from [89]).

atomically clean poly- and monocrystalline Al. In these electron spectra the following features have been observed (cf. fig. 1.29 from [89]):

- (a) At ca. 63 eV the Al-LMM Auger electron peak from sputtered neutral Al atoms, and minor peaks on its left side from Auger electron emission from sputtered Al^+ ions.
- (b) Doppler-shifting autoionisation lines from doubly excited projectiles (He^+ and Ne^{q+})
- (c) A broad peak at about 11 eV which is related to bulk-plasmon decay, and in some cases also a considerably weaker surface plasmon peak at 6.6 eV (c.f. fig. 1.4).

Fig. 1.22 shows electron spectra for impact of 5 keV H^+ (impact angle $\Psi = 5^0$ with respect to the surface) on poly- and monocrystalline Al.

For polycrystalline Al only a weak (indirectly excited [88]) bulk plasmon peak (c) can be recognized, whereas for the monocrystalline Al(111) surface a considerably more prominent peak is seen which however moves if the electron emission angle α is changed with respect to the surface, instead of staying at 11 eV as expected for bulk-plasmon decay (see above). Such "moving peaks" have already been observed in [86] and were there ascribed to plasmon decay. In contrast to this interpretation, we explain this feature by the diffraction of slow electrons from KE which have undergone multiple scattering near the surface of the monocrystalline target [20]. Variation of total electron yields when changing the target crystal orientation vs. the incident ion beam can be ascribed to changing conditions for projectile channelling (see, e.g., [90]). However, in the present measurements the ion incidence angle ψ was kept fixed and the angle of electron emission α varied. Similar diffraction features as for 5 keV H^+ have been found for 10 keV H_2^+ impact on Al(111) [20], but not for any other projectile ions. This can probably be understood from the fact that projectiles heavier than protons cause considerably stronger sputtering and thus more surface roughening than impact of H^+ or H_2^+.

As to a completely different explanation of the structured electron spectra arising from impact of H^+ on Al(111), one might regard autodetached electrons from doubly-excited H^{-**} [91]. In this explanation the shift of electron peaks with electron emission angle α can be interpreted as Doppler-shifted energies of autodetached electrons which originate from the moving projectile. However, considerably more experimental and theoretical studies will be needed to decide which of the two explanations is in fact correct or more relevant.

Another remarkable result concerns the relation between the importance of potential excitation of bulk plasmons and the projectile potential energy. Comparing relative plasmon-excitation probabilities for different primary ions

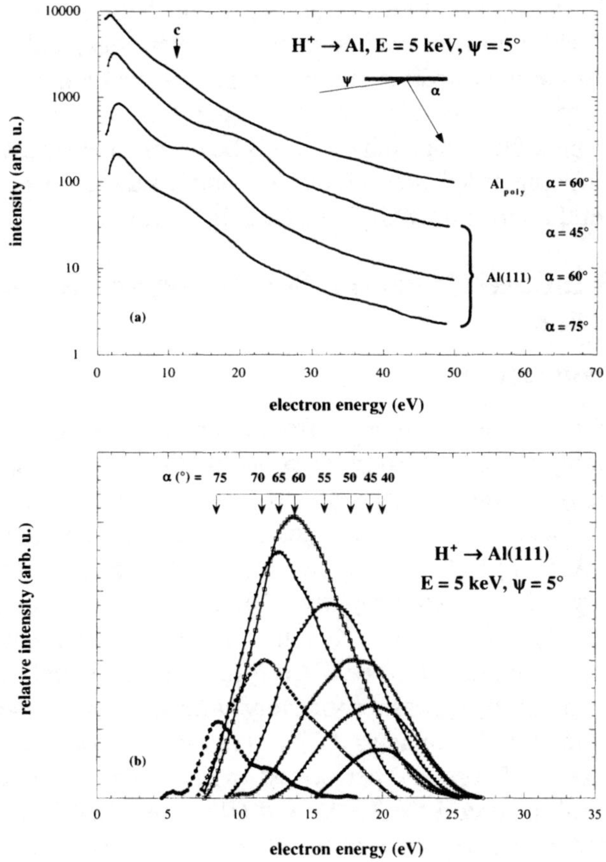

Figure 1.22. (top) Comparison of electron spectra (in arbitrary units) for impact of 5 keV H^+ on poly- and monocrystalline aluminum. Spectra have been set-off with respect to each other for better comparability.
(bottom) Relative size and electron-energy position at different electron emission angles α for the "moving peaks" in electron spectra induced by impact of 5 keV H^+ on Al(111) (from ref. [20]).

suggests an excitation process which is quasi-resonant with respect to the available potential energy [20]. PEP can of course not proceed for too low potential energy (e.g., for H^+ or Ar^+, where the observed plasmons are excited by fast electrons from KE). PEP is apparently most probable for Ne^+ where the ion neutralisation energy practically matches the potential energy which is required for bulk plasmon excitation (see above), and it is clearly less probable for higher potential energy (He^+, Ar^{2+}). In particular, the significance of bulk plasmon peaks in the electron spectra for Ne^+ and Ne^{2+} is not much different for poly- as well as monocrystalline Al surfaces, which can be explained in the following way [20]. The potential energy arising from the first neutralization step $Ne^{2+} \Rightarrow Ne^{+*}$ is too low for excitation of bulk plasmons, and the subsequent de-excitation of the intermediate singly charged excited ion $Ne^{+*} \Rightarrow Ne^{+}$ already provides too much potential energy. Therefore, for impact of Ne^{2+} the probability for bulk plasmon excitation will be dominated by the potential energy of the intermediate Ne^{+} ground state ion, and it is therefore not significantly more important than for impact of Ne^{+} (see fig. 1.21).

6. MCI Induced Potential Sputtering and Secondary Ion Emission

6.1 Introduction

Of all the phenomena which result from interaction of slow MCI with solid surfaces, probably the most intriguing concerns the removal (sputtering, desorption, ablation) of target particles due to the potential projectile energy. First experimental evidence for an ion-charge state dependent sputtering was found for impact of Ar^{q+} ($q \leq 5$) on silicon- and alkali halide surfaces where the secondary ion yields increased rapidly with q [21, 22]. Etching patterns on KCl bombarded by slow Ar- and Kr ions were larger for higher q [23]. On the other hand, for impact of 20 keV Ar^{q+} ($q \leq 9$) on Si only the secondary ion yield was found to increase noticeably with q whereas the total sputter yield, which is mainly due to neutral Si, did not measurably depend on q [24]. AFM (atomic force microscopy) performed on mica irradiated with comparably fast MCI (several hundred keV Xe^{44+} and U^{70+}) revealed blister-like defects which increased with q beyond $q \approx 30$ [8, 25]. For impact of Xe^{q+} ($q \leq 44$) and Th^{q+} ($q \leq 70$) at similar kinetic energies on SiO_2, TOF spectra for positive and negative secondary ions were dominated by single-atomic species, but molecular clusters have also been observed [8]. For impact of Th^{q+} absolute secondary ion yields start to increase with q beyond $q \approx 25$ to up to about 25 for positive- and about 5 for negative secondary ions for Th^{70+}, which indicated total sputtering yields (including neutral secondaries) far larger than the respective kinetic sputtering yields (about 2.5 for 500 keV Th^{+} impact on SiO_2). Further work performed under similar conditions showed that also GaAs, UO_2 and graphite surfaces and thin films are much more efficiently ablated by MCI like Th^{70+} than expected from the involved kinetic projectile energy [8]. These rather strong ablation effects were explained by a combination of effects as “Coulomb explosion” according to Bitenskii et al. [92], “electronic sputtering” (defect-mediated desorption, see below) and “ultrafast electronic excitation” [8].

6.2 Investigation of PS for polycrystalline target films on a quartz microbalance crystal; defect-mediated desorption

In a genuinely different experimental approach (MCI induced ablation of thin films covering the face of a quartz oscillator crystal, cf. fig. 1.13), strongly q-dependent total sputtering yields were demonstrated for impact of much slower ($\geq$ 100 eV) but also considerably less highly charged ions (only up to Xe^{28+}) on alkali halide (LiF, NaCl) [26, 27], SiO_2 [27] and Al_2O_3 [29] insulator surfaces (cf. figs. 1.23 - 1.24).

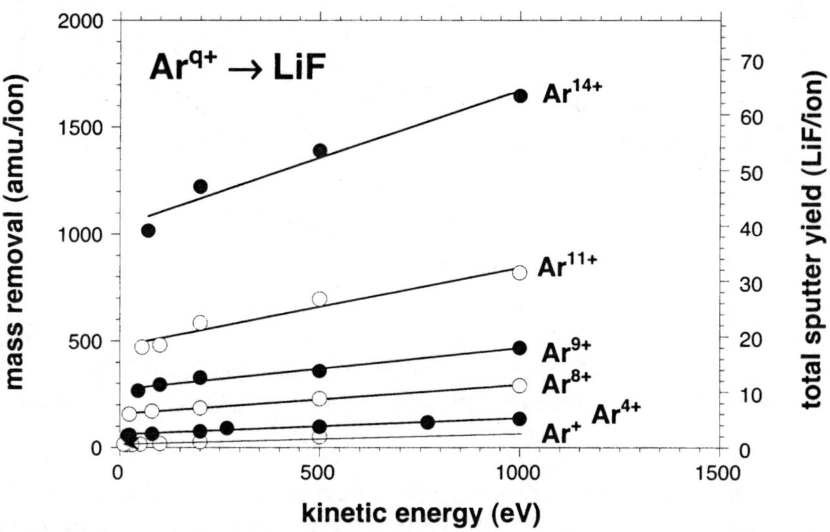

Figure 1.23. Mass removal for sputtering of LiF by highly charged Ar^{q+} (q = 1, 4, 8, 9, 11 and 14) ions as function of ion impact energy (data from ref. [27], lines for guidance only).

The related secondary ion yields were typically smaller by two orders of magnitude than the total sputtering yields. On the other hand, for MCI impact on Au, Si, and GaAs only kinetic sputtering independent of q was observed. This q-dependent sputtering is explained by "defect-mediated desorption", a process similar to electron- and photon-stimulated desorption (ESD, PSD). The mechanism proceeds by electronic defect production due to the MCI induced electron capture. Because of electron-phonon coupling in the insulator lattice, the electronic defects can become localized ("self-trapping of excitons and holes") and in further consequence give rise to the observed strong desorption. Such localization is not possible for many other insulators and definitely not for metals and semiconductors (for more detailed explanations see [93] and recent refinements in [28]).

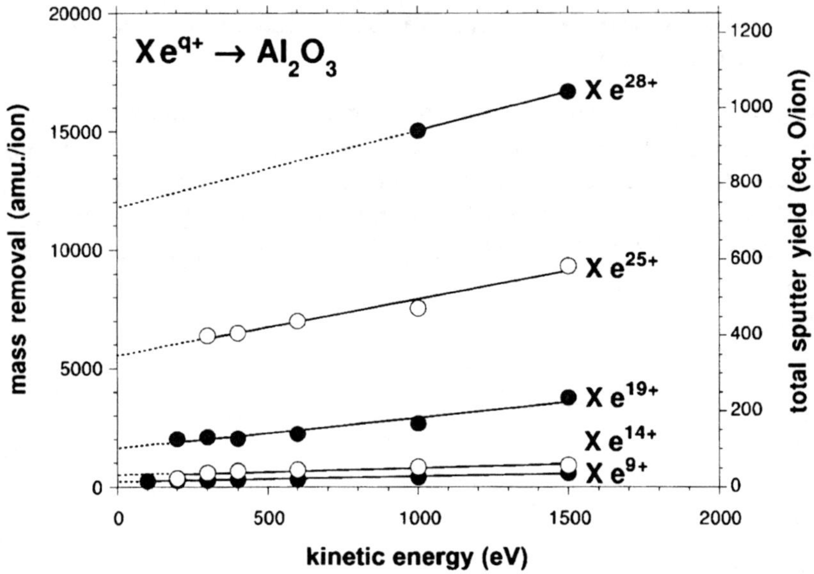

Figure 1.24. Mass removal for sputtering of Al_2O_3 by highly charged Xe^{q+} ($q = 9$, 14, 19, 25 and 28) ions as function of ion impact energy (data from ref. [29], lines for guidance only).

However, recent studies with an insulator target (MgO) for which no electron-phonon coupling is given [29] indicate that the localization of capture-induced electronic defects, which is necessary for desorption to proceed, could also be mediated or at least supported by the kinetic projectile energy (cf. fig. 1.25).

In this context, another example concerns the damage imposed on self-assembled monolayers of alkano-ethiolates on gold substrates by impact of 0.1 eV metastable Ar and 350 keV Xe^{44+} [94]. The authors of this study have argued that the much larger (by five orders of magnitude) damaging effect due to MCI impact is caused by a much larger (by a factor of ca. 5000) potential projectile energy. However, the kinetic projectile energies involved are also relevant, considering their ratio of more than one million! It is thus important to conduct such studies at sufficiently low MCI impact energies in order to clearly assess the role of the potential projectile energy.

This rather interesting observation requires further systematic studies, preferably by means of UHV-AFM (atomic force microscopy) techniques which are especially suited for investigating single ion-impact induced defects on well defined monocrystalline target surfaces. AFM measurements on insulator crystals (e.g. Al_2O_3) are currently in progress at TU Wien (as an example, cf. fig. 1.26):

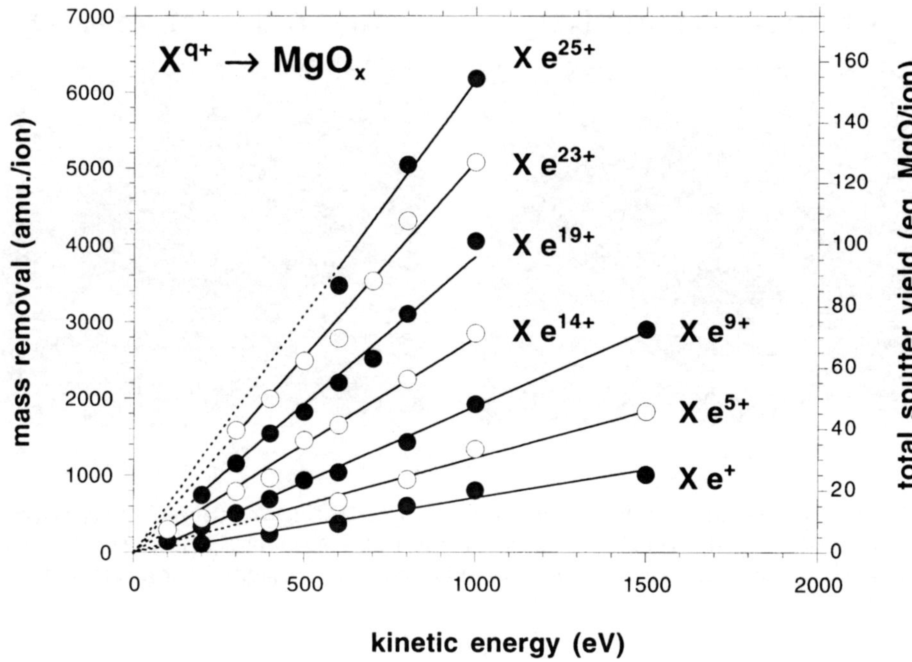

Figure 1.25. Mass removal for sputtering of MgO_x by highly charged Xe^{q+}(q = 1, 5, 9, 14, 19, 23, and 25) ions as function of ion impact energy (data from ref. [29]). Lines for guidance only.

Before bombardment, the Al_2O_3 c-plane (0001) single crystal surface is CO_2 snow cleaned and then annealed for three hours at 400° C in UHV. This preparation technique yields very flat crystal surfaces (roughness of 0.093±0.06 nm rms, left part of fig. 1.26). Bombardment with Ar ions of different charge states and kinetic energies result in hillock-like nanodefects (right part of fig. 1.26). However, the density of these nanodefects did not correspond to the applied ion dose (ca. 5×10^{12} ions/cm^2). 500 eV Ar^+ ion impact caused defects with about one nanometer height and some tens of nanometers width (left part of fig. 1.26), while defects produced by 500 eV Ar^{7+} ions were several nenometers high and had lateral dimensions of about 100 (!) nanometers (right part of fig. 1.26).

Al_2O_3 might therefore be a good candidate for nanostructuring via potential sputtering and therefore of relevance for attractive applications of HCI-surface interactions, in the rapidly emerging field which combines microelectronics with nanotechnology.

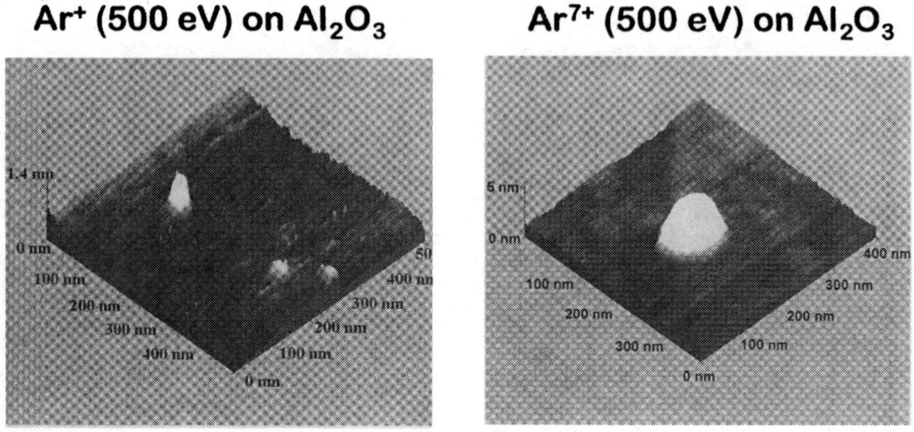

Figure 1.26. Atomic-force microscopic image of an Al_2O_3 single crystal surface bombarded with 500 eV Ar^+ (left) and Ar^{7+} ions (right) as seen in UHV AFM contact mode. The observed defect size (both height and lateral dimension) increase with the projectile charge state (from [95]).

7. Summary and Outlook

In this survey we have described experimental techniques and the state of knowledge on phenomena which are induced by inelastic interaction of slow ions with clean solid surfaces. We have focused on recent developments in the already well-established areas of potential- and kinetic electron emission which are of continuing practical importance. In addition, we have described progress in understanding of the more recently recognized phenomena of plasmon excitation in free-electron metal surfaces (PEP) and sputtering of insulator surfaces (PS), which are both induced by the potential energy of the projectile ions. Nowadays the common availability of compact multiply-charged ion sources and affordable technology for preparation and maintenance of clean solid surfaces in ultrahigh vacuum environment has contributed greatly to the status in these fields which, however, require and also deserve further work, to obtain a satisfactorily comprehensive understanding of the processes considered here and others closely related to them.

Acknowledgments

The research conducted at TU Wien which has been described in this survey in sections 1.3 - 1.6 has been supported by the Austrian Research Foundation FWF. Part of the work on PE was carried out within the 3. Framework Human Capital and Mobility Programme of the European Union under contract No. CHRT-CT93-0103.

References

[1] H. D. Hagstrum, Phys.Rev. **96**, 325 (1954)

[2] H. D. Hagstrum, Phys.Rev. **96**, 336 (1954)

[3] U. A. Arifov, E. S. Mukhamadiev, E. S. Parilis, and A. S. Pasyuk, Sov.Phys.Tech.Phys. **18**, 240 (1973)

[4] J. Burgdörfer, P. Lerner, and F. W. Meyer, Phys.Rev.A **44**, 5674 (1991)

[5] F. Aumayr and HP. Winter, Comments At.Mol.Phys. **29**, 275 (1994)

[6] A. Arnau, F. Aumayr, P. M. Echenique, M. Grether, W. Heiland, J. Limburg, R. Morgenstern, P. Roncin, S. Schippers, R. Schuch, N. Stolterfoht, P. Varga, T. J. M. Zouros, and HP. Winter, Surf. Sci. Reports **229**, 1 (1997)

[7] HP. Winter and F. Aumayr, J. Phys. B: At. Mol. Opt. Phys. **32**, R39 (1999)

[8] T. Schenkel, A. V. Hamza, A. V. Barnes, and D. H. Schneider, Progr. Surf. Sci. **61**, 23 (1999)

[9] HP. Winter and F. Aumayr, Physica Scripta **T92**, 15 (2001)

[10] J. D. Gillaspy, J. Phys. B: At. Mol. Opt. Phys. **34**, R93 (2001)

[11] N. Stolterfoht, chapter 2 in this book.

[12] M. Rösler and W. Brauer, in *Particle Induced Electron Emission I*, edited by G. Höhler (Springer, Berlin, 1991), Vol. 122.

[13] D. Hasselkamp, in *Particle Induced Electron Emission II*, edited by G. Höhler (Springer, Heidelberg, 1992), Vol. 123, p. 1

[14] R. Baragiola, in *Chap. IV in Low energy Ion-Surface Interactions*, edited by J. W. Rabalais (Wiley, 1993)

[15] H. Eder, F. Aumayr, and HP. Winter, Nucl.Instrum.Meth.B **154**, 185 (1999)

[16] J. Lörincik, Z. Sroubek, H. Eder, F. Aumayr, and HP. Winter, Phys. Rev. B **62**, 16116 (2000)

[17] H. Raether, *Surface Plasmons,Springer Tracts in Modern Physics 111, Springer, Berlin* (Springer, Berlin, 1988)

[18] R. A. Baragiola and C. A. Dukes, Phys. Rev. Lett. **76**, 2547 (1996)

[19] D. Niemann, M. Grether, M. Rösler, and N. Stolterfoht, Phys. Rev. Lett. **80**, 3328 (1998)

[20] H. Eder, F. Aumayr, P. Berlinger, H. Störi, and HP. Winter, Surf. Sci. **472**, 195 (2001)

[21] S. S. Radzhabov, R. R. Rakhimov, and P. Abdusalumav, Izv.Akad.Nauk SSSR Ser.Fiz. **40**, 2543 (1976)

[22] S. N. Morozov, D. D. Gurich, and T. U. Arifov, Izv.Akad.Nauk SSSR Ser.Fiz. **43**, 137 (1979)

[23] S. S. Radzhabov and R. R. Rakhimov, Izv.Akad.Nauk SSSR Ser.Fiz. **49**, 1812 (1985)

[24] S. T. de Zwart, T. Fried, D. O. Boerma, R. Hoekstra, A. G. Drentje, and A. L. Boers, Surf.Sci. **177**, L939 (1986)

[25] D. H. Schneider, M. A. Briere, J. McDonald, and J. Biersack, Rad.Eff.Def.Solids **127**, 113 (1993)

[26] T. Neidhart, F. Pichler, F. Aumayr, HP. Winter, M. Schmid, and P. Varga, Phys.Rev.Lett. **74**, 5280 (1995)

[27] M. Sporn, G. Libiseller, T. Neidhart, M. Schmid, F. Aumayr, HP. Winter, P. Varga, M. Grether, and N. Stolterfoht, Phys.Rev.Lett. **79**, 945 (1997)

[28] G. Hayderer, M. Schmid, P. Varga, H. Winter, F. Aumayr, L. Wirtz, C. Lemell, J. Burgdörfer, L. Hägg, and C. O. Reinhold, Phys.Rev.Lett. **83**, 3948 (1999)

[29] G. Hayderer, S. Cernusca, M. Schmid, P. Varga, H. Winter, F. Aumayr, D. Niemann, V. Hoffmann, N. Stolterfoht, C. Lemell, L. Wirtz, and J. Burgdörfer, Phys.Rev.Lett. **86**, 3530 (2001)

[30] B. Wolf, *Handbook of Ion Sources* (CRC Press, Boca Raton, New York, London, Tokyo, 1995)

[31] R. Geller, B. Jacquot, and M. Pontonnier, Rev.Sci.Instrum. **56**, 1505 (1985)

[32] P. Sortais, Nucl.Instrum.Methods B **98**, 508 (1995)

[33] E. D. Donets, in *The Physics and Technology of Ion Sources*, edited by I. G. Brown (John Wiley, New York, 1989), p. chapt. 12.

[34] E. D. Donets, in *HCI-92*, edited by P. Richard, M. Stöckli, C. L. Cocke and C. D. Lin (AIP Conf. Proc., Manhattan, Kansas, 1992), Vol. 274, p. 663.

[35] R. E. Marrs, M. A. Levine, K. D. A, and J. R. Henderson, Phys.Rev.Lett. **60**, 1715 (1988)

[36] R. E. Marrs, P. Beiersdorfer, and D. Schneider, Physics Today **10**, 27 (1994)

[37] I. G. Brown, *The Physics and Technology of Ion Sources* (John Wiley, New York, 1989)

[38] V. P. Ovsyannikov and G. Zschornack, Rev. Sci. Instrum. **70**, 2646 (1999)

[39] A. J. T. Holmes, in *The Physics and Technology of Ion Sources (Chap. 4)*, edited by I. G. Brown (John Wiley, New York, 1989)

[40] E. Taglauer, Appl.Phys. A **51**, 238 (1990)

[41] S. Speller, W. Heiland, and M. Schleberger, Exper. Meth. in Phys. Sci. **38**, 1 (2001)

[42] H. Winter, Progr. Surf. Sci. **63**, 177 (2000)

[43] H. Niehus, W. Heiland, and E. Taglauer, Surf. Sci. Reports **17**, 213 (1993)

[44] A. Niehof and W. Heiland, Nucl. Instrum. Meth. Phys. Res. B **48**, 306 (1990)

[45] L. Folkerts, S. Schippers, D. M. Zehner, and F. W. Meyer, Phys.Rev.Lett. **74**, 2204 (1995)

[46] C. Röthig, (University Osnabrück, Germany, 1994)

[47] R. M. Jaeger, H. Kuhlenbeck, H. J. Freund, M. Wuttig, W. Hoffmann, R. Franchy, and H. Ibach, Surf.Sci. **259**, 235 (1991)

[48] U. Diebold, J. M. Pan, and T. E. Madey, Surf.Sci. **331-333**, 845 (1995)

[49] G. Lakits, F. Aumayr, and HP. Winter, Rev.Sci.Instrum. **60**, 3151 (1989)

[50] F. Aumayr, G. Lakits, and HP. Winter, Appl.Surf.Sci. **47**, 139 (1991)

[51] H. Kurz, F. Aumayr, C. Lemell, K. Töglhofer, and HP. Winter, Phys.Rev.A **48**, 2192 (1993)

[52] F. Aumayr, H. Kurz, D. Schneider, M. A. Briere, J. W. McDonald, C. E. Cunningham, and HP. Winter, Phys.Rev.Lett. **71**, 1943 (1993)

[53] M. Vana, F. Aumayr, P. Varga, and HP. Winter, Europhys. Lett. **29**, 55 (1995)

[54] M. Vana, F. Aumayr, P. Varga, and HP. Winter, Nucl. Instrum. Meth. Phys. Res. B **100**, 284 (1995)

[55] F. Aumayr, T. D. Märk, and HP. Winter, Intern. J. Mass Spectrom. Ion Processes **129**, 17 (1993)

[56] M. Vana, F. Aumayr, C. Lemell, and HP. Winter, Intern. J. Mass Spectr. Ion Proc. **149/150**, 45 (1995)

[57] C. Lemell, J. Stöckl, J. Burgdörfer, G. Betz, HP. Winter, and F. Aumayr, Phys.Rev.Lett. **81**, 1965 (1998)

[58] C. Lemell, J. Stöckl, HP. Winter, and F. Aumayr, Rev. Sci. Instrum. **70**, 1653 (1999)

[59] W. Heiland and E. Taglauer, Meth.Exp.Phys. **22**, 299 (1985)

[60] G. Hayderer, M. Schmid, P. Varga, H. Winter, and F. Aumayr, Rev.Sci.Instrum. **70**, 3696 (1999)

[61] J. Burgdörfer, in *Fundamental Processes and Applications of Atoms and Ions*, edited by C. D. Lin (World Scientific, 1993)

[62] H. Winter, Europhys.Lett. **18**, 207 (1992)

[63] F. W. Meyer, L. Folkerts, H. O. Folkerts, and S. Schippers, Nucl.Instrum.Meth.Phys.Res.B **98**, 441 (1995)

[64] L. Hägg, C. O. Reinhold, and J. Burgdörfer, Phys.Rev.A **55**, 2097 (1997)

[65] C. Auth, T. Hecht, T. Igel, and H. Winter, Phys.Rev.Lett. **74**, 5244 (1995)

[66] Q. Yan and F. W. Meyer, in *Materials Science Forum* (Trans Tech. Publications, Switzerland, 1997), Vols. 239 - 241, p. 629.

[67] J. Ducree, H. J. Andrä, and U. Thumm, Phys. Rev. A **60**, 3029 (1999)

[68] M. Delaunay, M. Fehringer, R. Geller, D. Hitz, P. Varga, and HP. Winter, Phys.Rev.B **35**, 4232 (1987)

[69] Y. Yamazaki, M. Kakutani, and T. Azuma, J. Phys. Soc. Jap. **65**, 1199 (1996)

[70] K. Tökesi, L. Wirtz, C. Lemell, and J. Burgdörfer, Phys. Rev. A **61**, 020901(R) (2000)

[71] N. Stolterfoht, private communication (2001)

[72] J. P. Briand, G. Giardino, G. Borsoni, V. Le Roux, N. Bechu, S. Dreuil, O. Tüske, and G. Machicoane, Rev. Sci. Instrum. **71**, 627 (2000)

[73] A. Arnau, P. A. Zeijlmans van Emmichoven, J. I. Juaristi, and E. Zaremba, Nucl.Instrum.Meth.Phys.Res.B **100**, 279 (1995)

[74] J. Burgdörfer, C. Reinhold, and F. Meyer, Nucl.Instrum.Meth.Phys.Res.B **98**, 415 (1995)

[75] N. Stolterfoht, A. Arnau, M. Grether, R. Köhrbrück, A. Spieler, R. Page, A. Saal, J. Thomaschewski, and J. Bleck-Neuhaus, Phys.Rev.A **52**, 445 (1995)

[76] J. Mrogenda, J. Ducree, E. Reckels, J. Leucker, and H. J. Andrä, Appl. Surf. Sci. **136**, 269 (1998)

[77] J. Stöckl, C. Lemell, HP. Winter, and F. Aumayr, Phys. Scr. **T92**, 135 (2001)

[78] H. Kurz, K. Töglhofer, HP. Winter, F. Aumayr, and R. Mann, Phys.Rev.Lett. **69**, 1140 (1992)

[79] F. Aumayr, G. Betz, T. D. Märk, P. Scheier, and HP. Winter, Intern. J. Mass Spectrom. Ion Processes **174**, 317 (1998)

[80] R. A. Baragiola, E. V. Alonso, and A. Olivia-Florio, Phys. Rev. B **19**, 121 (1979)

[81] U. Fano and W. Lichten, Phys.Rev.Lett. **14**, 627 (1965)

[82] U. Wille and R. Hippler, Phys.Rep. **132**, 129 (1986)

[83] G. Spierings, I. Urazgil'din, P. A. Zeijlmans van Emmichoven, and A. Niehaus, Phys.Rev.Lett. **74**, 4543 (1995)

[84] J. A. Yarmoff, T. D. Liu, S. R. Qui, and Z. Sroubek, Phys. Rev. Lett. **80**, 2469 (1998)

[85] H. Eder, K. Mertens, K. Maass, H. Winter, HP. Winter, and F. Aumayr, Phys. Rev. A **62**, 052901 (2000)

[86] B. vanSomeren, P. A. Zeijlmans van Emmichoven, I. F. Urazgil'din, and A. Niehaus, Phys. Rev. A **61**, 032902 (2000)

[87] P. Apell, J.Phys.B:At.Mol.Opt.Phys. **21**, 2665 (1988)

[88] E. A. Sanchez, J. E. Gayone, M. L. Martiarena, O. Grizzi, and R. A. Baragiola, Phys. Rev. B **61**, 14209 (2000)

[89] HP. Winter, H. Eder, F. Aumayr, J. Lörincik, and Z. Sroubek, Nucl. Instrum. Meth. Phys. Res. B **182**, 15 (2001)

[90] N. Benazeth, Nucl. Instrum. Meth. Phys. Res. B **194**, 405 (1982)

[91] J. Lörincik, Z. Sroubek, F. Aumayr, and HP. Winter, Europhys. Lett. **54**, 633 (2001)

[92] I. S. Bitenskii, M. N. Murakhmetov, and E. S. Parilis, Sov.Phys.Tech.Phys. **24**, 618 (1979)

[93] F. Aumayr, J. Burgdörfer, P. Varga, and HP. Winter, Comments At.Mol.Phys. **34**, 201 (1999)

[94] L. P. Ratliff, E. W. Bell, D. C. Parks, A. I. Pikin, and J. D. Gillaspy, Appl. Phys. Lett. **75**, 590 (1999)

[95] I. C. Gebeshuber, S. Cernusca, F. Aumayr, and HP. Winter, Proc. 11th Int. Conf. on Highly Charged Ions (HCI-2002), Caen/France, 1.-6. 9. 20002, Nucl. Instrum. Meth. Phys. Res. B. *to be published*

Chapter 2

INTERACTION OF SLOW HIGHLY CHARGED IONS WITH SURFACES

Yasunori Yamazaki
Institute of Physics, University of Tokyo,
Komaba, Meguro, Komaba, Tokyo Japan
and
Atomic Physics Laboratory, RIKEN, Hirosawa, Wako,
Saitama, Japan
yasunori@phys.c.u-tokyo.ac.jp

Abstract Interaction of a slow highly-charged ions (HCIs) with metal and semiconductor surfaces is discussed with particular attention on findings uniquely obtained with microcapillary targets and hydrogen-terminated Si surfaces. The microcapillary targets enable for the first time extraction of hollow atoms (ions), *i.e.*, atoms (ions) in highly- and multiply-excited states, into vacuum. The intrinsic nature of hollow atoms (ions) was studied through X-ray and visible light measurements. X-ray measurements revealed that a considerable fraction of charge-changed ions was in meta-stable states keeping innershell holes with lifetimes of ∼ns or longer, *i.e.*, metastable multiply excited states were effectively formed. Spin-aligned states with less than half-filled innershells and/or highly excited high angular momentum states were proposed as possible candidates to explain such extreme metastability. The angular distributions of these metastable ions were as narrow as those of the incident HCIs, *i.e.*, a high quality beam of hollow atoms (ions) can be prepared with the microcapillary technique. Visible light measurements provide detailed information on electronic states formed at the beginning of the HCI-surface interactions. It was found that principal quantum numbers of the states were around the incident charge states, which is consistent with the prediction of the classical over barrier model. Proton sputtering from well-defined H-terminated Si surfaces were studied. It was found that the proton yields were proportional to $q^{\sim 5}$. Desorption induced with double electron transfer from a chemical bond was proposed as a principal mechanism of the proton sputtering with HCIs.

Keywords: slow highly charged ions, hollow atom (ion), microcapillary, high sensitivity surface analysis, potential sputtering

F.J. Currell (ed.), The Physics of Multiply and Highly Charged Ions, Vol. 2, 47-67.

1. Introduction

Interaction of slow highly-charged ions (HCIs) with metal, semiconductor, and insulator surfaces has been intensively studied in the last two decades [1–9] because of the exotic nature of the collision dynamics, which involves formation and evolution processes of hollow atoms on the HCI, sputtering induced with HCIs, abrupt deposition of very large potential energy in a localized area of the target, *etc*. The latter two processes are expected to be useful for high-sensitivity surface analysis and surface modification. Charge state evolutions above and below the surface with multiple electron transfers are the key process to govern these phenomena.

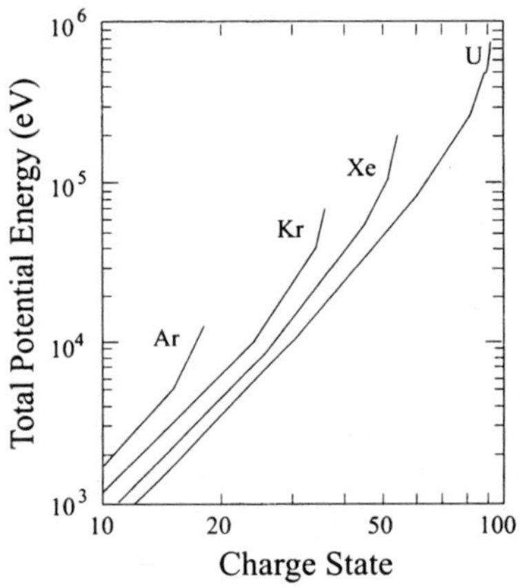

Figure 2.1. Potential energy of Ar, Kr, Xe, and U ions as a function of charge state q.

HCIs are characterized by two parameters, *i.e.*, the charge state q, and the potential energy ϵ_q, the energy to be released when the HCI is neutralized. The relation between q and ϵ_q is shown in fig. 2.1 for several HCIs. It is seen that ϵ_q depends not only on q but also on elements, which is particularly true when K-shell electrons are involved. For example, $\epsilon_q(Kr^{54+})$ is $\sim$3 times larger than ϵ_q (U^{54+}) although $\epsilon_q(Kr^{10+})$ is only $\sim$10% larger than $\epsilon_q(U^{10+})$. Very crudely, collision processes with large impact parameters are governed by q, but those with small impact parameters are by ϵ_q.

A general scenario on the interaction of a slow HCI with a surface is briefly discussed in section 2.2. It will be shown that a flat surface target has conceptual limitations in studying the interaction of HCI with a surface. Section 2.3 is devoted to a microcapillary technique, which enables study into the nature of isolated hollow atoms (ions) overcoming the difficulty involved in the flat surface experiment. In section 2.4, proton sputtering induced with HCIs is discussed, which is induced as a result of multiple electron transfer.

2. Fundamental Process in HCI-Surface Interaction - Classical Over Barrier Mechanism

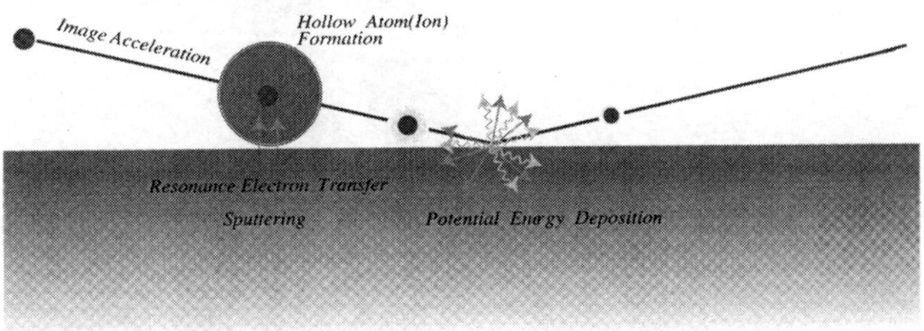

Figure 2.2. A schematic view of atomic processes taking place when a slow highly-charged ion approaches a metal/semiconductor surface.

When a slow HCI approaches a metal surface, it is accelerated toward the surface due to its image force, then at a certain distance, starts to capture target electrons, and finally jumps into the surface releasing all its potential energy through emissions of Auger electrons and X-rays. In the case of a glancing angle of incidence, a considerable fraction of ions are specularly reflected when the target surface is atomically flat. A schematic view of the processes involved is given in fig. 2.2. The active electron to be transferred is subject to the following interactions; the interaction with the HCI, with the image of the HCI, and with the image of the electron itself (see fig. 2.3(a)). The interaction potential is given by

$$V(\vec{r}_e, \vec{R}) = -\frac{q}{|\vec{r}_e - \vec{R}|} + \frac{q}{|\vec{r}_e + \vec{R}|} - \frac{1}{4|z|} \tag{2.1}$$

where $\vec{r}_e(x, 0, z)$ is the electron position with z-axis perpendicular to the surface, and $\vec{R}(0, 0, d)$ is the position of the HCI.

Fig. 2.3(b) is a 3D representation of $V(\vec{r}_e, \vec{R})$ with a HCI at $\vec{R} = (0, 0, 5)$. It is seen that a saddle point appears along the line connecting the HCI and its image charge, through which target electrons can be transferred. The position

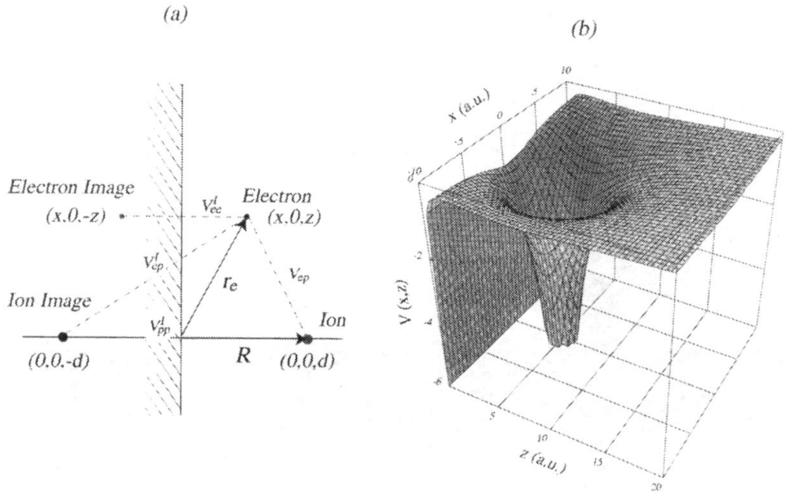

Figure 2.3. (a) Forces acting on an electron. (b) 3D representation of $V(\vec{r}_e, \vec{R})$. The HCI of q=10 is located at $d = 5a.u.$

of the saddle point and its depth are evaluated to be $d/(8q)^{1/2}$ and $-(2q)^{1/2}/d$, respectively, $i.e.$, the saddle point gets deeper as d gets smaller. When the potential at the saddle point is shallower than the workfunction of the metal, valence electrons can be transferred to the HCI only via a tunneling process. On the other hand, when it is equal to or deeper than the workfunction of the metal, electrons are transferred classically overcoming the barrier (which will be called classical over barrier (COB) condition, hereafter). Because the transition rate under the COB condition is much higher than that under tunneling condition [10], electron transfer phenomena in slow HCI-surface collisions are more or less governed by the period when the COB condition is satisfied [7]. Equating the depth of the saddle point to be the workfunction W, the critical distance d_c, where the first electron transfer takes place via the COB mechanism, is evaluated as

$$d_c \sim \frac{\sqrt{2q}}{W}. \tag{2.2}$$

For a typical metal having $W \sim 5eV (\sim 0.2)$, d_c is of the order of nm. Because the electron transfer takes place resonantly, the binding energy of the transferred electron is similar to the workfunction, $i.e.$, the electron is transferred to high

Rydberg states of the HCI. The principal quantum number n_c in which the first electron is transferred is given by

$$n_c \sim \frac{q}{\sqrt{2W(1+\sqrt{q/8})}}. \quad (2.3)$$

n_c for $W = 0.2$ is shown by the solid line in fig. 2.4 as a function of q. The dashed line corresponds to $n = q$, which more or less reproduces n_c for $q \lesssim 20$. The HCI captures the second electron as soon as $d \sim \sqrt{2(q-1)}/W$ (eq. 2.2) is satisfied, *i.e.*, electrons are successively transferred into high Rydberg states of the HCI as it gets closer to the surface. Such an atomic state with multiple electrons in Rydberg states still keeping inner shell hole(s) is called a hollow atom formed above the surface (HAA). It is noted that the radii of HAA are comparable to d_c, and accordingly the electrons can move back and forth between the ion and the target when they are in resonance. In this respect, the hollow atom in front of the surface had better be called a "dynamic hollow molecule".

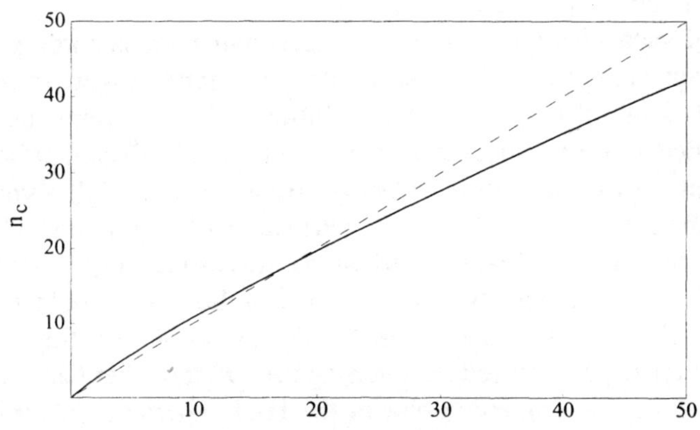

Figure 2.4. The solid line shows n_c predicted by eq. 2.3 for $W = 0.2$. The dashed line shows $n = q$, which is seen to be a good approximation of n_c for $q \lesssim 20$.

The charge state of the HCI in front of the surface evolves according to eq. 2.2 like

$$q(d) \sim \begin{cases} q & (d \geq d_c) \\ \frac{1}{2}W^2d^2 & (d < d_c) \end{cases} \quad (2.4)$$

Taking this charge state variation, the total kinetic energy of the HCI at d from the surface due to the image acceleration, $\Delta\epsilon_{im}(d)$, is given by

$$\Delta\epsilon_{im}(d) \sim -\int_{\infty}^{d} \frac{q(z)^2}{4z^2} dz = \begin{cases} \frac{q^2}{4d} & (d \geq d_c) \\ \frac{Wq^{3/2}}{3\sqrt{2}} - \frac{W^4}{48} d^3 & (d < d_c). \end{cases} \tag{2.5}$$

Accordingly, the time interval, τ_{HA}, from the first electron capture till the HCI jumps into the surface is estimated as

$$\tau \sim \int_0^{d_c} \frac{dz}{v} = \sqrt{\frac{m}{2}} \int_0^{d_c} \frac{dz}{\sqrt{\Delta\epsilon_{im}}} \sim \sqrt{\frac{3\sqrt{2}m}{W^3 q^{1/2}}} \tag{2.6}$$

or

$$\tau \sim 2.4 \times 10^{-14} \frac{m(\text{amu})^{1/2}}{q^{1/4}} (\text{sec}), \tag{2.7}$$

which is typically $\sim 10^{-13}$sec or shorter.

Innershell vacancies with lifetimes longer than τ are accordingly filled via quasi-resonant charge transfer and/or inter- and intra-Auger transitions after the HCI jumps into the surface. The evolution of hollow atoms near a surface is then divided into two successive stages; (1) a soft above-surface collision accompanying resonant multiple electron transfers into high Rydberg states, and (2) a hard collision at or below the surface involving innershell transitions (hereafter referred to as Stage I and Stage II, respectively). Hollow atoms formed at the Stage II will be referred to as hollow atom below the surface hereafter (HAB). As is evident from the discussion above, the major part of the potential energy is released during Stage II. On the other hand, in the Stage I, the area around the entrance point of the HCI experiences a strong charge-up, which is re-neutralized by valence electrons around the entrance point with a time constant τ_{re}. τ_{re} is expected to be $\sim \omega_{pl}^{-1}$ for metals, where ω_{pl} is the plasma angular frequency. For insulators, τ_{re} is $\sim$ 10, which is estimated from a typical band width or a hopping matrix element between the nearest neighbor atoms. It is expected that the Coulomb repulsion between charged particles formed on the surface during Stage I induces particle emission when reneutralization of the charged area is slow.

The effect of the image acceleration of HCI toward the surface was successfully monitored through measurements of glancing angle scattering [11]. The left top of fig. 2.5 visualizes the way to detect the image acceleration. The HCI is attracted toward the surface due to the image force until it is neutralized (see also fig. 2.2), and specularly reflected at the surface. Because the reflected ion

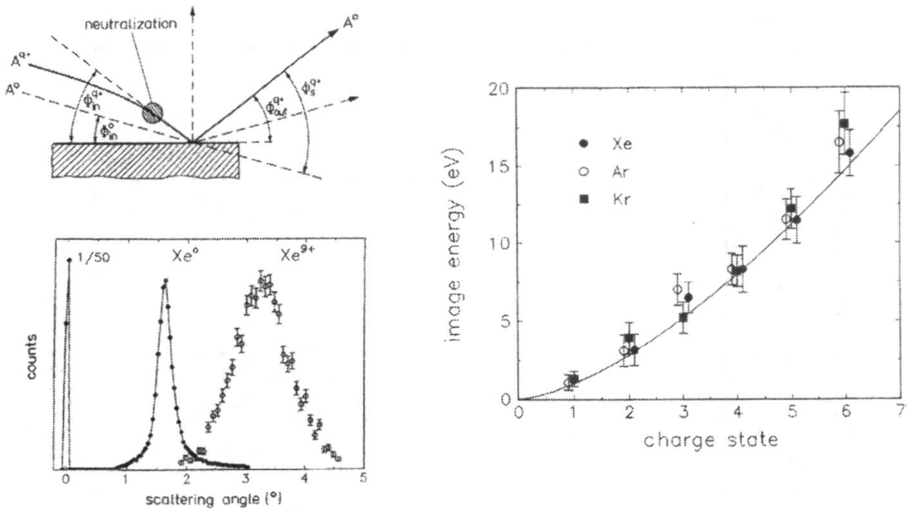

Figure 2.5. Left Top: A schematic drawing of a glancing scattering of a HCI taking into account the image acceleration and neutralization at the surface. Left Bottom: Angular distribution of 25keV Xe^0 and Xe^{9+} ions scattered from an Al(111) surface. Right: Energy gain of HCI as a function of the charge state q for Xe, Ar, and Kr ions.[11]

is almost neutralized (see fig. 2.6), the particle trajectory on their way out is not affected by the image force. The left bottom of fig. 2.5 shows the observed angular distributions of scattered Xe^0 and Xe^{9+} ions with the same incidence angle, which tells that the scattering angle of Xe^{9+} ions is much larger than that of Xe^0. The angular variation was converted to the energy gain of the HCI during the image acceleration, which is plotted in the right half of fig. 2.5. It is seen that the transferred energy is almost independent of atomic number of the ion but depends on q. The solid line in the right half of fig. 2.5 shows $\Delta\epsilon_{im}(0)$ for $W \sim 4.3$eV, which reproduced the experimental findings quite satisfactorily.

Fig. 2.6 shows a charge state distribution of 3.75keV/u O^{q+} ions ($q = 3 - 8$) specularly reflected from a Au(110) surface [12]. It is seen that the major fraction of the reflected ions were neutralized independent of q, *i.e.*, their innershell holes had been filled before they left the surface area. The time necessary to form the neutral ground state was estimated to be only 30fs or less. It seems that some very quick innershell filling processes take place when HCIs are at or very near the surface in addition to the COB process.

The fraction of the potential energy deposited to a solid target from HCIs was measured employing a silicon surface barrier detector as a target [13]. It

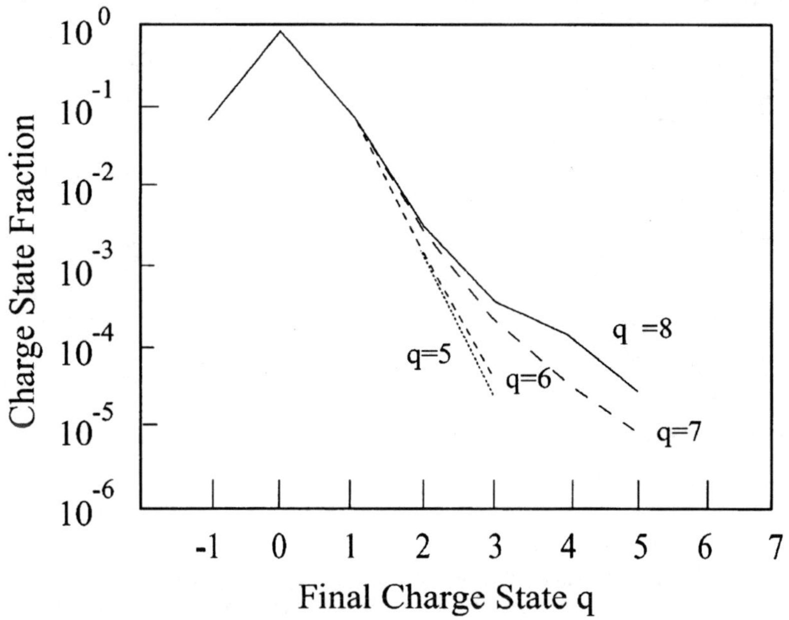

Figure 2.6. Charge state distribution of specularly reflected ions for 3.75keV/u O^{q+} on Au(110) [12].

was found that ~35% (or 60keV) of the potential energy of Au^{69+} ions was dissipated in the depletion layer of the target.

A part of the potential energy is also released via X-ray emission. High resolution X-ray spectroscopy was applied for HCI-surface collisions, where HABs were successfully observed [1].

Various experiments had been done to sort out processes taking place above or below the surface, *i.e.* the Stage I or Stage II, through *e.g.*, measurements of Auger electron spectra. Some Auger electron peaks were identified to have above-surface origin. However, some controversial observations were also reported. Fig. 2.7 plots the intensity of K-Auger electrons as a function of emission angle for 135eV Ne^{9+} bombarding Al(111) [6]. The open squares are the intensity of a K-Auger electron peak supposed to have above-surface origin and the solid squares are for the total K-Auger electrons. Although the above-surface component was expected to show no emission angle dependence, the intensity of the "above-surface peak" followed that of the total K-Auger peak which were emitted from ions already at or below the surface.

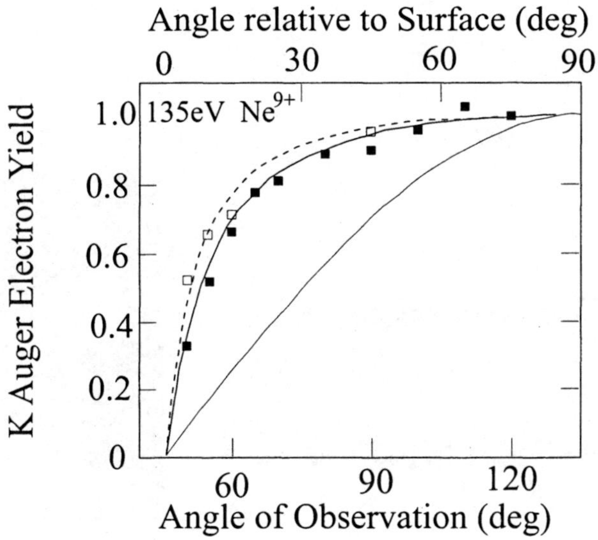

Figure 2.7. Angular distributions of the total K-Auger electron peak intensities (closed squares), and of the peak expected to be emitted above the surface (open squares) for 135eV Ne^{9+} [6].

3. Beam Capillary Spectroscopy

To get a more concrete idea on what is happening in the Stage I, a microcapillary foil [14, 15] was employed, which is a slice of a bundle of thin straight straws, into which HCIs are shot [16, 17]. Fig. 2.8(a) schematically shows three different "fates" of HCIs in the capillary depending on their radial positions in the capillary, *i.e.*, (1) when the distance of the HCI to the inner wall is always larger than the critical distance d_c, the HCI passes through the capillary keeping its initial charge state, (2) when the distance is shorter than d_c, a HAA is formed and then hits the inner wall and collapses, (3) if, however, a HAA is formed near the exit of the capillary, it can escape into the vacuum if the influence of the image acceleration which varies the ion trajectories is neglected. It is noted that the HAA so formed is isolated in vacuum and is conceptually different from the "dynamic hollow molecule" in front of a surface, which had been called a hollow atom. The ratio of the charge-changed particles out of all the transmitted HCIs, f, may be crudely given by

$$f \sim \frac{2\pi a d_c}{\pi a^2} = \frac{2(2q)^{1/2}}{aW}, \tag{2.8}$$

where a is the radius of the capillary.

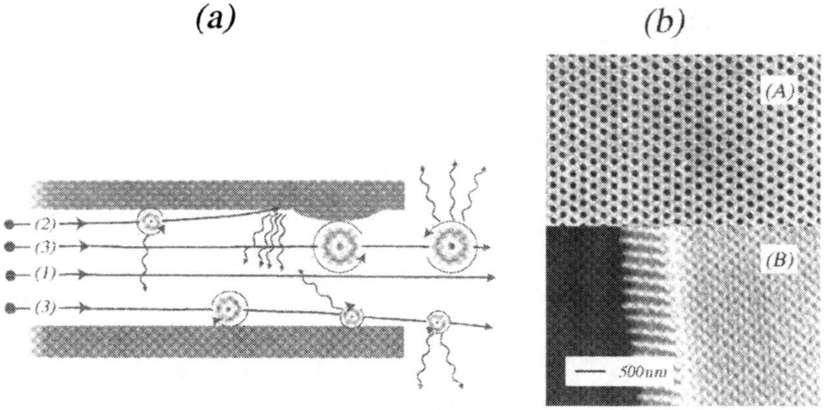

Figure 2.8. (a) Schematic view of ion-capillary collisions. (b) An SEM image of the Au capillary view from (A) the top and (B) 45° [15].

Fig. 2.8(b) shows an SEM image of a highly ordered microcapillary made of Ni, which has $\sim$ 1mm^2 in area with thickness of $\sim$ 1.5μm and has a multitude of straight holes of $\sim$ 100nm in diameter [15]. In this case, the charge-changed fraction f is expected to be a few % of the transmitted HCIs.

3.1 Primordial Stages of Hollow Atom Formation

In order to investigate the very beginning of the hollow atom (ion) formation, visible light spectroscopy has been introduced [20]. The transition energy $\Delta\epsilon(\Delta n)$ of a HCI with core charge q capturing an electron into a principal quantum number n_c is approximately given by

$$\Delta\epsilon(\Delta n) \sim -\frac{q^2}{2n_c^2} + \frac{q^2}{2(n_c - \Delta n)^2} \sim \frac{q^2 \Delta n}{n_c^3} \sim \frac{\Delta n}{q}, \quad (2.9)$$

where Δn is the difference of the principal quantum number before and after the transition. Fig. 2.9 shows the transition energy estimated with eq. 2.9 as a function of q, which indicates that visible light is observed when q is around 10 and a $\Delta n = 1$ transition is realized. It is expected that the combination of the capillary target and visible light spectroscopy provides a unique chance to directly determine n_c at the beginning of the hollow atom formation.

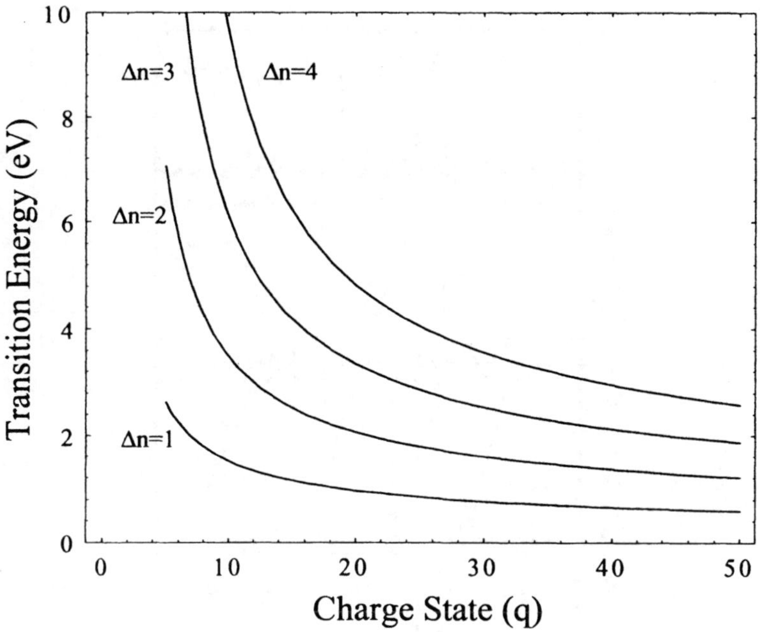

Figure 2.9. The transition energy $\Delta\epsilon$ for Δn =1,2,3 and 4 as a function of charge state q for $W = 4.5eV$ (for details see the text).

Fig. 2.10 shows visible light spectra observed when 2keV/u Ar^{q+} (q=6-11) ions transmitted through a Ni capillary [21]. Considering that the lifetimes of excited states which emit visible light are in the range of sub ns or even longer, such a measurement was never possible with a flat surface target. Most of the strong lines were successfully attributed to $\Delta n = 1$ transitions of ions having captured one electron in the capillary. The numbers in the parentheses, $(n, n-1)$, are the initial and the final principal quantum numbers so identified. The solid lines connect a series of transitions with the same n. It is seen that the series with $n \sim q+1$ was the strongest [21]. In order to get quantitative information on the initial state distribution, the spectra were measured with higher resolution so that transitions belonging to different angular momentum quantum number states were distinguished. The transition lines belonging to $l = n-1$ and $l = n-2$ states were successfully identified in the case of 2keV/u Ar^{7+} ions transmitted through a Ni capillary.

A cascade analysis revealed that $n = 8$ and $n = 9$ states were strongly populated for both $l = n-1$ and $l = n-2$ [21], *i.e.*, the distribution width of the principal quantum number of the initial states is quite narrow. Eq. 2.3 predicts that $n_c \sim 8.3$, showing that the average number of n is well reproduced with the

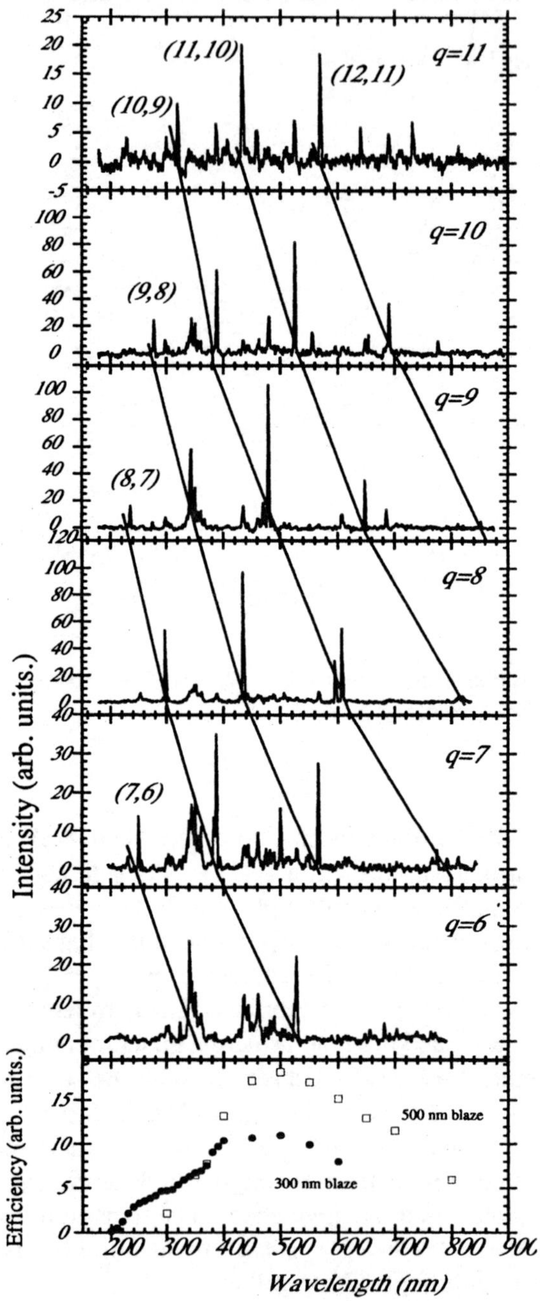

Figure 2.10. Visible light spectra for 2keV/u Ar^{q+} (q=6-11) ions transmitted through a Ni microcapillary [21].

COB model. Fig. 2.10 also shows that transitions with the same wavelengths are observed at two different charge states (*e.g.*, the transition at 480nm for q=9 is also clearly seen for q=10). This observation is consistent with a general expectation that Auger relaxation rates are rather high, and accordingly, only one electron can be left at high Rydberg states, which slowly decays via radiative transitions even when electrons are multiply transferred.

The broad band structure around 350nm and 450nm observed for all the spectra were attributed to transitions from sputtered Ni atoms, indicating that the inner wall of the capillary is somewhat sputter-cleaned continuously during the measurements.

3.2 Evolution of Hollow Atoms

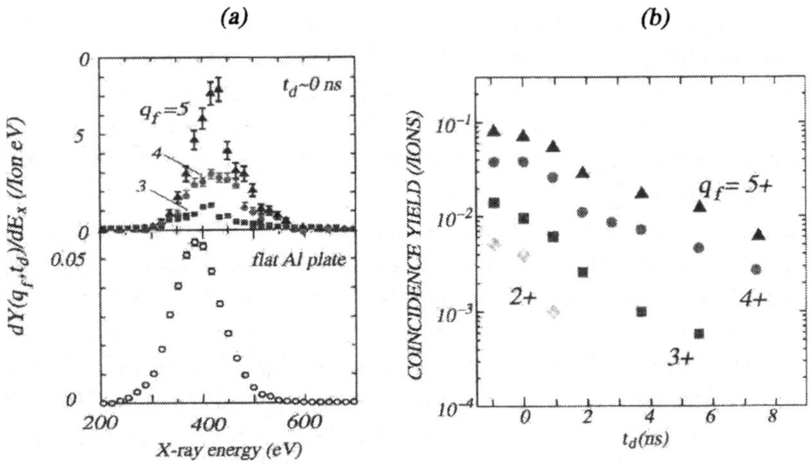

Figure 2.11. (a) Energy spectra of N x-rays for 2.1keV/u N^{6+} ions transmitted through a Ni microcapillary in coincidence with exiting charge state of 5 (▲), 4 (♦) and 3 (■). (b) The delayed x-ray yields normalized per one N ion with q_f; q_f=5 (▲), 4 (•), 3 (■), and 2 (♦) [8].

Soft x-rays emitted from ions transmitted through the capillary were also measured to study the inner shell filling processes [8, 16]. In contrast to the visible light emission, which probes the early stage of the hollow atom evolution, x-rays probe their later stage. Fig. 2.11(a) shows the energy spectra of x-rays measured with a Si(Li) detector in coincidence with exiting charge states q_f=3, 4, and 5 for 2.1keV/u N^{6+} ions transmitted through a Ni capillary. As a reference, an x-ray spectrum observed when the same ions hit a flat Al plate is shown in the lower part of fig. 2.11(a). Although the energy resolution of the Si(Li) detector is not enough to identify individual core configurations, the peak energies for the capillary transmitted ions were seen to be almost the same as each other and were about 30eV higher than that for the flat Al target, which indicates that the L-shell holes were not completely filled at the moment of X-ray emission independent of q_f. A similar behavior was also observed for Ne^{9+} ions transmitted through a Ni capillary [16, 17]. Fig. 2.11(b) shows the decay curve of the x-ray intensities for each q_f shown in fig. 2.11(a), which tells that the lifetimes of the N K-shell hole are of the order of ns depending weakly on q_f, which are several orders of magnitudes longer than a typical lifetime of an N K-shell vacancy. The observation that the K-shell vacancy is fairly stabilized and the L-shell is not much filled indicates that the spin of K- and L-shell electrons are aligned and accordingly the K-shell filling process is strongly suppressed. Another important observation is that the production ratio of the stabilized states is very large, which are seen from the coincidence yields ($\sim$ 8%, $\sim$ 4%, $\sim$ 1%, and $\sim$ 0.5% for 2,3, 4 and 5 electron systems, respectively). The lifetimes of innershell holes were found to be longer for higher charge states, *e.g.*, it is several tens ns for 3keV/u Ar^{13+} transmitted through a Ni capillary [17].

In order to identify the electronic configurations involved in the innershell filling processes with better accuracy, a grating soft x-ray spectrometer combined with a LN_2 cooled CCD was developed [22]. In the case of 5keV/u Ne^{9+} ions transmitted through a Ni capillary, the major core configurations during K x-ray emission were $1s2p^1P$, $1s2p^3P$, and $1s2s2p^4P$ states [17]. Fig. 2.12 shows an example of such high resolution x-ray spectra for 2.3keV/u $^{15}N^{7+}$ transmitted through a highly ordered Ni capillary. $np - 1s$ transitions with n as high as 8 were identified together with a $1s2p^1P - 1s^2$ transition [23]. Considering that the principal quantum number of the initially populated state is $n \sim q + 1$ (see fig. 2.10), the observed $8p - 1s$ transition is the transition filling the K-shell hole directly from the initially populated state. This observation tells that low angular momentum states like p-states are also populated although high angular momentum states are much more preferred as discussed in section 2.3.1.

3.3 Charge State Fraction

Eq. 2.2 tells that the charge state of the capillary transmitted ions are uniquely determined by the distance from the capillary wall. In other words, the charge state distribution $f(q_f)$ can also be estimated as

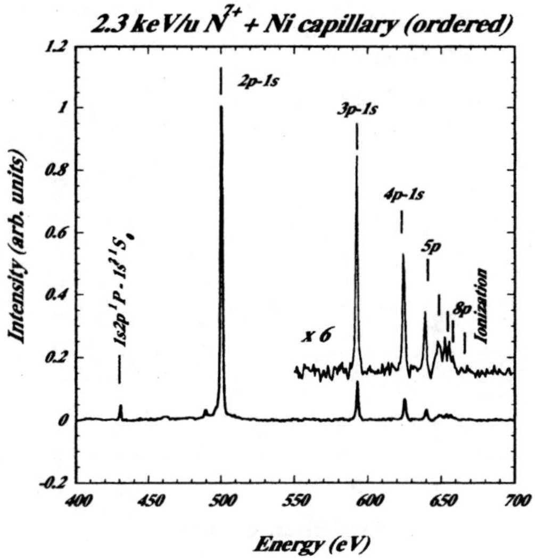

Figure 2.12. High resolution K X-ray spectrum measured downstream of the Ni capillary for 2.3keV/u $^{15}N^{7+}$ ions [23].

$$f(q_f) \sim 2^{3/2} \frac{q_f^{1/2} - (q_f - 1)^{1/2}}{aW} = \frac{2^{3/2}}{(q_f^{1/2} + (q_f - 1)^{1/2})aW}, \tag{2.10}$$

if the influence of the image acceleration is negligible, where q_f is the exiting charge state. Eq. 2.10 indicates that the charge state distribution is a monotonically decreasing function of q_f. The open circles in fig. 2.13 show the charge-changed fraction observed for 2.1keV/u N^{6+} ions transmitted through a Ni capillary [8], which is not in accord with the monotonic decrease predicted with eq. 2.10. Autoionization is supposed to be the key process to bridge the discrepancy. The transferred electrons are in highly excited states, and a considerable fraction of them autoionize, which results in reduction of lower charge state components and enhancement of higher charge state components. The solid circles show the results of a sophisticated Monte Carlo simulation taking into account such cascading processes and image acceleration of HCIs [18, 19]. As is seen, the observation was quite satisfactorily reproduced. It is also noted that the charge state distribution observed here is qualitatively different from that for the specular reflection case shown in fig. 2.6.

It has recently been found that an insulator capillary target acts like a beam guide, *i.e.*, HCIs travel along the capillary without changing their original

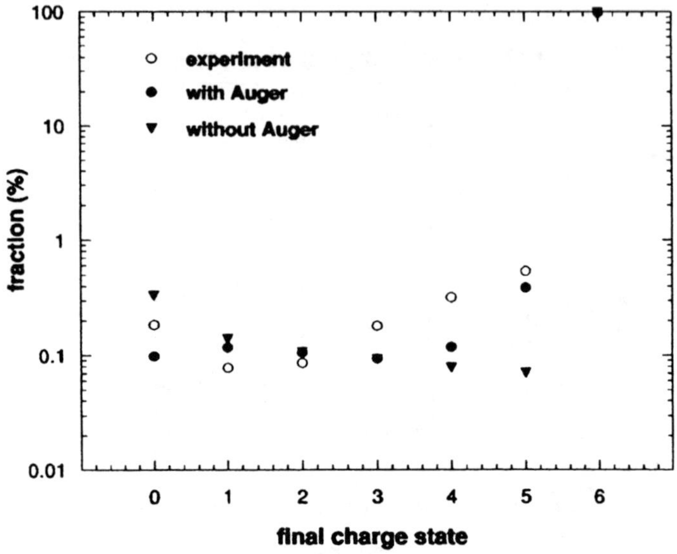

Figure 2.13. Final charge state distribution of 2.1keV/u N^{6+} ions transmitted through a Ni microcapillary. Open circles:experiment [8], solid circles:simulation [18], solid inverted triangles: eq. 2.10.

charge states, which could be an interesting direction of capillary applications [24].

4. Proton Sputtering - the Counterpart of Hollow Atom Formation

HCIs induce particle emission as a result of multiple electron transfer (Stage I) and potential energy deposition (Stage II). In this section, we discuss a new mechanism of proton sputtering induced during the Stage I. Proton yields for were found to be proportional to $\propto q^{\sim 5}$ for all the HCIs tested [25].

4.1 Sputtering of protons and Si^+

Fig. 2.14(a) shows the sputtering yield of proton and Si^+ from a Si(100)-(2$\times$1)H surface with Xe^{q+} ions as a function of the charge state q [28]. The q-dependence of the proton yield was as steep as that of untreated surface ($q^{\sim 5}$) observed for slow Ne^{q+}, Ar^{q+}, Kr^{q+} and Xe^{q+} ions [25]. On the other hand, the Si^+ ion yield was almost independent of q. These findings strongly indicate that the proton sputtering mechanism is (1) qualitatively different from Si^+ ion sputtering which is governed by cascading binary collisions (kinetic sputtering), and (2) the same between untreated- and well-defined surfaces.

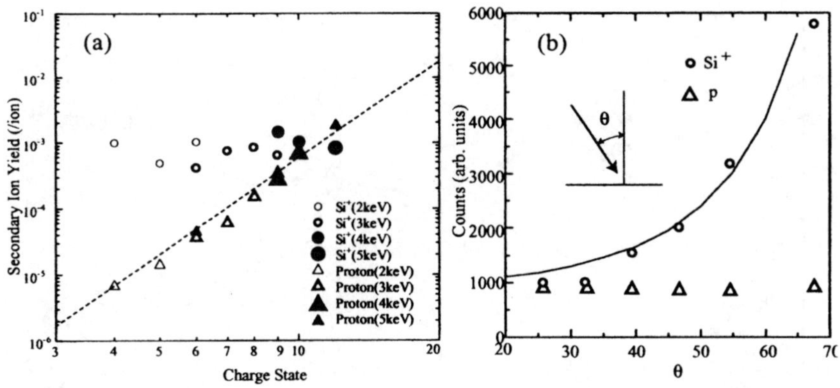

Figure 2.14. (a) Sputtering yields of proton and Si^+ ions from a Si(100)-(2×1)H surface as a function of q [28]. The dashed line shows $\propto q^5$. (b) Sputtering yields of proton and Si^+ as a function of incidence angle θ measured from a Si(100)-(2×1)H surface with 3keV Xe^{8+} [28]. The solid line shows $\cos^{-2}\theta$.

The absolute proton yield for Si(100)-(2×1)H, Si(100)-(3×1)H and Si(100)-(1×1)H surfaces for 3keV Xe^{8+} were $\sim 1.2 \times 10^{-4}$/ion, $\sim 2.0 \times 10^{-4}$/ion, and $\sim 1.3 \times 10^{-3}$/ion, respectively. It is noted that the yield variation among well-defined Si-H surfaces was much larger than the relative surface abundance of hydrogen atoms, which are 3, 4, and 6 for Si(100)-(2×1)H, Si(100)-(3×1)H and Si(100)-(1×1)H surfaces, respectively.

Fig. 2.14(b) shows sputtering yields of proton and Si^+ ions from a Si(100)-(2×1)H surface with 3keV Xe^{8+} as a function of incidence angle θ. The Si^+ yield shows strong dependence on the incidence angle θ, which is another particular feature of the kinetic sputtering. On the other hand, the proton yield has no dependence on θ, again proving that the proton sputtering takes place without momentum transfer from the incident ion [28].

4.2 Energy Distribution of Protons

The longitudinal energy distributions of sputtered protons normal to the surface were determined with a high resolution TOF spectroscopy, the results of which are given in fig. 2.15 for (a) 600eV $Ar^{6+}$and 500eV Ar^{12+} ions, and (b) 4.8keV Ar^{6+} and 4.8keV Ar^{12+} ions bombarding untreated surfaces [9]. It

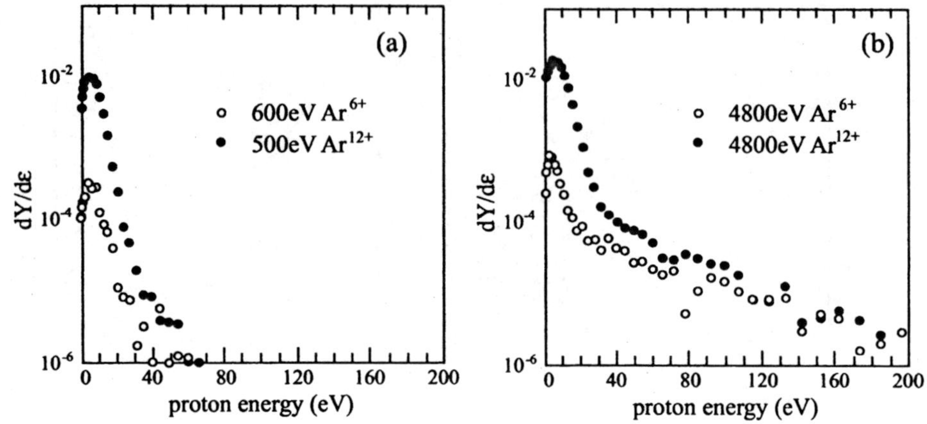

Figure 2.15. The energy distribution of sputtered protons for (a) 600eV Ar^{6+} and 500eV Ar^{12+}, and (b) 4.8keV Ar^{6+} and 4.8keV Ar^{12+} [9]

is seen that (1) the energy distribution was highly nonthermal with a peak at several eV, which depended neither on q nor on the kinetic energy, (2) the higher energy tail grew for higher kinetic energy, and (3) the yields for $q = 12$ was almost two orders of magnitudes larger than those for $q = 6$. The peak energy was consistent with the Coulomb energy between H^+ and X^+ ions which are produced as a result of double electron removal from a H-X bond. The fact that the peak energy depended strongly on q but barely on the kinetic energy indicates that the sputtering is induced by q.

Fig. 2.16 shows the 2-dimensional distribution of sputtered protons for (a) Si(100)-(2×1)H, (b) Si(100)-(3×1)H, and (c) Si(100)-(1×1)H surfaces. The distribution width gives a measure of the lateral kinetic energy of protons. According to the model that the proton sputtering is induced by the mutual repulsion between two atoms in the same bond, which is proton and Si^+ ions in this case, the proton energy reflects the reneutralization time of Si^+. On the other hand, the proton yield reflects the reneutralization probability of proton itself. It is seen that the distribution width increases in accord with the yield

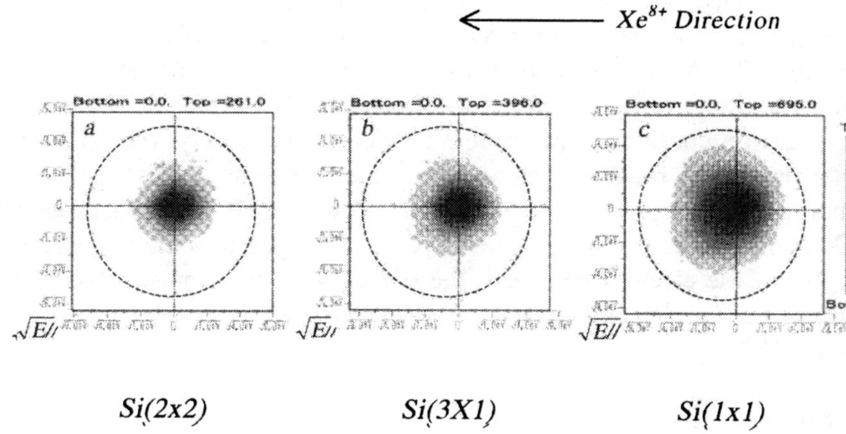

Figure 2.16. Two dimensional distributions of protons sputtered from (a) a Si(100)-(2×1)H surface (b) a Si(100)-(3×1)H surface and (c) a Si(100)-(1×1)H surface with 3keV Xe^{8+}. [28]

variation described in section 2.4.1, *i.e.*, the reneutralization probability of proton is strongly correlated with that of Si^+.

A scanning tunneling microscopy (STM) study has revealed that the Si(100)-(2×1)H and Si(100)-(3×1)H surfaces have atomically flat surfaces, and the Si(100)-(1×1)H surface is not flat despite the fact the surface shows (1×1) LEED pattern [29]. It is understood that a bond-breaking H-termination (H-Si=Si-H a H-Si-H, H-Si-H) and an etching reaction (H-Si-H+2H a SiH4) deteriorates the surface during the atomic hydrogen exposure at room temperature.

4.3 Model Consideration

A proton sputtering model based on the COB process [7] was proposed [27], which successfully reproduced the observed q-dependence. In this simulation, double electron transfer process discussed above are explicitly incoorporated. This model also predicts a saturation of the q dependence for higher charge states ($q > 10$), which was actually confirmed experimentally [25].

The proton sputtering mechanism discussed above has some similarities to the so-called Auger stimulated desorption (ASD), which is known to be effective in electron- and photon- induced desorptions (ESD, PSD), where production of innershell vacancies followed by Auger transitions plays an essential role [30]. Straightforward processes like excitation or ionization of a bond electron to a dissociative state, which is very important for dissociation of isolated molecules, is often ineffective for atoms or molecules on a surface because of quick reneutralization due to charge transfer from the substrate. In other words, suppression of reneutralization is very important for ESD/PSD to take place on a surface. The ASD process realizes this suppression via double ionization of anion atoms through an ionization of innershell electrons followed by an Auger electron emission. The reneutralization of the doubly ionized anion is usually suppressed because surrounding atoms have positive charges from the beginning, *i.e.*, a repulsive force lasts for a considerable time which induces the Coulomb sputtering. In the case of HCI induced proton sputtering observed here, the double electron transfer is the key to cause the repulsive force. It is noted that similar phenomena were also reported for F^+ emission for Ar^{2+} ion impact [31].

Acknowledgments

The author is deeply indebted to H.Masuda, S.Ninomiya, S.Thuriez, Y.Morishita, Y.Iwai, H.Torii, K.Kuroki, N.Okabayashi, and K.Komaki for their collaboration and fruitful discussions.

References

[1] J.P.Briand, L. de Billy, P.Charles, S.Essabaa, P.Briand, R.Geller, J.P.Desclaux, S.Bliman, and C.Ristori, Phys.Rev.Lett. **65**, 159 (1990)

[2] H.Winter, Europhys.Lett. **18**, 207 (1992)

[3] H.Kurz et al., Phys.Rev.Lett. **69**, 1140 (1992)

[4] J.Das and R.Morgenstern, Phys.Rev.A **47**, R755 (1993)

[5] F.W.Meyer et al., Phys.Rev.A **48**, 4479 (1993)

[6] M.Grether, A.Spieler, R.Koerbrueck, and N.Stolterfoht, Phys.Rev. **A52**, 426 (1995)

[7] J.Burgdoerfer, P.Lerner, and F.W.Meyer, Phys.Rev.A **44**, 5674 (1991)

[8] S.Ninomiya, Y.Yamazaki, F.Koiki, H.Masuda, T.Azuma, K.Komaki, K.Kuroki, and M.Sekiguchi, Phys.Rev.Lett. **78**, 4557 (1997)

[9] K.Kakutani, T.Azuma, Y.Yamazaki, K.Komaki, and K.Kuroki, Jpn. J. Appl. Phys. **34**, L580 (1995)

[10] S.Deutscher, X.Yang, and J.Burgdoerfer, Phys.Rev.A **55**, 466 (1997)

[11] C.Auth, T.Hecht, T.Igel, and H.Winter, Phys.Rev.Lett. **74**, 5244 (1995)

[12] L. Folkerts,S. Schippers,D.M. Zehner, and F.W.Meyer, Phys. Rev.Lett. **74**, 2204 (1995)

[13] T.Schenkel A.V.Barners, T.R.Niedermayr, M.Hattass, M.W.Newman, G.A.Machicoane, J.W.McDonald, A.V.Hamza, and D.H.Schneider, Phys.Rev.Lett. **83**, 4273 (1999)

[14] H.Masuda and K.Fukuda, Science **268**, 1466 (1995)

[15] H. Masuda, H. Yamada, M. Satoh, H. Asoh, M. Nakao, T. Tamamura, Appl. Phys. Lett. **71**, 2770 (1997)

[16] Y.Yamazaki, S.Ninomiya, F.Koike, H.Masuda, T.Azuma, K.Komaki, K.Kuroki, and M.Sekiguchi, J. Phys. Soc. Jpn. **65**, 1199 (1996)

[17] Y.Yamazaki, Int. J. Mass Spec. **192**, 437 (1999)

[18] K.Tokesi, L.Wirtz, C.Lemmell, and J.Burgdoerfer, Phys.Rev.A **61**, 020901 (2000)

[19] K.Tokesi, L.Wirtz, C.Lemmell, and J.Burgdoerfer, Phys.Rev.A **64**, 042902 (2001)

[20] Y.Morishita, S.Ninomiya, Y.Yamazaki, K.Komaki, K.Kuroki, H.Masuda, and M.Sekiguchi, Physica Scripta T **80**, 212 (1999)

[21] Y.Morishita *etal., Abstract book of egas conference and to be published*

[22] Y. Iwai, S. Thuriez, Y. Kanai, H. Oyama, K. Ando, R. Hutton, H. Masuda, K. Nishio, K. Komaki, Y. Yamazaki, RIKEN Rev. **31**,34 (2000)

[23] Y. Iwai, Y. Kanai, H. Oyama, K. Ando, H. Masuda, K. Nishio, M. Nakao, T. Tamamura, K. Komaki and Y. Yamazaki, Nucl.Instrum.Methods B *to be published*

[24] N.Stolterfoht, et al. Phys.Rev.Lett. **87**, 023201 (2001)

[25] K.Kuroki, T.Takahira, Y.Tsuruta, N.Okabayashi, T.Azuma, KKomaki, and Y.Yamazaki, Physica Scripta T **80**, 557 (1999)

[26] T. Neidhart,, F. Pichler, F. Aumayr,HP. Winter, M. Schmid, and P. Varga, Phys. Rev. Lett. **74**, 5280 (1995)

[27] J.Burgdoerfer and Y.Yamazaki, Phys.Rev.A **54** 4140 (1996)

[28] K.Kuroki, N.Okabayashi, H.Torii, K.Komaki, and Y.Yamazaki, Appl.Phys.Lett **81** *to be published*

[29] J.J.Boland, Phys.Rev.Lett. **65**, 3325 (1990)

[30] M.L.Knotek and P.L.Feibelman, Phys.Rev.Lett. **40**, 964 (1978)

[31] P.Varga, U.Diebold, and D.Wutte, Nucl.Instrum.Methods B **58**, 417 (1991)

Chapter 3

INTERACTION OF HOLLOW ATOMS WITH SURFACES

Auger electron emission and plasmon excitation

N. Stolterfoht
Hahn-Meitner-Institut, Bereich Strukturforschung, Glienickerstr. 100, D-14109 Berlin, Germany
stolterfoht@hmi.de

Abstract Experimental and theoretical methods for studying the formation of hollow atoms during the interaction of slow, highly charged ions with surfaces are reviewed. Particular attention is devoted to the emission of Auger electrons and the creation of plasmons near the surface. To exhibit the static aspect of hollow atoms, Hartree-Fock and Density Functional methods were used to evaluate electron charge densities above and below the surface. The dynamic properties of hollow atoms are revealed by means of cascade models which include the filling of inner shell orbitals via Auger transitions and collisional charge transfer. The theoretical results are compared with experiments of electron emission at energies from a few eV to a few hundred eV. In the electron spectra, Auger peak structures are shown to yield information about the filling of the hollow atom within the first atomic layers of the surface. Low-energy electrons ejected from an Al surface were measured to study spectral structures found near 11 eV attributed to the decay of plasmons. Absolute electron yields from the plasmon decay were primarily studied as a function of the incidence angle of the projectile. Finally, recent experiments revealing spectral structures near 6.5 eV which may be associated with the creation of surface plasmons.

Keywords: highly charged ions, hollow atoms, surfaces, low energy electrons, Auger lines, plasmon creation

1. Introduction

During the last two decades considerable interest has been devoted to studies of slow highly charged ions interacting with surfaces [1–3]. Early investigation by Hagstrum [4], Arifov [5], and Donets [6] have established the basic concepts

F.J. Currell (ed.), The Physics of Multiply and Highly Charged Ions, Vol. 2, 69-120.

of the field. In the late 1980s and early 1990s several groups [7–16] started to work in the field so that it became rapidly expanding. Hence, from the beginning, remarkable progress has been achieved in the understanding of the complex processes accompanying the interaction of highly charged ions with a surface.

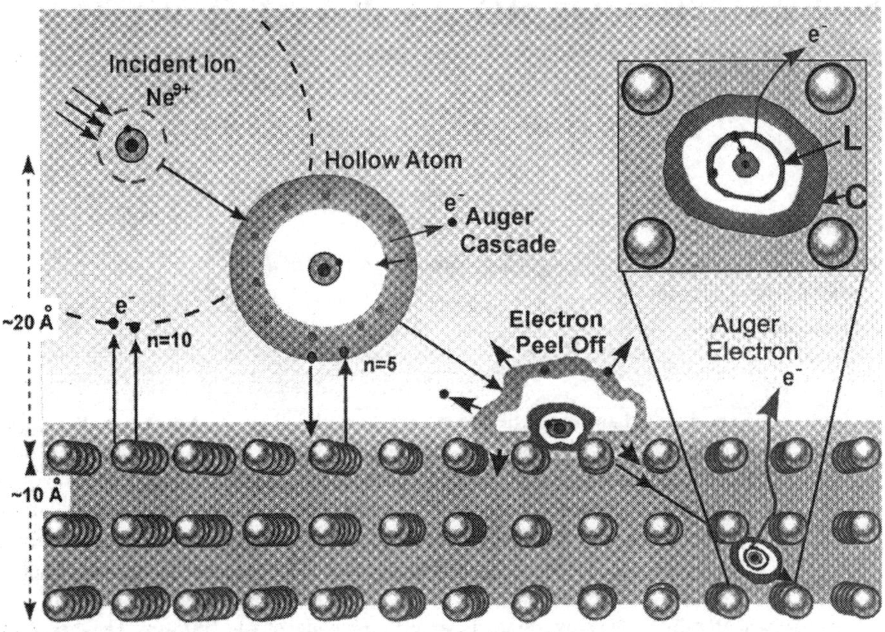

Figure 3.1. Pictorial view of the formation of a hollow atom at a surface [17]. In this example, a Ne^{9+} ion is approaching the surface where several electrons are captured into high Rydberg states. These electrons are "peeled off" upon entering the surface. Within the solid the strong screening cloud C is formed leaving empty orbitals in an inner region (see inset).

Fig. 3.1 provides an overview of the scenario commonly accepted within the community to date. When a highly charged ion approaches a surface, several electrons are resonantly transferred from the conduction band of the solid to the ion. Thus, in front of the surface, a "hollow" atom is produced with many electrons in higher orbitals and empty intermediate shells [6, 10, 18]. Hollow atoms are reduced in size by dielectronic processes such as autoionization where the electron-electron interaction leads to the relaxation of the Rydberg electrons accompanied by electron ejection. Another important reason for a shrinking of the electronic charge cloud is the flow of the Rydberg electrons back into the solid and the additional capture of electrons into lower orbitals. Hence, the diameter of the hollow atom decreases as it approaches the surface.

When the projectile hits the surface, the remaining Rydberg electrons are removed (peeled off) [10, 19]. Simultaneously, during the passage into the surface, the highly charged ion strongly attracts several electrons from the conduction band and, hence, it induces a negative charge cloud, denoted by C, which dynamically screens the nucleus [17, 20, 21]. The appearance of the intense cloud C is the outstanding property of a slow highly charged ion moving below the surface. This cloud gives rise to hollow atoms of the second generation whose dimensions are much smaller than those produced outside the solid (fig. 3.1).

For a quantitative description of the induced charge cloud C the electronic density of hollow atoms is evaluated. Since the induced electron density is relatively large, non-linear theories are required to describe their formation. Hence, the use of self-consistent field methods are adequate as applied within the framework of the Hartree-Fock method and the Density-Functional Theory [21, 22]. The calculations provide electron densities for the associated orbitals, which can be used to visualize the electronic clouds formed around the highly charged ions. In addition, the calculations yield Auger transition energies which can be compared with experiment.

Subsequently, as the hollow atom travels within the solid, the inner shell orbitals are successively filled by and collisional charge transfer [17, 23, 24]. This cascade-like filling of the hollow atoms represents the dynamic aspect of the interaction of highly charged ions with the surface. The processes associated with this filling cascade are rather complex involving a large number of sequential relaxation events. The modeling of the filling cascade requires substantial effort involving the solution of numerous rate equations. Because of the elaborate effort in the use of cascade models, their applications are still limited.

In this article, theoretical and experimental studies of heavy ions incident on a surface with a high charge state are reviewed. Particular attention is given to experiments with very slow ions of energies as low as a few tens of eV, which were primarily performed at the Hahn-Meitner-Institut Berlin. The main goal is to combine basic results from individual publications revealing common features for the production of Auger electrons and plasmons. The pronounced structures produced by Auger transitions provides information about the filling state of the hollow atom. While Auger electrons are generally ejected below the surface, specific examples are given where Auger emission take place well above the surface. Further emphasis is given to the creation of surface and bulk plasmons by highly-charged ions. Specific attention is devoted to the creation of surface plasmons of low momentum. The experimental results are compared to model calculations of the cascade-like filling of hollow atoms, which involve potential energy transfer to produce Auger electrons and plasmons.

2. Auger Electron and Plasmon Production

The mechanisms of Auger electron production can be understood as a scattering of two electrons within the attractive field of a hollow atom, where one of them is captured from the C cloud and the other one is excited (see fig. 3.2). Thus, potential energy of the hollow atom is converted to kinetic energy of the ejected Auger electron. In early years the analysis of K- and L-Auger-electrons has been performed under the assumption of above-surface emission [26–28] until Meyer *et al.* [13] and Burgdörfer and Meyer [18] concluded from detailed model calculations that the majority of the Auger electrons originate from below the surface.

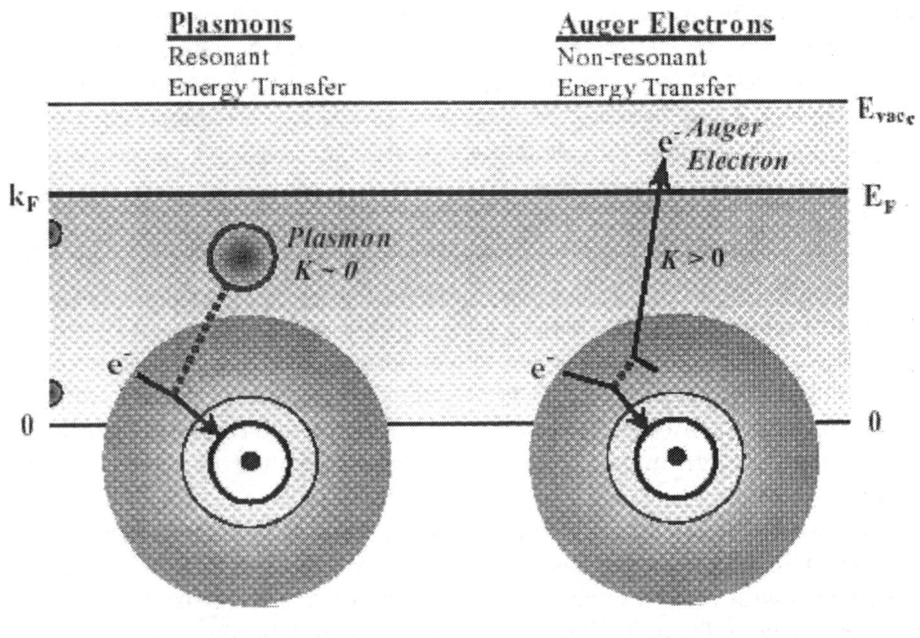

Figure 3.2. Schematic view of plasmon and Auger electron production by hollow atoms. While an electron is captured in an inner shell, the liberated potential energy is used for plasmon creation or excitation of a single electron. Note that plasmon production involves resonant energy transfer, i.e., the plasmon energy is about equal to the potential energy set free by the capture process.

Auger electrons from below surface emission have been examined in detail in recent years [29, 30]. This analysis of the hollow atoms suggested that several electrons are missing in the L shell during K-Auger transitions and these missing electrons are instead found in the M shell. Thus, it became common practice in calculations of Auger transitions near the surface that electrons missing in inner or valance shells are filled into the next higher-lying atomic orbital [28, 31, 32]. Later, it was recognized that this 'filling-up' procedure models the dynamic screening cloud C formed around the ion in the solid [31, 33].

Besides Auger electrons, hollow atoms traversing the solid can produce plasmons also shown in fig. 3.2. Plasmons are quantized collective oscillations of valence electrons in a metal, which are generally treated within the framework of the free-electron gas (jellium) model [34] and the Random Phase Approximation (RPA) [35].

Similar to a photon, a plasmon has a significant amount of energy (e.g., 15 eV for Al) whereas it generally has only little momentum. Hence, due to energy and momentum conservation there are constraints on the creation of plasmons in a direct process involving a binary collision, e.g., by an incident proton or electron as shown in fig. 3.3. In fact, it can readily be shown that plasmon creation in a binary interaction occurs only for momenta larger than the minimum momentum transfer of $K_{\min} \approx \omega_{pl}/v_p$, where ω_{pl} is the plasmon frequency and v_p is the velocity of the projectile. Hence, it follows that slow projectiles produce plasmons of relatively high momentum. Since $K_{\min}$ should not exceed the cut-off value K_c (fig. 3.3), the direct excitation process requires a threshold velocity corresponding to a minimum energy of about 33 eV for electron impact and 40 keV/u for heavy particles incident on Al [36].

In view of the kinematic constraints it was a surprise at first when plasmon creation was observed in electron emission spectra produced by ions with energies below a few keV [37–39]. Obviously, mechanisms different from binary excitation are important at such low energies. Controversial ideas have been put forward to interpret the creation of plasmons by slow ions. In principle, plasmons may be produced in direct collisions under participation of the lattice atoms (fig. 3.3) although this process has received less attention to date [40, 41]. For heavier projectiles, various groups [37, 38, 42–45] considered a *plasmon-assisted capture* process where the potential energy from the projectile produces a plasmon as indicated in fig. 3.2. It should be realized that in this case the projectile plays the role of a third body to balance the momentum so that lattice atoms are not required. For impact by protons, which have only little potential energy, the kinetic energy of the projectile may play a role adding to its potential energy. Also, high-energy electrons liberated in binary collisions with the projectiles may produce plasmons in secondary collisions, even at sub-threshold ion velocities [46].

Here, particular attention is devoted to the capture process producing plasmons by potential-energy effects. The energies involved in the transitions of an electron from the bottom of the band into the $2p$ shell of hollow neon are given in table 3.1 as function of the number of $2p$ vacancies. These energies

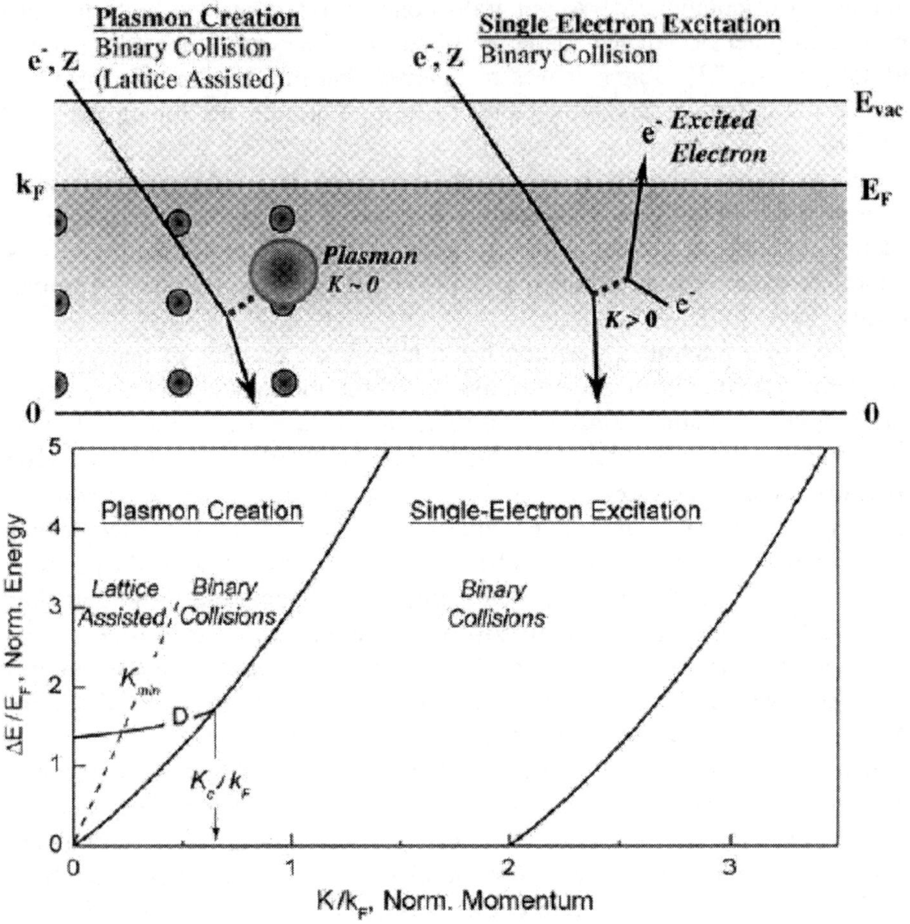

Figure 3.3. Diagrams to visualize plasmon production and single-electron excitation in ion-solid interactions. Plasmon production involves lattice assisted interactions and, more important, binary collisions depending on whether the momentum transfer is smaller or larger than $K_{\min} = \omega_{pl}/v_p$ (see text). Electron-hole pairs are produced in collisions of high-momentum transfer involving two-body interactions. The lower part of the figure shows the well-known momentum-energy diagram by Pines [35]. The curve labeled D represents the dispersion curve for bulk plasmons.

have been calculated by means of the Density Functional Theory [21]. The energy liberated in the capture process can be transferred into two branches involving either the creation of a bulk plasmon or an Auger electron (fig. 3.2). The branching ratio for plasmon excitation is also presented in table 3.1 [47]. It is important to note that the plasmon branch involves a resonance constraint as the liberated energy should be close to the energy of the plasmon. Therefore,

Table 3.1. Electronic transition energy from the bottom of the conduction band of Al to the $2p$ level of a hollow Ne atom which contains a number of $2p$ vacancies. The results are based on total energies evaluated using the Density Functional Theory [21]. The branching ratios for plasmon production are evaluated on the basis of the formalism in Ref. [48].

Number of 2p Vacancies	1	2	3	4	5	6
Transition Energy (eV)	12.8	23.6	33.4	45.6	56.7	68.8
Plasmon Branching Ratio	0.53	0.35	0.05	< 0.01		

for neon with one vacancy in the $2p$ shell the branching ratio is largest since the energy $\gtrsim 12.8$ eV for transitions from above the bottom of the band is close to the 15 eV required for the plasmon creation. For two $2p$ vacancies, i.e., for increasing energy gap, the branching ratio drops to 0.35 and vanishes rapidly when the number of $2p$ vacancy exceeds two. Then, the branching ratio for producing the corresponding Auger electrons approaches unity. This demonstrates that plasmon production by hollow atoms occurs in a rather selective manner.

Plasmon creation is followed by its decay where the plasmon energy is transferred into a single valence electron [49, 50]. Hence, electrons of characteristic energies are ejected from the metal providing a signature for plasmons which can experimentally be studied by means of electron spectroscopy. Previous experiments performed using electron impact [51, 52] and fast ion impact [53–55] have revealed clear evidence for plasmon decay via electron emission. For plasmon decay all constraints of momentum and energy conservation hold. In fact, it is impossible to annihilate a plasmon in a binary interaction where the plasmon energy is transferred to a single electron. To balance the momenta, plasmon decay needs the lattice atoms. Accordingly, for metals with weak lattice interactions, referred to as nearly-free electron metals, the lifetime of the plasmon is significant (of the order of fs) and the corresponding width of the plasmon resonance is relatively narrow (about an eV) [36].

The mechanisms for plasmon production by slow ions resulting in spectral structures at low-electron energies are still under debate. In particular, controversial ideas have been raised in view of structures observed for Al near 11 eV. Apart from bulk plasmons [37, 47, 56], the decay of surface plasmons [39, 43, 57] has been considered. The interpretation of the 11 eV structure as a surface plasmon appears problematic, since the energy of low-momentum surface plasmons is expected to be in the vicinity of 6.5 eV [54]. Recently, it was postulated that the structures near 11 eV include, apart from bulk plasmons, contributions of multipole surface plasmons [39, 58, 59]. Moreover, significant peak structures at energies $\gtrsim 11$ eV, first attributed to bulk plasmons [41], were recently recognized as being due to diffraction of low-energy electrons [60, 61]. In later sections more details will be given for the mechanisms of plasmon production.

3. Experimental Techniques

3.1 Ultra-high vacuum chamber

Most experiments discussed in this article were performed using the 14.5 GHz ECR source at the Ionenstrahl-Labor (ISL) of the Hahn-Meitner-Institut in Berlin. The ion source provides projectiles with energies up to $20q$ keV where q is the charge state of the extracted ions. The end of the beam line is equipped with a deceleration lens system to extract ions at energies as low as $1q$ eV. When the deceleration system is used, the beam line is set on a high-voltage so that the experimental apparatus can be operated at ground potential.

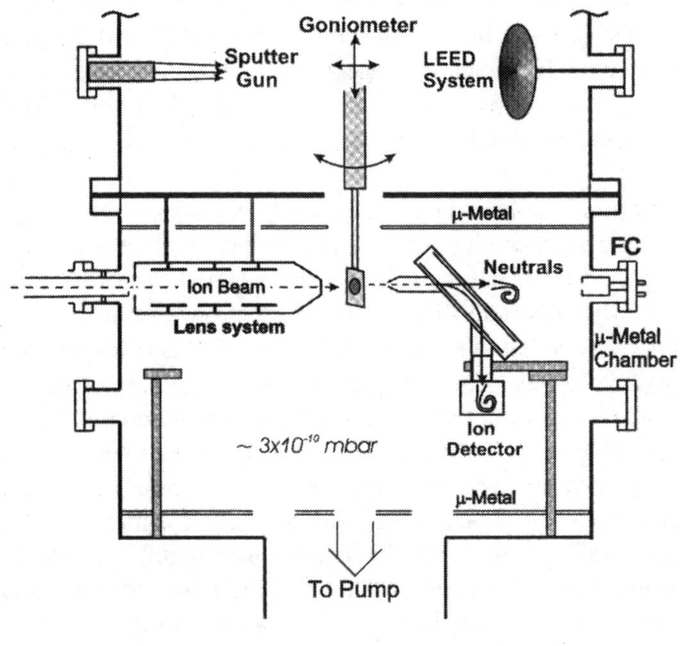

Figure 3.4. Ultra-high vacuum chamber used in the experiments. Ions are decelerated by a lens system and directed onto a solid target. The ejected electrons are analyzed by an electrostatic spectrometer. A μ-metal shield of the same dimension as the chamber is used for reducing the earth's magnetic field. In the upper part of the chamber the target sample is sputter cleaned and analyzed using a LEED system.

The main part of the deceleration (90 %) is performed outside the chamber whereas the remaining part (10 %) is done in a deceleration system in the chamber just in front of the target (fig. 3.4).

The apparatus consists of an ultra-high-vacuum (UHV) chamber including an electron spectrometer and facilities for surface preparation and examination (fig. 3.4). The apparatus has been described in detail previously [31, 33] so that only a few details shall be given. The base pressure during the measurements was a few 10^{-10} mbar. In the upper part of the UHV chamber the target can be cleaned by means of a sputter gun and the quality of the crystal can be verified by a LEED system. After surface preparation the target was moved to the lower part of the chamber where the measurements are carried out. Auger electron spectroscopy was used to verify the cleanliness of the surface. After careful cleaning no contaminations of the surface by C, N, and O could be observed.

In the experiments primarily Al(111) targets and Ne^{q+} ($q = 1-9$) projectiles were used. After the ions were accelerated, they were magnetically analyzed and collimated to a diameter of about $1 - 2$ mm at the position of the target. The emission of electrons from the target was measured using an electrostatic parallel-plate spectrometer. The electron spectra were normalized to an absolute scale taking acceptance angle, resolution, and transmission of the spectrometer and the efficiency of the channeltron into account [31]. The experimental set-up was optimized to accurately measure low-energy electrons [62, 63]. Experience shows that the spectrometer is capable of measuring reliable electron yields for energies as low as 2 eV.

3.2 Overview of electron spectra

Fig. 3.5 shows experimental results for electron spectra obtained with an Al(111) target bombarded by Ne^{q+} projectiles where q =1 to 4 [38]. The data represent double differential electron emission yield $d^2Y/d\Omega\, dE$ obtained at an angle of incidence of $\psi = 45^0$ and an observation angle of $\alpha = 15^0$ relative to the surface plane. (For the definition of the angles see also the inset in the figure.) Each electron spectrum exhibits a maximum at low energies in the range from 2 - 4 eV. Such a maximum is expected for kinetic electron emission as shown previously [64, 65]. Moreover, in fig. 3.5, one can see a significant increase of the electron yield with increasing charge state. This is due to the increase of potential electron emission [66] which is found to be roughly proportional to the corresponding potential energy.

The electron spectra exhibit various structures which can be attributed to emission of Auger electrons as well as the decay of plasmons produced within the surface and the bulk. Similar spectra indicating plasmon structures have been measured by Eder *et al.* [45] using Ne^+ and Ne^{2+}projectiles. The peaks labeled Ne^{2*}, Ne^{3*}, Ne^{4*} with centroid energies near 22 eV, 34 eV, and 46 eV are due to L-Auger transitions in hollow Ne with 2, 3, and 4 vacancies in the L shell, respectively. Apart from the projectile Auger electrons, the spectra in fig. 3.5 show distinct peaks near 63 eV due to the Auger decay of Al-2p vacancies [53] excited in binary collisions with a hollow Ne projectile [17] or in Al-Al collisions involving fast recoils [67]. A significant fraction of these electrons originates from sputtered Al decaying outside the solid.

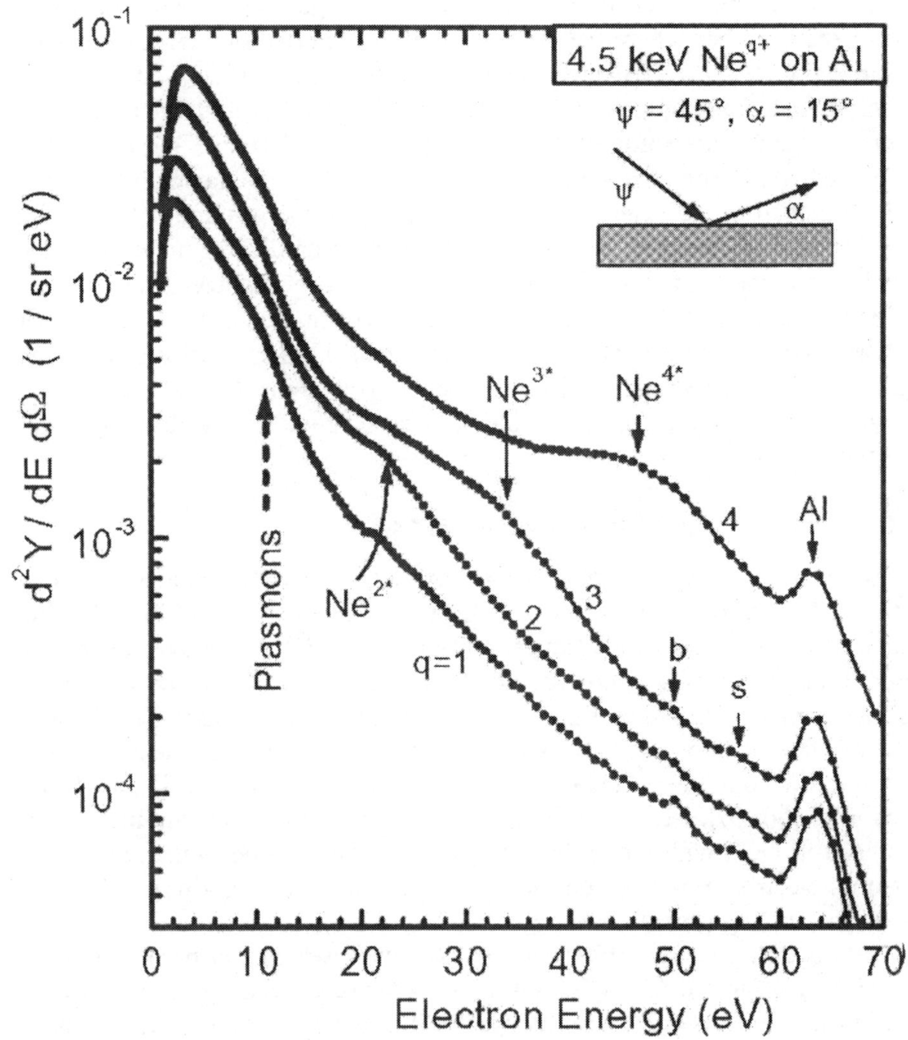

Figure 3.5. Double differential emission yields $d^2Y/dEd\Omega$ for electrons produced by 4.5 keV Ne^{q+} incident on an Al surface for the projectile charge states q = 1 - 4 [38]. The incident angle is $\psi = 45^0$ and the observation angle is $\alpha = 15^0$ relative to the surface plane. For the definition of the angles see also the inset in the figure.

The next examples refer to a spectral range of much higher energy where Ne K-Auger electrons occur. Several studies have been devoted to the measurements of K-Auger spectra produced by hydrogen-like ions [1]. These ions have a vacancy in the K-shell, which is filled in the last step of the cascade filling of the hollow atom. Therefore, the analysis of K-Auger transitions provides information about the whole filling cascade of a hollow atom. Thus, the spectroscopy of K-Auger spectra is a unique tool acquire details about the dynamic properties of hollow atoms formed above and below the surface.

Fig. 3.6 depicts K-Auger electron spectra obtained with an Al solid in comparison with previous results using a He gas target [68]. In the latter case, the projectile Ne^{6+} with an initial configuration $1s^2\,2s^2$ is ionized in the 1s shell so that the $1s\,2s^2\,{}^2S$ state is produced. It decays via KL_1L_1 Auger transitions into the final state $1s^2\,{}^1S$ giving rise to monoenergetic electrons with an energy of 652 eV. Another Auger maximum, observed at higher energies, is due to the configuration $1s\,2s2p$ produced by $1s$ ionization of the projectiles in the initial metastable state $1s^2\,2s2p\,{}^3P_0$ [68].

The spectrum obtained with Ne^{9+} incident on the solid target Al contain lines for the same $K\,L_1L_1$ Auger transition $1s\,2s^2\,{}^2S \rightarrow 1s^2\,{}^2S$, but shifted to the energy of 748 eV. This shift is due to the influence of the solid producing the screening cloud C around the ion. It is important to note that this energy shift produced by the solid is significant, i.e. 96 eV. It shows that the C cloud has a large influence on the orbital energies and, thus on the Auger transition energies. In the following section KL_1L_1 Auger transition energies for different electron clouds are shown to visualize the energy shift by screening effects.

4. Theoretical Methods

4.1 Calculations for hollow atoms

As noted in view of fig. 3.1, a hollow atom approaching the surface at large distances has an essentially different shape than after it has entered into the solid. To give an indication about their size, the profiles of hollow atoms were calculated [21, 22] using atomic structure programs. Hence, the shape of hollows atoms above and below the surface can be demonstrated by means of electron density plots.

When the ion approaches the surface, it is known from the classical over-the-barrier model [11] that resonant charge transfer takes place into orbitals whose outer boundary just touches the surface. For different distances of the ion from the surface, fig. 3.7 shows density plots of orbitals whose outer radii coincide with these distances [22]. For simplicity the orbitals are assumed to be spherically symmetric although they are likely to be polarized by the interaction with the surface [69, 70]. The wave functions underlying the density plots were obtained from Hartree-Fock calculations using the atomic structure code by

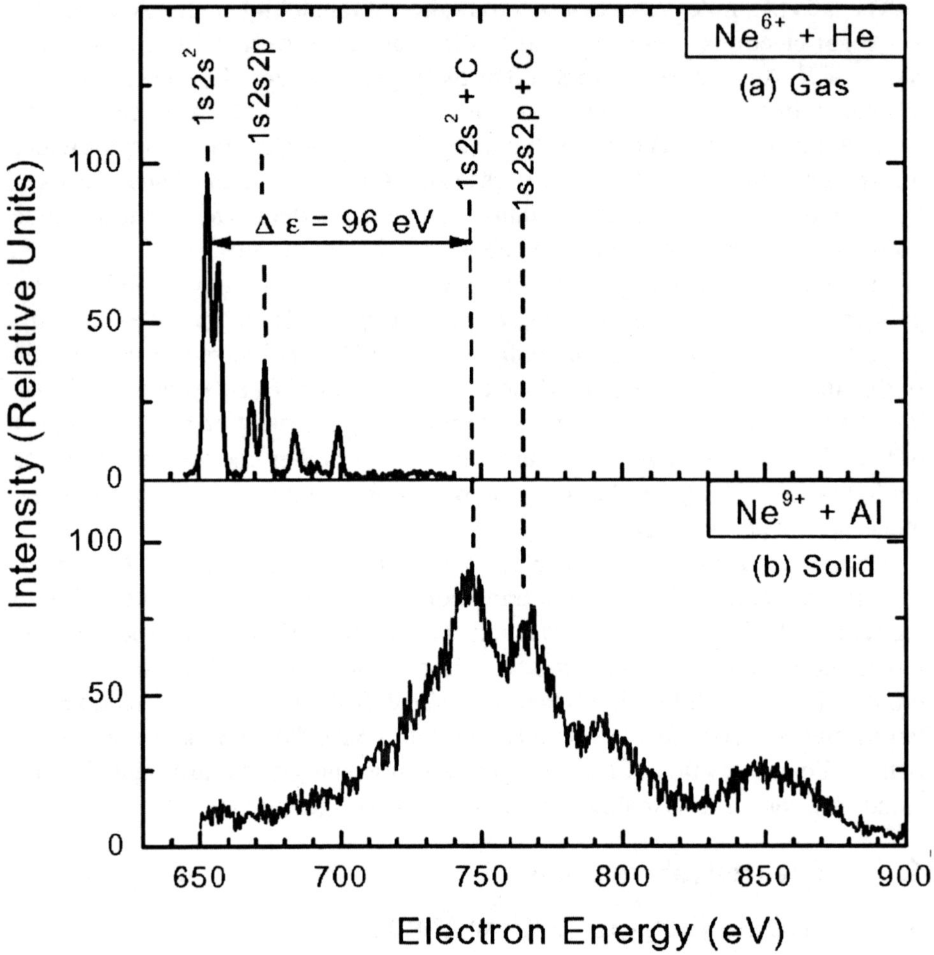

Figure 3.6. Comparison of electron spectra measured for Ne^{q+} impact on a He gas target and a solid target Al. In (a) the spectrum from the He target [68] exhibits a prominent Auger line at 652 eV which originates from the Auger transition $1s\,2s^2\,{}^2S \rightarrow 1s^2\,{}^2S$. In (b) the corresponding line is shifted by 96 eV for the Al target due to solid state effects manifested by the screening cloud C [22].

Cowan [71]. The calculations [22] were carried out for ions in the configuration $1s\,2s^2\,nl^7$ with core electrons in the $1s$ and $2s$ orbitals and 7 Rydberg electrons in outer orbitals with the principal quantum numbers n. It should be noted that 2s occupation and, hence, the consideration of the configurations $1s\,2s^2\,nl^7$ is unrealistic for large n. Nevertheless, it has been shown that the present picture provides useful information about hollow atom formation above the surface [22]. Essentially, the calculations explain the shift of the KL_1L_1 Auger transition energy by 96 eV observed in fig. 3.6.

The Hartree-Fock calculations were conducted for various Rydberg orbitals with the principal quantum number decreasing from $n = 9$ to 3 (fig. 3.7). Due to the fact that the diameter of the Rydberg orbitals scale with n^2, the orbital decreases strongly in size as the quantum number n decreases. The radius of the $n = 9$ orbital is as large as $\sim$ 35 a.u. (note the scale in the figure.) On the other hand, the $n = 3$ orbital has a relatively small diameter of about $\sim$ 4 a.u., which is smaller than the distance between the lattice atoms in the solid.

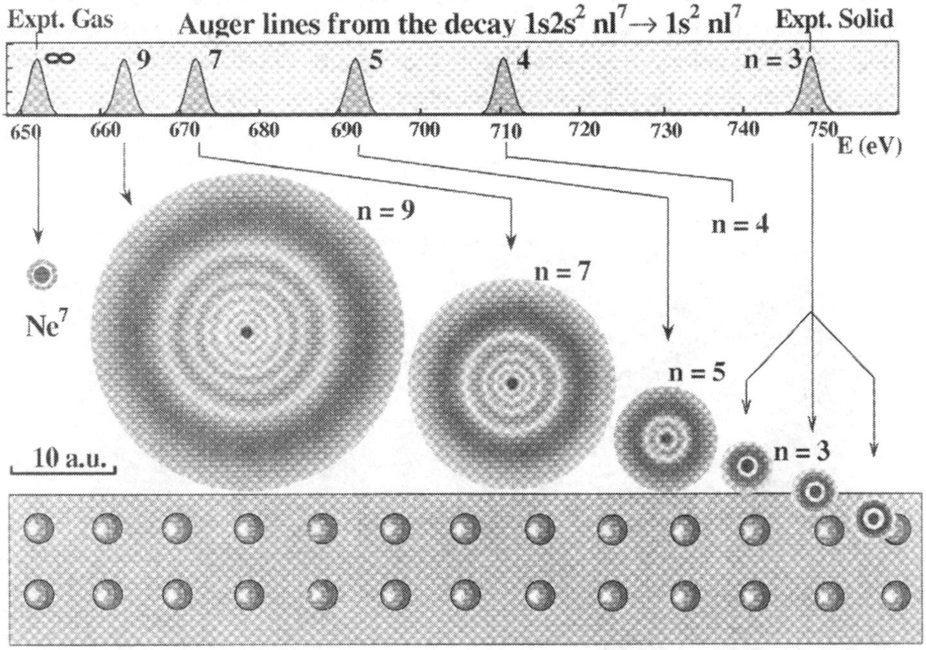

Figure 3.7. Density plot of Rydberg electrons with the principal quantum numbers $n = 9$, 7, 5, and 3. The plots are based on orbitals obtained from Hartree-Fock calculations of the configuration $1s\,2s^2\,(nl)^7$ where $l = 0$ and 1 using the Cowan code [71]. Note the length of 10 a.u. indicated in the figure. Auger line positions obtained with the Cowan code are shown at the top of the graph. From Ref. [22].

The electrons in outer Rydberg orbitals influence the K-Auger transition energy by screening effects. This can be seen in the upper part of fig. 3.7 which shows the energy of the Auger lines due to the KL_1L_1 transitions $1s\,2s^2\,nl^7 \rightarrow 1s^2\,nl^7$ where the quantum number n was varied in accordance with the density plots of the orbitals. The important result of the calculations is that significant energy shifts do not occur before the small values $n = 4$ or 3 are reached. The experimental value of the KL_1L_1 Auger transition energy is 748 eV for solid targets (fig. 3.6), which is reached for the smallest value $n = 3$ considered here. From the dimension of the $n = 3$ orbital it can be concluded that the ions have at least reached the surface, if not entered into the solid, when the 748 eV Auger electrons are ejected. In fact, detailed studies of the angular distributions of the Auger electrons have shown that these KL_1L_1 Auger electrons are ejected below the surface when the neon projectile energy is larger than a few hundred eV [31].

A theoretical study of hollow atoms inside the solid has been performed by Arnau *et al.* [21]. In the analysis, Density Functional Theory (DFT) was applied to the problem of a static charge impurity in jellium [72]. Thus, screening functions were determined by modeling the features of hollow projectile atoms. Fig. 3.8 shows results for electron densities for a hollow neon atom inside Al.

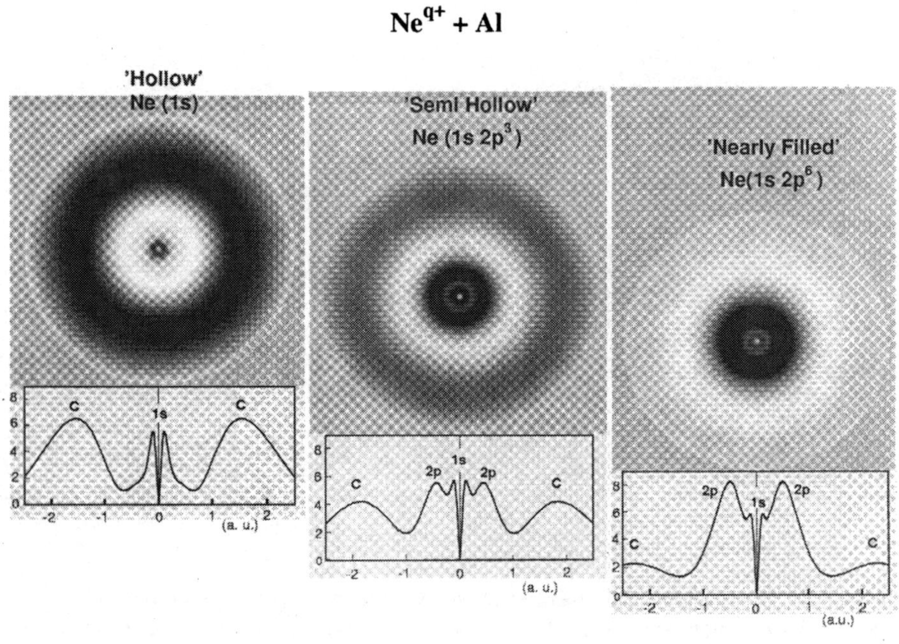

Figure 3.8. Density plots of hollow, semi-hollow, and nearly filled Ne atoms in Al. The data are calculated by means of the Density Functional Theory [72]. Note that the density is multiplied by the square of the distance r. From Ref. [22].

The bottom of the graph depicts the corresponding density functions. It is seen that the $1s$ electron density is clearly separated from the induced charge cloud that maximizes near 1.5 a.u. At about 0.8 a.u. the charge density exhibits a deep valley giving rise to a remarkable empty space which is a characteristic feature of the hollow atom.

When the hollow atom moves inside the solid it suffers binary collisions with individual target atoms. In these collisions the L shell ($n = 2$) becomes more and more filled due to charge transfer between the projectile and lattice atoms [17] as will be discussed in the following section. Consequently, when the L shell becomes increasingly filled, the induced charge cloud around the projectile decreases in intensity. Accordingly, as seen from fig. 3.8 the empty space diminishes as the hollow atom gets more and more filled in the L shell. When 6 electrons are located in the L shell, the induced electron density, i.e. the C shell, is barely visible.

To compare the formation of hollow Ne atoms in solids and vacuum, results of electron density calculations by means of the HF code [71] and the DFT code [72] are plotted in fig. 3.9. Since the hollow atoms are neutral, the number of electrons contained in the induced charge cloud is equal to the number of electrons missing in the core. For instance, a hollow atom with one K vacancy and an empty L shell contains 9 electrons in the induced charge cloud (fig. 3.9). It is seen that the charge cloud induced in the solid is not much different from the atomic charge cloud where 9 electrons are placed in the $n = 3$ orbital.

In accordance with the similarity of the charge clouds for the two cases, the energy of the KL_1L_1 transition in an atom with an occupied $n = 3$ orbital is found to be practically equal to the energy of the corresponding Auger transitions in the solid (figs. 3.6 and 3.7). It appears that the charge cloud induced in the solid is best modeled by electrons located in the lowest lying atomic projectile orbital that is not bound inside the solid. Although not fully understood, this rule was commonly used when energies for Auger transitions are determined for hollow atoms moving at the surface and inside the solid [31, 73, 32].

4.2 Cascade models for hollow atoms

In the past, various models have been developed describing the cascading decay of hollow atoms. The theoretical approaches can be divided into two groups: (i) models treating the mean number of electrons in a given shell [11, 74–77] and (ii) models treating the occupation probability for individual configurations [17, 20, 23–25]. An obvious conceptual difference in the two models can be seen from the calculated results: occupation numbers obtained from the mean charge model are generally larger than one, whereas occupation probabilities evaluated in the configuration model are always smaller than one. For the specific case of filling hollow atoms below the surface, an analytic evaluation [17] of the configuration model is obtained. This formalism exhibits the various parameters relevant for the creation of Auger electrons and plasmons by slow ions.

A more recent model [78], devoted to hollow Ar, combines the mean charge and configuration methods as will be described in the following. An important

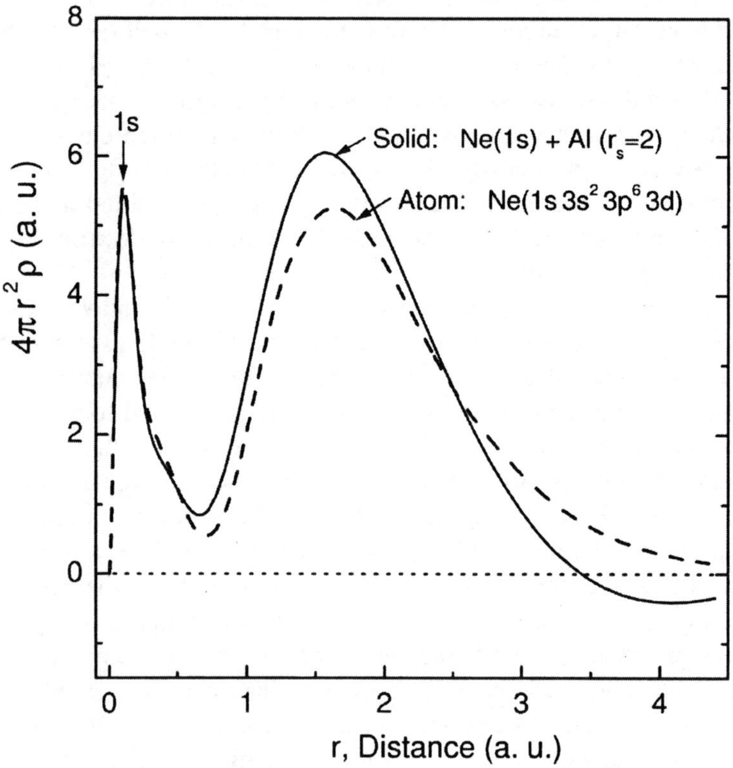

Figure 3.9. Electron densities of a hollow neon atom in Al ($r_s = 2$) calculated using the Density Functional Theory [72]. The results are for atoms with a vacancy in the K shell, an empty L shell, and an induced charge cloud C. The data for the solid are compared with Hartree-Fock results for an isolated atom with 9 electrons in the M shell [22]. The background density due to the jellium was subtracted. Note that the density is multiplied by $4\pi r^2$ where r is the electron distance from the neon nucleus.

feature of this model is that the N shell and higher shells are treated differently from the K, L, and M shells. For the lower shells the configuration method is applied [78], whereas for the higher shells we use parts of the over-the-barrier model by Burgdörfer *et al.* [11] and other similar models [74, 75, 77], also referred to as mean-charge method. It is noted that higher shells are only important for atoms moving above the surface. Within the solid higher shells do not exist and the N shell merges into the C shell.

4.2.1 Mean-charge model. Within the mean-charge model each shell of principle quantum number n is assumed to be occupied by a mean number of electrons corresponding to the charge q_n. The time-dependencies of these mean occupation numbers are determined by a set of coupled rate equations

$$\frac{dq_n}{dt} = I_n^{in}(t, \mathbf{q}) - I_n^{ex}(t, \mathbf{q}) \tag{3.1}$$

where the vector $\mathbf{q} = (...q_n...q_{\max})$ contains all information about the shell occupations. The ingoing current I_n^{in} and outgoing current I_n^{ex} correspond to the gain and loss of electrons for the n'th shell, respectively. Each current implies a contribution due to electrons transferred to (or from) the conduction band and a contribution due to electrons from Auger transitions. For more details the reader is referred to the original publication of the over-the-barrier model [11]. At this point it shall only be pointed out that transitions from the N shell into lower lying levels are excluded when this model is combined with the configuration model [78]. In fact, eq. (3.1) is solved for a fixed configuration $\mathbf{n}$ defined below. This fixing is released in the configuration model as discussed in the next subsection.

The time evolution of the system is governed by transition rates which are associated with the different electron transfer processes. The Auger transitions are designated by three labels, e.g., Γ_{KLL} the rate corresponds to the transfer of an electron from the L shell to the K shell while ejecting another L-shell electron into the continuum. To keep a consistent notation, two labels are used for x-ray transitions, e.g., Γ_{KL}, corresponds to the transfer from the L shell to the K shell (also denoted K_α in the literature). The collisional charge transfer is specified by a single label indicating the shell to which the transfer takes place, e.g., Γ_M.

Some effort is needed to proceed from the mean-charge model to the configuration model. The latter model involves rates associated with the transfer of electrons from the N shell. These rates require information about individual electrons occupying the N shell, which is not achieved when calculating the mean charge from eq. (3.1). For instance, when an occupation number $q_N < 1$ is obtained, it may seem as if, e.g., the radiative KN transition is not possible as it requires at least one electron in the N shell. However, it should be recalled that q_N stands for a mean value which involves a spectrum of occupation numbers.

Hence, to link the mean-charge model with the configuration model, an assumption about the number distribution associated with a given mean value has to be made. To evaluate the occupation probability P_ν^{bin} of the N shell by ν electrons we can use the well-known binomial distribution formula

$$P_\nu^{bin} = \binom{\nu_{\max}}{\nu} p^\nu (1-p)^{\nu_{\max}-\nu} \tag{3.2}$$

where $p = q_N/\nu_{\max}$. The maximum number of electrons in the N shell $\nu_{\max} = Z - \kappa - \lambda - \mu$ is obtained from the number of electrons κ, λ, and μ in the K, L, and M shell, respectively. The results from eq. (3.2) allows for the deduction of average transition rates, which were obtained as a sum of individual rates weighted by the corresponding probability P_ν^{bin}. It is noted that below the surface the binomial distribution is of less importance, since the C shell is readily filled with several electrons.

4.2.2 Configuration model for Ar. To study the dynamic properties of second-row atoms, such as argon, a cascade model was developed describing the stepwise filling of their empty inner orbitals near a surface [78]. For Ar^{17+} K, L, and M shells are considered, so that the filling dynamics is rather complex. Moreover, electron transfer processes above the surface are included in the analysis. The above-surface effects complicate the analysis so that the model equations have to be solved numerically. Nevertheless, the filling sequence of the hollow atom is determined by expressions that are similar to those known from the radioactive decay of nuclei [17, 23, 24]

Within the framework of the configuration method, we consider the time-dependent occupation of the configurations $\mathbf{n} = (\kappa, \lambda, \mu)$ where again κ, λ and μ are the number of electrons in the K, L and M shells, respectively. The main task of treating complex hollow atoms arises from the elaborate bookkeeping of the numerous configurations and decay processes occurring during the stepwise filling of its empty shells. The Ar^{17+} ion has empty L and M shells. During the filling of a hollow Ar atom, the configurations $\mathbf{n}$ are transiently produced. The transfer from one configuration to another takes place via different radiative and non-radiative electron transitions. Also, below the surface, collisional electron capture processes into the M shell are taken into account.

To facilitate the task of tracing the sequential filling of a hollow atom we use the configuration matrix shown in fig. 3.10. The diagram, introduced to visualize the cascade filling of Ar^{17+}, can be used for other second-row atoms, too. Each box in the matrix is associated with a configuration label $\mathbf{n} = (\kappa, \lambda, \mu)$. The upper level corresponds to ions with a K vacancy, whereas the lower level corresponds to ions with a filled K shell. In the configuration matrix, the stepwise filling of a hollow atom can be traced along a sequence of configurations $\{\mathbf{n}_1, \mathbf{n}_2, ..., \mathbf{n}_s\}$. The system follows a main configuration sequence, however, there are side paths which may be included in the analysis.

The Ar^{17+} system starts in the front at the upper-left corner associated with the configuration box (0,0) (where the $\kappa = 1$ label is omitted for brevity). The main transition which progresses the system to the (0,1) box corresponds to an MNN Auger process where an electron from the N shell is transferred into the M shell, while another N electron is ejected. A side path involves an LNN Auger process transferring the system into the (1,0) box. Furthermore a KLN transition may transfer the system from the (1,0) box into the (0,0) box with $\kappa = 2$ in the lower level. After sufficient time the system will be distributed over all configuration boxes. At the beginning, the system is centered in the

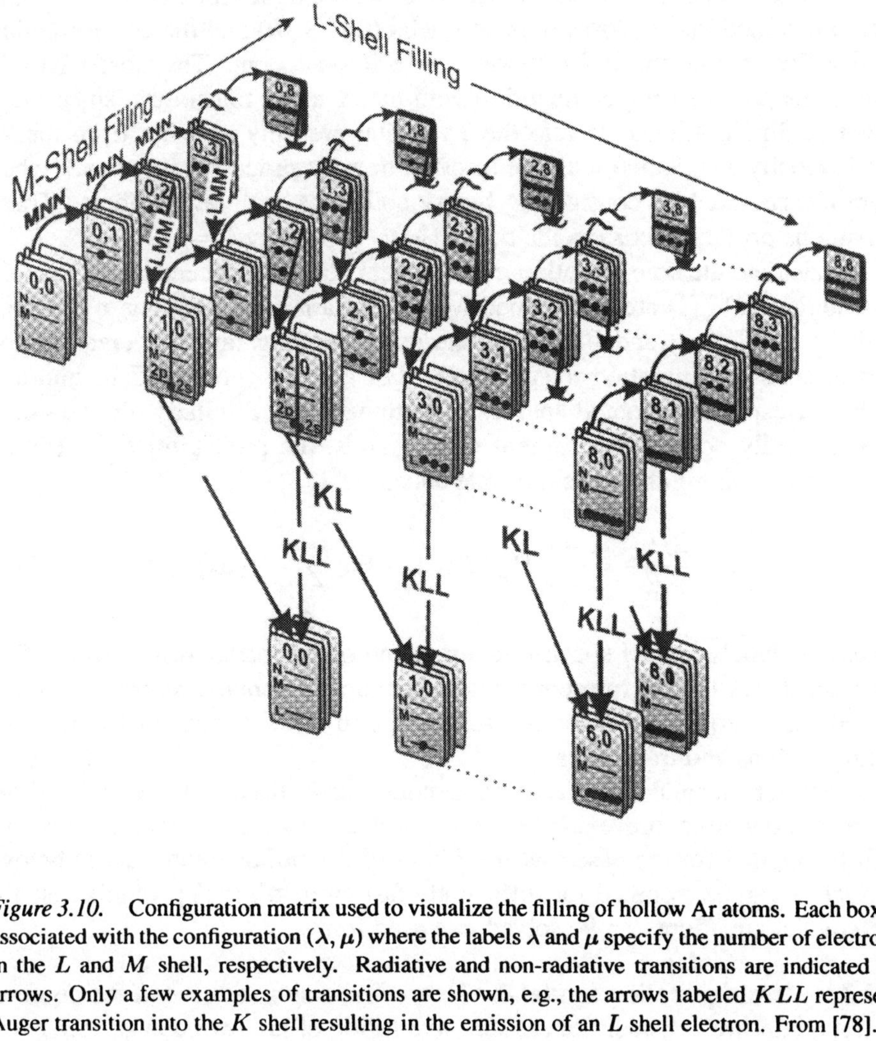

Figure 3.10. Configuration matrix used to visualize the filling of hollow Ar atoms. Each box is associated with the configuration (λ, μ) where the labels λ and μ specify the number of electrons in the L and M shell, respectively. Radiative and non-radiative transitions are indicated by arrows. Only a few examples of transitions are shown, e.g., the arrows labeled KLL represent Auger transition into the K shell resulting in the emission of an L shell electron. From [78].

left-front regions at the upper level, whereas with increasing time the system migrates into the right-back region at the lower level.

To visualize individual ingoing and outgoing fluxes consider an arbitrary configuration box (λ, μ) in the intermediate region of the upper level. The main ingoing flux is produced by an MNN Auger transition from the $(\lambda, \mu-1)$ box. Likewise, LMM, LMN, and LNN Auger transitions produce flux from the $(\lambda-1, \mu+2)$, $(\lambda-1, \mu+1)$, and $(\lambda-1, \mu)$ boxes, respectively. The outgoing flux involves all possible MNN, LXY, and KXY transitions where X and Y stand for the L, M, and N shells. Also, radiative KX transitions were included, whereas radiative transitions into higher shells (L, M etc.) are neglected [79].

It should be added that the steps involving the configurations with $\lambda = 1$ and 2 are processed via the $2p$ and $2s$ subshells (fig. 3.10) and the corresponding Coster-Kronig transitions LLN were taken into account. The Coster-Kronig transitions are in strong competition with the K x-ray transitions, since these radiative dipole transitions into the $1s$ orbital are only possible from the $2p$ level. Finally, single and multiple electron transfer processes into the M shell were incorporated by considering binary collisions with target atoms. These transitions provide flux from the boxes $(\lambda, \mu - i)$ where $i = 1, 2....$

To calculate the time evolution of the system, the time-dependent occupation probabilities $P_{\mathbf{n}_k}(t)$ are determined for all relevant configurations $\mathbf{n}_k$ labeled with k. Let $\Gamma_i(\mathbf{n}_i)$ and $\Gamma_j(\mathbf{n}_j)$ be the corresponding rates for creation and loss of such configuration, respectively. For a given configuration, multiple paths correspond to several sources of ingoing flux. Similarly, the outgoing flux generally occurs via different paths. Thus, the probabilities $P_{\mathbf{n}_k}(t)$ are obtained by solving the system of rate equations [78]

$$\frac{dP_{\mathbf{n}_k}}{dt} = \sum_i P_{\mathbf{n}_i}\Gamma_i(\mathbf{n}_i) - P_{\mathbf{n}_k}\sum_j \Gamma_j(\mathbf{n}_j) \tag{3.3}$$

where the labels i and j specify the input and output paths, respectively. This formula shows that the present calculations are elementary, however as mentioned, they require an elaborate selection and bookkeeping of the relevant configurations and transitions.

It should be recalled that the cascade model can be used to describe the filling of the hollow atom above and below the surface. In the following, the model will be applied for the case that the filling of the hollow atoms starts below-surface as can be assumed for sufficiently fast projectiles [17]. In this case the differential equations can be solved in closed form.

4.2.3 Configuration model for Ne. As a simpler case of the configuration model we consider the cascade-like filling of hollow neon as a first-row atom. For Ne^{9+} in Al the filling of the projectile L shell takes place via L-Auger transitions and collisional charge transfer governed by the L-Auger rate $\Gamma_{LCC}(\lambda)$ and the capture rate $\Gamma_L^c(\lambda)$, respectively. For $\lambda \geq 2$ the ensemble of atoms $N_\lambda(t)$ undergoes KLL-Auger transitions with the rate $\Gamma_{KLL}(\lambda)$. The time-dependent occupation probability $P_\lambda(t)$ with one vacancy in the K shell and λ electrons in the L shell is obtained from the rate equations [24, 17]

$$\frac{dP_\lambda}{dt} = \Gamma_L(\lambda - 1)\,P_{\lambda-1} - \Gamma_S(\lambda)\,P_\lambda \tag{3.4}$$

where the L-shell filling rate $\Gamma_L = \Gamma_L^c(\lambda) + \Gamma_{LCC}(\lambda)$ and the sum rate $\Gamma_S(\lambda) = \Gamma_L(\lambda) + \Gamma_{KLL}(\lambda)$.

For the intensity of the K-Auger electrons, ejected below the surface, we assume an exponential attenuation law

$$I_K^a(t) = I_K^0 e^{-\ell/\ell_K^a} \tag{3.5}$$

where I_K^0 is the original intensity, ejected from the hollow atom, ℓ is the travel distance of the electrons in the solid, and ℓ_K^a is the attenuation length. This well-known attenuation law may be transformed into a time-dependent expression $I_K^a(t) = I_K^0 e^{-\Gamma_K^a t}$ where Γ_K^a is the corresponding attenuation rate. It can readily be shown that the attenuation rate is obtained as $\Gamma_K^a = v_p/\ell_K^a$, where v_p is again the projectile velocity.

After time integration one obtains the *attenuated yield* of the Auger electrons in the elastic channel which may be evaluated analytically giving rise to the relatively simple expression

$$Y_K^a(\lambda) = \Gamma_{KLL}(\lambda) \frac{\prod_{\lambda'=0}^{\lambda-1} \Gamma_L(\lambda')}{\prod_{\lambda'=0}^{\lambda} \widetilde{\Gamma}_S(\lambda')} \tag{3.6}$$

Expression (3.6) implies the sum rate $\widetilde{\Gamma}_S(\lambda) = \Gamma_S(\lambda) + \Gamma_K^a$ modified by attenuation.

The Auger electrons lost by attenuation may be rescattered into the detector where the creation of Auger intensity is governed by the *build-up* intensity

$$I_K^b(t) = I_K^0 \left(1 - e^{-\ell/\ell_K^a}\right) e^{-\ell/\ell_K^b} \tag{3.7}$$

The first term describes the rescattering into the build-up channel and the second term governs its absorption which, in turn, is determined by the attenuation length ℓ_K^b. Similar to the primary channel, eq. (3.7) can be transformed in a time dependent expression. Then, the flux in the inelastic channel is integrated to obtain the *build-up yield* $Y_K^b(\lambda)$ which, in turn, is obtained as the difference of two terms each one analogous to that given in eq. (3.6). Since the Auger electrons are usually measured within a wide range of energies, covering most of the inelastic energy spectrum, we compare the experiment with the *transported yield* $Y_K(\lambda) = Y_K^a(\lambda) + Y_K^b(\lambda)$. More details can be obtained from Ref. [17].

Finally, electron refraction effects are included into the analysis. When the Auger electrons with energy ε' leaves the surface it experiences a potential step U so that its energy is reduced to $\varepsilon = \epsilon' - U$. Apart from the energy loss the electron experiences refraction effects where the emission angle β' inside the solid is altered to β outside the solid relative to the surface normal. The yield ratio of electrons passing from outside to inside the surface is obtained as

$$\frac{Y(\beta)}{Y'(\beta')} = \frac{\varepsilon \cos\beta}{(\varepsilon + U)\cos\beta'} = \frac{\cos\beta}{\left[\left(1+\frac{U}{\varepsilon}\right)\left(\cos^2\beta + \frac{U}{\varepsilon}\right)\right]^{1/2}} \tag{3.8}$$

For energies $\epsilon \gg U$ refraction effects are not important whereas for smaller electron energies these effect become significant. For instance, L-Auger electrons of Ne with energies of about 50 - 100 eV are significantly affected, whereas for K-Auger electron of about 748 eV the refraction effects are found to be negligible [17].

The cascade model requires a number of model parameters that have been determined by *a priori* methods in Ref. [17, 47]. Recent progress in calculating the Auger decay of hollow atoms located in a metal have been made by Deutscher et al. [80]. Moreover, the present cascade model requires cross sections for charge exchange between molecular inner-shell orbitals correlating with the L shells of the projectile and target atoms. These molecular orbitals were evaluated for the Ne + Al system using model matrix elements reported previously [81]. Then, the Landau-Zener model was utilized to determine related electron transfer probabilities which, in turn, were applied to evaluate the corresponding cross sections for electron capture.

5. Auger Electron Spectroscopy

Since the beginning of the field, experimental studies of Auger spectra played an important role to obtain information about the formation and decay of hollow atoms near a surface. Fig. 3.11 gives an overview of the locations where Auger emission occurs. Auger electrons may be ejected from the projectile before it hits the surface. This is particular true for autoionizing transitions between higher Rydberg states giving rise to electron emission of relatively low energies [22]. For potential emission, these electrons constitute a major contribution to the total electron emission. The mechanisms leading to potential emission of low-energy electrons have been discussed in detail by Winter and Aumayr [82]. Here, we shall focus the attention on the last steps of the filling cascade of the hollow atom, i.e., L- and K-Auger transitions.

Above-surface emission of K-Auger electrons has been searched for in various studies [1] using a *Doppler* shift analysis pioneered by Morgenstern and collaborators [12, 83, 84]. The analysis showed that Auger electrons are predominantly produced below surface. To date it is commonly accepted that K-Auger emission on the incident trajectory before hitting the surface occurs only in specific cases [84]. Hence, some effort is required to create suitable conditions for Auger electrons to be ejected above the surface [13].

This above-surface emission is expected when projectiles are used under grazing incidence so that they are reflected from the surface [85]. Thomaschewski *et al.* [86] have shown that K-Auger electrons are ejected between the topmost atomic layer and the jellium edge, denoted "at surface emission",

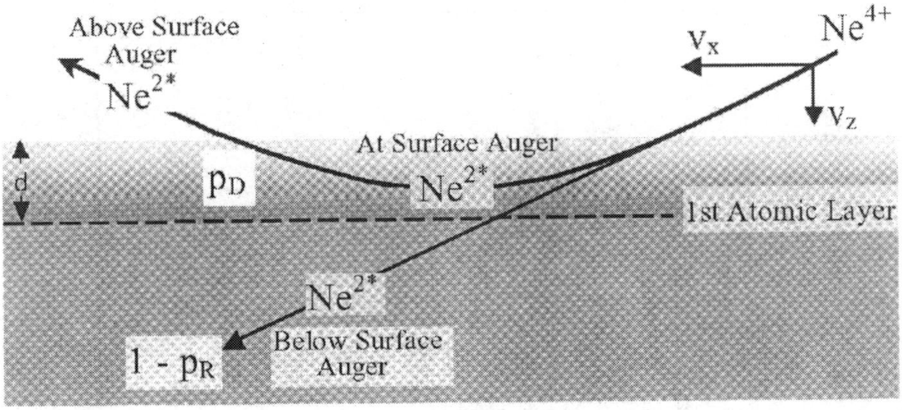

Figure 3.11. Diagram to visualize the 3 regions for Auger electron production: above, at, and below surface. Incident Ne^{4+} ions are neutralized and reflected at the surface with the probability p_R and, hence, they enter into the solid with the probability $1-p_R$. Reflected atoms Ne^{2*} survive the passage through the above surface region in a doubly excited state with the probability p_D.

where electron refraction effects can still play a role. A case where L-Auger electrons are ejected well above the surface on the outgoing part of a scattered projectile (fig. 3.11) is presented in a later subsection. Below-surface emission of K-Auger electrons is discussed first since it is most probable.

5.1 Below surface emission

5.1.1 Spectra of neon. In this subsection spectra of Auger electrons shall be interpreted by means of cascade model calculations. The Auger spectra refer to projectile velocities which are sufficiently high to ensure electron emission below the surface. Therefore, the analytic cascade model (eq. 3.6) is applied. In fig. 3.12 a typical K-Auger spectrum for 0.4-keV Ne^{9+} incident on Al is plotted. A similar spectrum was already shown in fig. 3.6.

The spectrum shows essentially 3 peaks that can be associated with the decay modes KL_1L_1, KL_1L_{23}, and $KL_{23}L_{23}$ of Ne with a K vacancy. The structure at higher energies labeled KLC is produced by Auger transitions where an electron from the induced electron cloud C is ejected. Each peak of the KLL group involves contributions from Ne projectiles with different numbers λ of electrons occupying the L shell. It is important to note that the relative intensities of the Auger peaks depend on the filling state of the hollow atom.

Fig. 3.12 also shows theoretical results from the cascade model displayed as bars below the experimental spectrum [33]. For each yield $Y_K(\lambda) = Y_K^a(\lambda) +$

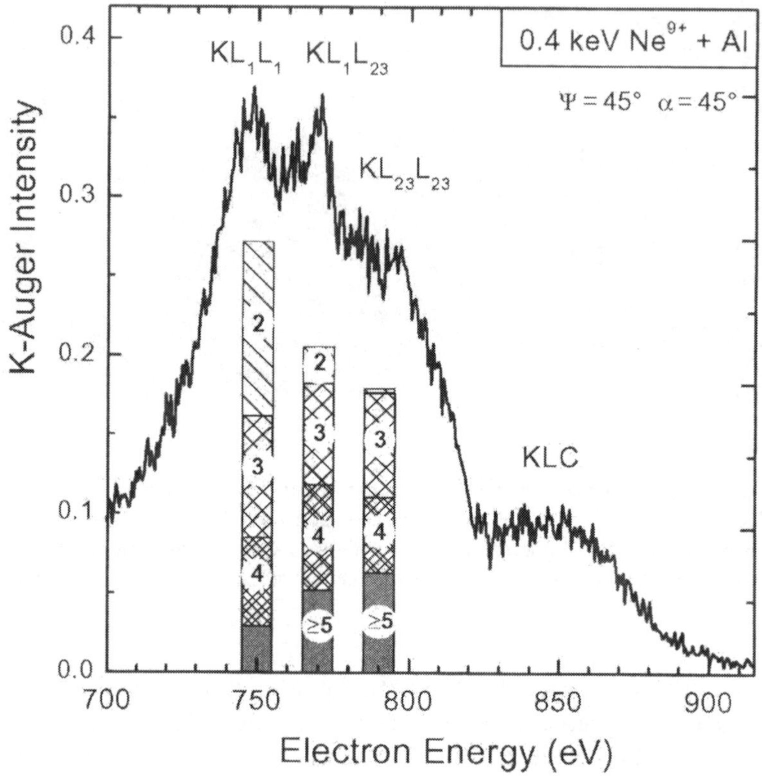

Figure 3.12. Experimental K-Auger spectra for Ne^{9+} incident on Al at the energy of 0.4 keV. The projectile angle of incidence is $\psi = 45^0$ and the observation angle relative to the surface plane is $\alpha = 45^0$. The bar diagram is evaluated using the cascade model discussed in the text. The numbers in the bars stand for the occupation number λ of the L shell. From Ref. [22].

$Y_K^b(\lambda)$, deduced from the cascade model, the branching into the decay channels KL_1L_1, KL_1L_{23}, and $KL_{23}L_{23}$ were evaluated using theoretical Auger rates [87]. As already noted the Auger peaks associated with these decay channels, include different contributions from the filling states λ.

The theoretical data in fig. 3.12 indicate that the first Auger peak at 748 eV is primarily composed of small $\lambda \lessapprox 2$ values, whereas the peak at higher energies near 790 eV has contributions from λ values as large as 8. Thus, the structures in the Auger spectrum yield information about the filling state of the hollow atom during the Auger transitions. In general, it follows that the low-energy part of the spectrum is a signature for a hollow atom in an early stage, whereas the high-energy part represents the hollow atom in a later stage. Thus, information about dynamic properties of the hollow atom can be obtained from the analysis of the Auger spectra.

In the pioneering period of the field there was a long-standing discussion about the question of whether the K-Auger electrons are ejected above or below the surface. Ultimate evidence for the location of the K-Auger decay has been achieved from the angular distributions of the Auger electrons. In the case of an above-surface emission the angular distribution is expected to be isotropic. Deviations from isotropy may occur outside the solid, however, the statistical average over a large number of states is likely to cancel such anisotropies.

Inside the solid, definite anisotropy effects are expected from the attenuation of the electrons during their transport to the surface. Electrons ejected at small angles with respect to the surface plane are subject to significant attenuation due to longer travel distances of the electron in the solid. On the contrary, attenuation effects are relatively small for observation angles perpendicular to the surface plane. Thus, attenuation effects produce anisotropy in the electron angular distribution in addition to possible refraction effects.

The attenuation method used by Köhrbrück *et al.* [31] revealed that K-Auger emission is strongly anisotropic for Ne^{9+} incident on Al with energies as low as 0.1 keV. This shows for the considered cases that the K-Auger electrons are essentially ejected below the surface. The same result was found by Grether *et al.* [33] for the KL_1L_1 spectral component observed at 748 eV (fig. 3.12). It should be added that Ne-K Auger electrons ejected within the solid are not much influenced by refraction effects [17].

5.1.2 Spectra of argon. In the early period of the field the projectile Ar^{17+} has been extensively used in studies of K- ray emission in ion-surface interactions by Briand and collaborators [10, 88, 89]. The method of x-ray spectroscopy is particular suited for argon since its K-shell fluorescence yield is significantly larger than that of neon. The studies of radiative transitions in Ar yielded detailed information about the occupation of the L and M shells during the K x-ray transition [90]. The high degree of information of the observed spectra represent a challenge for models devoted to the dynamic properties of hollow atoms.

In fig. 3.13 recent cascade model calculations are compared with experiments by Briand et al. [91]. The left column shows data for Ar^{17+} incident on SiH with the energies of 17 eV, 3.4 keV, and 170 keV. The spectra are composed of distinct peaks, each of which can be attributed to a specific number λ of electrons occupying the L shell during x-ray emission.

It is seen that the spectra change significantly as the projectile energy varies. At the lowest energy of 17 eV the spectrum consists of only a few prominent lines which are primarily due to the smallest L- shell occupation numbers $\lambda = 1$ and 2. Hence, at low projectile energies the L shell is barely occupied

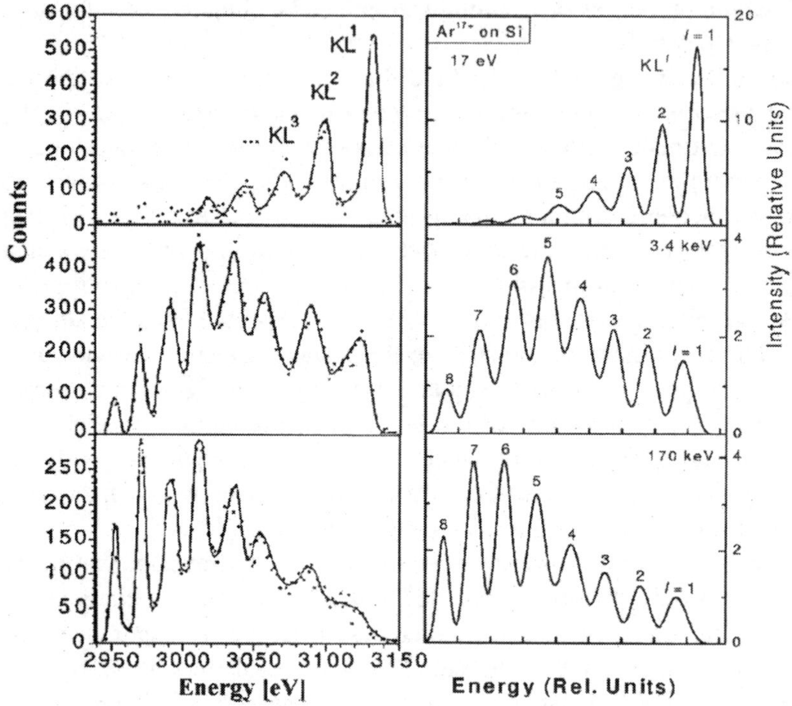

Figure 3.13. Experimental K x-ray spectra [91] produced by Ar^{17+} at normal incidence on SiH for energies of 17 eV, 3.4 keV, 170 keV (left column), in comparison with theoretical results obtained using the cascade model [78] (right column). The peaks are attributed to the number λ of electrons occupying the L-shell during the x-ray transition. For a given λ value a peak shift is observed due to a variation of the number μ of M-shell electrons. The experimental yields are plotted in relative units whereas the theoretical yields are given in terms of the total number of x-rays ejected per incident ion.

during K x-ray emission. At higher energies of 3.4 keV the spectrum exhibits intermediate occupation numbers for the L shell. Finally, when changing to the high energy of 170 keV the spectrum is still significantly altered indicating that the L- shell occupation is further enhanced.

In fig. 3.13 the right hand side shows the cascade model results for the emission yield [78]. For the lowest energy of 17 eV the incident ions are assumed to be accelerated by 80 eV due to image charge effects [19, 92]. It is recalled that the emission yield is evaluated for a given configuration denoting the occupation of the L and M shells. From fig. 3.13 it is seen that the intensities of the theoretical line spectra compare well with the corresponding experimental data. This provides confidence that the present model accounts for the essential features of hollow Ar atoms interacting with a surface.

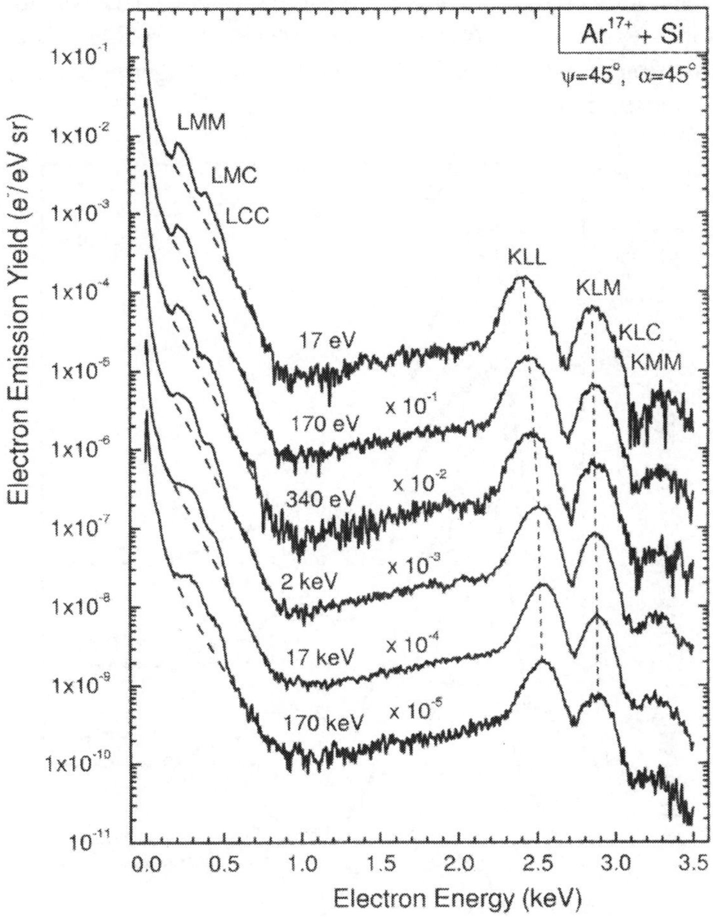

Figure 3.14. Electron spectra produced by Ar^{17+} incident at $\psi = 45^0$ on a Si(111) surface. The electron observation angle is $\alpha = 45^0$. Note that for graphical reasons the spectra are multiplied by the factors indicated in the figure. From Ref. [93].

In comparison with the x-ray studies, the number of electron spectroscopy experiments using highly charged Ar^{17+} ions are limited [93]. Fig. 3.14 shows electron spectra acquired within a wide range of Ar^{17+} energies. Compared to the x-ray data, the electron spectra exhibit less structures, however, K- and L-Auger electrons can simultaneously be measured. The Auger spectra exhibit different peaks which can be attributed to the shell structure of the hollow atoms formed near the Si surface. In particular, the spectra indicate peaks due to KLL- and LMM-Auger transitions in hollow Ar. These peaks show that both L and M shells exist in hollow Ar atoms when moving inside the solid. In addition, the highly charged Ar ion induces the negative charge cloud C, which coincides in shape with the N shell. The experiment provides clear evidence for the formation of the C shell (see for instance the Auger peaks due to LMC and KLC transitions).

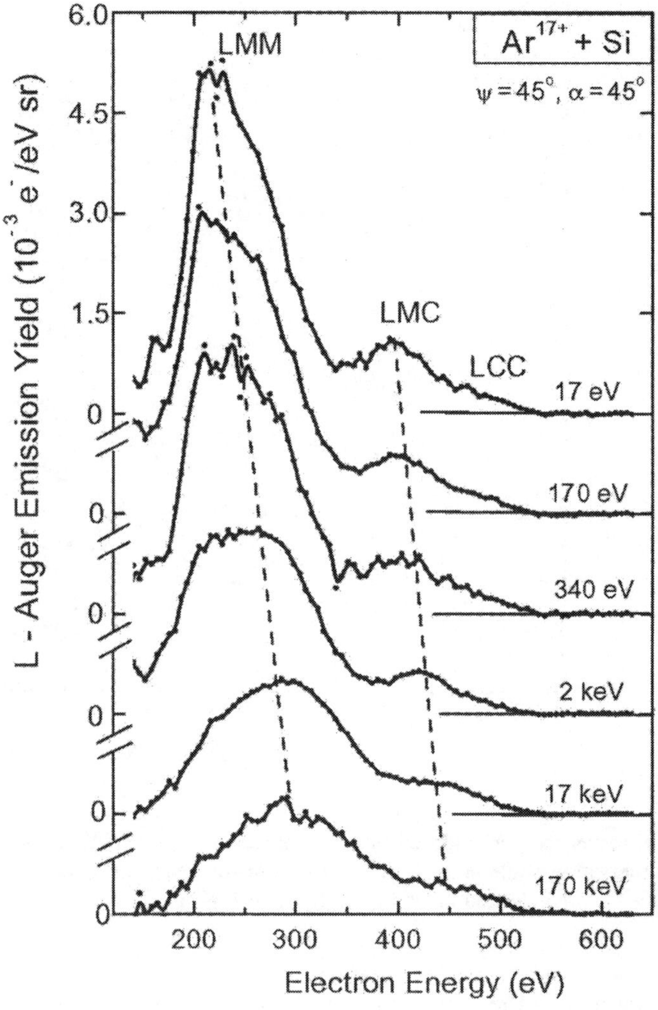

Figure 3.15. Emission yield for L-Auger spectra acquired for Ar^{17+} incident on a Si surface. The data are extracted from fig. 3.14. The continuous background was subtracted from the Auger spectra. Note the shift of the peak energies with varying projectile energy indicated in the spectra. From Ref. [93].

In fig. 3.14 the surprising results is that the electron spectra do not change significantly in shape and absolute value when the projectile energy is varied by 4 orders of magnitude. This demonstrates that the important quantity that governs the present emission spectra, is not the kinetic energy but the potential energy of the hollow argon atom. Nevertheless, closer inspection of the KLL and KLM Auger peaks shows a line shift with varying projectile energy. This line shift can be attributed to the decreasing population of the L shell during the K-Auger transition as discussed in conjunction with fig. 3.13. It is recalled that the number of L-shell electrons increases as the projectile energy increases. The increasing population of the L-shell is produced by an enhancement of LMM Auger transitions following an increasing population of higher orbitals [78]. When the number of L- shell electrons increases with the projectile energy, the centroid energy of the emission spectra is shifted to lower energies (contrary to the x-ray spectra).

Fig. 3.15 shows an enlarged portion of the electron spectra in an energy range covering the Ar L-Auger structures. The continuous background of the spectra was subtracted from the L-Auger intensities. It is seen that the centroid energy of the L-Auger peak varies in a similar manner as noted for the K-Auger electrons. This can be attributed to screening effects, since L-Auger transitions at low projectile energies take primarily place after the K-shell filling, whereas at high energies the L-Auger transitions occur before the K-shell filling [78, 91]. Furthermore, contrary to the K-Auger structures, the L-Auger peak decreases in intensity as the projectile energy increases. This decrease is attributed to attenuation effects on the Auger electrons ejected in deeper layers of the solid. In fact, for the lowest projectile energy of 17 eV, both K- and L-Auger electrons are found to be ejected above the surface [78]. This phenomenon will be discussed for other systems in the following section.

5.2 Above-surface emission

In this section, cases are presented where significant fractions of the L-Auger electrons are ejected outside the jellium edge at the outgoing part of a scattered neon projectile (fig. 3.11). Examples are given in fig. 3.16 [94]. The figure shows electron spectra observed at an emission angle of $\alpha = 20^0$ for different incidence angles of Ne^{4+} projectiles ($\psi = 3^0$, 7^0, and 20^0). The electron spectra exhibit various structures which can be attributed to emission of Auger electrons. The peaks labeled Ne^{2*}, Ne^{3*}, Ne^{4*} are due to L-Auger transitions in hollow Ne with 2, 3, and 4 vacancies in the L shell, respectively. The structures attributed to Ne^{3*} and Ne^{4*} are found to be relatively broad, indicating that these projectiles are likely to decay below the surface.

In contrast, the decay of Ne^{2*} atoms give rise to a rather distinct line at 22 eV, which can be associated with the initial state $1s^2 2s^2 2p^4 3s^2\ {}^1D$ decaying into $1s^2 2s^2 2p^5\ {}^2P$ under emission of an Auger electron [95]. The occurrence of an atomic Auger transition provides evidence that Ne^{2*} atoms decay well

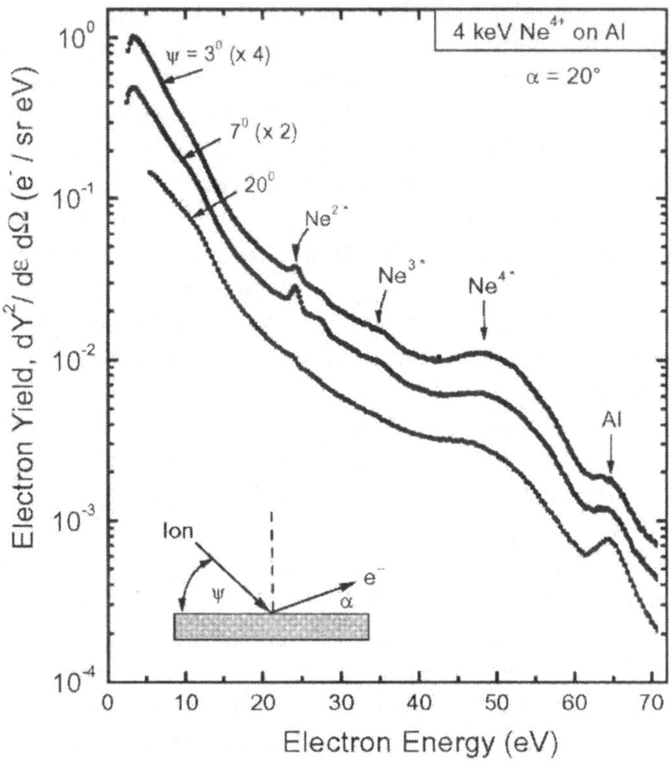

Figure 3.16. Electron yields measured for 4 keV Ne^{4+} impact on Al as a function of the electron energy. The incidence angles are $\psi = 3^0, 7^0$, and 20^0 as indicated. The peak labeled Al is produced by L-Auger transitions filling a vacancy in the L-shell of Al. The peaks labeled Ne^{2*}, Ne^{3*}, Ne^{4*} are due to L-Auger transitions in hollow Ne with, respectively 2, 3, and 4 vacancies in the L shell. Note that the spectra for 3^0 and 7^0 are multiplied by 4 and 2, respectively. From Ref [94].

above the surface. A more detailed analysis showed that the distinct Ne^{2*} peak is superimposed on a broader structure, indicating that part of the Ne^{2*} atoms decay also below the surface.

A *Doppler* analysis was performed to verify whether the L-Auger electrons are ejected by ions on its incident or outgoing trajectory, i.e., scattered from the surface [94]. This information can be extracted from the angular dependence of the energy of the ejected Auger electrons. Due to kinematic effects, the ejection energy sensitively depends on the observation angle relative to the direction of the moving emitter. Hence, in a careful analysis of the Auger electron energies yields information about a possible (specular) reflection of the projectiles.

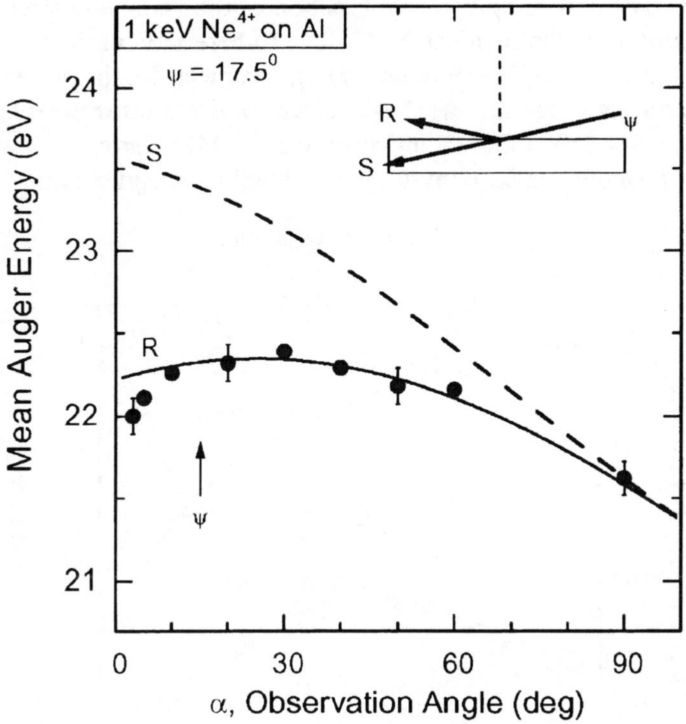

Figure 3.17. *Doppler* analysis of *L*-Auger electrons ejected from doubly excited Ne^{2*} atoms. The analysis is based on the data partially shown in fig. 3.16 [94]. The incident angle is ψ = 17.5°. The points represent experimental data. The curves labeled S and R refer to theoretical results for ions moving on a straight line trajectories and reflected ions, respectively.

Results of the *Doppler* analysis are given in fig. 3.17 showing theoretical curves evaluated from well-known kinematic formulas [96]. The curve labelled S is due to a trajectory where the projectile continues its incident direction on a straight line, whereas the curve labeled R is due to ions that are specularly reflected and have suffered energy losses in binary collisions with an Al target atom. It is seen that the uncertainties of the experimental data are much smaller than the energy difference between the S and R curves. Hence, the good agreement of the experimental with the curve labeled R unambiguously shows that the measured Auger electrons originate from reflected projectiles. In addition, as noted already, the sharpness of the Auger lines indicates that the electrons are ejected well out side the jellium edge.

Moreover, the analysis of the intensity of the distinct line at 22 eV allows for the determination of the fraction of Ne^{2*} atoms scattered from the surface. After integration of this line, absolute Auger electron yields were obtained as shown in fig. 3.18. The data are given as a function of the vertical projectile energy $E_z = E_p \sin^2 \psi$ of the incident Ne^{4+} ions (where ψ is again the incidence angle). It is seen that E_z is a suitable scaling parameter for the present electron yield. The results show that the Auger electron yields for different projectile energies of 1 to 4 keV follow a universal curve. This curve drops at low and high values forming a maximum in the intermediate energy region.

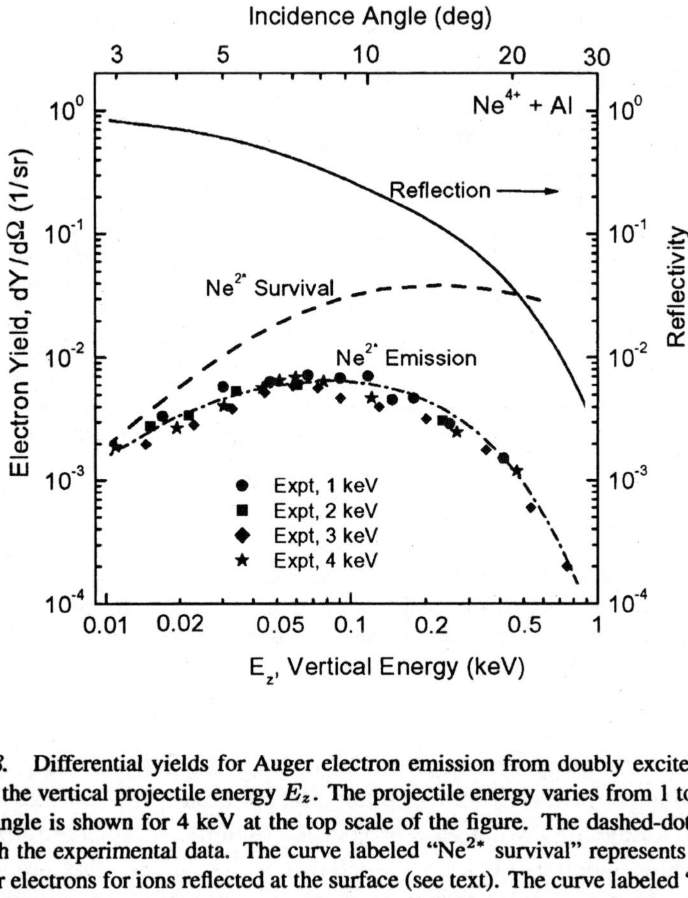

Figure 3.18. Differential yields for Auger electron emission from doubly excited Ne^{2*} as a function of the vertical projectile energy E_z. The projectile energy varies from 1 to 4 keV. The incidence angle is shown for 4 keV at the top scale of the figure. The dashed-dotted curve is a fit through the experimental data. The curve labeled "Ne^{2*} survival" represents the yield of Ne^{2*} Auger electrons for ions reflected at the surface (see text). The curve labeled "Reflection" referring to the right hand scale represents the fraction of reflected ions. From Ref [94].

To understand this energy dependence of the Auger electron yields, we consider two regions of the surface relevant for different trajectories of the projectiles. Looking back at fig. 3.11 these regions are located above the surface near the jellium edge and below the first atomic layer. The incoming ions are assumed to be reflected at the first atomic layer with the probability p_R and, hence, the ions enter into the bulk with the probability $1 - p_R$. When reflected, the ions emerge from the surface in the doubly excited state (Ne^{2*}) with the probability p_D. Then, the electron yield due to above-surface Auger emission from Ne^{2*} can be written as

$$\frac{dY_{Ne^{2*}}}{d\Omega} = \frac{dY^0_{Ne^{2*}}}{d\Omega} p_R p_D \tag{3.9}$$

where $dY^0_{Ne^{2*}}/d\Omega$ stands for the normalized angular distribution of the 22 eV Auger electrons. Assuming isotropic emission it follows that $dY^0_{Ne^{2*}}/d\Omega = 1/4\pi$. Expression (3.9) already explains the basic feature of the Ne^{2*} decay curve in fig. 3.18. The probabilities p_D and p_R are decreasing and increasing functions with E_z, respectively, and hence their product exhibits a maximum in an intermediate region.

For a more quantitative analysis, the cascade model (3.3) was used to describe the decay of hollow atoms in front of the surface and inside the bulk. The results are given in fig. 3.18 as the dashed line labeled "Ne^{2*} survival". Since there is no room to give a detailed description of the cascade model calculations, we provide a few explanations to make this curve plausible. After Ne^{2*} is produced close to the surface with nearly unity probability, it decays with an exponential behavior $\exp(-u/v_z)$ where u is an adjustable parameter and v_z is the vertical velocity of the projectile. It can readily be shown that $u \approx 2d\Gamma_L$ is obtained as the velocity needed for the double passage of the jellium edge region of thickness d within the lifetime Γ_L^{-1} of the hollow Ne atom. Thus, it may be understood that the energy dependence of the Ne^{2*} survival curve follows an exponential law with a v_z (or E_z) scaling.

After determination of the Ne^{2*} survival function, the reflection probability p_R was deduced from the present data. First, the experimental data were fitted by a smooth function (dashed dotted line in fig. 3.18) which, in turn, was divided by the Ne^{2*} survival function. The results are given as the solid line labeled "Reflection" referring to the right hand scale of fig. 3.18. The solid line represents the fraction of ions scattered at the first atomic layer. As expected this curve is close to unity at small scattering angles and it decreases monotonically with increasing scattering angle.

The present empirical method of determining the fraction of reflected ions is advantageous, as it takes into account the specific properties of the surface such as the surface roughness. The results of the present analysis will be used in the next sections dealing with the excitation of plasmons.

6. Creation and Decay of Plasmons

Structures due to the decay of plasmons can be observed in the low-energy range from about 5 to 14 eV [35, 54]. Looking back at fig. 3.5 in this energy range indications due to such plasmon structures can be seen. In particular, we shall focus on the plasmon structures near 6.5 eV and 11 eV which are generally associated with the decay of surface and bulk plasmons, respectively.

As in previous work [37, 38, 42], we consider excitation by potential energy transfer as a unique mechanism for plasmon production by slow heavy ions (fig. 3.2). As mentioned, this mechanism involves the capture of a valence electron into the L shell of the Ne projectile, which provides the energy for plasmon creation. However, other mechanisms for plasmon creation may be considered. Note first that Ne orbitals higher than the 2p shell cannot participate in the bulk plasmon creation, since those orbitals are not bound inside the solid [21]. However, energetic electrons produced directly in collisions as well as Auger electrons may excite plasmons when traveling through the solid [46, 47].

6.1 Spectral analysis

To enhance the visibility of the structures due to plasmon decay, which are superimposed on an intense background from other processes, it is common practice to differentiate the measured electron intensities $N(E)$ [35]. Furthermore, to obtain more information about the mechanisms for plasmon production, absolute values for the corresponding electron emission yield were extracted from the electron spectra. The principles of the procedure to obtain absolute plasmon yields from the spectral derivatives are shown in fig. 3.19 [47].

The energy ε_p liberated by the decaying plasmon is determined by the Lorentzian function $\mathcal{L}(\varepsilon_p, E_p, \Gamma_p)$ involving the plasmon energy $E_p = \hbar\omega_p$ and decay width Γ_p. In fig. 3.19(a) the Lorentzian, represented by the curve labeled *Initial Distribution*, is normalized to unit area. The energy distribution of the electrons excited from the conduction band is obtained as a convolution over the normalized density of states $\mathcal{D}(E) = N\sqrt{E}$ for $E \leq E_F$, where E_F is the Fermi energy. With the normalization factor $N = E_F^{-1/2}$ the integral of the $\mathcal{D}(E)$ function is obtained as $2E_F/3$. The convolution is performed by the integration [47]

$$F(\varepsilon) = \int_0^{E_F} \mathcal{L}(\varepsilon + U - E, E_p, \Gamma_p)\, \mathcal{D}(E)\, dE \qquad (3.10)$$

where U is the potential step at the surface (15.5 eV for Al) and ε is the energy of the plasmon-decay electron outside the solid. The result of the convolution is represented by the data labeled *Convoluted* in 3.19(b) for which the integral remains equal to $2E_F/3$. (The convoluted curve was shifted in energy so that a direct comparison is possible with the $\mathcal{D}(E)$ curve.)

The crucial point of the present method is that the derivative of the $F(\varepsilon)$ curve, also given in fig. 3.19(a), reproduces closely the initial energy distribution of the plasmon. Thus, the intensity of the plasmon-decay electrons is obtained from the integral of the *Derivative* peak multiplied by $2E_F/3$. It should be noted that this factor is independent of the shape of the initial plasmon energy distribution. To account for the transport of the electrons, attenuation and refraction effects

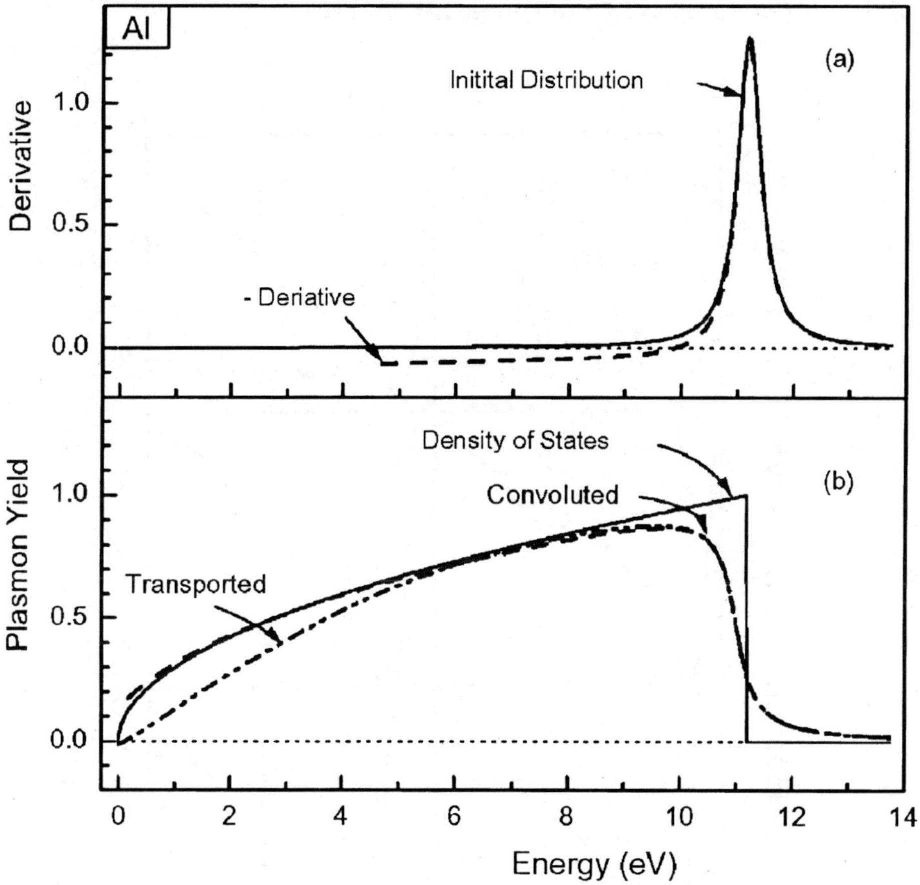

Figure 3.19. Method to determine absolute electron yields originating from plasmon decay. In (a) a Lorentzian function is shown with a width of $\Gamma_p = 0.5$ eV representing the *Initial Distribution* of the energy liberated by the plasmon decay (shifted to the Fermi energy $E_F = 11.2$ eV of Al). In (b) the *Density of States* in the Al conduction band normalized to unity at E_F is compared with the *Convoluted* curve obtained by convoluting the density of states with the initial energy distribution. The curve labeled *Derivative* represents the negative derivative of the convoluted curve. Also, (b) shows a curve labelled *Transported* which includes electron transport effects by the solid, such as attenuation and refraction. From Ref. [47].

have to be considered; this leads to the curve labeled *Transported* in fig. 3.19(b). When the *Transported* curve is normalized to the *Convoluted* curve near E_F, the transport effects are found to be small, i.e., the factor $2E_F/3$ was reduced by $\sim$ 20 % yielding the value 6.3 eV. It should be added that similar results were obtained by Sanchez *et al.* [97] using more empirical treatments of the plasmon dip structure.

6.2 Derivative of the plasmon spectra

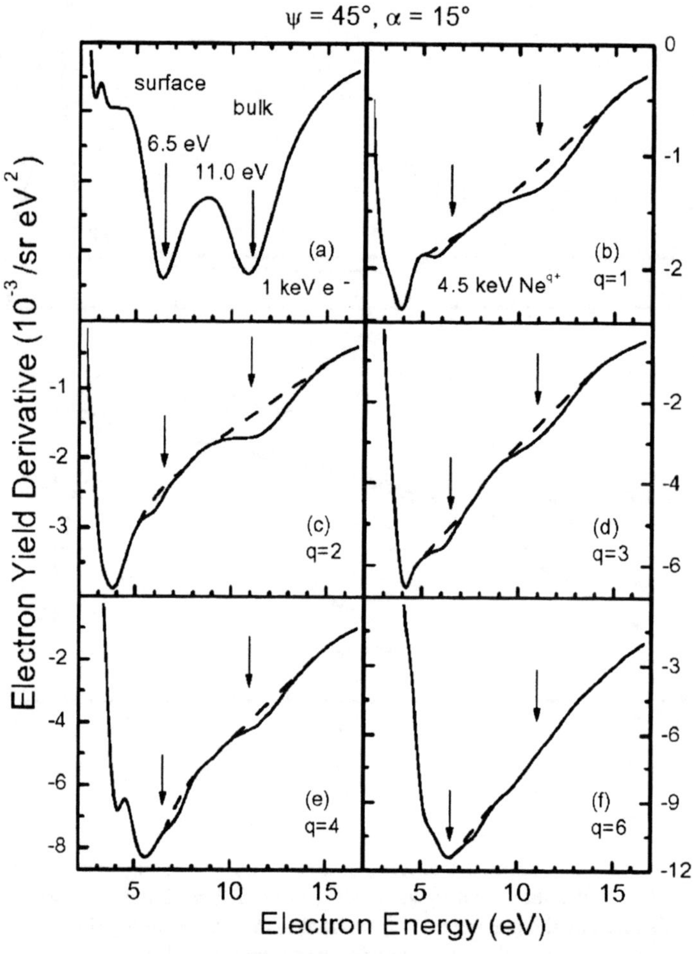

Figure 3.20. Derivative dN/dE of the double differential emission yields given in fig. 3.5. The results in (a) refer to 1 keV electron impact and those in (b) to (f) refer to 4.5 keV Ne^{q+} impact where $q = 1 - 4$ and 6. The dips near 6.5 eV and 11 eV are attributed to surface and bulk plasmons, which were also observed for $q = 5$. From Ref. [38]

In the following sections, derivatives dN/dE of the experimental electron spectra are discussed. Before evaluating the derivative, however, a Fourier transformation procedure was applied to smooth the spectra in fig. 3.5. This procedure has no significant effect on the energy resolution. The results are presented in fig. 3.20 [38]. As shown in the previous subsection 3.6.1, the plasmon structure appears as a negative peak (dip) when performing the derivative of the electron intensity [47]. This dip was separated from the electron background by fitting the intensities above and below the plasmon structure by a second-order polynomial (given in fig. 3.20 as dashed lines).

First, we consider fig. 3.20(a) showing auxiliary data for 1 keV electron impact which compare well with previous results [52]. The derivative dN/dE indicates a dip at 6.5 eV and at 11 eV which can be associated with the decay of surface and bulk plasmons, respectively. In figs. 3.20(b - f) the data for 4.5 keV Ne^{q+} impact with $q = 1 - 4$, 6 are given. The curves show structures which can also be associated with plasmon decay. For $q = 1$ the present data compare well with the results by Baragiola and Dukes [37]. The derivative dN/dE for $q = 1$ clearly shows the 11 eV structure commonly attributed to bulk plasmons. This structure is enhanced for $q = 2$, however, it becomes less pronounced in comparison with the background as the charge state further increases to $q = 5$. Finally, the bulk plasmon structure seems to disappear for $q = 6$.

The derivative dN/dE shows an additional structure which is shifted from about 6 eV to 7.5 eV as the incident charge state increases from $q = 1$ to 6. Hence, this structure is close to 6.5 eV where the signature for surface plasmons occurs in fig. 3.20(a). The 6.5 eV structure can be identified for all charge states so that one is tempted to take this structure as a real indication for surface plasmons. However, spectral features at low energies should be viewed with caution as they may be influenced by instrumental effects. Therefore, in the following, the attention is focused on bulk plasmons and, in the later subsection 3.6.5, improved spectra for surface plasmons will be discussed.

As an application of the foregoing theoretical analysis, the electron yields for bulk plasmons were determined from the data in fig. 3.20 as a function of the charge state q of the incident Ne^{q+} ion [47]. The results show that the plasmon decay yield increases significantly (nearly by a factor of 2) as the charge state q changes from 1 to 2. For higher charge states, however, the plasmon decay yield remains constant within the experimental uncertainties. This allows for the conclusion that the disappearance of the plasmon structures for higher charge states (fig. 3.20) is not caused by a reduction of the plasmon yield but by the increase of the background intensity.

Further results for the derivative $dN/d\varepsilon$ are extracted from fig. 3.16 for 4-keV Ne^{4+} impact on Al [94]. In fig 3.21 the graphs labeled (a), (b), and (c) on the left-hand side show $dN/d\varepsilon$ for incidence angles of $\psi = 3^0, 7^0$, and 20^0, respectively, relative to the surface plane. Each $dN/d\varepsilon$ curve clearly shows a structure near 11 eV which is primarily attributed to bulk plasmon decay [56, 47]. On the right hand side, the corresponding plasmon dip structures are shown as obtained after subtraction of the background intensity. It is seen that the centroid energy of the dip is slightly shifted as the incidence angle increases. This shift is larger than the uncertainties in the dip positions possibly caused by the background subtraction. Furthermore, it is found that the dip intensity for

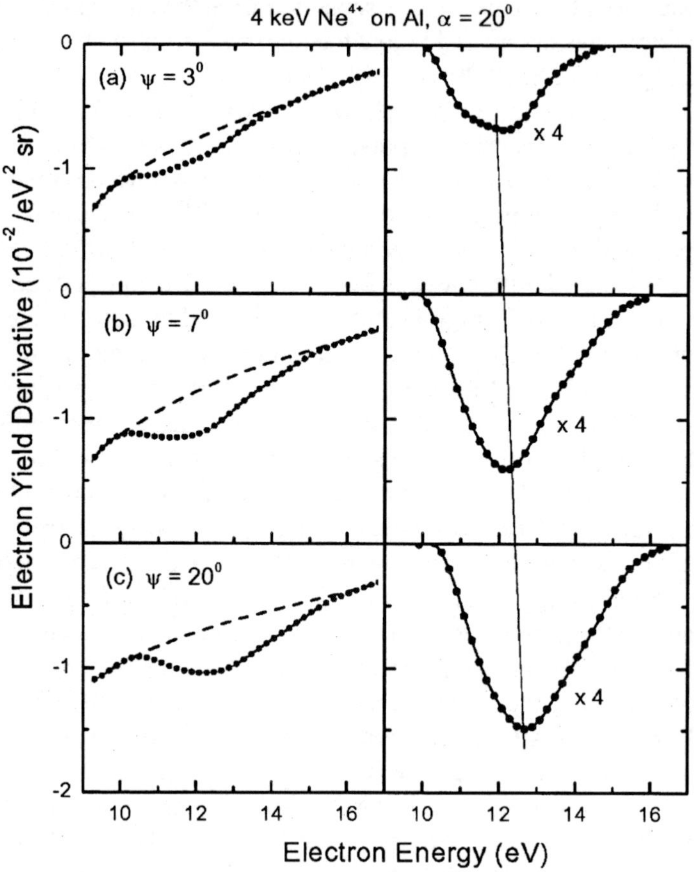

Figure 3.21. Derivative of electron yields measured for 4 keV Ne^{4+} impact on Al as a function of the electron energy. The observation angle of the electrons is equal to $\alpha = 20^0$ relative to the surface plane. In (a), (b), and (c) data are given for the incidence angles $\psi = 3^0$, 7^0, and 20^0, respectively, as shown in the left hand side graphs. On the right hand side the plasmon structures are shown after subtraction of the background. From Ref [94].

the grazing angle of $\psi = 3^0$ is significantly smaller than those for 7^0 and 20^0. The angular dependence of the plasmon yield shall be treated in more detail in the next section.

6.3 Variation of the incidence angle

Systematic studies of the plasmon structures near 11 eV were carried out as a function of the incident angle ψ [94]. In fig. 3.22 absolute values of the electron yield $dY/d\Omega$ are presented for projectile energies of 1, 2, and 4 keV. The $dY/d\Omega$ curves show strong variations with the incidence angle. After a region of nearly constant values for the larger angles, the yield decreases to a minimum and rises again at small angles. The location of the minimum changes with the projectile energy.

For an interpretation of the data, we consider fig. 3.23 recalling two relevant regions for plasmon production, i.e. above the surface near the jellium edge and below the surface defined by the first atomic layer. In these regions, it is likely that, respectively, surface and bulk plasmons are produced. One may argue, however, that the consideration of surface plasmons is inconsistent with the energy of the present plasmon structure. The 11 eV dip is significantly higher in energy than the Al surface plasmons found at 6.4 eV [54]. However, as emphasized previously [43, 39], the latter energy refers to plasmons of near zero momentum, whereas the decay of surface plasmons with larger momentum may produce electrons of higher energies.

Moreover, high momentum transfer may result in the production of surface plasmons of higher multipoles [39]. The occurrence of electrons from high momentum surface plasmons would be consistent with the shift of the plasmon dip observed in fig. 3.21. At small incident angles ($\psi = 3^0$) the electron spectrum might be composed of both surface and bulk plasmons whereas at higher incident angles ($\psi = 7^0$ and 20^0) the bulk plasmons dominate. Similar results have been obtained by Barone *et al.* [58] using Ne^+ projectiles.

In the following, the electron yields due to above- and below-surface processes are denoted $dY_a/d\Omega$ and $dY_b/d\Omega$, respectively. These yields are governed by the reflection probability p_R of the projectile at the surface, which strongly depends on the incidence angle ψ. In a simple model we set

$$\frac{dY_b}{d\Omega} = B\,(1 - p_R) \tag{3.11}$$

where B is a constant assuming that the ions penetrating into the solid produce bulk plasmons. Similarly, we set

$$\frac{dY_a}{d\Omega} = A\,p_R(1 - p_D) \tag{3.12}$$

The reflection probability p_R and survival probability p_D were determined from an independent analysis of the atomic Ne^{2*} Auger peak (fig. 3.18). Then, the sum $dY_a/d\Omega + dY_b/d\Omega$ was fitted to the experimental data treating A and B as adjustable parameters for each incident energy. The results are plotted in fig. 3.22 where apart from the sum (solid lines) also the individual yields $dY_a/d\Omega$ (dashed lines) and $dY_b/d\Omega$ (dashed dotted lines) are given. It is seen that the theoretical data compare well with experiment providing confidence for the essential features of the present model. For instance, the existence of a

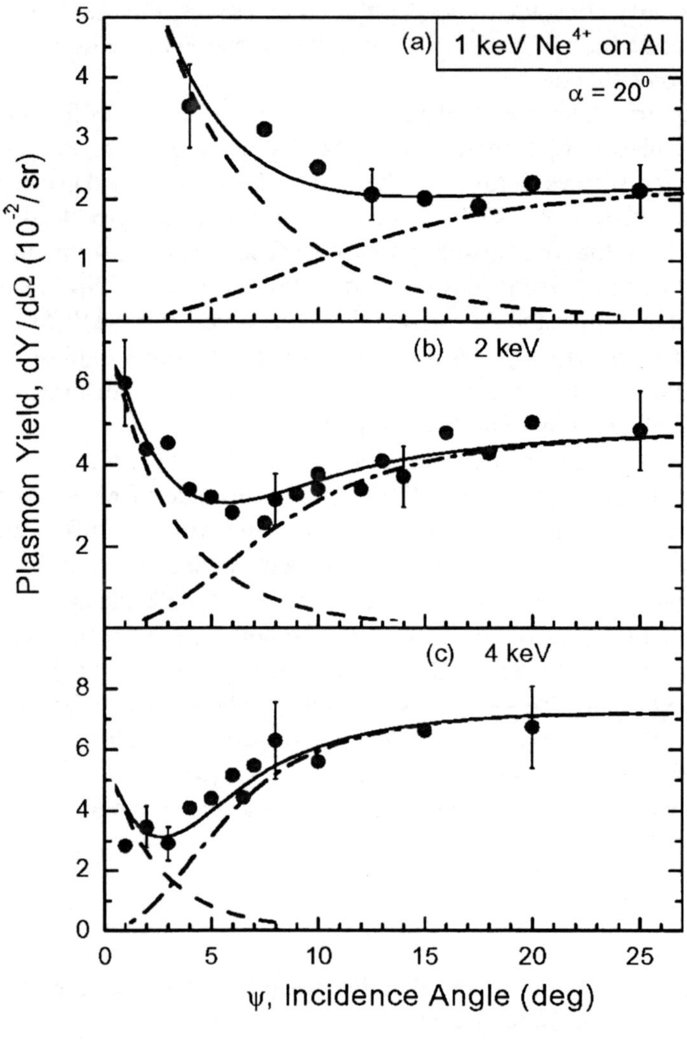

Figure 3.22. Plasmon yield from Al for Ne^{4+} impact as a function of the incidence angle ψ. The electron observation angle is $\alpha = 20^0$ relative to the surface plane. In (a), (b), and (c) data are given for impact energies of 1, 2, and 4 keV, respectively The curves are due to model calculations indicating the contributions from above and below the surface (see text). From Ref. [94]

minimum in the electron yields and its shift with the projectile energy can be associated with the onset of specular reflection of the projectiles at the surface.

The main goal of the present analysis is the separation of above and below surface contributions to the plasmon yield. We note that these contributions are governed by the parameters A and B, respectively, which are obtained as

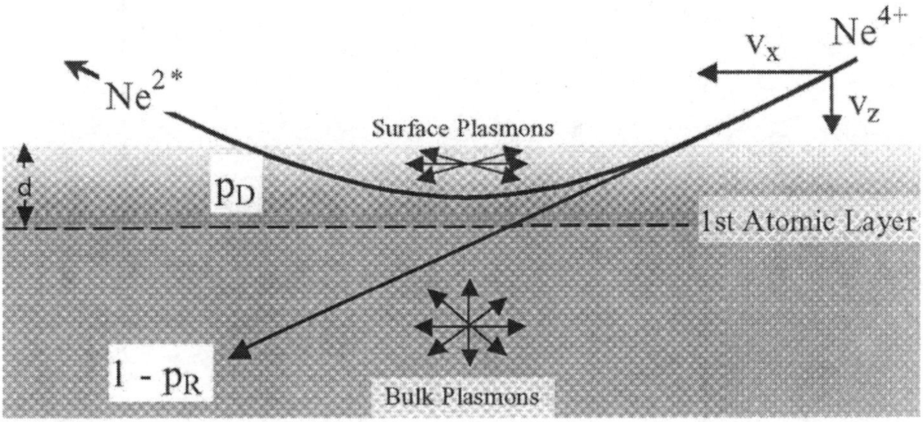

Figure 3.23. Diagram to visualize the two regions for plasmon production considered in the theoretical analysis. Details of the trajectories are described in fig. 3.11. From Ref [94].

asymptotic electron yields $A = dY_a(\psi \to 0)/d\Omega$ and $B = dY_b(\psi >> 0)/d\Omega$. Fig. 3.24 shows these parameters as a function of the projectile energy. It is found that the yields A and B have opposite energy dependencies, i.e., A decreases slightly, whereas B increases significantly with increasing projectile energy. It should be noted that Barone *et al.* [58] found the same opposite energy dependencies when verifying two components in the 11 eV structure produced by Ne^+ ions. The decrease of A with energy confirms potential energy effects being responsible for the above-surface processes. On the other hand, the increase of B with energy suggests that the below-surface processes are influenced by kinetic energy effects. This finding is consistent with the recent analysis of absolute plasmon yields for different incident charge states [47] suggesting that for charge states higher than 2 the plasmons are produced indirectly by secondary electrons.

6.4 Variation of the Observation Angle

It should be recalled that the data in the previous subsection refer to a single observation angle of the electrons α relative to the surface plane. Here, the information about the dependence on the electron observation angle is provided. The knowledge of this angular dependence is crucial for different reasons. First, it is well known that electrons originating in the bulk of the solid exhibit a cosine-like angular dependence for an observation angle $\beta = 90 - \alpha$ measured with respect to the surface normal [98]. On the other hand, electrons originating from shallow surface layers are expected to exhibit a more isotropic emission [31, 33]. Hence, the observed angular dependence of the

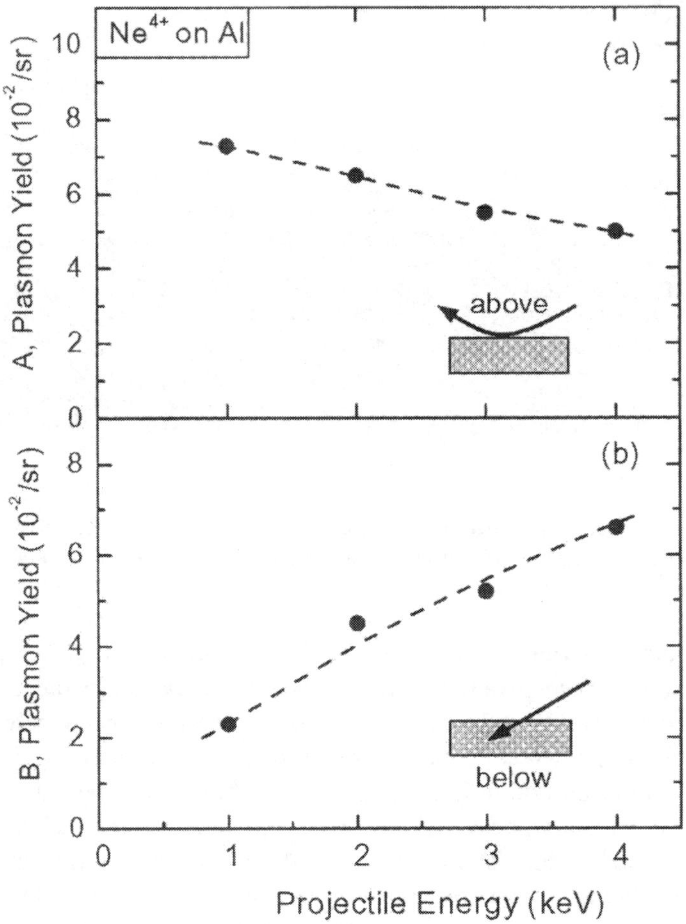

Figure 3.24. Electron yields $dY/d\Omega$ due to the decay of plasmons obtained from the results given in fig. 3.22. In (a) and (b) the data refer to the asymptotic values of $A = dY_a/d\Omega$ at $\psi \rightarrow 0$ and $B = dY_b/d\Omega$ at $\psi \rightarrow 25^0$ associated with plasmon production above and below the surface, respectively. From Ref [94].

plasmon-decay electrons provides information about the depth of the plasmon production. Furthermore, information about the angular dependence is required when total yields of electrons ejected into the hemisphere above the surface are evaluated.

For Ne^{4+} impact, electron emission yields have been measured for various observation angles [47]. Results of absolute plasmon yields are given in fig. 3.25 obtained using the procedure previously described in section 3.6.1. (Electron yields from plasmon decay were obtained by multiplying the integrated derivative peak by the quantity 6.3 eV.) From fig. 3.25 the experimental

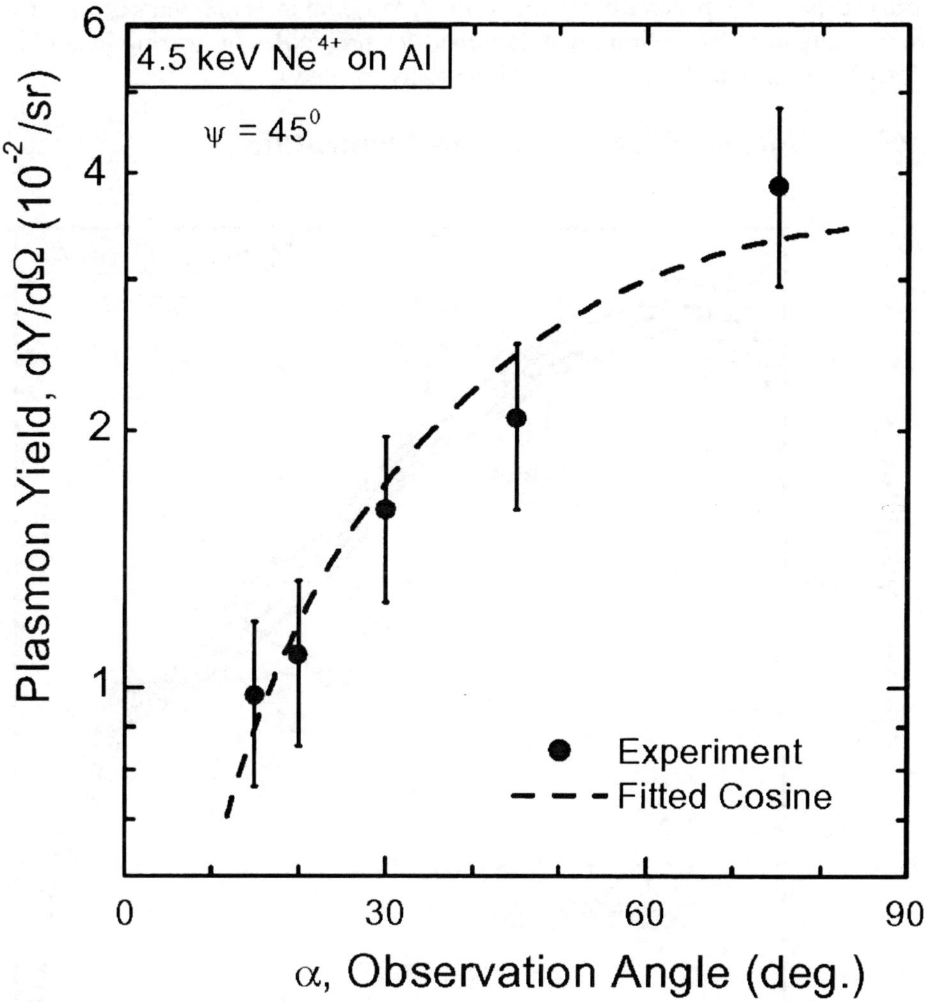

Figure 3.25. Angular distribution of electron yields for plasmon decay obtained by 4.5 keV Ne^{4+} impact on Al. The incident angle is $\psi = 45^0$. The observation angle $\alpha = 90 - \beta$ is measured relative to the surface plane. The experimental results (points) are compared with a fitted $\cos\beta$ function (dashed curve). From [47].

data are seen to closely follow a normalized $\cos\beta$ function, providing evidence that the 11 eV plasmon structure are produced well inside the solid. Similar results were found for other cases for which the incident angle was varied [47] and the projectile Ne^+ was used [99].

However, it should be realized that the angular dependencies, previously studied, refer to cases where the incident angle is rather large so that the projectile is expected to penetrate into the solid. Additional measurements are needed where angular distribution are determined for projectiles of grazing incidence for which surface plasmons are preferentially excited.

6.5 Surface Plasmons of Low Momentum

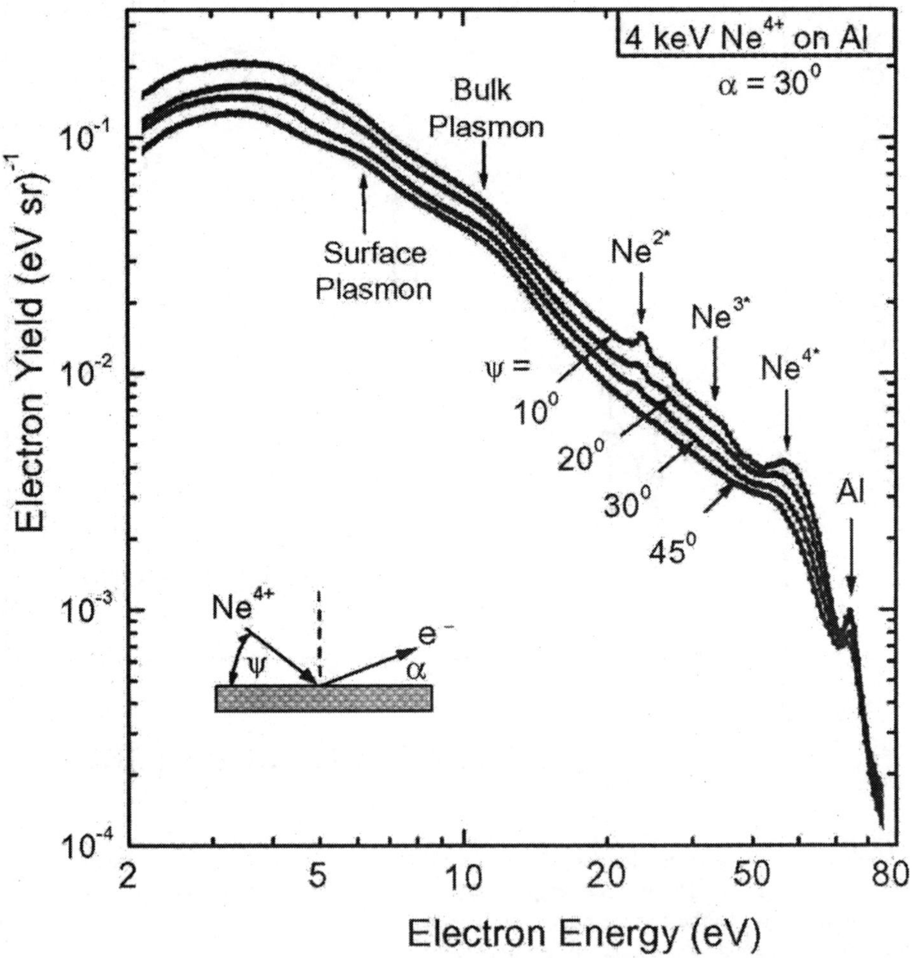

Figure 3.26. Electron yields measured for 4 keV Ne^{4+} impact on Al as a function of the electron energy. The incidence angles are $\psi = 10^0$, 20^0, 30^0, and 45^0 as indicated. The observation angle of the electrons is equal to a = 30^0 relative to the surface plane. The peak identification was already given in fig. 3.5. From [63].

This section is primarily concerned with the lower energy range where the 6.5 eV structure due to low-momentum surface plasmons is expected. Particular instrumental effort is needed to study this energy range that is influenced by various spurious effects [63]. Results obtained with an improved apparatus are given in fig. 3.26 referring to the incidence angles of $\psi = 10^0$, 20^0, 30^0, and 45^0. (Again, the angles are defined as the inset in the figure.) To enhance the visibility of the low-energy range, a logarithmic x-axis is chosen.

The electron spectra exhibit various structures which can be attributed to emission of Auger electrons as well as the decay of plasmons produced within the surface and the bulk. As discussed already in conjunction with fig. 3.16, the peaks denoted Al are produced by Auger transitions filling a vacancy in the L-shell of Al. The peaks labeled Ne^{2*}, Ne^{3*}, Ne^{4*} are due to L-Auger transitions in hollow Ne with 2, 3, and 4 vacancies in the L shell, respectively. Again, one observes a bump near 11 eV, which can be attributed to bulk plasmons possibly containing a component of multipole surface plasmons [94].

In addition, near 6.5 eV a further bump is visible which can well be identified on top of the background of secondary electrons. The 6.5 eV bump is likely to be associated with low-momentum surface plasmons. It should be recalled that in previous work the structure at 6.5 eV was not clearly visible (see fig.3.5). The recent experience showed that the 6.5 eV structure could only be detected in a clear manner after spurious instrumental effects were removed [63]. Nevertheless, it should be realized that uncertainties in the low-energy domain cannot fully be avoided.

In fig. 3.27 the results for the spectral derivative are given for 4 keV Ne^{4+} impact on Al. The graphs labeled (a), (b), (c), and (d) on the left-hand side show the derivative $dN/d\varepsilon$ for incidence angles of $\psi = 10^0$, 20^0, 30^0, and 45^0, respectively, relative to the surface plane. Each curve clearly shows two plasmon structures [38]. As discussed before, a plasmon structure appears as a negative peak (dip) when performing the derivative of the electron intensity. This dip was separated from the electron background by fitting the intensities above and below the plasmon structure by a second-order polynomial (given in fig. 3.27 (a - d) as dashed lines). On the right hand side of fig. 3.27 the corresponding plasmon dip structures are shown as obtained after subtraction of the background intensity. It is seen that the centroid energy of the dip is slightly shifted as the incidence angle increases in accordance with the results in fig. 3.21 [94].

For the further analysis of the data the peak structures were integrated after subtraction of the background [63]. The plasmon yields exhibit angular variations which are slightly outside the experimental uncertainties. The yield curve for bulk plasmons shows a maximum at about 20^0, whereas the surface plasmon data increases monotonically with decreasing incidence angle. For a better understanding of the data, we reconsider the two types of ion trajectories shown in fig. 3.11. The surface plasmons are preferentially produced by ions reflected at the surface while passing through the jellium edge. Moreover, when

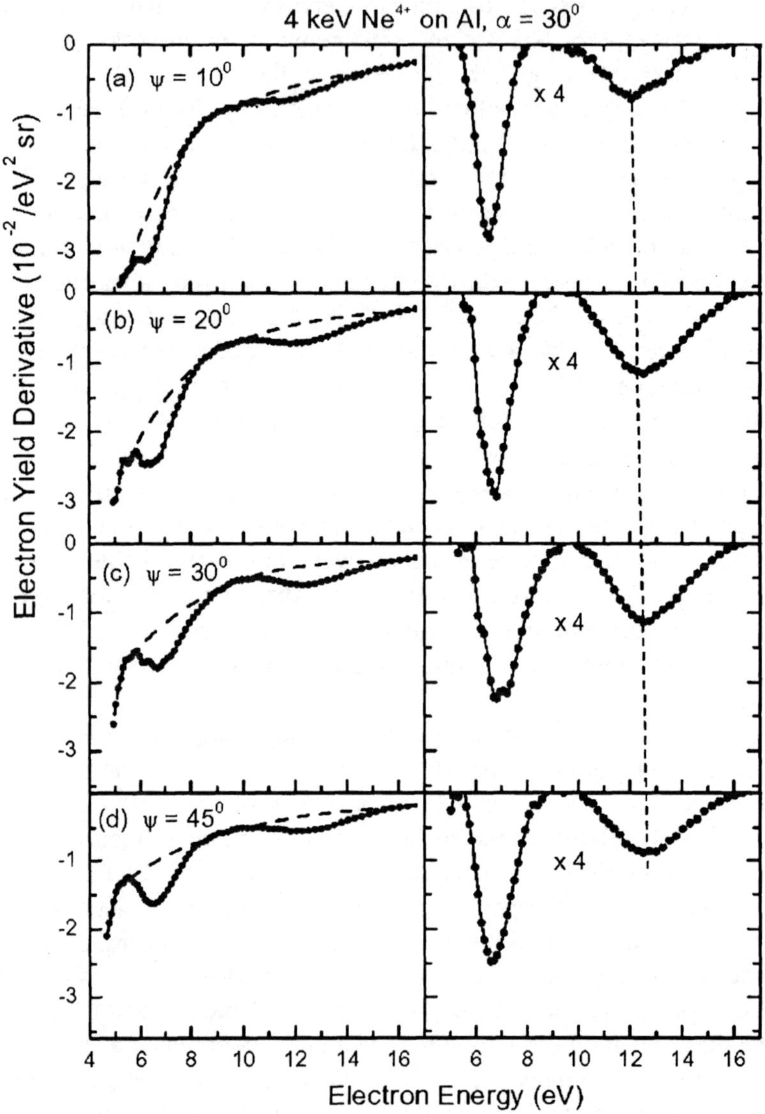

Figure 3.27. Derivative of electron yields measured for 4 keV Ne^{4+} impact on Al as shown in fig. 3.26. In (a), (b), (c), and (d) data are given for the incidence angles $\psi = 10^0$, 20^0, 30^0, and 45^0, respectively, as shown in the left hand side graphs. On the right hand side the plasmon structures are shown after subtraction of the background. From [63].

still above the surface, incident ions can create surface plasmons even if they subsequently enter into the bulk. This may explain the increase of the surface plasmon yield with decreasing incidence angle.

The bulk plasmons are expected to be produced uniquely by ions that enter into the surface. At large incidence angles the ions penetrate deeply into the bulk so that the electrons ejected by the plasmon decay are increasingly absorbed in the bulk. At small incidence angle the ions are increasingly reflected from the surface so that the production of bulk plasmons is suppressed. Thus, the maximum of the bulk plasmon curve at an intermediate angle near 20^0 is plausible.

From fig. 3.24 it was inferred that the 11 eV structures due to bulk plasmons are created by secondary electrons below the surface. This conclusion is consistent with a previous analysis of absolute plasmon yields for different incident charge states [47] as mentioned above. Furthermore it was inferred from fig. 3.24 that possible contribution of multipole surface plasmons are produced above the surface by potential energy effects. Similarly, the present results for low-momentum surface plasmons may provide indications for potential energy effects [47].

7. Conclusions

The present article is concerned with experimental and theoretical efforts to clarify mechanisms for Auger electron and plasmon creation by multiply-charged ions. Absolute values for the electron yield from Auger emission and plasmon decay have been evaluated. Primary attention is devoted to capture processes which provide the potential energy necessary for Auger electron and plasmon creation. This is done because the productions of plasmons and L-Auger electrons are based on the same potential energy effect.

In particular, for neon projectiles the analysis of the 11 eV plasmon structure is performed on the basis of results concerning projectile L-Auger transitions. An analytic expression was used to estimate the contribution of the below surface processes on the plasmon production. The comparison between theory and experiment shows that the contributions to plasmon production from below and above the surface change significantly when the angle of incidence reaches that of specular reflection. However, it is felt that the interpretation of the 11 eV structure, involving a contribution of high-momentum surface plasmons, remains to be debated. The different energy dependencies of the above and below surface contributions confirm potential energy mechanisms, however, point also to other mechanisms for plasmon production.

Specifically, when analyzing plasmon production below the surface, the contribution by secondary interactions with high-energy electrons should be taken into account. The comparison between theory and experiment suggests that for low incident charge states potential energy effects are responsible for plasmon production, whereas for higher charge states secondary collisions with high-energy electrons become important. Nevertheless, the discussed mechanisms are not fully understood at present so that further work is needed to clarify various questions about plasmon creation by slow multi-charged ions. Particularly, further work is required to verify the 6.5 eV structure in terms of surface plasmon production and related potential-energy effects using highly-charged ions.

Acknowledgments

I am indebted to Martin Grether, Volker Hoffmann, Dagmar Niemann, and Jan-Hauke Bremer for their great help in performing the experiments at HMI in Berlin. I am grateful to Max Rösler and Raul Baragiola for long-standing fruitful collaborations and to Bela Sulik and John Tanis for many helpful comments on the manuscript.

References

[1] A. Arnau, F. Aumayr, P. Echenique, M. Grether, W. Heiland, J. Limburg, R. Morgenstern, P. Roncin, S. Schippers, R. Schuch, N. Stolterfoht, P. Varga, T. Zouros, and H. Winter, Surf. Sci. Reports **27**, 113 (1997)

[2] J. Burgdörfer, *Review of Fundamental Processes and Applications of Atoms and Ions*, p. 517, World Scientific, Singapore, 1993.

[3] H. Winter and F. Aumayr, J. Phys. B **32**, R39 (1999)

[4] H. Hagstrum, Phys. Rev. **96**, 325 (1954)

[5] U. Arifov, E. Mukhamadiev, E. Parilis, and A. Pasyuk, Zh. Tekh. Fiz. **43**, 375 (1973), [Sov. Phys.-Tech. Phys. 18, 240 (1973)]

[6] E. Donets, Nucl. Instrum. Methods Phys. Res. Sect. B **9**, 522 (1985)

[7] M. Delaunay, M. Fehringer, R. Geller, P. Varga, and H. Winter, Europhys. Lett. **4**, 377 (1987)

[8] H. Andrä, Nucl. Instrum. Methods Phys. Res. Sect. B **43**, 306 (1989)

[9] G. Lakits, F. Aumayr, M. Heim, and H. Winter, Phys. Rev. A **42**, 5780 (1990)

[10] J.-P. Briand, L. D. Billy, P. Charles, S. Essabaa, P. Briand, R. Geller, J. P. Desclaux, S. Bliman, and C. Ristori, Phys. Rev. Lett. **65**, 159 (1990)

[11] J. Burgdörfer, P. Lerner, and F. Meyer, Phys. Rev. A **44**, 5674 (1991)

[12] L. Folkerts and R. Morgenstern, Europhys. Lett. **13**, 377 (1990)

[13] F. Meyer, S. Overbury, C. Havener, P. Zeijlmans van Emmichoven, and D. Zehner, Phys. Rev. Lett. **67**, 723 (1991)

[14] R. Köhrbrück, K. Sommer, J. Biersack, J. Bleck-Neuhaus, S. Schippers, P. Roncin, D. Lecler, F. Fremont, and N. Stolterfoht, Phys. Rev. A **45**, 4653 (1992)

[15] S. Schippers, S. Hustedt, W. Heiland, R. Köhrbrück, J. Kemmler, D. Lecler, J. Bleck-Neuhaus, and N. Stolterfoht, Phys. Rev. A **46**, 4003 (1992)

[16] D. Schneider, M. Briere, M. Clark, J. McDonald, J. Biersack, and W. Siekhaus, Surf. Sci. **294**, 403 (1993)

[17] N. Stolterfoht, A. Arnau, M. Grether, R. Köhrbrück, A. Spieler, R. Page, A. Saal, J. Thomaschewski, and J. Bleck-Neuhaus, Phys. Rev. A **52**, 445 (1995)

[18] J. Burgdörfer and F. Meyer, Phys. Rev. A **47**, R20 (1993)

[19] F. Aumayr and H. Winter, Comments At. & Mol. Phys. **29**, 275 (1994)

[20] J. Burgdörfer, C. Reinhold, and F. Meyer, Nucl. Instrum. Methods Phys. Res. Sect. B **98**, 415 (1995)

[21] A. Arnau, R. Köhrbrück, M. Grether, A. Spieler, and N. Stolterfoht, Phys. Rev. A **51**, R3399 (1995)

[22] N. Stolterfoht, D. Niemann, M. Grether, A. Spieler, A. Arnau, C. Lemell, F. Aumayr, and H. Winter, Nucl. Instrum. Methods Phys. Res. Sect. B **124**, 303 (1997)

[23] R. Page, A. Saal, J. Thomaschewski, L. Aberle, J. Bleck-Neuhaus, R. Köhrbrück, M. Grether, A. Spieler, and N. Stolterfoht, Phys. Rev. A **52**, 1344 (1995)

[24] H. Limburg, S. Schippers, I. Hughes, R. Hoekstra, R. Morgenstern, S. Hustedt, N. Hatke, and W. Heiland, Phys. Rev. A **51**, 3873 (1995)

[25] K. Suto and T. Kagawa Phys. Rev. A **58**, 5004 (1998)

[26] F. Meyer, C. Havener, K. Snowdon, S. Overbury, D. Zehner, and W. Heiland, Phys. Rev. A **35**, 3176 (1987)

[27] S. D. Zwart, A. Drentje, A. Boers, and R. Morgenstern, Surf. Sci. **217**, 298 (1989)

[28] H. Andrä, A. Simionovici, T. Lamy, A. Brenac, G. Lamboley, J. Bonnet, A. Fleury, M. Bonnefoy, M. Chassevent, S. Andriamonje, and A. Pesnelle, Z. Physik D **21**, S301 (1991)

[29] R. Köhrbrück, D. Lecler, F. Fremont, P. Roncin, K. Sommer, T. Zouros, J. Bleck-Neuhaus, and N. Stolterfoht, Nucl. Instrum. Methods Phys. Res. Sect. B **56/57**, 219 (1991)

[30] S. Schippers, S. Hustedt, W. Heiland, R. Köhrbrück, J. Bleck-Neuhaus, J. Kemmler, D. Lecler, and N. Stolterfoht, Nucl. Instrum. Methods Phys. Res. Sect. B **78**, 106 (1993)

[31] R. Köhrbrück, M. Grether, A. Spieler, N. Stolterfoht, R. Page, A. Saal, and J. Bleck-Neuhaus, Phys. Rev. A **50**, 1429 (1994)

[32] S. Schippers, J. Limburg, J. Das, R. Hoekstra, and R. Morgenstern, Phys. Rev. A **50**, 540 (1994)

[33] M. Grether, A. Spieler, R. Köhrbrück, and N. Stolterfoht, Phys. Rev. A **52**, 426 (1995)

[34] J. Lindhard, Kgl. Danske Videnskab. Mat-fys. Medd. **28**, Nr. 8 (1954)

[35] D. Pines, *Elementary excitations in solids*, W.A. Bejamin Inc., New York, 1964.

[36] M. Rösler and W. Brauer, *Particle Induced Electron Emission I, Springer Tracts in Modern Physics*, volume 122, Springer Verlag, Berlin, 1991.

[37] R. Baragiola and C. Dukes, Phys. Rev. Lett. **76**, 2547 (1996)

[38] D. Niemann, M. Grether, M. Rösler, and N. Stolterfoht, Phys. Rev. Lett. **80**, 3328 (1998)

[39] P. Riccardi, P. Barone, A. Bonanno, A. Oliva, and R. Baragiola, Phys. Rev. Lett. **84** (2000)

[40] N. Stolterfoht, Nucl. Instrum. Methods Phys. Res. Sect. B **154**, 13 (1999)

[41] B. Van Someren, P. A. Zeijlmans van Emmichoven, I. F. Urazgil'din, and A. Niehaus, Phys. Rev. A **61**, 032902 (2000)

[42] A. Almulhem and M. Girardeau, Surf. Sci. Reports **210**, 138 (1989)

[43] R. Monreal, Surf. Sci. **388**, 231 (1997)

[44] F. A. Gutierrez, H. Jouin, S. Jequier, and M. Riquelme, Surface Science **431**, 269 (1999)

[45] H. Eder, F. Aumayr, P. Berlinger, H. Störi, and H. Winter, Surf. Sci. **472**, 195 (2001)

[46] S. Ritzau, R. Baragiola, and R. Monreal, Phys. Rev. A **59**, 15506 (1999)

[47] N. Stolterfoht, D. Niemann, V. Hoffmann, M. Rösler, and R. Baragiola, Phys. Rev. A **61**, 52902 (2000)

[48] R. Díez Muiño, A. Salin, N. Stolterfoht, A. Arnau, and P. M. Echenique, Phys. Rev. Lett. **76**, 4636 (1996)

[49] M. Chung and T. Everhart, Phys. Rev. B **15**, 4699 (1977)

[50] M. Rösler, Scan. Micr. **8**, 3 (1994)

[51] L. Jenkins and M. Chung, Surf. Sci. **33**, 159 (1972)

[52] T. Everhart, N. Saeki, R. Shimizu, and T. Koshikawa, J. of Appl. Phys. **47**, 2941 (1976)

[53] C. Benazeth, N. Benazeth, and L. Viel, Surf. Sci. **78**, 625 (1978)

[54] D. Hasselkamp and A. Scharmann, Surf. Sci. **119**, L388 (1982)

[55] K. Groeneveld, R. Maier, and H. Rothard, Il Nuovo Cimento D **12**, 843 (1990)

[56] D. Niemann, M. Rösler, M. Grether, and N. Stolterfoht, Nucl. Instrum. Methods Phys. Res. Sect. B **135**, 460 (1998)

[57] P. Riccardi, P. Barone, M. Camarca, A. Oliva, and R. Baragiola, Nucl. Instrum. Methods Phys. Res. Sect. B **164**, 886 (2000)

[58] P. Barone, R. A. Baragiola, A. Bonanno, M. Camarca, A. Oliva, P. Riccardi, and F. Xu, Surf. Sci. **480**, L421 (2001)

[59] R. Baragiola, C. A. Dukes, and P. Riccardi, Nucl. Instrum. Methods Phys. Res. Sect. B **182**, 73 (2001)

[60] H. Winter, H. Eder, F. Aumayr, , J. Lörincik, and Z. Sroubeck, Nucl. Instrum. Methods Phys. Res. Sect. B **182**, 15 (2001)

[61] A. Niehaus, P. A. Zeijlmans van Emmichoven, I. F. Urazgil'din, and B. V. Someren, Nucl. Instrum. Methods Phys. Res. Sect. B **182**, 1 (2001)

[62] D. Niemann, M. Grether, A. Spieler, N. Stolterfoht, C. Lemell, F. Aumayr, and H. Winter, Phys. Rev. A **56**, 4774 (1997)

[63] N. Stolterfoht, J. H. Bremer, V. Hoffmann, M. Rösler, and R. A. Baragiola, Nucl. Instrum. Methods Phys. Res. Sect. B, in print (2002)

[64] D. Hasselkamp and A. Scharmann, Vak. Tech. **32**, 9 (1983)

[65] P. Zeijlmans van Emmichoven, C. Havener, and F. Meyer, Phys. Rev. A **43**, 1405 (1991)

[66] H. Kurz, F. Aumayr, C. Lemell, K. Töglhofer, and H. Winter, Phys. Rev. A **48**, 2182 (1993)

[67] R. A. Baragiola, E. Alonso, and H. Raiti, Phys. Rev. A **25**, 1969 (1982)

[68] A. Itoh, D. Schneider, T. Schneider, T. Zouros, G. Nolte, G. Schiwietz, W. Zeitz, and N. Stolterfoht, Phys. Rev. A **31**, 684 (1985)

[69] S. Deutscher, X. Yang, J. Burgdörfer, and H. Gabriel, Nucl. Instrum. Methods Phys. Res. Sect. B **115**, 152 (1996)

[70] A. Borisov and U. Wille, Nucl. Instrum. Methods Phys. Res. Sect. B **115**, 137 (1996)

[71] R. Cowan, *The Theory of Atomic Structure and Spectra*, University of California Press, Berkeley, 1981.

[72] E. Zaremba, L. Sander, H. Shore, and J. Rose, J. Phys. F **7**, 1763 (1977)

[73] H. Andrä, A. Simionivici, T. Lamy, A. Brenac, and A. Pesnelle, Europhys. Lett. **23**, 361 (1993)

[74] S. Winecki, C. Cocke, D. Fry, and M. Stöckli, Phys. Rev. A **53**, 4228 (1996)

[75] W. Huang, H. Lebius, R. Schuch, and N. Stolterfoht, Phys. Rev. A **56**, 3777 (1997)

[76] J. Ducrée, F. Casali, and U. Thumm, Phys. Rev. A **57**, 338 (1998)

[77] J. Ducrée and H. Andrä, Phys. Rev. A **60**, 3029 (1999)

[78] N. Stolterfoht, J. Bremer, and R. Díez Muiño, Int. J. Mass Spectrom. **192**, 425 (1999)

[79] R. Díez Muiño, A. Salin, N. Stolterfoht, A. Arnau, and P. M. Echenique, Phys. Rev. A **57**, 1126 (1998)

[80] S. A. Deutscher, R. Díez Muiño, A. Arnau, A. Salin, and E. Zaremba, Nucl. Instrum. Methods Phys. Res. Sect. B **182**, 8 (2001)

[81] N. Stolterfoht, *Progress in Atomic Spectroscopy*, volume Part D, p. 415, Plenum Publishing Corp., New York, 1987.

[82] H. Winter and F. Aumayr, this book, 2002.

[83] J. Das, L. Folkerts, and R. Morgenstern, Phys. Rev. A **45**, 4669 (1992)

[84] J. Das and R. Morgenstern, Phys. Rev. A **47**, R755 (1993)

[85] L. Folkerts, S. Schippers, D. Zener, and F. Meyer, Phys. Rev. Lett. **74**, 2204 (1995)

[86] J. Thomaschewski, J. Bleck-Neuhaus, M. Grether, A. Spieler, and N. Stolterfoht, Phys. Rev. A **57**, 3665 (1998)

[87] M. Chen and B. Craseman, Phys. Rev. A **12**, 959 (1975)

[88] J.-P. Briand, L. de Billy, P. Charles, J. Desclaux, P. Briand, R. Geller, S. Bliman, and C. Ristori, Europhys. Lett. **15**, 233 (1991)

[89] J.-P. Briand, L. de Billy, P. Charles, J. Desclaux, P. Briand, R. Geller, S. Bliman, and C. Ristori, Z. Physik D **21**, S123 (1991)

[90] J. D. Gillaspy, J. Phys. B **34**, R93 (2001)

[91] J.-P. Briand, G. Giardino, G. Borsoni, M. Froment, M. Eddrief, C. Sébenne, S. Bardin, D. Schneider, J. Jin, H. Khemliche, Z. Xie, and M. Prior, Phys. Rev. A **54**, 4136 (1996)

[92] C. Auth, T. Hecht, T. Igel, and H. Winter, Phys. Rev. Lett. **74**, 2523 (1997)

[93] N. Stolterfoht, V. Hoffmann, D. Niemann, and J. H. Bremer, American Institute Physics Proceedings (AIP, New York, N. Y.) **500**, 646 (2000)

[94] N. Stolterfoht, J. H. Bremer, V. Hoffmann, M. Rösler, and R. A. Baragiola, Nucl. Instrum. Methods Phys. Res. Sect. B **182**, 89 (2001)

[95] F. Xu, R. Baragiola, A. Bonanno, P. Zoccali, M. Camarca, and A. Oliva, Phys. Rev. A **50**, 4048 (1994)

[96] N. Stolterfoht, Phys. Rep. **146**, 315 (1987)

[97] E. A. Sánchez, J. E. G. M. L. Matiarena, O. Grizzi, and R. A. Baragiola, Phys. Rev. A **61**, 14209 (2000)

[98] P. Sigmund and S. Tougaard, Springer Ser. Chem. Phys. **17**, 2 (1981)

[99] P. Riccardi, P. Barone, M. Camarca, N. Mandarino, F. Xu, A. Oliva, and R. Baragiola, Nucl. Instrum. Methods Phys. Res. Sect. B **182**, 84 (2001)

Chapter 4

INTERACTIONS OF HIGHLY CHARGED IONS WITH C_{60} AND SURFACES

U. Thumm

J. R. Macdonald Laboratory, Department of Physics, Kansas State University Manhattan, Kansas 66506–2604, USA

Uwe.Thumm@phys.ksu.edu

Abstract Slow collisions between highly charged ions and many-electron targets, such as large atoms, molecules, clusters, or surfaces, usually lead to the transfer of several electrons from the complex target to the projectile. The efficient capture of target electrons is related to the relatively long (on an atomic time scale) interaction time of typically several femtoseconds and the strong Coulomb force of the highly charged projectile that acts like a vacuum cleaner for loosely bound target electrons while offering a large number of excited projectile states into which electrons can be captured. This chapter will cover part of the interesting physics involved in collisions with two particular types of complex targets: gaseous C_{60} (as a representative for easily mass-selected carbon-cage molecules called "fullerenes") and solid surfaces. The theoretical description of these collisions is presented as simply as possible, with a strong emphasis on the modeling of basic electronic interaction mechanisms. For C_{60} targets, a variety of observables recently have been measured in laboratories around the world against which existing models can be scrutinized. While for C_{60} targets modern coincidence experiments allow for the selection of distant projectile trajectories that do not result in the destruction of the target's carbon cage, even in the most grazing collisions between highly charged ions and surfaces, close encounters cannot be avoided. This complicates the study of collisions with surfaces to the extent that the interaction mechanisms which dominate while the ion is close to (and possibly inside) a surface are not yet understood in full detail. These mechanisms are a matter of intense ongoing research to which the reader will be introduced in this chapter.

Keywords: solid surfaces, fullerenes, C_{60} *many-electrontargets, electronicinteractionmechanisms*

F.J. Currell (ed.), The Physics of Multiply and Highly Charged Ions, Vol. 2, 121-165.

1. Introduction

Highly charged ions (HCIs) carry a large amount of potential energy. This energy equals the sum of successive ionization energies required to remove a given number of electrons in order to generate an ion in a particular charge state. A significant fraction or all of this energy may be released whenever an HCI gets in contact with matter and may lead to the fragmentation of a molecular or cluster target, to the creation of blisters and craters on the surface of solid targets, to the emission of a large number of electrons in collisions with complex targets, and to the "sputtering" of atoms and ions during the impact on solid surfaces. For the past few years, the high energy density of HCIs and the fact that very fast HCIs traveling through matter (e.g. water or organic tissue) deposit most of their total energy over a well localized volume have allowed for the successful treatment of certain cancers in otherwise inaccessible parts of the human body [1]. Other applications include the use of HCIs for ion lithography [2, 3], which is of particular interest to the semiconductor industry [4]. These technological advances are unthinkable without prior basic research in atomic, molecular, solid state, and surface physics. They serve as a motivation for the detailed investigations of HCI–matter interactions reviewed in this chapter.

The basic physical processes that occur during the interactions of HCIs with gaseous matter, i.e. atoms or molecules, have been studied in great detail for over two decades. A comprehensive body of work, both experimental and theoretical, exists for a large variety of projectile ions, targets, and kinetic energies of the incident projectile. Next to spectroscopy, atomic collisions constitute one of the pillars on which our understanding of atoms and molecules rests. The basic collision processes investigated for gaseous atoms and molecules are usually classified by the elementary electronic processes of interest. For atomic targets, these are excitation and ionization processes of either collision partner and combinations thereof, e.g., the simultaneous transfer of one electron from one collision partner to the other accompanied by electronic excitation (so–called "transfer excitation"). For molecular targets, collision processes become more complex as the relative nuclear motion within the molecules gets affected by the collision. This may result in collisionally induced vibrational and rotational excitation or fragmentation of the molecule.

Clusters are large molecules and, therefore, all mechanisms that are operative in interactions between ions and simple molecules also will appear in collisions with clusters, complicated by their larger number of constituents (electrons and nuclei). For example, while the simplest molecule (H_2) can fall apart only into a few fragment combinations (H+H, p+H, p+p), a C_{60} fullerene might undergo multiple fragmentation into vibrationally excited fragments, which subsequently split into smaller fragments and yield a wide distribution of fragment masses.

With respect to increasing complexity of the target, ion–cluster collisions fall intermediate between ion–atom and ion–surface interactions. Cluster targets, C_{60} in particular, combine many features of charge transfer from atoms in the gas phase and from surfaces. In the early nineties, gaseous C_{60} became a popular target for collision experiments with ions owing to the convenience with which it is handled in the laboratory. While metal clusters are usually generated with a broad mass (size) distribution, C_{60} is available in pure form in macroscopic amounts as a target of specific mass (720 amu). This eliminates the need of either mass–selecting charged target clusters before the collisions take place (at the expense of reducing the target density and count rate in the experiment) or, alternatively, of carrying out experiments in which detailed information gets lost due to the inherent averaging over cluster masses. The truncated icosahedral molecular structure and its large number of vibrational degrees of freedom make the highly symmetrical C_{60} cluster unusually stable [5]. Due to the large number of vibrational degrees of freedom, a perturbation to the cage structure of the cluster can quickly equilibrate over the cluster. This tends to avoid the accumulation of enough vibrational energy over the volume of individual chemical bonds to induce fragmentation by breaking carbon–carbon bonds. Thus, due to its relatively large thermal stability, C_{60} can be evaporated easily and serve as a gaseous target for incoming electrons [6, 7], ions [5, 8], and charged clusters [9]. The cage-structure of C_{60} was observed to withstand the reflection from a surface [10], and photoionization and collision experiments with HCIs have produced C_{60}^{i+} ions in positive charge states up to i=9 [11] that do not dissociate ("Coulomb explode") for at least several microseconds.

Highly charged projectile ions are particularly suitable for investigations of the dynamic electronic response of C_{60} clusters to a strong external perturbation that may significantly distort the electronic charge distribution of the easily polarizable target and lead to the capture and emission of a large number of electrons. In this context, "dynamic" means that the cluster–electron distribution continuously adjusts itself to the Coulomb force of the moving projectile. In recent coincidence experiments the deflection of slow Ar^{8+} projectiles was found to depend sensitively on the dynamic polarizability of the target [12]. The analyses and theoretical modeling of such collisions provide not only an important tool for examining properties of fullerenes and their basic interaction mechanisms with highly charged projectile ions, but also allow for the study of the complicated electronic dynamics involved in the collisional creation and post-collisional decay of unstable and multiply excited projectiles (hollow ions) due to Auger electron and X-ray emission[11, 13–15]. So–called "Auger electrons" are emitted in relaxation steps that involve two active projectile electrons.

Within a simplified picture of this Auger transition, one active electron transits to a lower electronic projectile shell while transferring the energy released in this transition to the second active electron that is emitted. In contrast, the emission of X rays requires only one active electron. This simplified explanation in terms of one active electron neglects collective electron readjustments ("configuration interactions") that happen in ions with at least two electrons in response to the primary radiative transition Interactions with ions contribute to our understanding of the electronic response of delocalized fullerene electrons in the initial phase of a chemical reaction that leads to the binding of reactive groups. In this way, the study of fullerene interactions with ions may contribute to the analysis of reactions in the new field of fullerene chemistry. More complex than aromatic chemistry, which is based on the two–dimensional ring structure of benzene, fullerene chemistry is based on three–dimensional carbon–cage molecules and may allow for the synthesis of new chemicals and pharmaceuticals of unforeseen properties and efficiency.

The emerging field of slow HCI–C_{60} interaction studies was significantly stimulated by experiments conducted at Kansas State University and at the University of Giessen [16]. In rapid succession, fullerenes were used in atomic collision laboratories around the world in the nineties. These experiments first probed the interaction between slow, highly charged ions and gaseous C_{60} targets by measuring the final charge–states of target and projectile in coincidence. One of the striking features of these coincidence measurements is the distinction between hard collisions and non–destructive, soft collisions. Hard collisions occur at relatively small impact parameters and lead to fragmentation of the target–carbon cage. In contrast, soft collisions occur at relatively large impact parameters (always larger than the C_{60} radius), can result in the capture of several electrons, and do not lead to the immediate fragmentation of the target. To some extent, in soft, more distant collisions, a fullerene may be viewed as a spherically bent monolayer of a graphite surface. Details of the target electron distribution are less important for the interaction with a distant projectile. Soft HCI—C_{60} collisions allow for the investigation of "above surface" effects, which are disguised by the more violent interaction dynamics in hard collisions with C_{60} and for small ion–surface distances in surface collisions [17]. Very recently, triple coincidence measurements have been performed [15] in which the ejected number of electrons is detected along with the final charge state of the projectile and with the mass and charge of the recoiling target or target fragments. These experiments resolve the many–electron interaction dynamics in unprecedented detail and allow, for example, the discrimination between the number of electrons captured into excited projectile states and the number of "stabilized" projectile electrons that are kept after the projectile has relaxed into its ground state.

Collisions of HCIs with metal surfaces have been a subject of intense experimental and theoretical interest for more than a decade, and several characteristics of the electron exchange and emission processes are becoming well established [17–32]. In contrast to soft collisions with buckyballs (C_{60}), collisions with surfaces always involve close encounters with target atoms. The attractive interaction between the positively charged projectile and the negative image charge it induces in the surface bends the projectile trajectory towards the surface, even at the most grazing angles of incidence, and leads to a large overlap of target and projectile electronic states.

Insulator surfaces have been added to the list of target materials more recently, and interesting new phenomena are being investigated [33–38]. Recent experiments have measured the final charge state distribution of the surface–scattered projectile [19, 30, 32], the projectile deflection angle [33, 39], and the emission of electrons [17, 25, 26, 29, 34, 40] and photons [18, 23, 24, 37] during and after the projectile–surface interaction.

The remainder of this chapter is organized in the following way. Section 4.2 reviews resonant electron exchange and the emission of projectile Auger electrons during and after soft collisions of an HCI with C_{60}. In section 4.2.1, we give an overview of the dynamical classical over–barrier model (COM). Section 4.2.2 illustrates the dynamic shift of projectile and target energy levels during the interaction. Section 4.2.3 includes remarks on the electronic structure of C_{60}. Numerical results for charge–state evolutions and emitted electron yields follow in section 4.2.4. In the subsequent subsections, theoretical predictions of the dynamical COM are discussed and compared with recently measured observables, such as cross sections for the capture of a specific number of target electrons (section 4.2.5), final projectile charge states (section 4.2.6), the projectile kinetic image energy gain (section 4.2.7), and the projectile scattering angle (section 4.2.8).

Collisions with surfaces are reviewed in section 4.3. In section 4.3.1, we discuss the forces that determine the motion and kinetic energy change of the projectile ion along its classical path. In section 4.3.2 we describe the basic elements in the modeling of HCI–surface collisions by extending the COM of section 4.3.1 in order to include interactions that become dominant close to (and inside) the target surface. Section 4.3.3 summarizes these interaction mechanisms in the form of a set of rate equations for the occupation numbers of projectile shells. In section 4.3.4 simulations done within the extended COM are compared with measured final charge–state distributions of reflected ions and emitted electron yields. A brief summary follows in section 4.4. Unless otherwise stated, atomic units are used throughout this chapter. In these units, the elementary charge, the electron mass, Planck's constant divided by 2π, and the Bohr radius are equal to one ($e = m = \hbar = a_0 = 1$).

2. Collisions of Highly Charged Ions with C_{60}

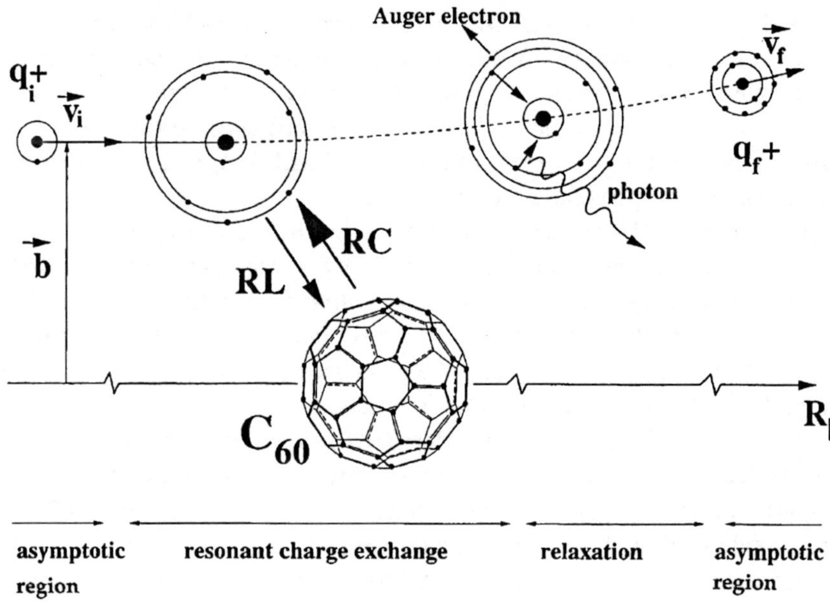

Figure 4.1. Sketch of the collisions scenario (not to scale). A highly charged ion with initial charge q_i, velocity $\vec{v}_i$, and impact parameter $\vec{b}$ carries one tightly bound electron into the collision. At distances of the order of 10 to 100 a.u. from the target center it resonantly captures (RC) and (to a lesser extent) loses (RL) electrons. This leads to a positively charged target and a multiply excited, hollow projectile. Downstream and past the resonant interaction region, the unstable hollow projectile relaxes by emitting Auger electrons and X-ray photons. It may then be detected with final charge state q_f and velocity $\vec{v}_f$. The projection of the target–projectile distance R onto the incident–beam direction is denoted by $R_{||}$.

The basic scenario for a distant collision between an HCI and a C_{60} cluster is sketched in fig. 4.1. On an incident trajectory that is determined by the impact parameter $\vec{b}$ and the asymptotic velocity $\vec{v}$ of the incident projectile, starting at a distance of typically ten to 50 atomic units from the target center, an HCI captures resonantly several electrons into excited states. The distance at which the capture sequence starts primarily depends on the charge state of the incident ion and the ionization energy of the cluster. "Resonant" capture means that the active electron is transferred to a projectile electronic state that has about the same energy as its original target electronic state. The capture sequence typically ends shortly after the projectile has passed the point of closest approach to the fullerene target and results in the formation of a multiply excited ("hollow") ion. These collisionally formed hollow ions are electronically unstable

Table 4.1. Time scales relevant for the formation and relaxation of hollow ions (of core charge q_{core}). Typical values for 80 keV Ar^{8+} interacting with C_{60}. The plasmon response time is estimated as $\sqrt{\frac{2\pi}{n_{el}}}$, where n_{el} is the average electron density of C_{60}.

collision time for resonant exchange:	5 fs
orbiting time of first active projectile level (n=7):	0.8 fs
plasmon response time of C_{60}:	0.2 fs
average time between successive electron capture events:	0.3 fs
projectile Auger transitions:	> 0.1 fs
projectile radiative transitions:	$q_{core}^{-4} \cdot 10^{-8}$ s

and release their excitation energy by emitting electrons and photons. Compared to the time it takes to form the hollow ion by multiple electron capture (typically a few femtoseconds) , some relaxation steps are slow, and the full relaxation of the excited projectile can be completed only during a relatively long time, downstream from the target. In typical experiments, the time of flight of the projectile between the interaction region and the detector is of the order of microseconds and long enough to allow for complete relaxation. Relevant time scales for a typical collision system are assembled in table 4.1.

2.1 The formation of hollow ions in ion–C_{60} collisions

The main aspects of the interaction between slow highly charged ions and complex atomic or molecular targets can be described by using mostly classical model assumptions. In COMs the electronic interaction with highly charged ions is modeled within the independent electron picture and is based on the effective potential to which an active electron, i.e. an electron that might be captured or lost, is subjected. An important feature of this effective potential is the potential barrier located between target and projectile. Position and height of the barrier change during the collision due to the relative motion and changing electronic structure of the collision partners. The COM allows for resonant transitions if the motion of a target or projectile electron across the potential barrier is classically possible, if the initial electronic state is at least partly occupied, and if the final state is not fully occupied in order to prevent Pauli blocking, i.e., the violation of Pauli's exclusion principle that excludes two electrons from occupying the same quantum state. The basic idea of classical capture over a potential barrier is show in fig. 4.2. In the past, various versions of over–barrier models have been applied successfully to slow collisions of ions with atoms [41–43], surfaces [21, 38], and clusters[12, 14, 16, 27, 44, 45]. The model of Bárány and Setterlind [27] represents C_{60} as a dielectric sphere. Radius a and dielectric constant ϵ of the sphere are free parameters.

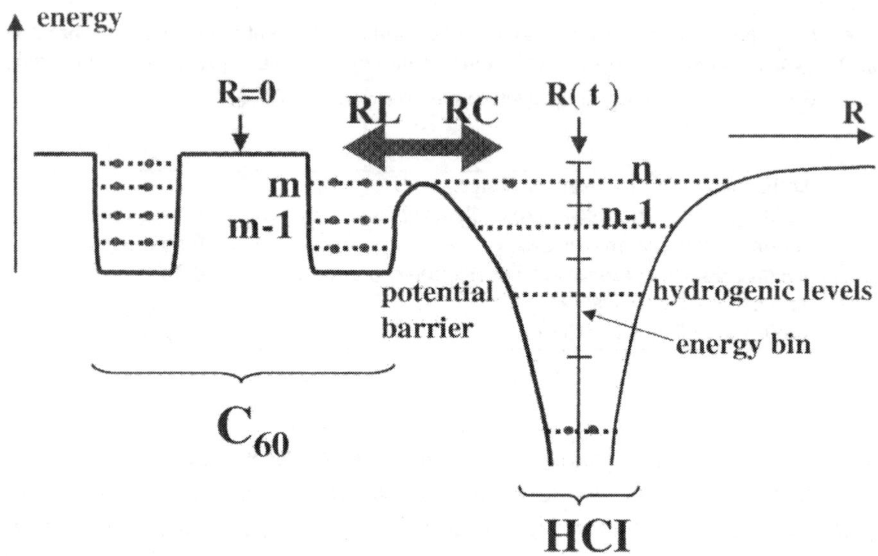

Figure 4.2. Basic idea of the classical over–barrier model applied in this chapter. The effective potential governs the motion of an active electron between the C_{60} target (centered at $R = 0$) and a highly charged ion (HCI) at a distance $R(t)$. Resonant capture (RC) and resonant loss (RL) of electrons occur across the potential barrier

For applications to collisions of slow Ar^{8+} with C_{60}, this model agrees well with experiments and other theories [16, 44]. An attractive feature in all COMs is that basic ideas of classical dynamics and electrostatics yield reasonable estimates for charge–transfer cross sections, charge–state distributions, and other observables, the computation of which is beyond the technical feasibility of full quantum calculations. Details of the best specific formulation of the model remain under investigation. For very slow projectiles, ionization of the target occurs slowly with a small transfer of electronic energy, leaving the C_{60} vibrationally and electronically cold with a high probability of surviving the ionization intact. In the experiment of Walch *et al.* [16] it was shown that C_{60}^{i+} up to i=6 could be produced in this manner with slow Ar^{8+} ions incident on C_{60} with relevant impact parameters in the range of typically 10 to 30 a.u.. Jin *et al.* [11] used a more highly charged projectile, Bi^{44+}, to produce C_{60}^{9+}. For charge states $i < 9$, Jin *et al.* found C_{60}^{i+} ions with lifetimes of at least 5μsec.

During the ion–cluster interaction, energy levels, level occupations, transition rates, and total charges of target and projectile change as a function of R, the distance between the centers–of–mass of target and projectile. For the slow collisions considered in this chapter, R does not change on the time scale of resonant electronic transitions, and an adiabatic approximation is generally justified. In order to be captured or recaptured, the active electron is required to overcome the potential barrier V_B between target and projectile that is formed by the total electronic potential

$$V(q_p, q_t, R, z) = -\frac{q_p}{|R-z|} - \frac{q_t}{z} + V_{im}(q_p, R, z) \tag{4.1}$$

where q_p and q_t are the charges of projectile and target acting on the electron in transition. The electron coordinate along the "inter-nuclear axis" is denoted by z. This axis joins the C_{60} center and the projectile nucleus with the origin at the buckyball center of mass. The image potential, V_{im}, includes the active electron's interaction with its own image charge and with the image of the effective projectile charge q_p in the target. The barrier height V_B is found numerically for any distance R as the maximum of $V(q_p, q_t, R, z)$, considered to be a function of z.

As the projectile approaches the target, the first resonant capture (RC) of an electron becomes possible at the distance R_1^* between the projectile and target centers–of–mass, when V_B energetically moves below the highest occupied target level. Similarly, as R decreases, a second, third, etc. electron may be captured at critical distances $R_2^* > R_3^* \ldots$ on the incoming trajectory. Note that for the purpose of investigating electron capture mechanisms, the heavy projectile may be assumed to move along a straight–line trajectory (projectile scattering angles in typical experiments are of the order of a few mrad, cf. section 4.2.8 below). The critical distances R_i^* are then equal to critical impact parameters at which the trajectory becomes tangent to a sphere of radius R_i^* about the target center. Since the *dynamical* COM discussed in this chapter treats the electronic charge as a continuous parameter, some assumption has to be made as to when a complete elementary charge has been transferred. This leaves some freedom in the precise definition of critical radii. We define R_1^* as the impact parameter at which charge begins to flow from the target to the projectile and R_2^* as the impact parameter at which one unit of charge has left the target, etc.

The projectile may be described within an independent electron approach, based on hydrogenic shells n with energy levels, occupation numbers, and degeneracies denoted by $\epsilon_n^p(R)$, $a_n(R)$, and $A_n = 2n^2$, respectively. Angular momentum sub-levels are not resolved. The time evolution of the occupations $a_n(t)$ and $b_m(t)$ of projectile shells n and target levels m are obtained by integrating classical rate equations of the form

$$\frac{d}{dt}a_n = \Theta(A_n - a_n)\Gamma_n^{RN} - \Gamma_n^{RL}\,a_n + \sum_{n'>n}\Gamma_{n',n}^{AI} - 2\sum_{n'<n}\Gamma_{n,n'}^{AI} \tag{4.2}$$

$$\frac{d}{dt}b_m = \Gamma_n^{RL}\,a_n - \Gamma_n^{RN}, \tag{4.3}$$

with the known initial occupations of projectile and target, a_n^0 and b_m^0. Θ is the unit step function. Analytical expressions for the RC rates Γ_n^{RN} and RL rates Γ_n^{RL} can be derived as classical transfer currents [44, 46]. Resonant transition rates and occupation numbers depend on $R(t)$, and the above equations are solved numerically, together with Newton's equation for the projectile motion. In particular, the resonant neutralization rate Γ_n^{RN} depends on the populations b_m of all (energetically shifted) target levels m that lie within a small interval around the (shifted) energy of the resonant projectile level n. Within the dynamical COM this "energy binning" is necessary in order to relate classical transfer currents to discrete quantum levels [44, 46]. Simplified illustrations of the elementary electronic processes in eqs. (4.2) and (4.3) are given in fig. 4.3.

For the two last terms in eq. (4.2), it is sufficient to include only fast Auger transitions for which the two active electrons start in the same shell and which may, to a small extent, deexcite a multiply excited projectile while competing resonant electron transfer occurs. An analytical approximation to these fast Auger rates,

$$\tilde{\Gamma}^{AI}_{n_{\text{ini}},n_{\text{fin}}} = \frac{5.06 \cdot 10^{-3}}{(n_{\text{ini}} - n_{\text{fin}})^{3.46}}, \tag{4.4}$$

was obtained by Burgdörfer *et al.* [20] by fitting atomic structure calculations [20, 47]. In addition to $\tilde{\Gamma}^{AI}_{n_{\text{ini}},n_{\text{fin}}}$ in eq. (4.4), the Auger transition rates $\Gamma^{AI}_{n_1,n_2}$ in eq. (4.2) include statistical weights. These empirical statistical factors, $1/(1+1.5a_n)$ and $\frac{1}{2}a_n(a_n-1)$, correct for the decrease of Auger transition rates for increasing populations a_n of the final shell and take the equivalence of electrons in the initial shell into account, respectively [44, 48]. Slow Auger relaxation channels are not included in eq. (4.2) because they can be neglected *during* the collision (cf. table 4.1). Further downstream, however, when resonant transfer processes are classically forbidden, slow Auger processes become crucial in determining the final charge–state of the projectile. Downstream Auger and radiative relaxation steps determine the observable final charge state of the projectile and can be accounted for by enhancing the dynamical COM with a post–collisional relaxation scheme [14] (see section 4.2.6 below).

2.2 The shift of target and projectile electric levels during the collisions of highly charged ions with C_{60}

During the collision, projectile energy levels shift due to image–charge effects, Stark shifts induced by a charged target, and the dynamical change in screening induced by varying level populations. Target energy levels $\epsilon_m^t(R)$ are shifted downward in the strong attractive electric field of the positive projectile. After the capture of target electrons, positive charge accumulates on the target, which results in an additional downward shift of the target and projectile spectra.

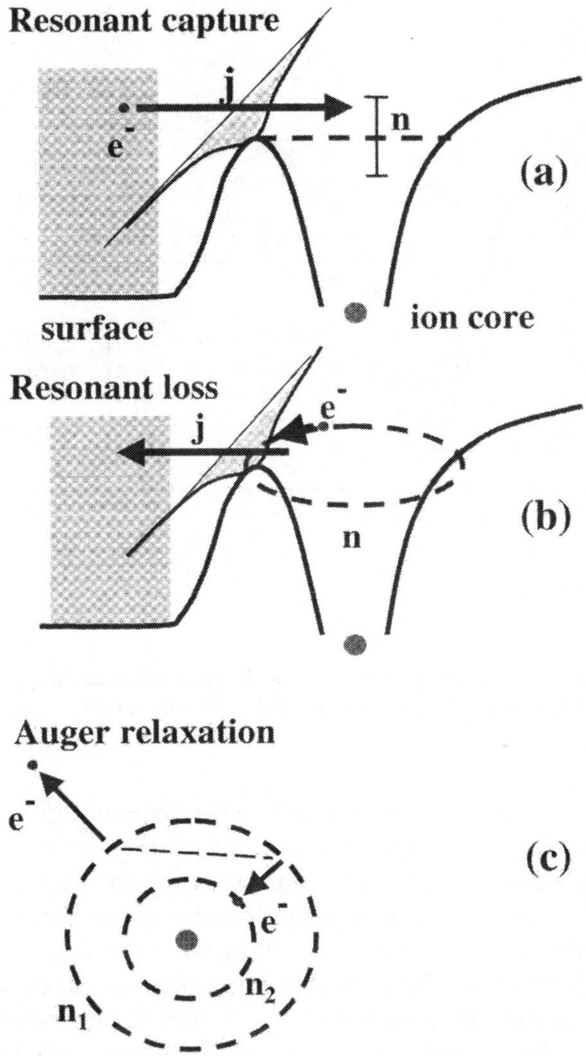

Figure 4.3. Illustration of resonant capture by (a) and resonant loss from (b) a highly charged ion, modeled as a classical current across a potential barrier. The Auger process (c) involves two electrons, one of which is emitted.

In order to understand how the occupation of target and projectile levels changes in response to resonant electron transfer processes, it is instructive to look at the energies of target and projectile levels relative to the top of the potential barrier as a function of $R_{||}$. With $R_{||}$ we denote the projection of the distance between the target center–of–mass and the projectile onto the incident

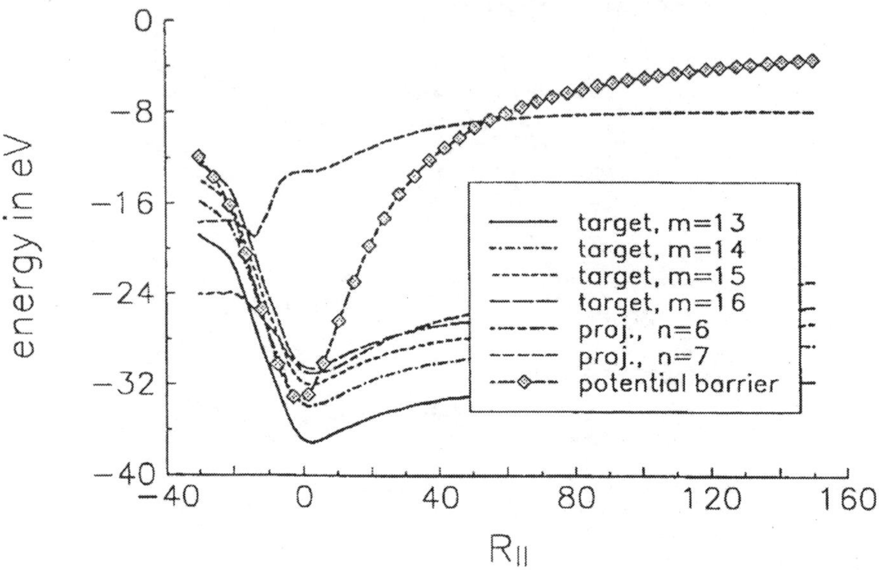

Figure 4.4. Shift of target and projectile energy levels relative to the potential barrier for classical over–barrier capture as a function of the distance $R_{\|}$ for 80 keV Ar^{8+} ions colliding with C_{60}.

beam direction (cf. fig. 4.1). For 80 keV (corresponding to a speed v=0.28) Ar^{8+} ions colliding with C_{60} with impact parameter $b = 15$, fig. 4.4 shows that projectile level $n = 6$ crosses the highest occupied target level (the Fermi level), labeled '$m = 15$'. The crossing happens above the potential barrier and near the point of closest approach. This results in a large current of target electrons from level $m = 15$ that are classically allowed to transfer to projectile level $n = 6$. Similarly, projectile level $n = 7$ crosses the same target level on the incident trajectory and above the barrier. The corresponding rapid resonant filling of projectile levels $n = 6$ and $n = 7$ is seen in fig. 4.5a (see section 4.2.4, below).

2.3 On the electronic structure of C_{60}

Large metallic clusters and C_{60} have band structures similar to metal surfaces with regard to a large portion of unoccupied states above the Fermi level. Smaller clusters tend to form narrowly spaced levels, rather than bands and, with respect to capture, may resemble a large atomic target.

The ground–state electronic structure of neutral C_{60} is well understood from first–principles calculations [46, 49–52]. In the calculation underlying fig. 4.4, the asymptotic energies of the target levels were taken from the local density

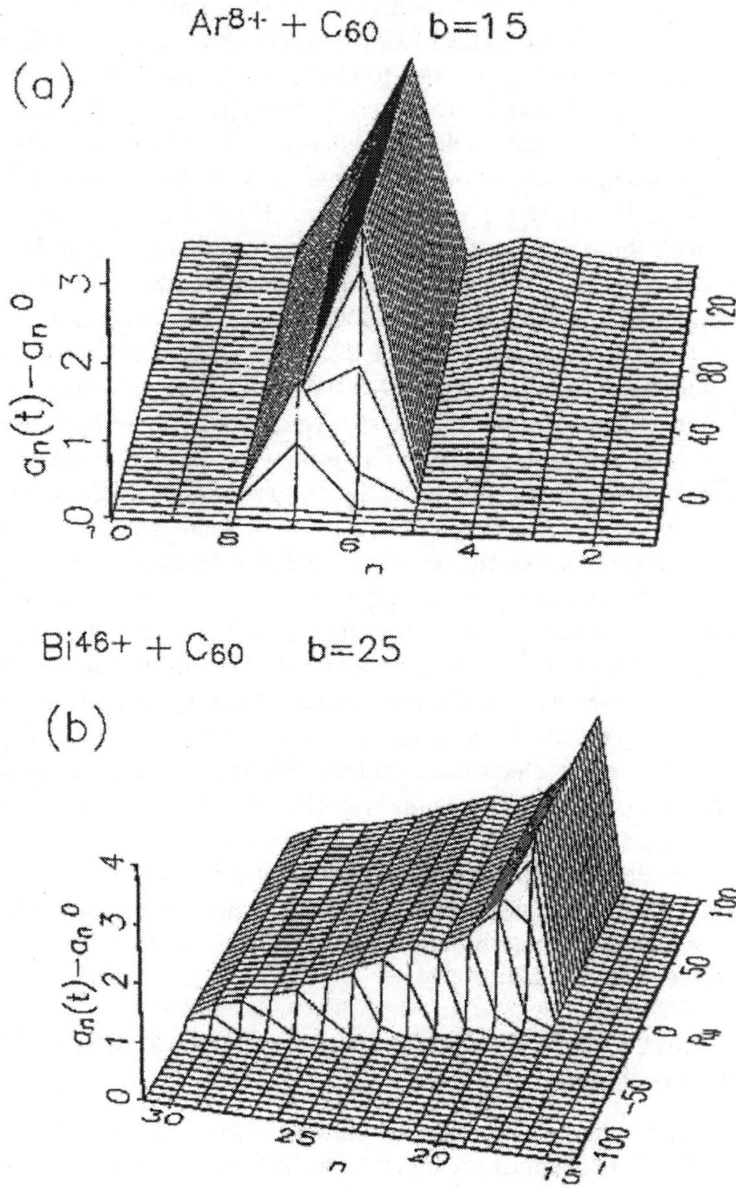

Figure 4.5. Evolution of projectile level occupations in collisions with C_{60}. The point of closest approach is at $R_{\|} = 0$. (a) For incident 80 keV Ar^{8+} projectiles and impact parameter $b = 15$. The graph shows the dynamical change $a_n(t) - a_n^0$ in the population of projectile shells $n = 1$ to $n = 10$. The initial shell population is denoted as a_n^0. $a_n(t)$ is the population of shell n at the distance $R_{\|}(t)$. (b) For 830 keV Bi^{46+} and $b = 25$.

functional approximation (LDA) calculation of Puska and Nieminen [49] in which the occupied valence states of C_{60} are represented by 15 highly degenerate levels. In all other applications to charge exchange and electron emission in soft ion–C_{60} collisions discussed in this chapter, multi–center self–consistent Dirac–Fock–Slater (DFS) calculations [46, 52] were employed to obtain the ground–state electronic structures of neutral C_{60} and its positive ions $C_{60}^{i+}, i = 0...6$. These calculations assumed that the cage structure does not change with the charge state i. The charge–state independent buckyball radius was taken as $a = 6.7$ [49, 50]. The calculations yield DFS single particle energies of C_{60}^{+i} for $i = 0 \ldots 7$ and the sequence of ionization potentials $I_{i=1...7} = 7.24, 10.63, 14.01, 17.44, 20.78, 24.24$, and $27.54\ eV$, in good agreement with other calculated [46, 49–52] and experimental data [6, 53, 54].

In agreement with the simple electrostatical picture that represents the charged C_{60} cluster as a uniformly charged sphere, the incremental increase of ionization energies is due only to the net charge of the cluster. Therefore, the sequence of ionization potentials increases linearly with the net cluster charge. Thus, higher, not reliably measured or exactly calculated ionization potentials I_i, $i > 7$ can be approximated by taking into account the work necessary to remove an eighth, etc., electron from the surface of a conducting sphere, $I_i = I_1 + (i-1)/a$ [44].

The LDA calculation of Puska and Nieminen [49] and the DFS calculation show noticeable differences in the calculated valence spectra of neutral C_{60}. However, the comparison of scattering calculations, based on the two different descriptions of the target–electronic structure [46], shows that cross sections for the production of specific target–charge states in soft collisions and charge–state evolutions of target and projectile agree within 10 per cent. This relative insensitivity of total capture cross sections to details of the C_{60} electronic structure is consistent with the primary importance of the sequence of target ionization energies I_i and the matching of (shifted) energy levels of target and projectile, as illustrated in section 4.2.2.

2.4 Charge–state evolution and electron emission during the collision

fig. 4.6 shows typical charge–state evolutions in HCI–C_{60} interactions, as a function of $R_{||}$. The changes are equal to the difference of the instantaneous occupation and the initial occupation of a particular projectile shell. The figure shows results for incident Ar^{8+} at a kinetic energy of 80 keV on a trajectory with impact parameter $b = 15$ (fig. 4.5a) and Bi^{46+} at 830 keV with $b = 25$ (fig. 4.5b). For the Ar^{8+} projectile, the $n = 6$ and $n = 7$ shells get resonantly populated on the incoming trajectory (cf. fig. 4.4 in section 4.2.2). This agrees with experimental evidence for capture into the $n = 7$ shell [16, 55]. Auger relaxation of the projectile on the outgoing trajectory ($R_{||} > 0$) leads to the

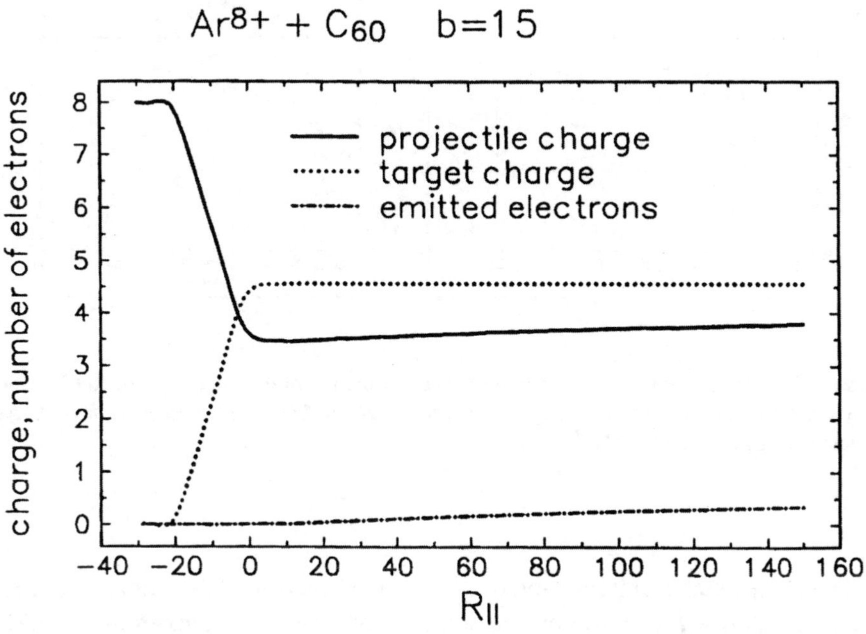

Figure 4.6. Projectile and target charge–state evolution and accumulated projectile Auger emission for incident 80 keV Ar^{8+} projectiles colliding with C_{60} at impact parameter $b = 15$.

partial depletion of projectile shells $n = 6$ and 7 and increases the population in projectile levels with $n = 3$ and $n = 4$.

The degree of population inversion achieved with incident Bi^{46+} projectiles is rather spectacular. Fig. 4.5b shows that resonant transitions first populate projectile shell $n = 31$ and, as the projectile further approaches the target, eventually lead to the population of projectile shells with principal quantum numbers between 19 and 31. The dynamical COM predicts that the large current of resonantly captured electrons originates in nearly degenerate levels near the Fermi level of C_{60} [14].

The charge–state evolutions in fig. 4.6 follow directly from the time–dependent occupations in eqs. (4.2) and (4.3). The projectile continues to relax by emitting electrons as it moves further away from the target and after charge exchange has become impossible (see section 4.2.6 below). However, in the average, Auger transitions are too slow to significantly depopulate excited projectile levels within the short collision time interval of about 15 fs covered in figs. 4.5a and 4.6. Therefore, the accumulated current of emitted Auger electrons displayed in fig. 4.6 and the increase of the net projectile charge on the part of the outgoing trajectory shown in the figure are very small.

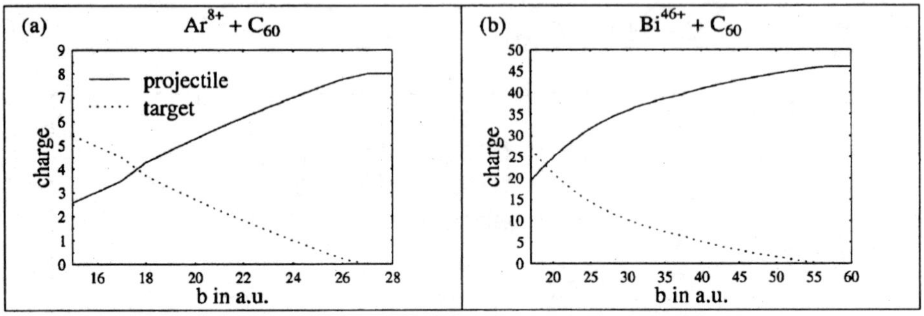

Figure 4.7. Projectile and target charge states as a function of the impact parameter b immediately after resonant electron exchange has ceased at $R_{||} = 150$. (a) For incident 80 keV Ar^{8+} projectiles. (b) For 830 keV Bi^{46+}.

The charge states of target and projectile as a function of the impact parameter and at a distance $R_{||} = 150$ on the outgoing trajectory are shown in fig. 4.7. At this distance all resonant electron transfer has stopped. The comparatively slow projectile Auger transitions had no time to relax the projectile. A lower limit for the impact–parameter range of non–destructive collisions is given by the highest charge state the target ions support without falling apart while interacting with the projectile. At the smallest impact parameter $b = 15$ in fig. 4.7a, the incident Ar^{8+} ion captures five electrons and thus maintains the carbon cage of the target [16]. At the smallest impact parameter in fig. 4.7b, the incident Bi^{46+} has captured a large number of electrons, which eventually leads to fragmentation of the target. In this case, the dynamical COM is applicable under the realistic assumption that the multiply charged fullerene is stable *during* the collision, i.e. at least for several femtoseconds.

2.5 Cross–sections for multiple electron capture

Similar to a solid surface, C_{60} provides a large reservoir of electrons with nearly the same binding energies and thus the cross section for the transfer of many electrons falls slowly with the number of electrons captured. With regard to an insulating target surface or an atomic target, electron capture by an HCI results in a multiply charged target. Figs. 4.5 and 4.6 illustrate the capture of several electrons by an HCI on its incident trajectory. The sequence of critical radii R_i^* for the capture of i electrons can be extracted from the impact parameter dependence of the final target charge state shown in fig. 4.7. The critical distances R_i^* for sequential over–barrier capture are related to geometrical cross sections for the production of specific charge states, $+i$, of

C_{60} by

$$\sigma_i = \pi({R_i^*}^2 - {R_{i+1}^*}^2) \tag{4.5}$$

and to the total geometrical cross section for charge exchange in non–destructive collisions by $\sigma_{tot} = \pi {R_1^*}^2$. This method for the calculation of cross sections has been applied to 50 keV N^{5+} [45], 3.3q keV Ar^{q+}, $q = 8, 15$ [56], and 830 keV Bi^{46+} [14] colliding with C_{60}.

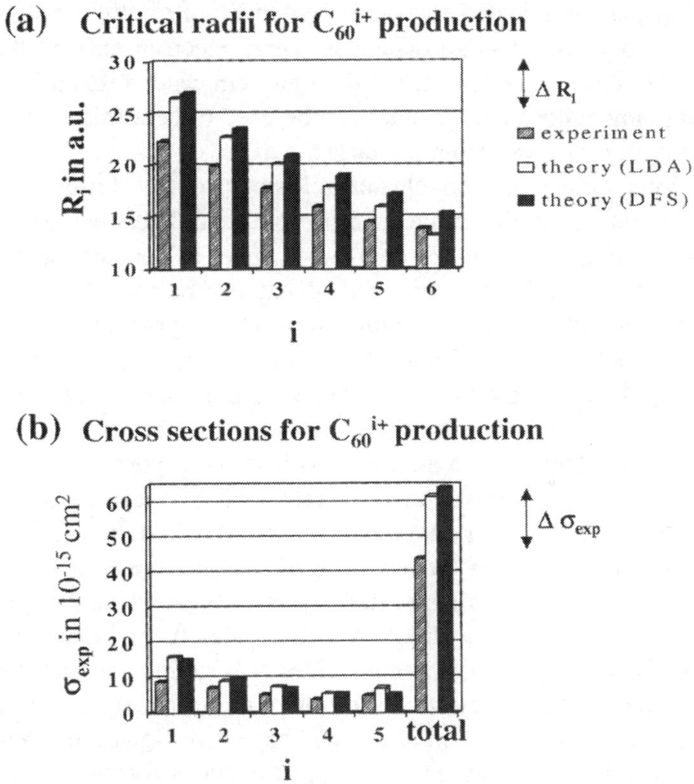

Figure 4.8. Critical radii R_i for sequential over–barrier capture (a) and corresponding cross sections (b) for capture of one to five electrons in collisions of 80 keV Ar^{8+} with C_{60}. Experimental results in comparison with simulations based on both LDA and DFS calculations of the C_{60} electronic structure.

Fig. 4.8 shows results for the system 80 keV Ar^{8+} on C_{60} discussed earlier in this chapter (cf. figs. 4.4, 4.5a, 4.6). The cross sections in fig. 4.8b are based on the critical radii in fig. 4.8a. The double arrows indicate experimental errors, ΔR and $\Delta\sigma_{exp}$, for critical radii and cross sections. Theoretical results are depicted in fig. 4.8 for simulations based on both, LDA [49] and DFS [46] calculations for the electronic structure of C_{60}. The calculated COM total cross sections agree with the absolute measurements for incident Argon ions performed by Walch *et al.* [16, 44] and by Selberg *et al.* [55, 56].

2.6 On the relaxation of hollow ions

In contrast to HCI–surface collisions, for cluster targets the hollow ion formed in the sequence of capture processes is not necessarily destroyed, but is likely to survive and deexcite far downstream. This offers the possibility of examining the complicated relaxation dynamics of multiply excited ions on the exit trajectory and at large distances where electron capture has ceased. Collisions with clusters are similar to the ion-atom case in that an intact recoil ion often remains whose charge state can be used to determine the number of electrons initially removed from the target.

Hollow ions, created in ion–cluster collisions, decay during the typically several microseconds of flight time the projectile needs to cover the macroscopic distance between the collision (resonant–exchange) region and the detector (cf. fig. 4.1 and table 4.1). This downstream decay can be modeled as a sequence of auto–ionizing and radiative relaxation steps [14]. Typically the HCI starts to relax via a cascade of Auger transitions. During these transitions lower lying levels are populated. Later, when the Auger relaxation cascade has populated lower–lying shells, radiative transitions may start to compete for subsequent relaxation steps. This is in agreement with known typical branching ratios (radiative versus non–radiative transition rates) which put increasing weight on radiative transitions as lower shells are populated. Radiative relaxation steps may then proceed along the "Yrast" line of maximal angular momentum of the active electron. This is supported by the statistical dominance of high angular momentum states, the dipole selection rule ($\Delta l = 1$), and the resonant population of high angular momentum states at large impact parameters.

Attempts to model the intricate relaxation of multiply excited ions have been made in the past. Benoit–Cattin *et al.* [57] have investigated the relaxation of doubly, triply, and quadruply excited projectile states formed in collisions of 70 keV N^{7+} ions with argon. Their discussion of possible relaxation paths is based on measured electron spectra in conjunction with predictions of the COM of Niehaus [42]. The emitted electron spectra are dominated by doubly excited lines, which are traced to excited states formed by either direct double capture or auto–ionizing cascades. Radiative stabilization following double electron capture of 10 keV/amu Ar^{q+}, $q \leq 17$ and Kr^{q+}, $q \leq 34$ ions colliding with argon has been discussed by Ali *et al.* [58]. Radiative stabilization was found to be of importance for the case of "asymmetric" doubly excited states, where the two excited electrons populate shells of different principal quantum number, $n << n'$. Hansen and collaborators [59] and Karim *et al.* [60] have recently calculated radiative and Auger decay rates for selected configurations of multiply excited ions. The *ab–initio* calculations of Vaeck and Hansen [59] predict that for increasing asymmetry of a doubly excited state, auto–ionization becomes less important and radiative transitions become possible in agreement with reference [58]. The close–coupling calculations of Chen and Lin [61] suggest a noticeable amount of radiative transitions (i.e. relatively large fluorescence yields) in certain quasi–symmetric configurations of high–lying, doubly excited Ar^{16+} states.

Very highly excited projectiles, such as generated in the case of incident Bi^{46+} ions (fig. 4.5b), offer a large number of auto–ionizing transitions between the many excited states of the hollow ion. These transitions can be combined to form an even larger number of possible relaxation cascades such that a rigorous theoretical treatment of the relaxation process is currently not possible. A simple relaxation scheme, based on intuition, basic features of emitted electron spectra, and wave function overlap arguments, has led to realistic final charge states of the relaxed projectile [14]. It consists of prioritizing possible relaxation steps and does not rely on detailed transition rates. For any configuration of the relaxing ion, the most likely next (Auger or radiative) transition is assumed to happen instantaneously and with unit probability.

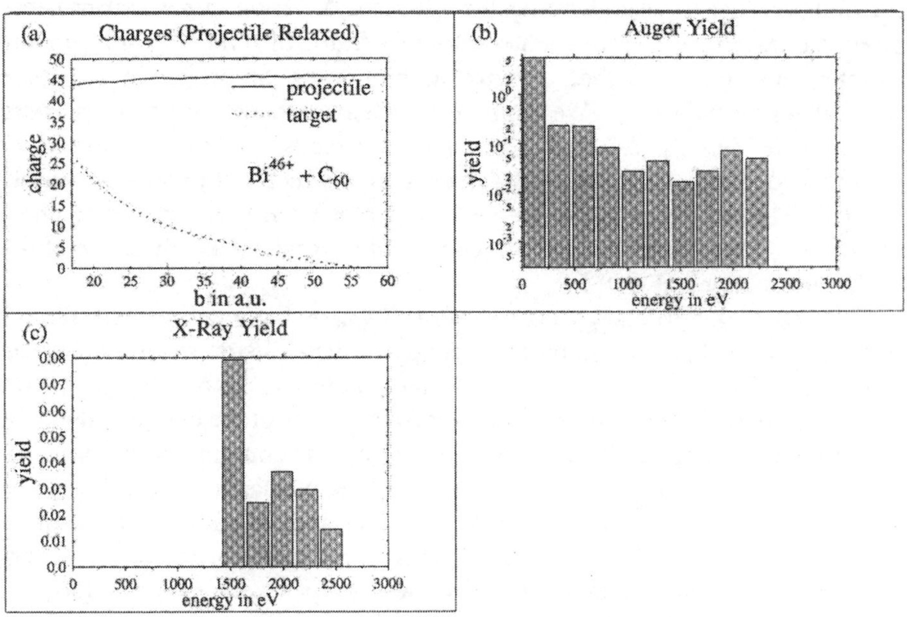

Figure 4.9. Results for 830 keV Bi^{46+} colliding with C_{60}, including downstream projectile relaxation. (a) Projectile and target charge states as a function of the impact parameter *b*. (b) Simulated Auger electron yield. (c) Simulated X-ray yield.

Results for incident 830 keV Bi^{46+} ions, including the downstream relaxation of the projectile are shown in fig. 4.9. As a consequence of the downstream relaxation process, most of the captured electrons get auto–ionized. This is easily seen by comparing the projectile charge states in fig. 4.9a with fig. 4.7b. The incident projectile charge effectively changes by no more than a few units. Auto–ionization is practically restricted to deexcitations that happen after the collision, as suggested earlier within a simple order–of–magnitude comparison of the collision time and typical Auger transition times (table 4.1). The simulated yields of emitted Auger electrons and X-rays in figs. 4.9b and 4.9c are accumulated over 234 eV–wide intervals of emitted electron and photon energies. These yields are normalized to the area perpendicular to the incident beam direction that the projectiles intersects with impact parameters between $b_{min} = 17$ and $R_1^* = 57.0$. For collisions in this impact parameter range, the total X-ray yield divided by total Auger yield amounts to 0.03.

2.7 Projectile kinetic energy gain

During a collision, a certain fraction of potential energy of the entire collision system may be transferred into kinetic energy of the projectile due to charge exchange processes. This energy can be calculated in two different ways. One possibility consists in integrating the net force between target and projectile along the trajectory. For the center–of–mass frame of reference, this amounts to integrating the force that governs the motion of the projectile of reduced mass along its trajectory. We define this systemic change in potential energy as *"nuclear" energy defect* Q_{nuc}, which is directly related to the motion of the reduced–mass projectile considered as an unstructured particle of variable charge. The net force is the sum of the direct Coulomb and image charge interactions between target and projectile and is provided as a function of time within the dynamical COM.

Obviously, due to energy conservation, Q_{nuc} is identical to the difference between the total electronic binding energy of the collision system before and after the collision. This *"electronic" energy defect* is denoted by Q_{el}. Since any COM–based simulation includes approximations of the complex dynamics of the multi–particle collision system that affect the coupling between nuclear and electronic degrees of freedom, the calculated values for Q_{el} and Q_{nuc} are expected to differ. The difference $|Q_{nuc} - Q_{el}|$ is indicative for the reliability of simulated translational kinetic energy gains. The electronic and nuclear energy defects for incident 46.2 keV Ar^{14+} ions are compared in reference [56].

In order to compare theory with measured projectile energy–gain spectra, the critical radii for the sequential capture of electrons may be related to energy defect values and to the number of electrons that are captured for a particular impact parameter. For impact parameters $b_i = R_{i+1}^*$, i electrons have been captured and the corresponding energy defects are $Q_{nuc}(b_i)$ and $Q_{el}(b_i)$. The simulated, discrete energy defects are folded with a Gaussian distribution G in order to correct for the finite experimental energy resolution. The full width at half maximum of G is adjusted to the resolution of the experiment. The simulated energy gain spectrum, differential in the projectile kinetic energy gain ΔE, is now given by the cross section

$$\frac{d\sigma}{d\Delta E} = \sum_i \sigma_i \, G(\Delta E - Q_i). \tag{4.6}$$

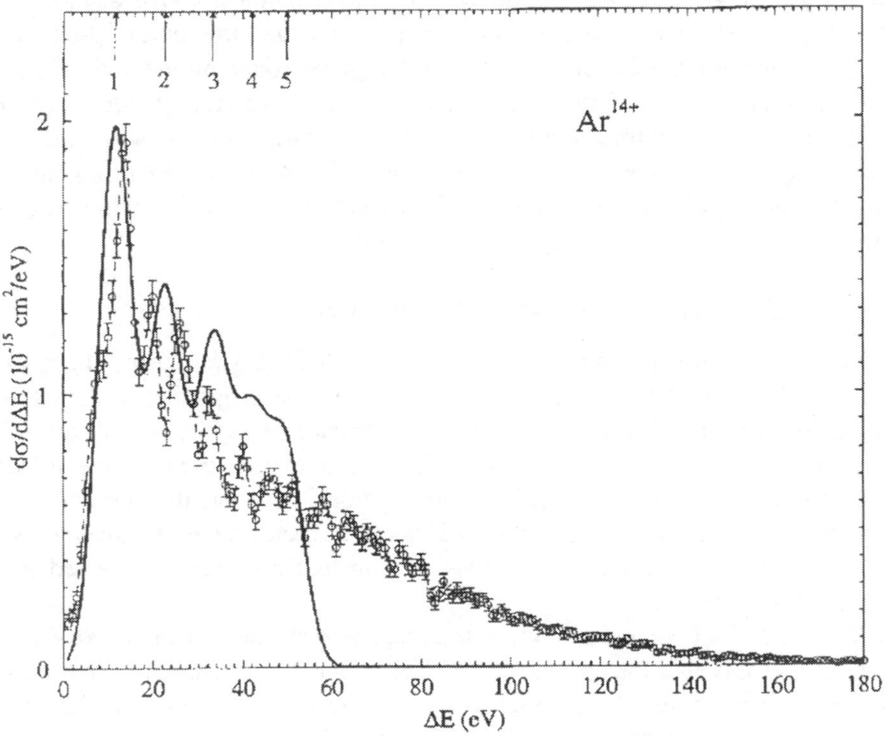

Figure 4.10. Simulated and measured projectile energy gain spectra for 46.2 keV Ar^{14+}–C_{60} collisions. The measured [55] and calculated [56] energy gain values and peak heights (cross sections) are absolute. The experimental errors in peak positions are typically ±0.5 eV. The arrows point to the calculated nuclear energy defects for the capture of 1, 2, 3, etc., electrons.

This method allows for the interpretation of peaks in measured spectra in terms of the corresponding number of captured electrons. In conjunction with the simulated projectile occupation changes, it also allows for the assignment of final projectile shells into which capture occurs at particular energy gains.

Fig. 4.10 contains an example for a measured differential energy gain spectrum from Selberg *et al.* [55] together with the simulated spectrum for the removal of a specific number of electrons from C_{60} [56]. Experimental and calculated spectra are absolute, both in intensity (peak heights) and energy gain. The calculated energy gains include the five lowest energy–gain peaks in fig. 4.10. In the overall trend, the lower part of the measured spectrum, which yields the dominant contribution to the total cross section, agrees well with the simulation. The calculated nuclear energy defects are indicated by numbered arrows and correspond to capture into specific projectile shells (1: capture into the $n = 12$ shell, 2: into $n = 11$, 3: into $n = 11$, 4: into $n = 10$, 5: into $n = 10$). The high energy gain region cannot be explained by the dynamical COM discussed in this chapter. Investigations are underway to understand this part of the kinetic energy gain spectra [62]. Similarly, even though reproducing the main features of the low–energy part of measured energy gain spectra on an absolute scale, the COM needs further refinement in order to more accurately agree with measured gains, e.g., for the capture of two electrons in fig. 4.10. A rewarding step in this direction could be the inclusion of projectile sub-shells nl, which are not resolved within the current version of the COM.

2.8 Projectile angular distributions

For slow highly charged ions colliding with C_{60}, simulations have predicted [45, 46] that the deflection function, i.e. the projectile scattering angle as a function of impact parameter, is characterized in its overall trend by two broad maxima that originate in the competition between attractive induced polarization and repulsive Coulomb interactions between the (charged) target and projectile. These maxima lead to strong measurable enhancements in the angle–differential scattering cross section that are usually referred to as 'Rainbow–Scattering'.

Walch *et al.* [12] have measured the angular distributions of 2.5 keV Ar^{8+} projectiles following the capture of 1 to 5 electrons from C_{60}. They determined the number of captured electrons by measuring simultaneously the scattered projectile and a charge-state-analyzed intact C_{60}^{i+} recoil ion. An experimental angle resolution of ≈ 0.01 degrees pointed to a potentially measurable prominent structure in the angle–differential cross section. The observed angular distributions in fig. 4.11 show a strong increase of the deflection angle with the number of electrons removed from C_{60}, due to the increasing Coulomb repulsion between positively charged collision partners.

A comparison with calculations based on the dynamical COM [12] shows good agreement only if the influence on the projectile trajectory by the large polarizibility of the C_{60} target is taken into account. This means that, due to the large polarizability of C_{60}, the trajectory of a highly charged ion captur-

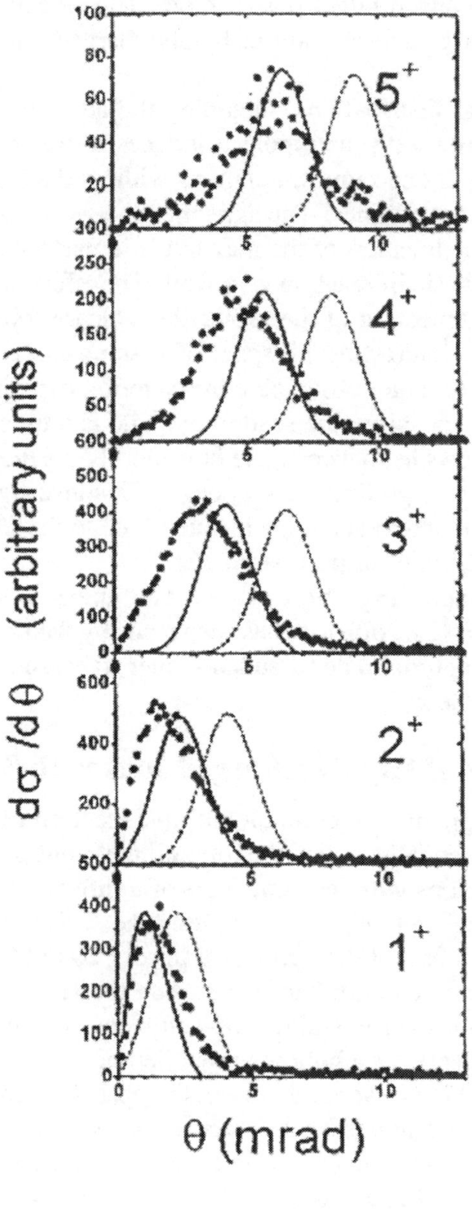

Figure 4.11. Angular distributions for the capture of i=1...5 electrons from C_{60} by 2.5 keV Ar^{8+} ions [12]. Each figure is labeled by $i+$. Θ is the deflection angle. The solid curves are dynamical COM calculations including target polarization. The dotted curves are calculated without this polarization.

ing electrons in the "soft" over-barrier region of impact parameters is affected measurably by the induced target polarization. This effect is a truncated analog of the image charge acceleration of highly charged ions incident on solid surfaces [39].

Calculated angular distributions, including the attractive self–image interaction of the projectile with the dipole it induces in the target, are shown in fig. 4.11 as solid lines. The same simulations without the induced polarization effects are shown as dotted lines. The agreement between the COM simulation and experiment in the location of the maxima is remarkable, showing that the model describes the basic interaction very well. The effect of the induced target polarization on the deflection of the projectile is clearly observed in all cases shown. It offers the interesting prospect of making the collective dielectric response of the cluster target observable in scattering experiments.

Since, classically speaking, the scattering angle is a function of the impact parameter, it also allows us to investigate how closely the projectile can pass by the target without disintegrating the C_{60} cage. This investigation is similar to the question of how much total energy is transferred to the C_{60} internal degrees of freedom as a function of impact parameter.

An alternative, interesting effort to calculate angle–differential scattering cross sections in HCI–C_{60} collisions has been made by Sakurai and Bárány [63]. This approach uses approximated Landau–Zener transition rates and classical potential scattering theory.

3. Collisions of Highly Charged Ions with Surfaces

The ideas of the previous section on multiple electron capture in soft collisions with C_{60} apply equally well to the formation of multiply excited projectile ions in surface collisions with incident HCIs of a sufficiently small component of the asymptotic incident velocity $|v_\perp^0|$ along the surface normal $\hat{e}_z$. As for collisions with fullerenes, at distances of typically 20 to 50 atomic units from the target surface, the potential barrier between the HCI and the target drops below the target Fermi level and initiates the efficient capture of target electrons, resulting in the formation of a hollow projectile ion.

However, due to the interaction between the positively charged HCI and the negative image charge it induces in the surface, the ion is accelerated towards the surface and hits the surface at a speed in excess of $|v_\perp^0|$. Even for the idealized case $v_\perp^0 \rightarrow 0$, that is for a projectile that starts out with no perpendicular velocity component at a very large distance from the surface, the image charge acceleration brings the projectile ion in very close contact with the surface and results in the "destruction" of the HCI. Thus, in contrast to large–impact parameter collisions with gaseous targets, for collisions with surfaces the COM needs to be extended in order to include those interaction mechanisms that dominate close to the surface. In this context, "close" can be specified by ion–surface distances at which the hollow ion starts getting destroyed due to significant overlap of the outermost populated states of the multiply excited ion and the electronic charge density of the surface.

3.1 Projectile motion and electron transfer at large distances from the surface

3.1.1 Projectile motion. The motion of the scattered projectile can be described by solving Newton's equation,

$$\Delta\vec{v} = \frac{\vec{F}(q_p, \vec{R})}{m_{\text{nuc}}}\Delta t\,, \tag{4.7}$$

where $\vec{R}$ denotes the position of the projectile nucleus with respect to a fixed point in the top–most lattice plane of the target. The force acting on the projectile,

$$\vec{F}(q_p, \vec{R}) = -\left(\frac{q_p(R)}{2(R - z_{\text{im}})}\right)^2 \hat{e}_z + \vec{F}_{TFM}(\vec{R}), \tag{4.8}$$

consists of two terms, the long–range interaction of the charged projectile with the image charge it induces in the surface and a short–range force $\vec{F}_{TFM}$ that represents all interactions of the projectile with individual target atoms in terms of a sum over binary forces between the projectile and target atoms localized at the target lattice points. The binary interactions are modeled as atomic Thomas-Fermi-Molière (TFM) potentials [64] and do not depend on the net projectile charge state $q_p = q_{nuc} - \sum_n a_n$. The unit vector $\hat{e}_z$ points along the surface normal, which is parallel to the z axis. The electronic coordinate $z = 0$ coincides with the top–most lattice plane. $z = z_{im}$ denotes the position of the image plane. Due to its dependence on $q_p(\vec{R})$, the projectile motion is coupled to its occupation evolution $\{a_n(t)\}$. Mass and charge of the projectile nucleus are designated as m_{nuc} and q_{nuc}. Recoil effects in close encounters with individual target atoms can be included by switching to a binary collision mode at distances below one–half of a lattice constant [17]. The inclusion of target recoil leads to a closer approach of the projectiles to the first bulk layer of the target as compared to a rigid crystal.

3.1.2 Kinetic energy gain. As discussed earlier for the case of HCI - C_{60} collisions, the self image interaction accelerates the projectile on the incident trajectory and thus increases its kinetic energy. Simulations as well as recent experiments [19] show that the neutralization of the HCI is completed prior to its reflection for a wide range of initial projectile charge states, leading to primarily neutral projectiles. Under the assumption that all reflected projectiles are neutral, and therefore do not experience any further image charge interactions that could change their kinetic energy close to the surface, the kinetic energy increase of the projectile ion is determined by the neutralization steps on the incident trajectory. This means that measurements of the kinetic energy gain are primarily sensitive to the distant interactions with the surface. In fact, an incident ion has accumulated over 70 % of its net kinetic energy gain by the time it reaches the critical distance R_1^* for over–the–barrier capture of the first target electron. Knowing that the complicated interaction dynamics at close distances to the surface is of little influence on the projectile kinetic energy gain, it is reasonable to first compare any model for HCI - surface interactions with measured kinetic energy gains, before proceeding to observables that are determined by the more intricate interaction dynamics at close distances.

Image energy gains of an HCI impinging on metal surfaces are characterized by an approximate $q^{3/2}$–increase with the initial projectile charge state q [65–67]. Within the COM, a lower limit for the kinetic energy gain can be deduced by assuming that the projectile is instantaneously and completely neutralized at the first critical distance $R_1^* \simeq \sqrt{8q+2}/(2W)$ [68], where W is the target work function. The energy gain for large q is then given by the expression

$$\Delta E = \frac{q^2}{4R_c} = \frac{W q^{3/2}}{4\sqrt{2}}. \tag{4.9}$$

We refer to this estimate as "*simple COM*". According to this result, the ratio $\Delta E/W$ does not depend on the target material.

The simple COM can be improved by letting one electron transfer to the HCI each time the over–barrier condition is fulfilled at consecutive critical radii, R_i^*, for the first, second, etc. capture. This version of the COM is called the "*staircase model*" [65]. In contrast to the dynamical COM (cf. section 4.2.1), in the simple and staircase model the charge transfer current is quantized. For energy gains of Xe^{q+} projectiles on an Al surface, the staircase model almost coincides with the dynamical COM for all initial charges q, and the simple model predicts, as expected, lower energy gains for all q (fig. 4.12). Except for the highest charge states, where the measured kinetic energy gains merge into a "plateau", both, the staircase and dynamical COMs agree with the experimental kinetic energy gains of Winter *et al.* [69], even though the more elaborate dynamical COM employs transition rates that depend on the width and depth of the potential saddle. Winter *et al.* deduced the energy gains from the difference of the measured asymptotic angles of incidence and reflection of the ion beam [67]. The simple model underestimates the measured energy gain, except for the highest charge states, where agreement with the experiment is fortuitous. As far as we know, the initial charge state q at which the experimentally observed plateau appears has not yet been reliably calculated. Fig. 4.12 also shows that the staircase COM calculation of Lemell *et al.* [65] agrees with our dynamical COM results [38].

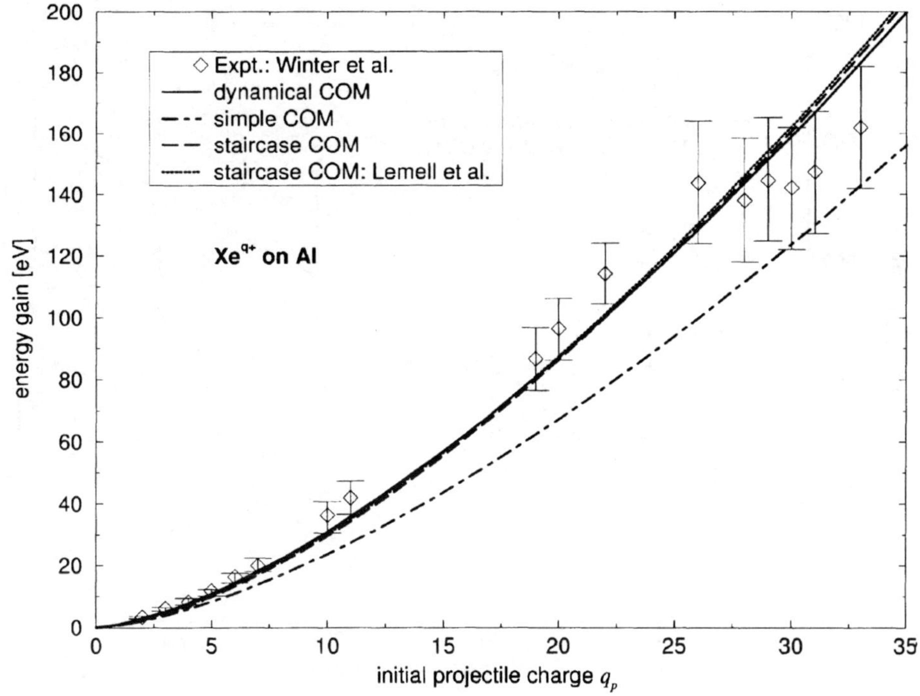

Figure 4.12. Experimental [69], simulated staircase COM results [65], and simulated energy gains from the dynamical COM [38] for Xe^{q+} (3.7 q keV, 1.5°) on an Al surface. The simple model assumes instantaneous complete neutralization at the first critical distance R_1^* and sets a lower boundary for projectile energy gains. The staircase COM instantaneously transfers one charge unit each time the over–barrier condition is fulfilled. In the dynamical COM continuous charge currents flow between projectile and surface.

Kurz *et al.* [26] have analyzed total electron yields for very high charge states as a function of the inverse projectile velocity. The data of Kurz *et al.* for $Xe^{q+}, q = 34 \ldots 50$ and $Th^{q+}, q = 61 \ldots 79$ on gold surfaces under perpendicular incidence provide, if at all, weak evidence for a deviation of the energy gain from the $q^{3/2}$–proportionality (fig. 4.13).

3.2 Electron transfer and emission at smaller distances from the surface

As the HCI moves closer to the topmost surface layer, electrons that were previously captured into highly excited states get increasingly disturbed and eventually lost. Lacking a detailed *ab-initio* description, models have been developed in order to identify and represent the most prominent features of the interaction scenario at small distances. Fig. 4.14 illustrates the various interaction

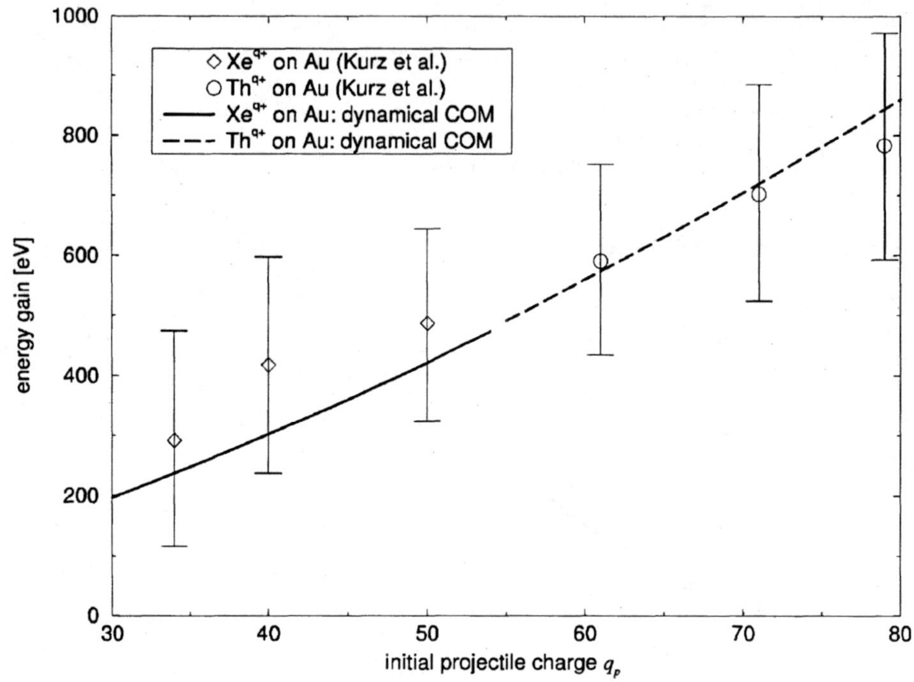

Figure 4.13. Experimental [26] and simulated dynamical COM data [38] for very high charge state ions impinging on polycrystalline gold.

stages during the ion's approach to the surface. Interactions in the near-surface zone are strongly influenced by target conduction band electrons that pour into the Coulomb well around the projectile core. The most prominent near–surface interaction mechanism is the direct transfer of electrons from target states into inner shells of the HCI (often referred to as "side feeding (SF)" [22, 70–72]) and the loss of loosely bound projectile electrons due to additional screening enforced by the tightly packed induced charge cloud (so called "peeling off" (PO)) [20, 73]. Simple illustrations of these basic interactions mechanism are shown in fig. 4.15. We emphasize that the separate investigation of largely simplified interaction rates for intuitive interaction modes, such as SF, PO, RC, and RL is justified only by the complexity of the many–electron dynamics involved. Since all of these mechanisms share a common cause, namely the strong perturbation of target (projectile) electrons by the projectile (target), modeling the intuitive mechanisms separately bears the risk of over-counting. The quality of these models therefore can be assessed only in comparison with experimental data.

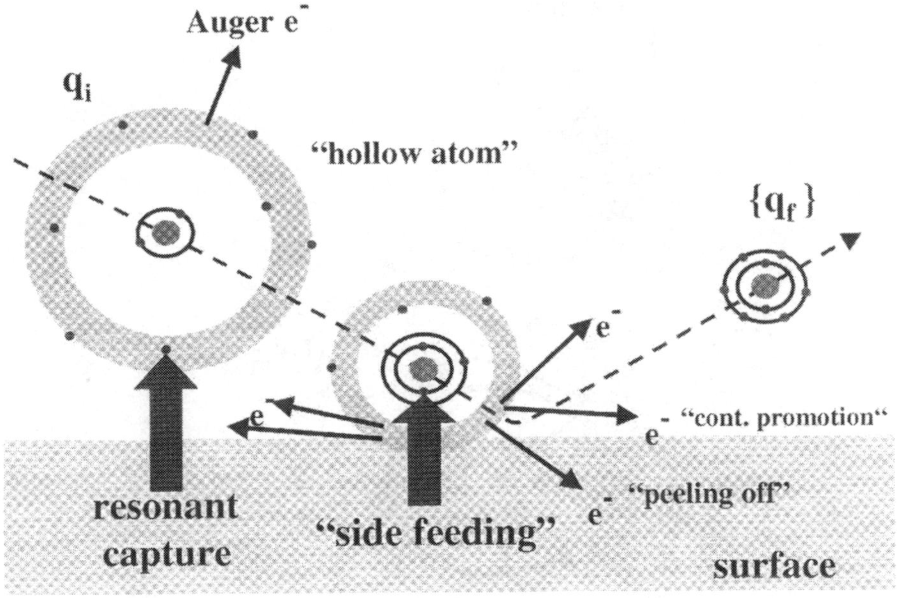

Figure 4.14. Sketch of the interaction scenario for the neutralization and reflection of an incident highly charged ion near a metal surface (see text).

3.2.1 Side feeding. For incident ion energies of up to several 100 keV and for a wide range of initial ion charge states and target materials, experiments on the final charge distribution of reflected projectiles have shown that the vast majority of the projectiles emerges in a neutral charge state [74–76]. Furthermore, it has been well known for more than a decade that the above-surface auto–ionization cascade does not allow for the full relaxation of the projectile on the incoming part of its trajectory. Simple estimates for Auger transition rates have suggested that this "bottle neck" originates in individual Auger relaxation steps of the hollow projectile that are slow compared with the time available between the second electron capture by the projectile and its close contact with the surface. This has led to the suggestion that tightly bound projectile levels are predominantly and very rapidly filled with target electrons in a region of strong overlap with the target electron distribution [17, 19, 22, 71, 77, 78]. This fast electron transfer mechanism was termed "SF", and was vaguely associated with the resonant transfer of localized, tightly bound target electrons.

Soon afterwards, it also became clear, that the SF mechanism leads to complete neutralization even in cases where the energies of projectile and target inner shells do not match and where, in consequence, an interpretation in terms of the resonant electron transfer fails. This stimulated the investigation of rapid

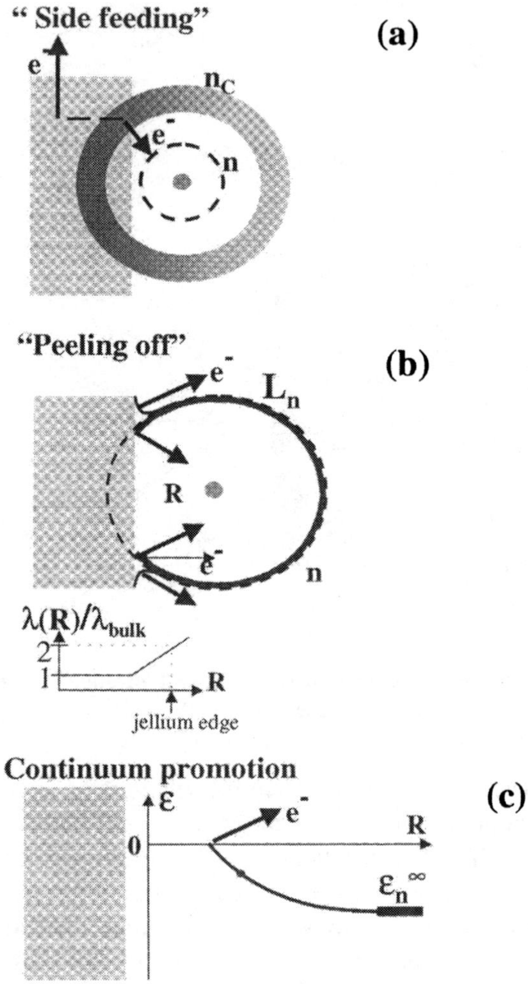

Figure 4.15. Illustrations of the near–surface electronic interaction mechanisms side feeding (a), peel–off (b), and continuum promotion (c). (b) also shows the assumed variation of the screening length λ as a function of the ion–surface distance R with respect to the bulk screening length λ_{bulk}.

inner shell transfer mechanisms that involve more than one active electron. In such a transfer, non-resonant projectile states become accessible into which an active electron can be captured. The active electron can restore the overall energy balance by transferring energy to a second electron (or an ensemble of other electrons, so-called "plasmons"). During this process, the second electron is excited or emitted. We introduced such an inner-shell population mechanism as an XCC-like process [48]. As a matter of convenience, we refer to it as SF. X designates an inner shell of the projectile ($K, L, M, \ldots$). This two-electron process is similar to regular Auger processes. However, the participating two electrons initially belong to the induced valence–band charge cloud (C) surrounding the projectile ionic core near and below the surface. LCC rates Γ_L^{SF} have been approximated by LCV processes [70] where a charge cloud electron (C) fills the L-vacancy while exciting a plasmon or an electron-hole pair in the valence band (V). Since C electrons remain localized about the ionic core, approximated XCC rates Γ_n^{SF} can be calculated in analogy to ordinary intra-atomic Auger rates [79]. In our dynamic simulation, we found that the LCV rates specified in reference [70] for N embedded into Al are too slow to explain measured final charge state distributions of reflected projectiles.

Ion-surface interaction models that include SF and other interactions close to the surface (in addition to the basic assumptions of the dynamical COM for distant interactions) include a variety of assumptions and adjustable parameters. In order to enable a meaningful comparison with experimental results, model calculations need to be performed with a fixed set of adjusted parameters for a variety of measured observables and of as many collision systems as possible. For this reason, model XCC SF rates were constructed in an attempt to include as much plausible physics as possible in their analytic form. These rates may then be fine–tuned by fitting experimental results [17, 48]. For all collision systems and all localized atomic levels $1 \leq n \leq n_{loc}$, we assume a single base rate $\Gamma_0^{SF} = 0.01 \approx 4 \cdot 10^{14}\, 1/s$. For each shell n, Γ_0^{SF} is multiplied by the number of n-shell vacancies N_n^{vac} and a factor that models the spatial variation of Γ_n^{SF},

$$N_n^{ol}(R) = \begin{cases} \frac{V_n^{ol}(R)}{V_n} & \text{if } R > z_j - \langle r \rangle_n \\ 1 & \text{otherwise} \end{cases} . \tag{4.10}$$

We account for the strong $(1/\Delta n)^{3.46}$-scaling of Auger rates $\Gamma_{n',n}^{AI}$ with the difference Δn between participating levels [20] (eq. (4.4)) and arrive at the LCV rate

$$\Gamma_n^{SF}(R) = \Gamma_0^{SF} q_c\, N_n^{vac} N_n^{ol}(R) \left(\frac{n_{\text{loc}} + 1 - n_L}{n_{\text{loc}} + 1 - n} \right)^{3.46} \tag{4.11}$$

for side feeding into projectile shell n. V_n^{ol} is the part of the orbital volume of level n, $V_n = 4\pi/3\ \langle r \rangle_n^3$, that overlaps with the metal electron distribution. q_c denotes the charge of the ionic core. For calculating this overlap, the metal electron distribution is supposed to extend to the (assumed planar) jellium edge located a distance z_j in front of the topmost layer of lattice points. At the jellium edge, the density of conduction electrons is 1/2 of the bulk electron density. According to this model $N_n^{ol}(R)$ vanishes for distances $R > \langle r \rangle_n + z_j$. If the projectile has penetrated the jellium edge by more than the orbital radius $\langle r \rangle_n$, we assume $N_n^{ol}(R)$ to remain constant at the value 1.

3.2.2 Peeling off. Outer orbitals that were resonantly populated at large ion–surface distances with typical orbital radii $\langle r \rangle_n \simeq R_{crit}$ are increasingly disturbed as the HCI approaches the bulk. By modeling PO as the instantaneous loss of an electron as soon as a certain fraction of the HCI orbital volume overlaps with the surface electron distribution, our simulation results for the final charge-state distributions of the projectile disagreed with experiment [74, 75, 80], since peeled off electrons are immediately replenished by RC. In contrast to previously implemented *instantaneous* PO mechanisms which become effective at the moment the HCI enters the bulk region [73, 81], we examined the influence of a *dynamic* PO on the speed of the electron transport from outer projectile levels into inner levels during the entire projectile–surface interaction [48].

We modeled PO by interpolating smoothly between the (remote) vacuum region and the bulk limit. We assumed that near the surface, for $R < \langle r \rangle_n + z_j - \lambda_{scr}$, when the electron has "lost touch" with the ionic core due to screening, the outermost orbital is likely to move to the valence band continuum if its radius $\langle r \rangle_n$ exceeds the screening length

$$\lambda_{scr}(R) = \lambda_{scr}^0 \left(\frac{\max(R, 0)}{z_j} + 1 \right) . \tag{4.12}$$

The screening length is assumed to reach its bulk value λ_{scr}^0 at $R = 0$. Above the first bulk layer, $\lambda_{scr}(R)$ increases linearly in R and equals $2\lambda_{scr}^0$ at the jellium edge $R = z_j$. Due to the nonlinear response of the surface electron distribution to the nearby HCI, the linear scaling in R, the neglect of a variation with the incident projectile charge q, and the particular choice of the slope, Eq. (4.12) can represent only a crude estimate for the dependence of λ_{scr} on q, n, and R.

Our model PO rate,

$$\Gamma_n^{PO}(R) = a_n \frac{N_n^{ol}(R)}{T_n} \frac{2\pi \langle r \rangle_n}{L_n(R)} \Theta(\langle r \rangle_n - \lambda_{scr}(R)) , \tag{4.13}$$

is composed of several parts. Similar to the derivation of the RL rate in [20], our base rate is given by the inverse orbiting time T_n of an electron in an unperturbed orbital. As in eq. (4.11), we reduce T_n by a volume factor $N_n^{ol}(R)$. The term $L_n(R)/(2\pi \langle r \rangle_n)$ corrects T_n to yield the "reaction time" for an atomic electron. We assumed that an electron which is captured at $R - z_j = \langle r \rangle_n$ and enters an atomic orbital does not get perturbed by the target electron gas until it has covered the distance $L_n(R)$. This period decreases with the ratio of the vacuum section $L_n(R)$ of the classical orbital above z_j and its circumference $2\pi \langle r \rangle_n$.

The unit step function Θ in eq. (4.13) disables PO for levels with shell radii $\langle r \rangle_n$ smaller than the screening length $\lambda_{scr}(R)$.

In the present simulation, PO rates were introduced to replace RL rates if $\langle r \rangle_n < \lambda_{scr}(R)$. Both, PO and RL, represent resonant electron flow into empty band states and could not be distinguished in a precise theory.

3.2.3 Continuum promotion (CP). As the HCI approaches the surface, due to the action of the repulsive projectile image potential $V_{im,p}$ and the mutual screening of projectile electrons, atomic levels are shifted upwards with respect to their asymptotic values ε_n^∞. As the orbital energies $\varepsilon_n = \varepsilon_n^\infty + V_{im,p}$ reach the ionization threshold, electrons in shell n are detached from the projectile, i.e., have effectively been promoted (and lost) to the continuum. We assume immediate electron loss due to CP as soon as $\varepsilon_n > 0$.

The energies $\{\varepsilon_n^\infty\}$ for a given instantaneous occupation $\{a_n(\vec{R})\}$ of projectile shells can be evaluated by using a standard atomic structure program [47].

3.3 Evolution of projectile level populations

3.3.1 Rate equations for the projectile population. During the HCI – metal surface interaction the populations a_n of projectile shells with principal quantum number n vary as a function of time. They are given as solutions of a system of rate equations of the form [48]

$$\begin{aligned} \frac{da_n}{dt} &= \theta(A_n - a_n)\Gamma_n^{RC} - a_n\Gamma_n^{RL} \\ &+ \sum_{n'>n} \Gamma_{n',n}^{AI} - 2\sum_{n'<n} \Gamma_{n,n'}^{AI} \\ &+ \theta(A_n - a_n)\Gamma_n^{SF} - a_n\Gamma_n^{PO} + \Gamma_n^{CP} \\ &- \delta_{n,M}\Gamma^{CK} - \delta_{n,L}\Gamma^{sCK} \\ &+ (\delta_{n,K} - \delta_{n,L} - \delta_{n,M})\, a_L a_M \Gamma^{KLM}\,. \end{aligned} \tag{4.14}$$

We have encountered the processes in the two first lines of eq. (4.13) before (cf. section 4.2.1). For ion–surface interactions, these two lines describe the interaction scenario at intermediate and large distances R. They correspond to the "traditional" COM [20] and include the rates for resonant capture, Γ_n^{RC}, resonant loss, Γ_n^{RL}, and Auger transition rates $\Gamma_{n',n}^{AI}$. As in section 4.2.1, $\Gamma_{n',n}^{AI}$ describes only fast Auger relaxation steps that require at least two equivalent active electrons in an outer shell. Slow Auger processes do not contribute to the relaxation of the excited projectile prior to its impact on the surface. The Auger rates in eq. (4.13) include statistical weights in order to take the variable number of electrons in the initial and final active shells, n and n', into account [48]. θ is the unit step function and $\delta_{n,n'}$ the Kronecker symbol.

For the simulations discussed below, we have added the terms in the third, fourth and fifth line, including the rates for SF, PO, and CP, Γ_n^{SF}, Γ_n^{PO}, and Γ_n^{CP}. In general, we do not resolve the populations in particular projectile subshells. For the L-shell and the applications to very slow collisions discussed below, however, we calculate the $2s$ and $2p$ binding energies and keep track of the respective subshell populations. We assume that the $2p$ level is preferentially populated via Auger ionization (AI) and side–feeding processes, as suggested by its higher degeneracy and its lower binding energy. With increasing occupation, L-shell relaxation via LLM Coster–Kronig (CK) and LLL super CK (sCK) transitions becomes energetically possible and proceeds by one and two orders of magnitude faster, respectively, than other Auger processes [82]. The rates for CK, sCK, and KLM Auger transitions are included in the last line of eq. (4.13).

3.3.2 Monte-Carlo Sampling. The time integration of eqs. (4.7) and (4.13) for the level occupations $\{a_n(t)\}$ and the projectile trajectory $\vec{R}(t)$ is performed by Monte-Carlo sampling over a large number (of the order of 5000) incident projectiles. Starting at a random position just above the ($R = R_{crit}$)-plane, the HCI moves along its trajectory from a position $\vec{R}(t_i)$ to $\vec{R}(t_{i+1})$ within a short time interval Δt_i. At each time t_i, separate values Δt_i for each transition type $X \in \{$AI,RC,RL,SF,PO$\}$ are drawn from an exponential random number distribution

$$\Xi(\Delta t) = [\int_0^\infty \exp(-\Gamma^X \tau) d\tau]^{-1} \exp(-\Gamma^X \Delta t). \tag{4.15}$$

The physical process X supplying the smallest Δt_i is chosen to take place, and all variables, such as configuration energies and occupations, are updated according to the change in $\{a_n(t)\}$ during Δt_i. The projectile, in an electronic configuration given by eq. (4.13), is then moved from $\vec{R}(t_i)$ to $\vec{R}(t_{i+1})$ by the force given in eq. (4.8). Next, the same procedure starts over again leading to time step t_{i+2}, etc. Should the smallest drawn Δt_i be large (we assumed larger than one), the projectile is moved from $\vec{R}(t_i)$ to $\vec{R}(t_{i+1})$ without any electronic transition taking place.

3.4 Comparison with Measurements

Having outlined the basic model assumptions in the extended dynamical COM, we proceed by comparing our simulated results with various experiments. We have already confirmed in section 4.3.1.2 that our model reproduces measured projectile energy gains over a large range of charge states of the incident ion and for different projectile and target species. However, we have only included distant ion–surface interactions (corresponding to the two first lines in eq. (4.13)) and simply assumed complete neutralization of the projectile at the surface. With respect to the significant modifications given by the third and fourth lines in eq. (4.13), this agreement cannot necessarily be expected for simulations of projectile kinetic energy gains that include the *entire* trajectory of the reflected projectile [38, 48].

The comparison with experiments in section 4.3.1.2 was facilitated by the fact that the kinetic energy gain is mainly accumulated at large ion-surface separations. The present calculation of energy gains within the extended COM includes the entire trajectory of the reflected projectile and, in particular, describes the rapid neutralization of the projectile near the surface. Changes in the SF or PO rate were found to sensitively affect the dominant charge state of the reflected ion and, therefore, the net kinetic energy gain of the projectile that accumulates along its reflected trajectory. The results discussed below were obtained with model transition rates and adjustable parameters that also reproduce the measured kinetic energy gains.

3.4.1 Final charge state distributions. After the reflection of highly charged ion beams on surfaces, high fractions of completely neutralized projectiles, typically well above 90%, have been observed, even on insulating targets [74–76]. The remaining charged fraction of the measured final projectile charge-state distributions $\{q_{final}\}$ overwhelmingly consists of singly charged positive and negative ions. The traditional COM [20], does not allow for the efficient neutralization of the incident HCI, due to the "bottle–neck" problem mentioned above. Burgdörfer *et al.* [22] have included a resonant L-shell filling mechanism to comply with measured final charge states. The extended COM discussed in this chapter reproduces the strong trend towards neutrality in the q_{final}-distribution while, in addition, keeping agreement with other observables.

Fig. 4.16 shows the simulated charge fractions $q_{final} = 0, 1$, and 2 of the final charge–state distributions for ground state (gs) H-like ions and metastable (mt) He-like ions C^{q+}, N^{q+}, and O^{q+} in $(1s, 2s)$ configurations, impinging with $E_{kin} = 13q$ eV and a grazing angle of $\Theta = 5°$ on Al(111). The simulated fractions are recorded for reflected projectiles which have passed the first capture distance R_1^*. After this point, less than 0.1% of the beam still exhibits the original K-shell vacancy which eventually causes re-ionization. Also shown are measurements by Folkerts *et al.* [75] for O^{q+} $(3 \leq q \leq 8)$ at $E_{kin} =$ 3.75 keV/amu on Au(110) under surface channeling conditions. Note that the simulations were applied to different projectile types containing a single K-shell hole while the experiment was performed only for O^{q+}-projectiles with K-shell vacancies for $q \geq 7$. Simulations for incident projectile ions with a filled K-shell leads to slightly higher degrees of neutralization.

Neglecting the influence of the projectile kinetic energy, the measured charge-state distributions in fig. 4.16 agree well with our simulation results. Shifting the final charge state fractions of the metastables by $\Delta q = +1$ towards the

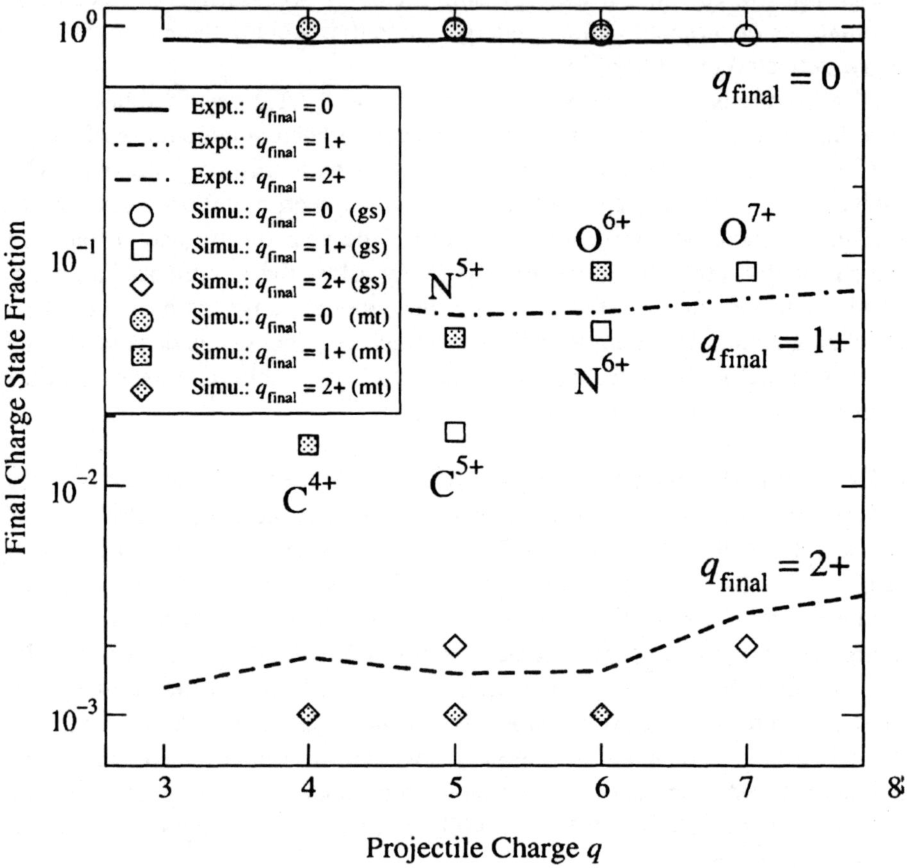

Figure 4.16. Final charge state fractions. The lines represent experimental data by Folkerts *et al.* [75] for O^{q+} impinging at $E_{kin} = 3.75$ keV/amu on Au(110) under surface channeling conditions. The simulation results [48] are given for H-like (gs) and He-like ($1s2s$) metastable (mt) C^{q+}, N^{q+}, and O^{q+} ions scattering with $E_{kin} = 13q$ eV and $\Theta = 5°$ off an Al(111) surface.

right, the data points for the H-like and He-like counterparts almost coincide for $q_{final} \leq 1$. This means that the distribution of q_{final} depends strongly on the nuclear charge and is rather insensitive to the initial L-electron.

3.4.2 Low-energy electron emission. Electron emission originates from various processes. Apart from auto–ionization, SF and CP may contribute to the emitted electron yield. In fig. 4.17a we show experimental and simulated electron emission spectra for N^{6+} interacting with an Al(111) surface, incident under $\Theta = 45°$ with kinetic energies E_{kin} of either 80 eV or 10 eV. For

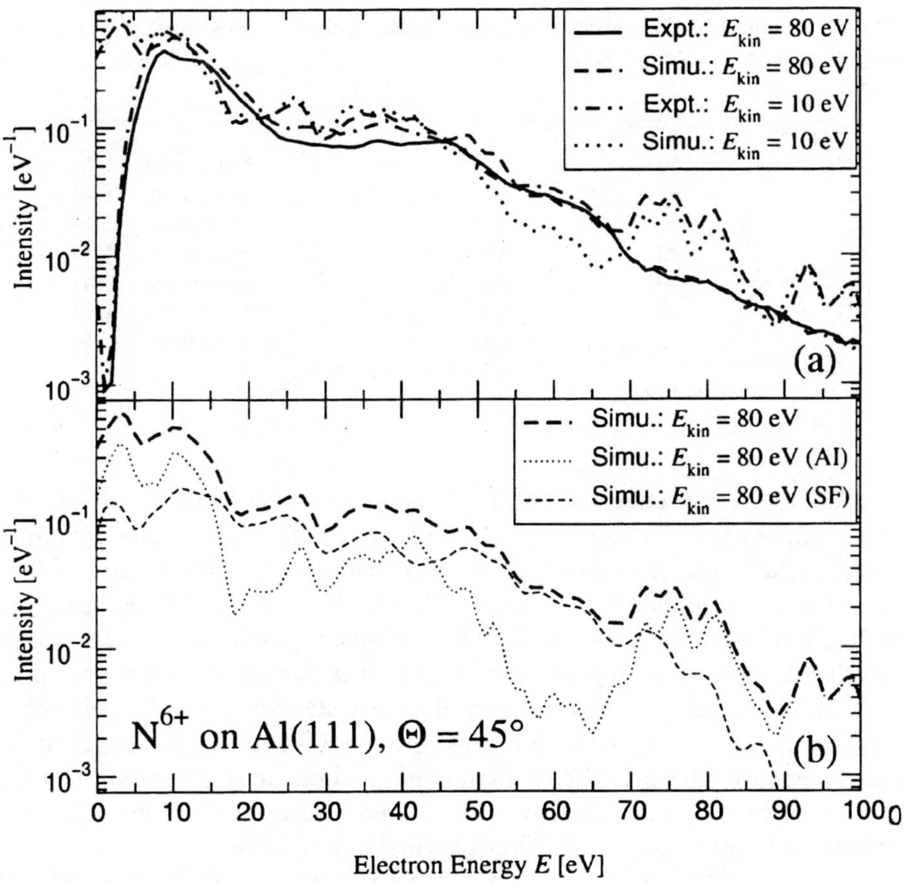

Figure 4.17. Low-energy electron spectra of N^{6+} incident under $\Theta = 45°$ on an Al(111) surface. Experimental and simulated spectra for incident energies $E_{kin} = 80$ eV and 10 eV (a). (b) SF and AI contributions to the $E_{kin} = 80$ eV spectrum.

these low incident kinetic energies, the simulated image energy gains amount to $E_{gain} = 17.1$ eV and 16.3 eV, respectively. The simulated data have been convoluted with the spectrometer resolution of 0.7% [48].

Vanishing spectrometer transmission and stray magnetic fields aggravate the detection of electrons at the lowest displayed energies, and only the experimental results for emitted electron energies above $E > 20$ eV are reliable. The spectra in fig. 4.17 and all the following plots are normalized to the integral K-Auger intensity. At electron energies $E > 10$ eV, the simulated spectra exhibit reasonable agreement with the experiment for both energies of the incident projectile. Additional structures in the simulated spectra are reminiscent

Table 4.2. Number of electrons emitted per projectile, γ, for N^{6+} ions, incident with kinetic energy E_{kin} on Al and Au surfaces.

system	E_{kin}/eV	γ		
N^{6+} + Al	10	4.9	experiment	[48]
	10	5.7	simulation	[48]
	80	4.3	experiment	[48]
	80	5.8	simulation	[48]
N^{6+} + Au	90	8.6 ±3.4	experiment	[81]
	90	9.8	experiment	[85]
	90	15.4 (AI:7.6, SF:6.6, CP:1.2)	simulation	[48]

of our simplified evaluation of transition energies which considers only ground state configurations for shells $n > 2$, neglects angular momentum coupling, and the perturbation and hybridization of ionic levels near the surface [83, 84]. Fig. 4.17b displays the contributions of AI and SF to the simulated spectrum for 80 eV incident ions. While the SF mechanism produces a comparatively smooth spectrum in the region $E < 90$ eV, AI transitions generate structures below 20 eV, which we associate with the early stage of projectile relaxation above the surface, where small Δn steps between Rydberg states prevail. In the same interaction phase, highly-excited configurations may also emit L-Auger electrons, which enhance the emission of more energetic electrons. CP does not contribute noticeably to the electron yield in fig. 4.17a.

The estimated total number γ of electrons emitted per incident ion is obtained by integration over all emitted electron energies. Integrating the spectral yields in fig. 4.17a above 20 eV leads to yields of $\gamma = 5.7$ and 5.8 emitted electrons per incident ion for the simulated spectra and to $\gamma = 4.9$ and 4.3 for the experiments with $E_{kin} = 10$ eV and 80 eV, respectively (table 4.2).

Low-energy Auger spectra for N^{6+} colliding with an Au surface at perpendicular incidence with E_{kin} ranging from 90 eV to 60 keV recently have been published by Niemann *et al.* [81]. Their total emission yield for $E_{kin} = 90$ eV, $\gamma = 8.6 \pm 3.4$, agrees well with their simulated yield of $\gamma_{AI} = 9.8$ for mere auto–ionization. With a different technique, Eder *et al.* [85] measured a yield of $\gamma = 9.8$ under similar scattering conditions. For the same collision system, the extended COM simulation including dynamic PO, CP, and SF mechanisms provides electron yields $\gamma_{AI} = 7.6$ for Auger emission, $\Gamma_{SF} = 6.6$ due to SF, and $\Gamma_{CP} = 1.2$ due to CP [48]. These values add up to a total yield, $\gamma = 15.4$, including contributions from $E < 20$ eV (table 4.2). The discrepancy with Eder *et al.* and Niemann *et al.* might be rooted in the experimental difficulty to measure low-energy electrons ($E < 20$ eV), which produce the greatest contribution to γ, as well as in the necessary simplifications embedded in the simulation.

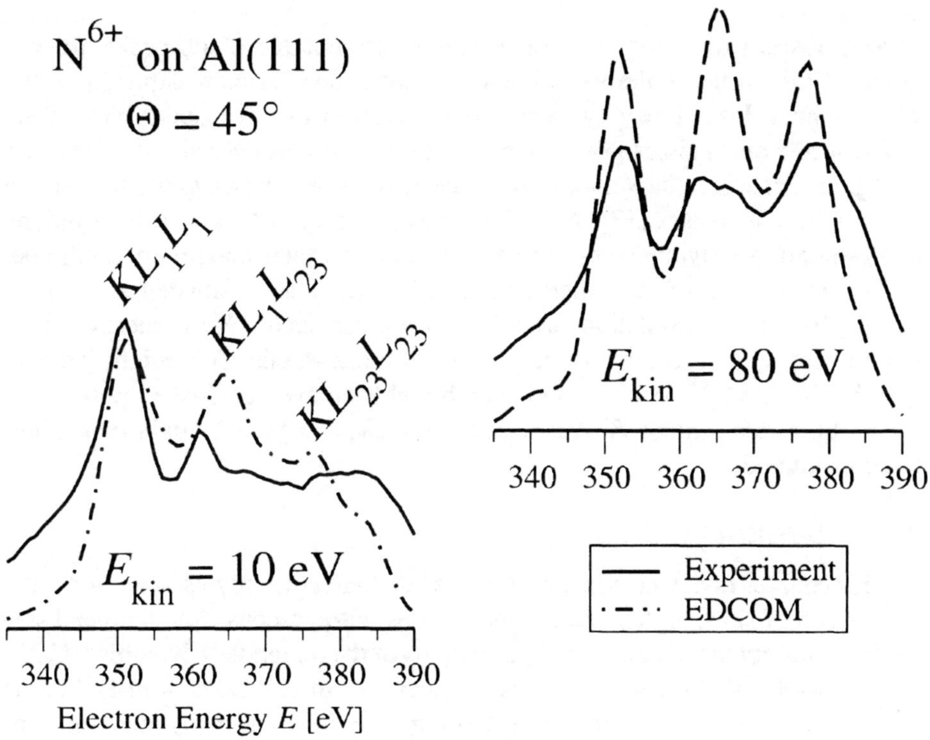

Figure 4.18. KLL spectra of N^{6+} incident under $\Theta = 45°$ on an Al(111) surface. Experimental and simulated spectra for the incident energies $E_{kin} = 10$ eV and 80 eV.

3.4.3 K-Auger spectra. K-Auger spectra for H-like and $(1s2s)$-metastable He-like incident ions can be subdivided into a well-structured KLL region, a broad, less intense peak consisting of KLM and KLC transitions and small contributions from KXY transitions with $X, Y \in \{M, N, \ldots\}$. Fig. 4.18 shows measured and simulated KLL electron spectra for N^{6+} colliding with an Al(111) surface under $\Theta = 45°$ with $E_{kin} = 80$ eV (upper right part) and $E_{kin} = 10$ eV (lower left part). For both projectile energies, the KLL region extends between the KL_1L_1 peak at $E = 352$ eV and the $KL_{23}L_{23}$ peak at $E = 378$ eV. The peak widths reflect the spread in initial L-shell populations at the time of K-Auger decay. The broad KLM/KLC peak is situated on the high-energy side of the KLL region.

In general, the KLL sub-peak intensities sensitively depend on the ratio between the L-shell filling rate $\Gamma_L^{fill} = \Gamma_{n,L}^{AI} + \Gamma_L^{SF}$ (eqs. (4.4) and (4.11)) and KLL decay rates Γ_K^{AI}. Only crude estimates for Γ_n^{fill} are available in the relevant interaction region, and Γ_K^{AI} is known only for free ions [86, 87] (cf. section 4.3.2.1). For shells $n > 2$, we have neglected the fine-structure in our simulation.

With respect to intensity ratios between different KLL sub-peaks, the extended COM follows the experimental trend: towards increasing E_{kin}, the KL_1L_1 peak loses intensity, which is transferred into the upper part of the KLL spectrum. This can be understood in view of stronger side feeding into the $2p$-orbital when the vertex of the trajectory moves closer to the first lattice layer as E_{kin} increases [17]. Near the vertex, the projectile is very slow and the exponentially decaying SF rates in eq. (4.11) reach their maximum amplitude.

The upper edge of the experimental $KL_{23}L_{23}$ peak is situated at a higher energy than in the simulation. In order to establish such a KLL energy, all six neutralizing electrons have to be present in the L-shell. This might indicate that the SF rate Γ_n^{SF} in eq. (4.11), which yields an average L-shell population $a_L = 4.6$ at the time of K-Auger decay for $E_{kin} = 80$ eV, might be slightly underestimated.

4. Summary

This chapter reviewed some of the ideas that are currently used to model the interaction of slow, highly charged projectiles with gaseous C_{60} and metal surfaces. It has put together several applications of the (extended) dynamical COM to observables that recently have been measured in collisions with HCI. Due to the complexity of the collision system and the inherent many–electron processes, *ab-initio* calculations, based on quantum mechanical matrix elements, are currently out of reach.

For soft collisions with C_{60} targets, we discussed the formation and decay of hollow projectile ions within the dynamical COM, supplemented by a simple downstream relaxation scheme. This scheme allows for the simulation of energy differential and total yields of post–collisionally emitted projectile Auger electrons and photons. The close agreement between the dynamical COM and various measured quantities confirms the picture of large-impact-parameter capture from C_{60} as an over-barrier process, very similar to corresponding processes in both ion-atom and ion-surface collisions.

For collisions of slow multiply-charged ions with solid surfaces, the dynamical COM needed to be extended to take into account interactions at small ion–surface distances, such as electron peel–off, side feeding, and continuum promotion. We calculated the population dynamics of the projectile by Monte-Carlo sampling along the entire ion trajectory over a large number of trajectories. For the classical motion of the projectile we included all relevant binary interaction potentials between the projectile and individual surface atoms. Our results are in reasonable agreement with various experimental observables for different combinations of projectiles, target types, incident angles and beam energies. This has been achieved *without* adapting the free parameters involved in the simulation to a particular collision system.

Future investigations, both experimental and theoretical, are necessary to refine these simulations in order to better understand the exciting life of a hollow ion during and after its interaction with complex targets, such as fullerenes and surfaces.

Acknowledgments

This work was supported by the Division of Chemical Sciences, Geosciences, and Biosciences, Office of Basic Energy Sciences, Office of Science, U.S. Department of Energy. The author gratefully acknowledges stimulating discussions and fruitful collaborations with J.J. Ducrée, A. Bárány, F. Casali, H. Cederquist, C.L. Cocke, B. Fricke, L. Hägg, and B. Walch.

References

[1] M. Krämer, J. Radiat. Res. **42**, 39 (2001); M. Scholz, A.M. Kellerer, W. Kraft–Weyrather, G. Kraft, Radiat. Environ. Biophys. **36**, 59 (1997)

[2] J.D. Gillaspy, D.C. Parks, and L.P. Ratliff, J. Vac. Sci. Technol. B **16 (6)**, 3294 (1998)

[3] L.P. Ratliff, R. Minniti, A. Bard, J.D. Gillaspy, D. Parks, A.J. Black, and G.M. Whitesides, Appl. Phys. Lett **74 (4)**, 590 (1999)

[4] E. Lerner, *The Industrial Physicist*, **5 (3)**, 18 (June 1999); **5 (5)**, 18 (October 1999)

[5] D.C. Lorentz, Comments At. Mol. Phys. **33**, 125 (1997)

[6] P. Scheier, B. Dünser, R. Wörgötter, S. Matt, D. Muigg, G. Senn, and T.D. Märk, Int. Rev. in Phys. Chem. **15**, 93 (1996)

[7] D. Hathiramani, V. Schäfer, K. Aichele, U. Hartenfeller, F. Scheuermann, M. Steidl, M. Westermann, and E. Salzborn, p. 653 in *"Photonic, Electronic, and Atomic Collisions"* ed. F. Aumayr, G. Betz and H.P. Winter, World Scientific, Singapore (1998)

[8] A. Bárány, p. 641 in *"Photonic, Electronic, and Atomic Collisions"* ed. F. Aumayr, G. Betz and H.P. Winter, World Scientific, Singapore (1998)

[9] H. Shen, P. Hvelplund, D. Mathur, A. Bárány, H. Cederquist, N. Selberg, and D.C. Lorentz, Phys. Rev. A **52**, 3847 (1995)

[10] H.G. Busmann, Th. Lill, B. Reif, and I.V. Hertel, Surf. Sci. **272**, 146 (1992)

[11] J. Jin, H. Khemliche, M.H. Prior, Z. Xie, Phys. Rev. A **53**, 615 (1996)

[12] B. Walch, U. Thumm, M. Stöckli, C.L. Cocke, and S. Klawikowski, Phys. Rev. A **58**, 1261 (1998)

[13] J.P. Briand, L. de Billy, J. Jin, H. Khemliche, M.H. Prior, Z. Xie, M. Nectoux, and D. Schneider, Phys. Rev. A **53**, R2925 (1996)

[14] U. Thumm, Phys. Rev. A **55**, 479 (1997)

[15] S. Martin, J. Bernard, L. Chen, A. Denis, and J. Désesquelles, Eur. Phys. J. D **4**, 1 (1998)

[16] B. Walch , C.L. Cocke, R. Voelpel, and E. Salzborn, Phys. Rev. Lett. **72**, 1439 (1994)

[17] J. Thomaschewski, J. Bleck-Neuhaus, M. Grether, A. Spieler, and N. Stolterfoht, Phys. Rev. A**57**, 3665 (1998)

[18] M. Schulz, C.L. Cocke, S. Hagmann, and M. Stöckli, Phys. Rev. A **44**, 1653 (1991)

[19] S. Winecki, C.L. Cocke, D. Fry, and M.P. Stöckli, Phys. Rev. A **53**, 4228 (1996)

[20] J. Burgdörfer, P. Lerner, and F.W. Meyer, Phys. Rev. A **44**, 5674 (1991)

[21] J. Burgdörfer, p. 517 in *Review of Fundamental Processes and Applications of Atoms and Ions*, ed. C. D. Lin, World Scientific, Singapore, (1993)

[22] J. Burgdörfer, C. Reinhold, and F.W. Meyer, Nucl.Instrum.Methods Phys. Res. B **98**, 415 (1995)

[23] M.W. Clark, D. Schneider, D. Dewitt, J.W. McDonald, R. Bruch, U.I. Safranova, I.Y. Tolstikhina, and R. Schuch, Phys. Rev. A **47**, 3983 (1993)

[24] J.P. Briand, B. d'Etat-Ban, D. Schneider, M.A. Briere, V. Decaux, J.W. McDonald, and S. Bardin, Phys. Rev. A**53**, 2194 (1996)

[25] F.W. Meyer, C.C. Havener, and P.A. Zeijlmans van Emmichoven, Phys. Rev. A **48**, 4476 (1993)

[26] H. Kurz, F. Aumayr, HP. Winter, D. Schneider, M.A. Briere, and J.W. McDonald, Phys. Rev. A **49**, 4693 (1994)

[27] A. Bárány and C.J. Setterlind, Nucl.Instrum.Methods Phys. Res. B **98**, 407 (1995)

[28] A.G. Borisov, R. Zimny, D. Teillet–Billy, and J.P. Gauyacq, Phys. Rev. A **53**, 2457 (1996)

[29] J. Limburg, S. Schippers, I. Hughes, R. Hoekstra, R. Morgenstern, S. Hustedt, N. Hatke, and W. Heiland, Phys. Rev. A. **51**, 3873 (1995)

[30] W. Huang, H. Lebius, R. Schuch, M. Grether, and N. Stolterfoht, Phys. Rev. A**56**, 3777 (1997)

[31] U. Thumm, J.Ducreé, P. Kürpick, and U. Wille, Nuc. Instrum. Methods Phys. Res. B**157**, 11 (1999); B. Bahrim, P. Kürpick, U. Thumm and U. Wille, ibid **164–165**, 614 (2000)

[32] Q. Yan, D.M. Zehner, and F.W. Meyer, Phys. Rev. A **54**, 641 (1996)

[33] C. Auth, and H. Winter, Phys. Lett. A **217**, 119 (1996)

[34] J. Limburg, S. Schippers, R. Hoekstra, R. Morgenstern, H. Kurz, F. Aumayr, and HP. Winter, Phys. Rev. Lett. **75**, 217 (1995)

[35] F. Aumayr, p. 631 in *Proceedings of the XIX'th Int. Conf. on the Physics of Electronic and Atomic Collisions*, ed. L. Dubé et al., AIP Conf. Proc. 360, AIP press, New York (1995)

[36] L. Hägg, C.O. Reinhold, and J. Burgdörfer, Phys. Rev. A**55**, 2097 (1997); Nucl.Instrum.Methods Phys. Res. B **125**, 133 (1997)

[37] J.P. Briand, S. Thuriez, G. Giardino, G. Borsoni, V. Le Roux, M. Froment, M. Eddrief, C. de Villeneuve, B. d'Etat-Ban, and C. Sébenne, Phys. Rev. A **55**, R2523 (1997)

[38] J.J. Ducrée, F. Casali, and U. Thumm, Phys. Rev. A**57**, 338 (1998)

[39] H. Winter, Europhys. Lett. **18** 207 (1992); J. Phys.: Condens. Matter **8**, 10149 (1996)

[40] J.J. Ducrée, J. Mrogenda, E. Reckels, M. Rüther, A. Heinen, Ch. Vitt, M. Venier, J. Leuker, H.J. Andrä, and R. Díez Muiño, Phys. Rev. A**57**, 1925 (1998)

[41] H. Ryufuku, K. Sasaki and T. Watanabe, Phys. Rev. A**21**, 745 (1980)

[42] A. Niehaus, J. Phys. B**19**, 2925 (1986)

[43] A. Bárány, p. 246 in *Proceedings of the XVI Int. Conf. on the Physics of Electronic and Atomic Collisions*, ed. Dalgarno et al, AIP Conf. Proc. 205, AIP, New York, (1990)

[44] U. Thumm, J. Phys. B **27**, 3515 (1994) In this reference, a factor 1/2 is not printed in the self-image potentials. This factor, however, was included in all computations. In addition, the present work uses a different sign convention in the definition of interaction potentials.

[45] U. Thumm, J. Phys. B **28**, 91 (1995)

[46] U. Thumm, T. Baştuğ, and B. Fricke, Phys. Rev. A **52**, 2955 (1995)

[47] R. D. Cowan, *The Theory of Atomic Structure and Spectra*, University of California Press, Berkeley, (1981)

[48] J.J. Ducrée, H.J. Andrä, and U. Thumm, Phys. Rev. A**60**, 3029 (1999)

[49] M.J. Puska and R.M. Nieminen, Phys. Rev. A**47**, 1181 (1993)

[50] J.L Martins, N. Troullier, and J.H. Weaver, Chem. Phys. Lett. **180**, 457 (1991)

[51] C. Yannouleas, and U. Landman, Chem. Phys. Lett. **217**, 175 (1994)

[52] T. Baştuğ, P. Kürpick, J. Meyer, W.–D. Sepp, B. Fricke, and A. Rosen, Phys. Rev. B **55**, 5015 (1997)

[53] C. Lifshitz, M. Iraqi, T. Peres, and J.E. Fischer, Rapid Commun. Mass Spectrom. **5**, 238 (1991)

[54] I.V. Hertel, H. Steger, J. de Vries, B. Weisser, C. Menzel, B. Kamke, and W. Kamke, Phys. Rev. Lett. **68**, 784 (1992)

[55] N. Selberg, A. Bárány, C. Biedermann, C.J. Setterlind, H. Cederquist, A. Langereis, M.O. Larsson, A. Wännström, and P. Hvelplund, Phys. Rev. A **53**, 874 (1996)

[56] U. Thumm, A. Bárány, H. Cederquist, L. Hägg, and C.J. Setterlind, Phys. Rev. A **56**, 4799 (1997)

[57] P. Benoit–Cattin, A. Bordenave–Montesquieu, M. Boudjema, A. Gleizes, S. Dousson, and D. Hitz, J. Phys. B **21**, 3387, (1988)

[58] R. Ali, C.L. Cocke, M.L.A. Raphaelian, and M. Stöckli, J. Phys. B **26**, L117, (1993); Phys. Rev. A **49**, 3586 (1994)

[59] N. Vaeck and J.E. Hansen, J. Phys. B **28**, 3523, (1995), and refs. therein.

[60] K.R. Karim, S.R. Grabbe, and C.P. Bhalla, J. Phys. B **29**, 4007 (1996)

[61] Z. Chen and C.D. Lin, J. Phys. B **26**, 957, (1993)

[62] H. Cederquist, private communication.

[63] P. Sakurai, *M.Sc. Thesis, Univ. Stockholm*, unpublished (1997)

[64] D. S. Gemmell, Rev. Mod. Phys. **46**, 129 (1974)

[65] C. Lemell, H. P. Winter, F. Aumayr, J. Burgdörfer, and F. Meyer, Phys. Rev. A **53**, 880 (1996)

[66] J. Burgdörfer, C. Reinhold, L. Hägg, and F. Meyer, Aust. J. Phys. **49**, 527 (1996)

[67] H. Winter, J. Phys: Condens. Matter **8**, 10149 (1996)

[68] J. Burgdörfer and F. Meyer, Phys. Rev. A **47**, R20 (1993)

[69] H. Winter, C. Auth, R. Schuch, and E. Beebe, Phys. Rev. Lett. **71**, 1939 (1993)

[70] R. Díez Muiño, N. Stolterfoht, A. Arnau, A. Salin, and P. M. Echenique, Phys. Rev. Lett. **76**, 4636 (1996)

[71] L. Folkerts and R. Morgenstern, Europhys. Lett. **13**, 377 (1990)

[72] N. Stolterfoht, R. Köhrbrück, M. Grether, A. Spieler, A. Arnau, R. Page, A. Saal, J. Thomaschewski, and J. Bleck-Neuhaus, Nucl. Instrum. Methods Phys. Res., Sect. B **99**, 4 (1995)

[73] C. Lemell, H. P. Winter, F. Aumayr, J. Burgdörfer, and C. Reinhold, Nucl. Instrum. Methods Phys. Res., Sect. B **102**, 33 (1995)

[74] F. W. Meyer, L. Folkerts, H. O. Folkerts, and S. Schippers, Nucl. Instrum. Methods Phys. Res., Sect. B **98**, 441 (1995)

[75] L. Folkerts, S. Schippers, D. M. Zehner, and F. W. Meyer, Phys. Rev. Lett. **74**, 2204 (1995), erratum: Phys. Rev. Lett. **75**, 983 (1995)

[76] S. Winecki, M. P. Stöckli, and C. L. Cocke, Phys. Rev. A **56**, 538 (1997)

[77] R. Díez Muiño, A. Arnau, and P. M. Echenique, Nucl. Instrum. Methods Phys. Res., Sect. B **98**, 420 (1995)

[78] M. Grether, A. Spieler, R. Köhrbrück, and N. Stolterfoht, Phys. Rev. A **52**, 426 (1995)

[79] R. Díez Muiño, A. Salin, N. Stolterfoht, A. Arnau, and P. M. Echenique, Phys. Rev. A **57**, 1126 (1998)

[80] S. Winecki, M. P. Stöckli, and C. L. Cocke, Phys. Rev. A **55**, 4310 (1997)

[81] D. Niemann, M. Grether, A. Spieler, N. Stolterfoht, C. Lemell, F. Aumayr, and H. P. Winter, Phys. Rev. A **56**, 4774 (1997)

[82] J. Limburg, J. Das, S. Schippers, R. Hoekstra, and R. Morgenstern, Phys. Rev. Lett. **73**, 786 (1994)

[83] P. Kürpick, U. Thumm, and U. Wille, Nucl. Instrum. Methods Phys. Res., Sect. B **125**, 273 (1997)

[84] P. Kürpick and U. Thumm, Phys. Rev. A **58**, 2174 (1998); U. Thumm, *Book of invited papers, XXII Internat. Conf. on the Physics of Photonic, Electronic, and Atomic Collisons*, Santa Fe, NM, ed. S. Datz et. al. (Rinton Press, 2002) p. 592

[85] H. Eder, M. Vana, F. Aumayr, H. P. Winter, J. I. Juaristi, and A. Arnau, Physica Scripta **T 73**, 322 (1997)

[86] J. Hansen, O. Schraa, and N. Vaeck, Physica Scripta **T41**, 41 (1992)

[87] S. Schippers, J. Limburg, J. Das, R. Hoekstra, and R. Morgenstern, Phys. Rev. A **50**, 540 (1994)

II

INTERACTIONS WITH GASEOUS TARGETS

Chapter 5

PHOTON EMISSION SPECTROSCOPY OF ELECTRON CAPTURE AND EXCITATION BY MULTIPLY CHARGED IONS

R. Hoekstra and R. Morgenstern

K.V.I., Atomic Physics, RijksUniversiteit Groningen,
Zernikelaan 25, 9747 AA Groningen, The Netherlands

hoekstra@kvi.nl

Abstract Photon Emission Spectroscopy (PES) is one of the experimental techniques which has contributed appreciably to our present understanding of the physics of highly charged ions, in particular electron capture and excitation processes. This chapter describes in detail most aspects of PES in the spectral range encompassing the vacuum ultraviolet and visible light regimes. It reviews some selected results obtained for a variety of collision systems, ranging from real one-electron systems, multiply charged ions colliding on atomic hydrogen, to systems with more active electrons to illustrate the versatility of PES method.

Keywords: highly charged ions, charge transfer, photoemission spectroscopy, electron capture

1. Introduction

The field of highly-charged ion physics has grown rapidly since the mid-eighties. Although its importance for astrophysical and man-made plasma was realized much earlier, the growth was initiated by the advent of powerful sources for multicharged ions, such as so called electron cyclotron resonance ion sources (ECRIS [1]) and later on also electron beam ion sources (EBIS [2]). In 1982 the KVI obtained the first ECRIS outside the laboratory of its inventor Geller [1]. Ever since we took our first ECRIS into operation we have been investigating the interaction of highly charged ions with matter, ranging from atomic hydrogen via molecules to solid-state surfaces. One of the main branches of research has been one-electron transfer and target excitation in collisions

F.J. Currell (ed.), The Physics of Multiply and Highly Charged Ions, Vol. 2, 169-192.

of highly charged ions on ground state and laser excited atoms. One of the experimental techniques we have used to study one-electron capture is Photon Emission Spectroscopy (PES). PES exploits the fact that the transfer of one electron in collisions of highly charged ions, A^{q+}, on neutral atoms, B populates excited states. This is given by:

$$A^{q+} + B \longrightarrow A^{(q-1)+}(nl) + B^{+} + \Delta E \longrightarrow A^{(q-1)+}(n'l') + hv + B^{+} + \Delta E \tag{5.1}$$

with n and l the principal and angular quantum numbers of the state into which the electron is captured. Because the electron is transferred almost resonantly, i.e. without a considerable change of binding energy, it is predominantly captured into an excited state. Subsequent to the capture process the excited particle $A^{(q-1)+}(nl)$ will decay to a lower lying state $A^{(q-1)+}(n'l')$ under the emission of a photon (hv). The wavelength of the emitted photon is characteristic for a specific $A^{(q-1)+}(nl \longrightarrow n'l')$ transition and can therefore be used to measure charge transfer into different $A^{(q-1)+}(nl)$ states. We have used this experimental method of Photon Emission Spectroscopy (PES) to study state selective charge transfer indicated by the above equation. Another method commonly used is Translational Energy Spectroscopy (TES) which exploits the kinetic energy gain or loss (ΔE) of the projectile ions [3, 4].

The fact that one-electron capture in collisions of highly-charged ions on neutral targets leads to excited states which decay radiatively can be nicely illustrated within the framework of simple classical models such as the over-the-barrier (COB) model [5–7]. Figure 5.1 depicts the potential $V(r)$ experienced by the electron in the joint Coulomb field of the ions A^{q+} and B^{+}. Along the internuclear axis this potential is given by (in a.u.):

$$V(r) = -\frac{1}{r} - \frac{q}{R-r} \tag{5.2}$$

where r is the distance of the electron from the target ion B^{+} and R is the internuclear distance. Classically speaking the electron can transit from the target to the multicharged ion if the barrier height, $V_{\max}$, drops below the binding energy of the electron, I_{BR}, which at an internuclear distance R is just the binding energy I_B at infinite seperation shifted by q/R. This condition yields a critical distance R_c that has to be reached for electron transfer:

$$R_c = \frac{2\sqrt{q}+1}{I_B} \tag{5.3}$$

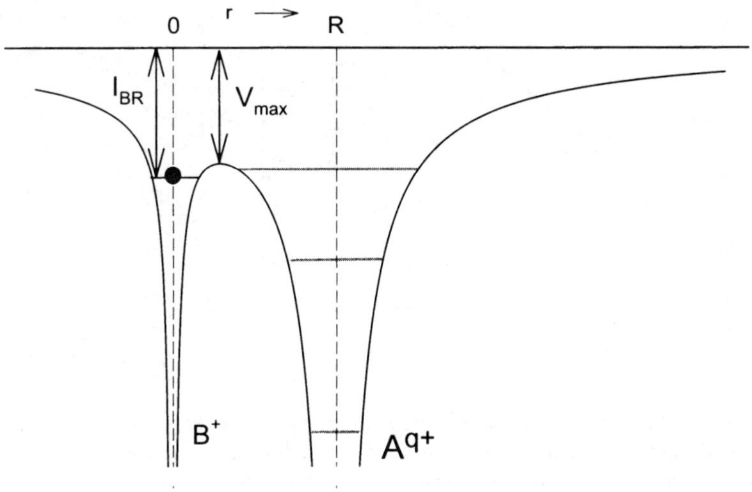

Figure 5.1. Schematic representation of the classical over-the-barrier description of one-electron transfer.

Assuming unit probability for the electron transfer inside the distance R_c (absorbing sphere approximation [8]), the total one-electron transfer cross section σ_{tot} in atomic units is given by

$$\sigma_{tot} = \pi R_c^2 \tag{5.4}$$

For a typical collision system such as C^{6+}-H we find a capture distance of 11.8 (6.2 Å) and a cross section of 437 (122 Å^2). In the keV/amu energy range the real cross section [9] is about half of the one calculated via eqs. 5.3 and 5.4. This is more or less a common ratio between COB model and experiment [10, 11]. The total cross section is basically constant at low energies and therefore it does not present a critical test of sophisticated theories. A more stringent test is the measurement of the final-state population in the multicharged ion. Predominantly excited states are populated as can be inferred from the COB model. The electron is transferred resonantly, at infinite internuclear distance this corresponds to a final binding energy I_q in the multicharged ion of

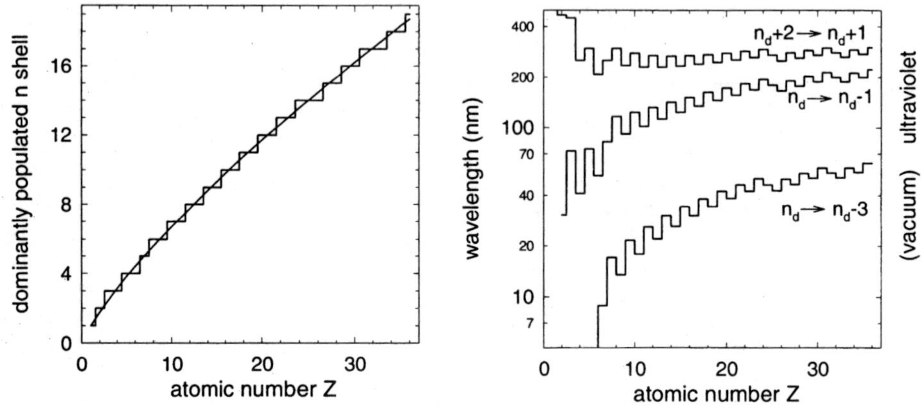

Figure 5.2. a) Principal quantum numbers n of the dominantly populated n shell in collisions of fully stripped ions on atomic hydrogen. The smooth curve represent the classical principal quantum number n_{cl}, while the stepped curve represents a binning on to discrete principal quantum numbers n. b) Wavelengths of some selected transitions starting out of the dominantly populated n shells (n_d) and out of a higher excited series ($n_d + 2$), see text.

$$I_q = I_B + \frac{q-1}{R_c}. \tag{5.5}$$

Using hydrogenic binding energies ($I_q = \frac{1}{2}Z^2/n^2$) the binding energies calculated with equation 5.5 can be converted into non-discrete, so-called classical principal quantum numbers n_{cl}. The results are depicted in figure 5.2a together with the results of binning the classical quantum numbers on to real, discrete quantum numbers. To do so, we used the following relation [12]:

$$[(n-1)(n-\frac{1}{2})n]^{\frac{1}{3}} \leq n_{cl} \leq [(n+1)(n+\frac{1}{2})n]^{\frac{1}{3}} \tag{5.6}$$

In particular for larger values of n_{cl} the relation is almost equal to rounding off n_{cl}. For example, for the C^{6+}-H system a final state binding energy of 0.92 (25 eV) is calculated. This corresponds to n_{cl}=4.4 which by eq. 5.6 is binned to the $n = 4$ level in C^{5+}. The strong dominance of this $n = 4$ shell is indeed observed in experiments (cf. figure 5.3). From figure 5.2a it is seen that the dominantly populated n shell shifts from n=1 for protons to n=19 for fully stripped Kr (Z=36). So highly excited states are predicted to be populated which in their turn will relax through photon emission. Although the dominantly populated n shell (n_d) changes strongly with Z the subsequent photon emission

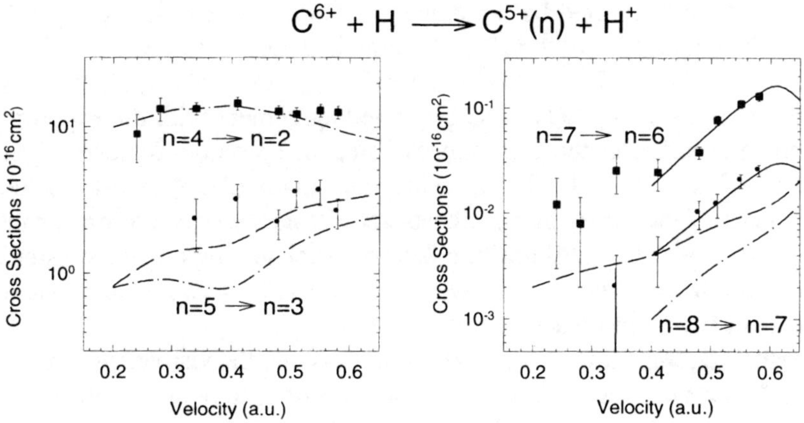

Figure 5.3. Cross sections for line emission following charge transfer in C^{6+} - H collisions. The experimental data are from Hoekstra *et al* [14]. The theoretical curves represent results of 3 AO-type of calculations with increasing basis set sizes (cf. text): dashed-dotted line - AO1 ([15]), long-dashed line - AO2 ([16]) and solid line - AO3 ([17]). The dominant line emission channels are also well predicted by other theoretical methods (Ref. [14] and references therein).

falls mainly in the vacuum ultraviolet to ultraviolet spectral range (10 - 350 nm). This is indicated in figure 5.2b which depicts as examples the wavelengths of transitions from n_d to lower-lying shells with principal quantum numbers $n_d - 1$ and $n_d - 3$. It is of note that because within a specific n shell the higher angular momenta states are most often the strongest populated ones [13], the strongest transitions are the ones with $\Delta n = 1$, followed by $\Delta n = 2, 3, 4$, etc. Transitions from higher lying levels and with small values of Δn shift towards longer wavelengths. Their intensity is normally orders of magnitude smaller because the higher lying levels are only weakly populated.

This is illustrated for charge transfer in collisions of C^{6+} on H in figure 5.3 which compares experimental and theoretical CVI($n \to n'$) line emission cross sections. Experiments were performed for various emission lines including the dominant one from the $n = 4$ shell and others from higher shells up to the non-dominantly populated $n = 8$ shell. The corresponding line emission cross sections differ by orders of magnitude, see figure 5.3. This indicates the extreme resonant nature of charge transfer. The determination of absolute cross sections for line emission from the non-dominant high-n states by means of crossed-beam laboratory experiments is on the edge of experimental possibilities. The theoretical modelling requires the most extensive computer resources available. The latter is nicely documented by the evolution of atomic orbital (AO) calculations [15–17] in which the electronic wavefunction of the collision complex is expanded in a basis of atomic wavefunctions of the separate collision partners. Limited by computer memory, the first attempts (AO1) to calculate the capture into the $C^{5+}(8l)$ states included only the $C^{5+}(4l, 5l, 8l)$ states in the basis set. In the AO2 calculations all the $C^{5+}(6l, 7l)$ states were added to the basis set of AO1. The improved agreement with the experiments (see figure 5.3) indicates the importance of stepwise promotion processes in the population of high-n levels. The final step was made (AO3) by also including excited states of the target into the basis set. Basis set sizes (computer memory) are to a large extent still limiting factors in calculating charge transfer into non-dominant high-n levels for heavier elements (for example Ne^{10+}) and for systems with two or more active electrons [18].

Nevertheless, if the transition wavelength falls into the visible spectral range, as for example the weak CVI($8 \to 7$) line emission, they are of practical use for fusion-plasma diagnostics which requires long-distance detection via fiber optics systems [19, 20]. PES in the (vacuum) ultraviolet and visible spectral range thus serves both fundamental and applied needs. The latter category encompasses a whole variety of astrophysical topics (e.g. [21–23]), ranging from the Jovian atmosphere, via comets to interstellar space.

2. Photon Emission Spectroscopy

The basic lay-out of a PES experiment is depicted in figure 5.4. A well collimated beam of ions, A^{q+}, is crossed with a neutral target. Photons originating from a certain region along the ion beam, the observation length L, and emitted towards the spectrometer are detected. In the earlier experiments the neutral target was a small gas-filled cell, while nowadays neutral target beams as shown in figure 5.4 are favoured. By this photon count rates can be enhanced by a factor of 2 - 5 because with a target beam it is possible to increase the target density in the region from where the photons are observed without a significant increase of density elsewhere, so maintaining single collision conditions.

Of main concern in PES experiments is the photon flux, because of the low detection efficiency due to the small solid angle of observation. In particular at shorter wavelengths the detection efficiency may be further reduced by the low quantum efficiency of photon detectors. However, in our opinion the low detection efficiency is more than balanced by the unsurpassed resolution of the PES technique and the possibility of polarization measurements yielding information on the level of magnetic substate population.

The number of photons S detected in a period T relates to the emission cross section $\sigma_{em}(i \to f)$ for a transition between state i and f as follows:

$$\sigma_{em}(i \to f) = \frac{4\pi}{\omega K_q(\lambda)} \frac{qe}{IT} \frac{S}{\int_L n(z)dz} PF_{obs}^{-1}(i) \tag{5.7}$$

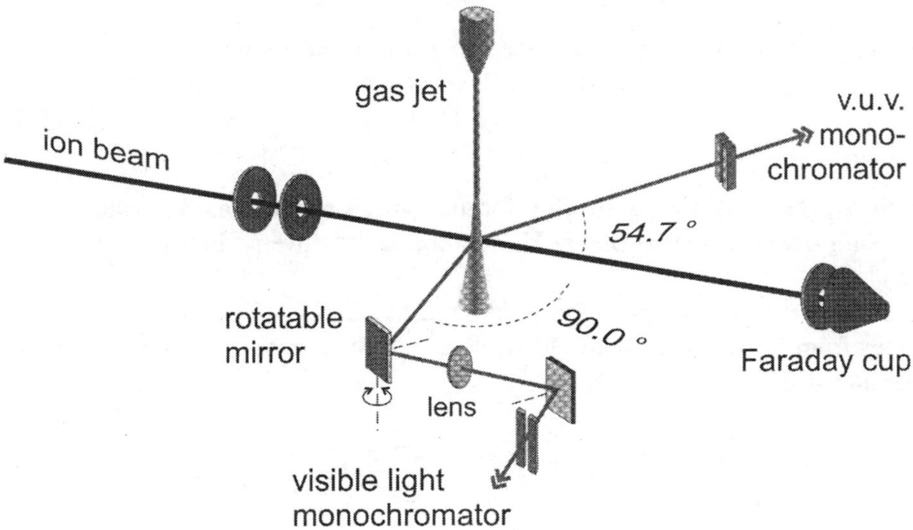

Figure 5.4. Lay-out of a typical Photon Emision Spectroscopy experiment.

where ω is the solid angle of observation, $K_q(\lambda)$ the quantum efficiency of the optical detection system at the wavelength λ of the transition between state i and f, q the charge state of the ions, e the elementary charge, I the electrical ion beam current, T the measuring time, $\int_L n(z)dz$ the neutral particle density in the observation region L, z the position along the ion beam, P a correction factor for polarization effects and F_{obs} the fraction of excited particles that decays within the observation region L of the spectrometer.

There are three factors in eq. (5.7) that need further consideration. Firstly, there is the group of parameters, ω, $K_q(\lambda)$, and $\int_L n(z)dz$, which are of importance for the absolute calibration of the system and usually thought to be the bottleneck of the PES method. Secondly, the parameters P and F_{obs} have to be discussed, since although sometimes complicating the interpretation of experimental results they can be used beneficially to extract extra information from the raw experimental data. Before treating these factors separately in sections 5.2.1, 5.2.2 and 5.2.3, the relations between state selective electron capture and excitation cross sections and measured emission cross sections are discussed.

If there is no appreciable contribution from cascades from higher states to the population of state i then the state selective cross section $\sigma(i)$ is directly calculated from the emission cross section via

$$\sigma(i) = B_{if}^{-1}\sigma_{em}(i \to f) \tag{5.8}$$

where the branching ratio B_{ij} of the i to f transition is given by

$$B_{if} = A_{if}\,(\Sigma_j A_{ij})^{-1} \tag{5.9}$$

with A_{if} the transition probability for the transition from state i to state f, and the sum over all final states j, $\Sigma_j A_{ij}$, equals the inverse lifetime of state i, $(\tau_i)^{-1}$.

If there is also a significant contribution to the population of state i by cascades from higher lying states k, the state selective cross section $\sigma(i)$ must be calculated from

$$\sigma(i) = B_{if}^{-1}\sigma_{em}(i \to f) - \sum_k B_{ki}\frac{F_{obs}(k,i)}{F_{obs}(i)}\sigma(k) \tag{5.10}$$

where $F_{obs}(i)$ and $F_{obs}(k,i)$ are the lifetime correction factors of the observed $i \to f$ transition and the $k \to i \to f$ cascade respectively. These lifetime factors are discussed in section 5.2.2 in more detail. But as a rule of thumb, it is justified to neglect the lifetime correction factor of a transition from a state j if the decay length of this state, i.e., the product of its lifetime and the velocity of the moving ion, is much smaller than the observation length of the spectrometer.

2.1 Calibration procedures

The systematic error in absolute emission cross sections determined via eq. (5.7) is governed by the parameters ω, $K_q(\lambda)$, and $\int_L n(z)dz$. Determining these parameters separately, their combined uncertainties range typically at best from 15 - 20% in the visible spectral range to values up to 30% at wavelengths in the Vacuum Ultra Violet (VUV).

Instead of the cumbersome separate determination of the above quantities it is often more efficient to determine their product *in situ* by measuring ion and electron impact processes with well known cross sections. Assuming for the moment that the transition rates are so high that $F_{obs} = 1$ and that the detection geometry is such that $P = 1$, then eq. (5.7) reduces to

$$\sigma_{em}(i \to f) = \frac{C}{K(\lambda)}\frac{q}{Q}\frac{S}{p_{eff}} \tag{5.11}$$

where C is the absolute calibration constant which includes the geometry of the photon detection system and the following conversion factors: accumulated charge, Q ($Q = IT$), to the number of projectile ions, relative sensitivity, $K(\lambda)$, to the quantum efficiency $K_q(\lambda)$, and the effective beam pressure, p_{eff}, to the target density $\int_L n(z)dz$.

Comparison of a measurement on a static target (known pressure) with one on a beam target yields a fair impression of an apparent, effective beam pressure, p_{eff}. An experimentally well-usable indicator for the effective pressure of a target beam is the pressure on the high-pressure side of the capillary from which the target beam effuses.

In the visible spectral range ($\sim$ 300 - 700 nm), the relative sensitivity can be measured by placing inside the collision chamber a tungsten-ribbon lamp [24] or a quartz-iodine lamp [25] of which the wavelength dependent emissivities are well known. In comparison to the experiments the lamps are extremely intensive light sources, so care has to be taken of straylight and pile-up effects. Especially at wavelengths below 350 nm where the lamp intensities, in particular of the tungsten-ribbon lamp, are very low relative to the total intensity of the light source, straylight may dominate the signals. For a procedure to determine and correct for the straylight contribution see for example Winter and Bloemen [26]. Normally the count rates in a PES experiment are well below 1 kHz, so most of the (photon multiplier) electronics is not designed for count rates exceeding 1 MHz. The latter is a count rate easily accessible with lamps, which may lead to pile-up in the detector electronics. To circumvent such pile-up problems neutral density filters may be used to reduce the light intensity around its maximum. In this way the relative wavelength-dependent sensitivity of the whole spectrometer system can be calibrated with an uncertainty of approximately 5%.

Thereafter the relative calibration curve, $K(\lambda)$, can be put on an absolute scale by using known cross sections for line emission in the visible induced by electron impact. Such benchmark cross sections are the ones measured by Van Zyl *et al.* [27] for excitation of the HeI($4s^1S \rightarrow 3p^1P$), HeI($5s^1S \rightarrow 3p^1P$) and HeI($6s^1S \rightarrow 3p^1P$) transitions at 505, 444 and 417 nm by the impact of 500 eV electrons on He. Furthermore the emission cross sections determined by Möhlmann *et al.* [28] for Balmer-β (486 nm) and Balmer-γ (434 nm) excitation by 400 eV electron impact on molecular hydrogen may be used.

Leaving synchrotons out off consideration, at wavelengths $\ll$300 nm there are no light sources which can be used for *in situ* determination of the relative sensitivity of the spectrometers. Here the sensitivity is determined from interpolation between absolute sensitivity points obtained by collisional excitation of atoms and molecules. Detailed discussions regarding the quality of electron impact cross sections can be found elsewhere e.g. refs. [29]. Using such electron impact processes one can calibrate down to about 30 à 50 nm. But the strongest emission lines of low-Z multiply charged ions appear at even shorter wavelengths. To extend down to these wavelengths one can use branching ratios in Li-like ions as N^{4+}, O^{5+} and F^{6+}. For these ions the $4l \rightarrow 3l'$ transitions radiate between 35 and 65 nm, while the $4l \rightarrow 2l'$ transitions yield photons with wavelengths between 8 and 20 nm.

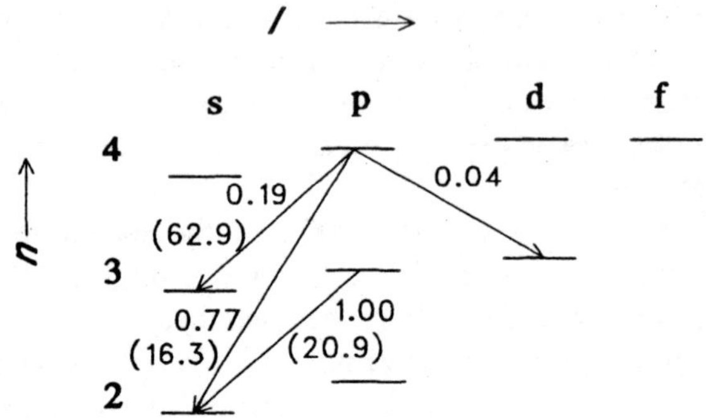

Figure 5.5. Part of the NV decay scheme, indicated are the relevant branching ratios and wavelengths in nm (in brackets).

Because the method is not as trivial as it may sound, it will be illustrated for decay from $N^{4+}(4p)$. The relevant wavelengths and branching ratios are given in figure 5.5. From this figure it is seen that the sensitivity at 16.3 nm $K(16.3)$ is related to the known sensitivity at 62.9 nm $K(62.9)$ by:

$$K(16.3) = \frac{0.19\ S(16.3)}{0.77\ S(62.9)}\ K(62.9) \tag{5.12}$$

where S(16.3) and S(62.9) are the signal rates at 16.3 and 62.9 nm, respectively. Note that the signal at 62.9 nm includes the third order of the $3p \rightarrow 2s$ transition ($3 \times 20.93 = 62.8$ nm). The partial blending of first order lines by higher order components of other lines is encountered in almost all Li-like low-Z ions. Since the excited N^{4+} ions are produced in charge changing collisions the excited state population can be steered by choosing the appropriate targets. Collisions of N^{5+} on H_2 populate both $3l$ and $4l$ states while collisions on He populate the $3l$ level only [10]. So from collisions on He we can deduce the strength of the third order of the $3p \rightarrow 2s$ line, allowing us to subtract its contribution to the $4p \rightarrow 3s$ when performing the branching ratio measurements on the N^{5+} - H_2 system. At typical impact energies of a couple of keV/amu the third order contribution is about 50% of S(62.9).

The result and the accuracy of such a calibration procedure can be assessed from figure 5.6 which shows a typical sensitivity curve of our VUV spectrometer. Note the branching ratio point at 53.7 nm, from the HeI($3p \rightarrow 1s$) transition which is connected to the HeI($3p \rightarrow 2s$) line at 501.6 nm. So it links the calibration of a VUV spectrometer to a visible light spectrometer.

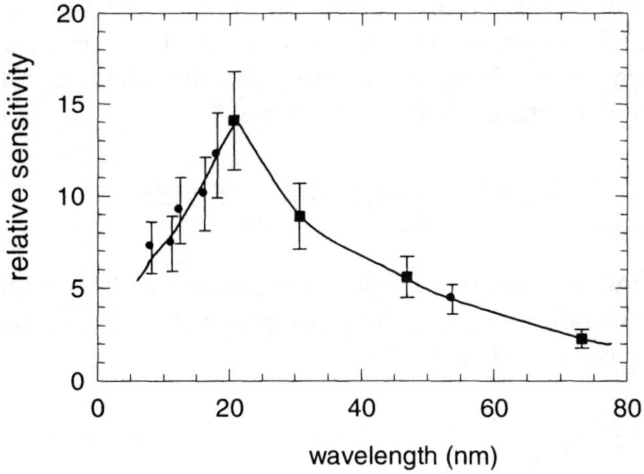

Figure 5.6. Example of an "absolute" calibration curve of a spectrometer for vacuum ultraviolet radiation. The squares stem from direct sensitivity measurements while the circles are obtained by means of the branching ratio method, see text.

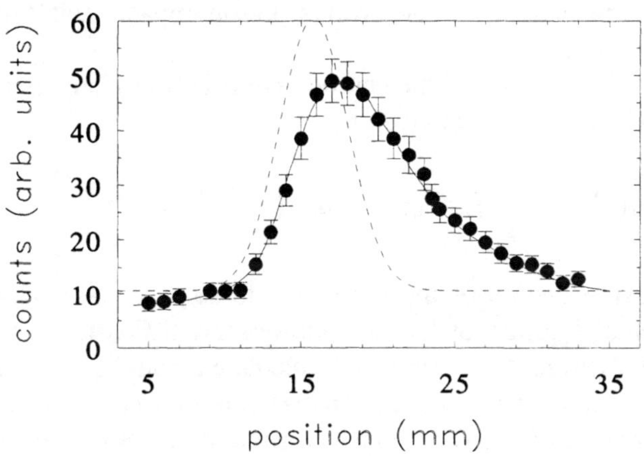

Figure 5.7. Photon emission of the OIII ($2p3p\ ^1P \rightarrow 2p3s\ ^1P^o$) transition along the ion beam axis. The dashed curve indicates the density of the target beam.

2.2 Lifetime effects

The lifetime correction factor F_{obs} becomes of importance in the determination of emission cross sections (see eq. 5.7) if the lifetime τ_i of the upper state i of the transition from $i \rightarrow f$ is so long that a significant fraction of these states, produced within the observation region, decays further downstream the beam and outside the observation region, cf. fig 5.7. This loss is normally not compensated for by decay of excited particles produced upstream from the observation region. The factor $F_{obs}(i)$ describes the effective loss of signal within the observation region L and is given by:

$$F_{obs}(i) = \int_L P_i(z)dz \,/ \int_L T(z)dz \tag{5.13}$$

with P_i(z) the emission profile of state i as a function of the position z along the beam, T(z) the target density profile along the beam. The spatial emission profile P_i(z) is given by [31, 32]:

$$P_i(z) = \frac{1}{v\tau_i} \int_{z_0}^{z} T(z')\exp[-(z - z')/v\tau_i]dz' \tag{5.14}$$

with v the velocity of the ions and z_0 the position of the last beam-collimating diaphragm in front of the target. Except for a gas-cell experiment with a constant target density profile [30], eqs. 5.13 and 5.14 cannot be solved analytically. The value of $F_{obs}(i)$ depends heavily on an accurate knowledge of T(z). Target density profiles can be obtained by electron or proton impact excitation of non-resonant target lines [32–34].

The spatial emission profile of the cascade contribution from state k to the observed transition from $i \rightarrow f$ is given by

$$P_{ki}(z) = \frac{1}{v\tau_i} \int_{z_0}^{z} P_k(z')\exp[-(z - z')/v\tau_i]dz' \tag{5.15}$$

with P_k, the emission profile of the first cascade step, $k \rightarrow i$, as given by eq. 5.14. The actual fraction of the cascade observed $F_{obs}(k, i)$ is obtained from eq. 5.13 with P_i(z) replaced by P_{ki}(z) calculated from eq. 5.15. From the equations it is clear that the form of the spatial emission profile is defined by the decay length $v \times \tau_i$ of the state under consideration. So reversely it is in principle possible to determine the lifetime of a state from a measured spatial emission profile, as done in beam-foil experiments e.g ref. [35].

For example, from the emission profile of the OIII($2s^2 2p3p\,^1$P $\rightarrow 2s^2 2p3s\,^1$P°) transition shown in figure 5.7 we determined the lifetime of the state $2s^2 2p3p\,^1$P to be $\sim$9 ns. The theoretical, inverse transition rate of the aforementioned transition was at the time of our measurement (1987) tabulated to be 27 ns [36]. Later theoretical work [37] confirmed the older results for the OIII($2s^2 2p3p\,^1$P $\rightarrow$ $2s^2 2p3s\,^1$P°) transition. But it also showed that 'two-electron' transitions of the type $2s^2 2p3p\,^1$P $\rightarrow 2s2p^3\,^1$L° (observable in VUV spectra) are important decay channels of the OIII($2s^2 2p3p\,^1$P) state, effectively reducing the lifetime

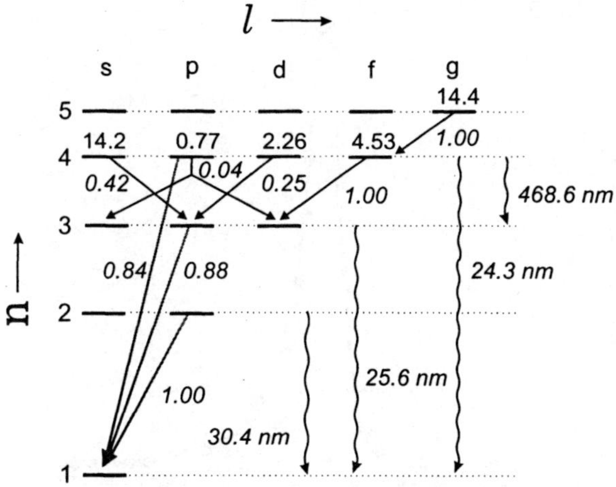

Figure 5.8. HeII level scheme with relevant transition wavelengths and branching ratios. The lifetimes (in ns) of the l and $5g$ states are given directly above these levels.

by more than a factor of 3 to almost 8 ns, in fair agreement with our experimental finding.

Emission profiles can also be used to disentangle contributions from different (degenerate) states to a specific emission line. This possibility, often used in collision experiments on static targets (gas cell and beam-foil experiments) was used in combination with a target beam to determine the cross sections for capture into the different H($3l$) states in H^+ - Li collisions [31] and by our group to determine $He^+(4l)$ capture cross sections resulting from collisions of He^{2+} on different targets [32, 34]. As an example let us discuss capture into $He^+(4l)$ states in collisions of He^{2+} on He. From the level scheme of hydrogenic He^+ ions (figure 5.8), it is clear that the different $4l \to 3l'$ transitions cannot be resolved spectroscopically. To extract the contributions of these transitions from the total HeII($n = 4 \to n = 3$) signals one can use the fact that the lifetimes of the HeII($4l$) states are different and relatively long, i.e. in the ns range. The longer the lifetime the further the emission extends downstream from the atomic target along the ion beam. The measured spatial HeII($n = 4 \to n = 3$) signal, S(z), is just equal to the sum of all the HeII($4 \to 3l'$) profiles, given by eq. 5.14 weighted by their branching ratios, $B_{4l,3l}$, and their cross sections, $\sigma(4l)$, and is thus given by, cf. eq. 5.11,

$$S(z) = \frac{K(\lambda)}{C}\frac{Q}{q}\sum_{l} B_{4l,3l'}P_{4l}(z)\sigma(4l) \qquad (5.16)$$

By measuring S(z) at some 30 positions along the beam axis (this is achieved by rotating the mirror in the light path of the visible light system, see figure

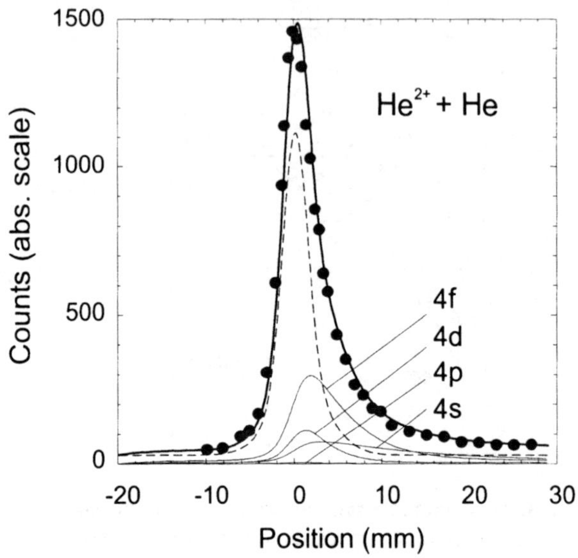

Figure 5.9. Spatial HeII($n = 4 \rightarrow n = 3$) emission profile for 4.5 keV/amu He^{2+} ions colliding with He. The broken line indicates the target-excitation component and the solid lines indicate the separate $He^{+}(4l)$ electron capture contributions.

5.4) it is possible to extract the state selective HeII($4l$) cross sections from the spatial emission profiles by a weighted least squares fit.

Figure 5.9 shows as an example the spatial HeII($n = 4 \rightarrow n = 3$) emission profile for 4.5 keV/amu He^{2+} ions colliding with He. Note that already at this relatively low impact energy the maxima of the profiles of the four $He^{+}(4l)$ projectile states, indicated in figure 5.9 by the thin full curves, are shifted downstream. An energy of 4.5 keV/amu corresponds to a velocity of 0.42 a.u., which is almost 1 mm/ns. The projectiles' emission profiles alone can never reproduce the measured profile which exhibits a maximum at $z = 0$, coinciding with the target center. This implies that for the symmetric He^{2+} - He system a fifth component is needed and is associated with $He^{+}(4l)$ target ions. Since the velocity of the target particles is negligibly small, the spatial emission profiles of all $He^{+}(4l)$ target-ion states collapse to the target profile, T(z) (eq. 5.14 with v approaching 0). The target ions are produced by two-electron processes in which one-electron is removed from the target and the other one is excited. Due to the symmetry of the He^{2+} - He system [38] the target and projectile-ion contributions are expected to be equal at low energies, i.e. energies well below 10 keV/amu as indeed found from deconvoluting spatial emission profiles [34].

The separate electron capture components derived from the deconvolution turn out to be sensitive to the exact shape of the neutral He target density,

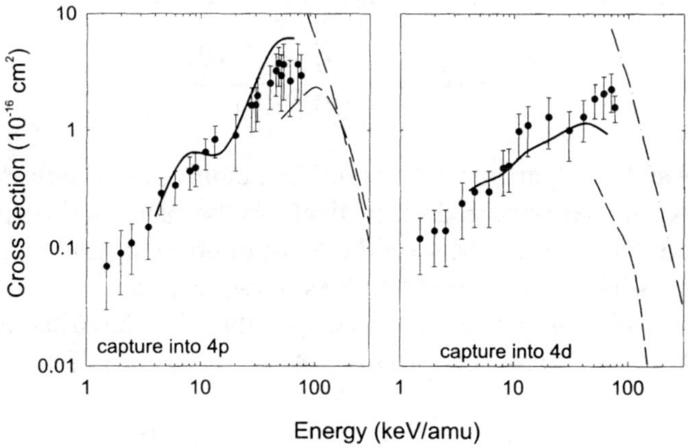

Figure 5.10. Cross sections for electron capture into $He^+(4p)$ and $He^+(4d)$ in collisions of He^{2+} on He. Experimental and CTMC (broken curve) results are from Folkerts *et al* [34]. The solid and dashed-dotted curves are the results of AO [38] and continuum distorted wave approximation [40] calculations, respectively.

especially the $4p$ scatters severely. This is due to its small contribution to the HeII($n = 4 \rightarrow n = 3$) transition (branching ratio of only 4%, see figure 5.8). But the cross section for capture into $He^+(4p)$ can be determined from the HeII($4p \rightarrow 1s$) transition in the VUV. This reduces the number of fit parameters by one and so it improves the quality of the fit. We could succesfully apply the method to H, Li and He targets [32–34], but not to Na because the Na target beam was too broad [39].

As an illustration of the state-selective results, figure 5.10 summarizes data for capture into $He^+(4p)$ and $He^+(4d)$ in collisions of He^{2+} on He. As the figure demonstrates, the largest differences between experiment and theory are observed around energies of 30 - 60 keV/amu, which is the high energy side of the applicability of AO calculations and the low-energy end of theoretical approaches suited for higher energies such as classical trajectory Monte-Carlo (CTMC) and continuum distorted wave (CDW) calculations. On the applied side, this is just the energy range of prime interest for diagnostic applications at large tokamaks such as JET [34]. The experiments bridge the gap between the predictions of the AO [38] and the higher energy calculations. More recent continuum distorted wave eikonal initial state calculations [41] show an improved agreement with the experiments.

2.3 Polarization effects

Photon emission by atoms or ions resulting from dipole transitions is in principle polarized [42]. The degree of polarization is defined as

$$\Pi(\theta) = \frac{I_{\|}(\theta) - I_{\perp}(\theta)}{I_{\|}(\theta) + I_{\perp}(\theta)} \tag{5.17}$$

where $I_{\|}(\theta)$ and $I_{\perp}(\theta)$ are the light intensities, emitted at an angle θ with respect to the ion beam, and polarized respectively in the plane and perpendicular to the plane defined by the ion beam and the line of observation. Polarized photon emission can affect the PES measurements in two ways:

(i) The radiation is not emitted isotropically. The intensity emitted at an angle θ is related to the polarization $\Pi_{\perp}$ at 90° by (see e.g. [43])

$$I(\theta) = I(90°)(1 - \Pi_{\perp}\cos^2\theta) \tag{5.18}$$

The polarization correction factor $P_1(\theta)$ by which the experimental data taken at a specific angle θ have to be multiplied with to account for the anisotropy of the emitted radiation is

$$P_1(\theta) = \frac{3 - \Pi_{\perp}}{3(1 - \Pi_{\perp}\cos^2\theta)} \tag{5.19}$$

From eq. 5.19 it is seen that if $\cos^2(\theta) = \frac{1}{3}$, P_1 is equal to 1 and therefore independent of the polarization. This occurs at the so-called magic angle of 54.7°. Nevertheless the majority of the PES experiments has been performed at 90° and must be corrected by the following factor

$$P_1(90°) = 1 - \frac{1}{3}\Pi_{\perp} \tag{5.20}$$

However, since $\Pi_{\perp}$ is normally not known an error of up to 33% ($-1 \leq \Pi_{\perp} \leq 1$) may be introduced by neglecting the possible anisotropy of the emitted radiation.

(ii) The spectrometers may have different sensitivities $K_{\|}(\lambda)$ and $K_{\perp}(\lambda)$ for parallel and perpendicularly polarized light respectively. These influences can be eliminated by tilting the spectrometers around the observation axis such that the entrance slit is inclined under 45° with the beam [44]. Otherwise a correction factor P_2 is needed to account for this instrumental effect. P_2 is given by [45]:

$$P_2 = \frac{(B + 1)(C + 1)}{2(BC + 1)} \tag{5.21}$$

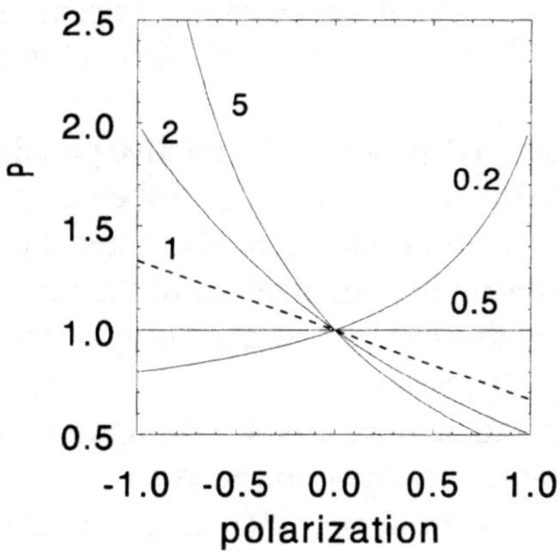

Figure 5.11. The total polarization correction factor P as a function of the degree of polarization for B=0.2, 0.5, 1 (dashed line), 2 and 5.

with

$$B = \frac{K_{\|}(\lambda)}{K_{\perp}(\lambda)} \qquad\qquad C = \frac{I_{\|}}{I_{\perp}} = \frac{1 + \Pi(\theta)}{1 - \Pi(\theta)} \tag{5.22}$$

In general B may deviate strongly from 1 and change rapidly with wavelength. For example for our visible light spectrometer equipped with a grating of 1800 lines/mm and blazed at 500 nm B increases from 0.1 at 400 nm to 3 at 600 nm. The total correction factor P appearing in the formula for the emission cross section, eq. 5.7, is given by the product of P_1 and P_2:

$$P = P_1 P_2 \tag{5.23}$$

To demonstrate that depending on the values of B polarization effects can be very appreciably, figure 5.11 shows the magnitude of the total correction factor P, for different sensitivity ratios B. For B=1 there is no instrumental effect and the total correction factor equals P_1, the correction factor for the anisotropic nature of the polarized photon emission (eq. 5.19). For values of $B \geq 1$ the effect of the anisotropy is amplified, while for $B \leq 1$ the instrumental effect counteracts the anisotropy effect. For the special case of B=0.5 the instrumental and anisotropy effects exactly balance each other and the measurements become independent of the polarization. For $B \ll 0.5$ the correction factor becomes large again. In conclusion a priori neglecting polarization effects may lead to significant errors. So if one hasn't the intention to measure polarization effects as described in the next section it is advised to perform the measurements under the *double magic angle*, i.e. with the spectrometer positioned under 54.7° with respect to the ion beam (P_1=1) and tilted 45° around the line of observation (P_2=1).

The degree of polarization depends on the relative population of the magnetic substates m of the decaying state i (see e.g. Sobelmann [42] and references therein). This implies that information on the magnetic substate population can be obtained by measuring the polarization of the radiation. The linear and circular degrees of polarization can be described fully by four parameters, the so-called Stokes parameters. In particular in the visible spectral range the complete set of Stokes parameters can be measured by the appropriate use of a $\frac{1}{4}\lambda$-plate and a linear polarizer [46]. In ion beam experiments, due to the cylindrical symmetry around the ion beam direction (z-axis), circular polarization is not possible and so the radiation can only be linearly polarized.

The linear polarization $\Pi_\perp$ of a transition between two states i and j depends on $I_{||}$ and $I_\perp$ (eq. 5.17) which on their turn depend on the magnetic substate populations, σ_{LSJM_j}, of the upper state i (L, S, J and M_j are the orbital angular momentum, the spin, the total angular momentum, and the total magnetic quantum number) and depend on the specific quantum numbers L', S', J' and M_j' of the final state f. Since polarization degrees are independent of the principal quantum number we will drop n from all state representations. Here we assume for the moment that the nuclear spin I is 0. If the spin is conserved during light emission then the intensities are given by [42, 47]:

$$I_{||} = C\sum_J\left\{\sum_{J'}\left[\frac{\mathbf{S}(LSJ,L'SJ')}{\sum_{J'}\mathbf{S}(LSJ,L'SJ')}\sum_{M_j}\sigma_{LSJM_j}(2J+1)\times \begin{pmatrix} J' & 1 & J \\ M_j & 0 & -M_j\end{pmatrix}^2\right]\right\} \tag{5.24}$$

$$I_\perp = \frac{1}{2}C\sum_J\left\{\sum_{J'}\left[\frac{\mathbf{S}(LSJ,L'SJ')}{\sum_{J'}\mathbf{S}(LSJ,L'SJ')}\sum_{M_j}\sigma_{LSJM_j}(2J+1)\times \left(\begin{pmatrix} J' & 1 & J \\ M_j-1 & 1 & -M_j\end{pmatrix}^2+\begin{pmatrix} J' & 1 & J \\ M_j+1 & -1 & -M_j\end{pmatrix}^2\right)\right]\right\} \tag{5.25}$$

where C is a constant depending on L and L' only, $\mathbf{S}(LSJ, L'SJ')$ is the line strength of the transition. From the 3J-symbols in eqs. 5.24 and 5.25 it is seen that $I_{||}$ and $I_\perp$ arise from transitions with $\Delta M = 0$ and $\Delta M = \pm 1$, respectively. The line strength $\mathbf{S}$ depends on J and J' as follows

$$\mathbf{S}(LSJ, L'SJ') \propto (2J+1)(2J'+1)\begin{Bmatrix} L & J & S \\ J' & L' & 1 \end{Bmatrix}^2 . \tag{5.26}$$

The sums over J and J' reflect the fact that for light ions transitions from different J states can not be resolved spectroscopically. What can be resolved are transitions from different LS states, except of course for the degenerate states in hydrogenic ions. The σ_{LSJM_j} distribution needed to calculate the intensities can be found by coupling LM_L and SM_S states and taking an equal population of the M_S magnetic substates. The latter is justified since normally captured and projectile-core electrons are not spin polarized. In this way the JM_J distribution relates to the LM_L population cross sections σ_{LM_L} as follows

$$\sigma_{LSJM_j} = \frac{1}{(2S+1)} \sum_{M_L+S_Z=M_J} |\langle LM_L SM_S|JM_J\rangle|^2 \sigma_{LM_L}. \tag{5.27}$$

Note that this implies that the J-state distribution does not depend on parameters of the collision. This is justified since the energy splitting between the different J states within one L multiplet is very small. In an one-electron system σ_{LM_L} is equal to the state selective capture cross section σ_{lm}. If the ion capturing the electron has a core with an angular momentum $L_c \neq 0$ then σ_{LM_L} is found from

$$\sigma_{LM_L} = \alpha_L \frac{1}{(2L_c+1)} \sum_{M_{L_c}+m=M_L} |\langle L_c M_{L_c} lm|LM_L\rangle|^2 \sigma_{lm} \tag{5.28}$$

where in contrast to the "J" case of eq. 5.27 α_L is a factor depending on experimental conditions, such as the collision velocity, due to the appreciable energy difference between nlm terms with different L quantum numbers. In the stepwise manner described above by eqs. 5.24 to 5.28 it is possible to calculate the polarization $\Pi_\perp$ of every dipole transition. In general extra degrees of freedom lead to a reduction of the maximum and minimum values of $\Pi_\perp$. This is shown in table 5.1 which gives the formulae for $\Pi_\perp$ and their maximum and minimum values $\Pi_\perp^{max}$ and $\Pi_\perp^{min}$ respectively for $p \to s$ transitions. The maximum value corresponds to the sole population of the $m = 0$ substate while the minimum value corresponds to situation that only the $|m| = l$ magnetic substates are populated.

Although instructive, the above stepwise calculations are rather lengthy and tedious, in particular if also a further step to hyperfine levels needs to be included. A compact formula for the polarization $\Pi_\perp$ can be derived directly by means of the density matrix formalism. By using the state multipoles $\langle T(l)^\dagger_{KQ}\rangle$ of the density matrix [48] the polarization is given by:

Table 5.1. Formulae for the polarization $\Pi_\perp$ and its maximum and minimum values for different $p \to s$ transitions. It is of note that for symmetry reasons the σ_{l,m_l} and $\sigma_{l,-m_l}$ are equal and therefore the results are expressed in terms of $\sigma_{|m_l|} = 0.5 \times (\sigma_{l,m_l} + \sigma_{l,-m_l})$ with the index l dropped for clarity.

transition	$\Pi_\perp$	$\Pi_\perp^{min}$	$\Pi_\perp^{max}$
$sp\,^1\mathrm{P^o} \to ss\,^1\mathrm{S}$	$\frac{\sigma_0 - \sigma_1}{\sigma_0 + \sigma_1}$	-1	1
$p\,^2\mathrm{P^o} \to s\,^2\mathrm{S}$	$\frac{3\sigma_0 - 3\sigma_1}{7\sigma_0 + 11\sigma_1}$	-0.27	0.43
$sp\,^3\mathrm{P^o} \to ss\,^3\mathrm{S}$	$\frac{15\sigma_0 - 15\sigma_1}{41\sigma_0 + 67\sigma_1}$	-0.22	0.37
$pp\,^3\mathrm{P} \to ps\,^3\mathrm{S^o}$	$\frac{15\sigma_0 - 15\sigma_1}{149\sigma_0 + 283\sigma_1}$	-0.05	0.10

$$\Pi_\perp(l \to l') = \frac{(-1)^{l+l'} \begin{Bmatrix} 1 & 1 & 2 \\ l & l & l' \end{Bmatrix} G(l)_2 \, (3/2)^{1/2} \, \langle T(l)^\dagger_{20} \rangle}{\frac{2}{3(2l+1)^{1/2}} \langle T(l)_{00} \rangle + (-1)^{l+l'} \begin{Bmatrix} 1 & 1 & 2 \\ l & l & l' \end{Bmatrix} G(l)_2 \frac{1}{6^{1/2}} \langle T(l)^\dagger_{20} \rangle} \tag{5.29}$$

with $\langle T(l)_{00} \rangle$ given by:

$$\langle T(l)_{00} \rangle = \frac{1}{(2l+1)^{1/2}} \sum_{m=-l}^{l} \sigma_{lm} \tag{5.30}$$

Apart from a constant factor $\langle T(l)_{00} \rangle$ represents the total population of the l-state and

$$\langle T(l)^\dagger_{20} \rangle = 5^{1/2} \sum_{m=-l}^{l} (-1)^{l-m} \begin{pmatrix} l & l & 2 \\ m & -m & 0 \end{pmatrix} \sigma_{lm} \tag{5.31}$$

is a measure of the charge cloud's alignment. The factor $G(\ell)_2$ accounts for depolarization effects due to fine- and hyperfine interactions. Its absolute values range from 1 (no depolarization) down to 0 in case of complete depolarization. If both fine- and hyperfine effects have to be accounted for and the observation time is long as compared to the lifetime of the state the depolarization factor for a (quasi-) one-electron atom becomes [48]:

$$G(l)_2 = \frac{1}{(2S+1)(2I+1)} \sum_{\substack{J\,J' \\ F\,F'}} (2J+1)(2J'+1)(2F+1)(2F'+1) \times$$

$$\begin{Bmatrix} J & F & I \\ F' & J' & 2 \end{Bmatrix}^2 \times$$
$$\begin{Bmatrix} l & J & S \\ J' & l & 2 \end{Bmatrix}^2 \frac{\gamma_l^2}{\gamma_l^2 + [(E_{J,F} - E_{J',F'})/\hbar]^2}. \qquad (5.32)$$

$\gamma_l = (2\pi\tau_l)^{-1}$ is the radiation line width of a given nl-level and $E_{J,F} - E_{J',F'}$ are its fine- and hyperfine splittings. The lm selective populations σ_{lm} in Eq. 5.30 and 5.31 are often provided by theoretical calculations. It is therefore convenient to use these cross sections to evaluate the above given equations for the transitions under study. Also minimum and maximum values for the polarizations can easily be derived.

From the last factor on the right-hand side of eq. 5.32 it is clear that there will be a strong depolarization ($G(l)_2$ towards 0) if the (hyper-) fine splittings are large compared to the linewidth γ_l. As an example we consider a $p\,^2\mathrm{P}^\circ \rightarrow s\,^2\mathrm{S}$ transition. From table 5.1 it is seen that for cases where $\gamma_l \gg E_{J,F} - E_{J',F'}$ the validity range of the polarization extends from -0.27 to 0.43. This holds for example for the HeII($4p \rightarrow 3s$) transition. Considering the NaI(3p→3s) transition it is found that the splittings range from 16 to 119 MHz, while the radiation width is only 10 MHz. Therefore, a strong depolarization due to hyperfine couplings is expected. Inserting all fine- and hyperfine-splittings $E_{J,F} - E_{J',F'}$ into eq. 5.32, the mininum and maximum polarization degrees are found to be reduced to [49] -0.08 and 0.14. A reduction by more than a factor of 3.

The number of PES experiments exploiting polarization degrees to obtain information on magnetic substate distributions is very limited, e.g. [50–54]. The central finding of all these keV/amu experiments is the increase of the polarization with impact energy up to the maximum degree of polarization. This implies that the $m = 0$ magnetic substate becomes predominantly populated, which means that the electronic charge cloud gets aligned along the ion beam axis.

In conclusion PES is an experimental technique that allows for studying on the most fundamental level (nlm state population) charge transfer and excitation processes in collisions of highly charged ions on neutral targets.

References

[1] R. Geller, *Electron Cyclotron Resonance Ion Sources and ECR Plasmas*, IOP Publishing, Bristol, (1996)

[2] E. D. Donets, 245 in: *The Pysics and Technology of Ion Sources*, ed. I.G. Brown, Berkeley, (1988)

[3] F.G. Wilkie, R.W. McCullough and H.B. Gilbody, J.Phys. B: At. Mol. Phys. **19**, 239 (1986)

[4] U. Jellen-Wutte, J. Schweinzer, W. Vanek and HP. Winter, J. Phys. B: At. Mol. Phys. **18**, L779 (1985)

[5] N. Bohr and J. Lindhard, K. Dan.Vidensk. Selsk. Mat. Fys. Medd. **28**, 7 (1954)

[6] A. Bárány, G. Astner, H. Cederquist, H. Danared, S.Huldt, P. Hvelplund, A. Johnson, H. Knudsen, L. Liljeby and K.-G.Rensfelt, Nucl. Instrum. Meth. B **9**, 379 (1985)

[7] A. Niehaus, J. Phys. B: At. Mol.Phys. **19**, 2925 (1986)

[8] R.E. Olson and A. Salop, Phys. Rev. A. **14**, 579 (1976)

[9] F.W. Meyer, A.M. Howald, C.C. Havener and R.A.Phaneuf, Phys. Rev. A. **32**, 3310 (1986)

[10] D. Dijkkamp, D. Ciric, E. Vlieg, A. de Boer and F.J. de Heer, J. Phys. B: At. Mol. Phys. **19**, 4763 (1985)

[11] H. Ryufuku, K. Sasaki, T. Watanabe, Phys. Rev. A. **21**, 745 (1980)

[12] R.C. Becker and A.D. MacKellar, J. Phys. B: At. Mol. Phys. **17**, 3923 (1984)

[13] R.K. Janev and HP Winter, Phys. Rep. **117**, 265 (1985)

[14] R. Hoekstra, D. Ciric, F.J. de Heer and R. Morgenstern, Phys. Scr. T. **28**, 81 (1989)

[15] W. Fritsch, Phys. Rev. A. **30**, 3324 (1984)

[16] W. Fritsch, J. Physique. Coll. **50**, 87 (1989)

[17] W. Fritsch, in *The Physics of Highly Charged Ions* ed. P. Richard, M. Stöckli, C.L. Cocke and C.D. Lin AIP Conf. Proc. **274**, 24 (1993)

[18] W. Fritsch and C.D. Lin, Phys. Rep. **202**, 1 (1991)

[19] A. Boileau, M. von Hellermann, L.D. Horton and H.P. Summers, Plasma Phys. Control. Fusion **31**, 779 (1989)

[20] R.C. Isler, Plasma Phys. Control. Fusion **36**, 171 (1994)

[21] D.E. Shemansky, in *Atomic Processes in Plasmas* ed. Y.-K. Kim and R.C. Elton AIP Conf. Proc. **206**, 163 (1989)

[22] C.M. Lisse, K. Dennerl, J. Englhauser, M. Harden, F.E. Marshall, M.J. Mumma, R. Petre, J.P. Pye, M.J. Ricketts, J. Schmitt, J.Trümper and R.G. West, Science **274**, 205 (1996)

[23] V.A. Krasnapolsky, M.J. Mumma, M. Abbott, B.C. Flynn, K.J. Meech, D.K. Yeomans, P.D. Feldman, C.B. Cosmovici, Science **277**, 1488 (1997)

[24] J. Van den Bos, G.J. Winter and F.J. de Heer, Physica **40**, 357 (1968)

[25] R. Stair, W.E. Schneider and J.K. Jackson, Appl. Opt. **11**, 1151 (1963)

[26] HP. Winter and E.W.P. Bloemen, J. Phys. E: Sci. Instr. **15**, 1002 (1982)

[27] B, Van Zyl, G.H. Dunn, G. Chamberlain and D.W.O. Heddle, Phys. Rev. A **22**, 1916 (1980)

[28] G.R. Möhlmann, F.J. de Heer and J. Los, 1977 Chem. Phys. **25**, 103

[29] J.M. Ajello, D.E. Shemansky, B. Franklin, J. Watkins, S. Srivastava, G.K. James, W.T. Sims, C.W. Hord, W. Pryor, W. McClintock, V. Argabright and D. Hall, Appl. Opt. **27**, 890 (1988); W. Jans, B. Möbius, M. Kühne, G. Ulm, A. Werner and K.-H. Schartner, Appl. Opt. **34**, 3671 (1995); P.J.M Van de Burgt, W.B. Westerveld, J.S. Risley, J. Phys. Chem. Ref Data. **18** 1757 (1989)

[30] E.W.P. Bloemen, D. Dijkkamp and F.J. de Heer, J. Phys. B: At. Mol. Phys. **15**, 3489 (1982)

[31] F. Aumayr, M. Fehringer and HP. Winter, J. Phys. B: At. Mol. Phys. **17**, 4185 (1984)

[32] R. Hoekstra, F.J. de Heer and R. Morgenstern, J. Phys. B: At. Mol. Opt. Phys. **24**, 4025 (1991)

[33] R. Hoekstra, E. Wolfrum, J.P.M. Beijers, F.J. de Heer, HP. Winter and R. Morgenstern, J. Phys. B: At. Mol. Opt. Phys. **25**, 2587 (1992)

[34] H.O. Folkerts, F.W. Bliek, L. Meng, R.E.Olson, R. Morgenstern, M. von Hellermann, H.P. Summers and R. Hoekstra, J. Phys. B: At. Mol. Opt. Phys. **27**, 3475 (1994)

[35] S. Bashkin, D. Burns, L.J. Curtis, L.J. Heroux, J. Macek, R. Marrus, I. MArtinson, I.A. Sellin, O. Sinanoglu, W. Whaling and W. Wiese *Beam foil spectorscopy* ed. S. Bashkin, Springer Verlag (1976)

[36] W.L. Wiese, M.W. Smith and B.M. Glennon, *Atomic transition probabilities*, US Govt Printing Office, Washington DC, (1966)

[37] D. Luo, A.K. Pradhan, H.E. Saraph, P.J. Storey and Yan Yu, J. Phys. B: At. Mol. Opt. Phys. **22**, 389 (1989)

[38] W. Fritsch, J. Phys. B: At. Mol. Opt. Phys. **27**, 3461 (1994)

[39] A.R. Schlatmann, R. Hoekstra, H.O. Folkerts and R. Morgenstern, J. Phys. B: At. Mol. Opt. Phys. **25**, 3155 (1992)

[40] D. Belkić and R. Gayet, private communication

[41] H.F. Busnengo, A.E. Martinez, R.D. Rivarola and H. Tawara, J. Phys. B: At. Mol. Opt. Phys. **30**, L805 (1997)

[42] I.I. Sobelmann, *Atomic Spectra and Radiative Transitions*, Springer, Berlin, (1979)

[43] B.L. Moiseiwitsch and S.J. Smith, Rev. Mod. Phys. **40**, 238 (1968)

[44] P.N. Clout and D.W.O. Heddle, J. Opt. Soc. of Am. **59**, 715 (1969)

[45] L. Wolterbeek Muller and F.J. de Heer, Physica **48**, 345 (1970)

[46] D. Clarke and J.F. Grainger, *Polarized Light and Optical Measurement* Pergamon Press, Oxford, (1971)

[47] I.C. Percival and M.J. Seaton, Phil. Trans. Roy. Soc. London A **251**, 113 (1958)

[48] K. Blum, *Density Matrix Theory and Applications* Plenum, New York, (1996)

[49] S. Schippers, P. Boduch, F.W. Bliek,R. Hoekstra, R. Morgenstern and R.E. Olson, J. Phys. B: At. Mol. Opt. Phys. **28**, 3271

[50] R. Baptist, J.J. Bonnet, G. Chauvet, J.P. Desclaux, S. Dousson and D. Hitz, 1984 J. Phys. B: At. Mol. Phys. **17**, L417 (1995)

[51] L.J. Lembo, K. Danzmann, Ch. Stoller, W.E. Meyerhof and T. Hänsch, Phys. Rev. A. **37**, 1141 (1988)

[52] R. Hoekstra, M.G. Suraud, F.J. de Heer and R. Morgenstern, J. Physique. Coll. **50**, 387 (1989)

[53] C. Laulhé, E. Jacquet, P. Boduch, M. Chantepie, G. Cremer, N. Gherardi, X. Husson, D. Lecler and J. Pascale, J. Phys. B: At. Mol. Opt. Phys. **30**, 1530 (1997)

[54] H. Tanuma, T. Hayakawa, C. Verzani, H. Kano, H. Watanabe, B.D. De-Paola, N. Kobayashi, J. Phys. B: At. Mol. Opt. Phys. **33**, 5091 (2000)

Chapter 6

LOW ENERGY ELECTRON CAPTURE MEASUREMENTS USING MERGED BEAMS

C. C. Havener
Physics Division, Oak Ridge National Laboratory, Oak Ridge, TN 37831
havenercc@ornl.gov

Abstract Low energy electron capture is an important process in plasmas where multicharged ions and neutral atoms or molecules can exist at eV/amu energies and below. At eV/amu collision energies there are no scaling laws to characterize the electron capture cross section and both theory and experiment are difficult. The capability of the merged-beams technique to address low energy electron capture is discussed.

Keywords: merged beams, low energy, electron capture, experiment

1. Introduction

Electron capture by a multicharged ion, X^{q+} with charge q, from a neutral atomic or molecular target, T is defined by

$$X^{q+} + T \rightarrow X^{(q-1)+}(n, l) + T^{+} \tag{6.1}$$

where one (or more) electron is transferred from the neutral target to the multicharged ion. (n, l) denotes the principal and angular quantum number, respectively, of the final (typically excited) state of the product ion. For low energy collisions, the single electron transfer process typically dominates, with a cross section for single capture that is typically large, on the order of 10^{-16} cm^2 – 10^{-14} cm^2. The electron capture process is important in many environments where multicharged ions and neutrals exist, e.g., in magnetically confined fusion plasmas, semiconductor processing, and photo-ionized nebulae and has received a significant amount of experimental and theoretical effort. For veloc-

F.J. Currell (ed.), The Physics of Multiply and Highly Charged Ions, Vol. 2, 193-217.

ities in the range $0.1 < v < 1$ a.u. (1 a.u. in velocity corresponds to a particle with energy of 24.97 keV/amu) the behavior of the total capture cross section (sum over (n, l) in eq. 6.1) has been successfully parameterized in terms of ionic charge, target binding energy, and collision velocity [1]. For lower collision velocities, less is known about electron capture by multicharged ions colliding with neutral atoms and molecules. The relative nuclear motion between collision partners is slow compared to the orbital motion of the active electrons in the system. Electrons of the temporary quasi-molecule formed in the collision have sufficient time to adjust to the changing interatomic field as the nuclei approach and separate. Fully quantum coupled-channel molecular-orbital calculations are considered the most accurate but are difficult to perform. The status and a review of current theory is addressed in another chapter.

Most of the experimental work on electron capture has been based on ion-beam gas-target methods using conventional ion sources which are typically operated at acceleration voltages above $\sim$ 1000eV. For this technique, a 1000 eV acceleration potential would result in a collision energy of $(q/m \times 1000)$ eV/amu where m is the mass (a.u.) of the accelerated ion of charge q. At lower energies the beam-gas technique becomes difficult due to the energy spread and emittance of the primary ion beam. Complete collection of the primary and product ions (or photons) is limited both by the divergence of the primary ion beam and the angular scattering [2] that takes place during the collision. Measurements with a H target are performed in a gas cell configuration using a thermal-dissociation atomic hydrogen target (e.g., [3]) or in a crossed-beams configuration using an effusive H beam from a radio-frequency (RF) discharge source (e.g., [4]). State-selective techniques include photon emission spectroscopy (e.g., [4]) and translational energy spectroscopy (e.g., [5, 6]). Special low energy ion sources incorporate a "cool" source of ions extracted from, e.g., a laser produced plasma [3, 7] or ions from recoils [8]. Such sources are able to produce ion beams typically as low as $q \times 10$ eV. However, beam-gas or crossed-beam techniques, which utilize these and other sources, still convert a significant portion of the beam energy into collision energy and are then limited to the energy spread, emittance, and intensity of the ion beams.

Special techniques are needed to measure electron capture cross sections at eV/amu collision energies. At the University of Nevada, Las Vegas, an ion trap technique is used to measure reaction rates (i.e. product of cross section times velocity integrated over velocity distribution) of stored ions interacting via electron capture with target gas [9]. With the introduction of an RF ion beam guiding technique to multicharged ions [10], conventional beam-gas techniques have been extended to eV/amu energies. At the Tokyo Metropolitan University (TMU) in Japan, a mini-EBIS multicharged ion source [11] in combination with an RF ion beam guide allows total electron capture cross sections to be measured to energies less than an eV/amu [12]. At KVI Groningen, The Netherlands, a recently constructed cross-beams apparatus uses an RF ion beam guide and decelerated ions from an ECR ion source to measure state-selective electron capture cross sections using photon spectroscopy [13] down to 5eV/amu. At Oak Ridge National Laboratory (ORNL) the merged-beams technique is currently being used at to study electron capture by multicharged ions from neutral H or D from 0.02 eV/amu to 5000 eV/amu [14]. In the merged-beams technique, relatively fast (keV) beams are merged producing low energy collisions in a moving center-of-mass frame.

The use of merged beams to access very low interaction energies has a long history. Almost forty years ago it was realized [15, 16] that collisions in the center-of-mass frame could be performed with an energy resolution significantly lower than the energy spread of each beam. The experimental feasibility of the technique was first demonstrated [17] at the Space Science Laboratory, General Dynamics/Convair, San Diego, California, where the cross section for the symmetric charge exchange of Ar^+ + Ar at a 5 eV/amu collision energy was successfully measured. With its two ion sources needed for operation along with a de-merging magnet, retarding grid assembly, hemispherical analyzer, and a beam chopping scheme to separate signal from noise, the technique was found to be technically challenging. Another early merged-beams apparatus [18] was developed to look at symmetric resonant electron capture in H^+ + H or D^+ + D collisions. A clever "overtaking" beam technique was used in which a fraction of the ion beam was first neutralized by a charge exchange cell and then the remaining ions electrostatically accelerated in order to overtake the neutral beam, producing collisions in the energy range between 2.5 and 50 eV/amu. From the success of these early experiments, the basic merged-beams technique showed promise. There was intense interest in the collision physics made possible by merged beams and several apparati were developed mostly during the late 1960s and 1970s which studied collisions between atomic or molecular ions and neutrals (see references in [14]. Merged-beam technology developed and measurements were performed with excited state beams, with detailed energy and angular scattering analysis, and state-selective photon spectroscopy. The early experiments were technically rather difficult and were typically limited to collision systems with relatively large cross sections using beams with low charge states. At Louvain-la-Neuve, Belgium, the development of the merged-beams technique continued and includes recent measurements with coincidence techniques [19] and with animated beams [20].

With the development of the Electron Cyclotron Resonance (ECR) ion source [21] intense beams of multicharged ions became available and a new opportunity arose to study heavy particle collisions in merged beams. A merged-beams apparatus was developed at ORNL [22] in conjunction with the ORNL ECR ion source [23] and is currently used to access collisions between multicharged ions and ground state H or D from 20 meV/amu to 5000 eV/amu. The merged-beams technique is unique in its ability to measure absolute cross sections for fundamental collisions. Reviews of the merged-beams technique can be found in the literature [14, 24, 25].

It is important to point out that the merged-beams technique is also used in electron-ion collisions. Early work on dissociative recombination of molecular ions was reported by Auerbach *et al* [26]. Development of the technique has continued with measurements of dielectronic recombination [27] and electron-impact excitation of ions [28, 29]. Recent dramatic improvements have been realized in going from a single-pass to a multi-pass merged-beams configuration at heavy-ion storage rings (see e.g., [25]. Because of beam expansion techniques [30] the low energy limit and resolution in an ion-storage ring experiment surpass the performance of a single-pass merged-beams configuration with electrons. These storage ring experiments have become a powerful tool to study electron-ion collisions. No multipass merged-beams configurations exist for heavy particle interactions.

2. Merged-Beams Technique

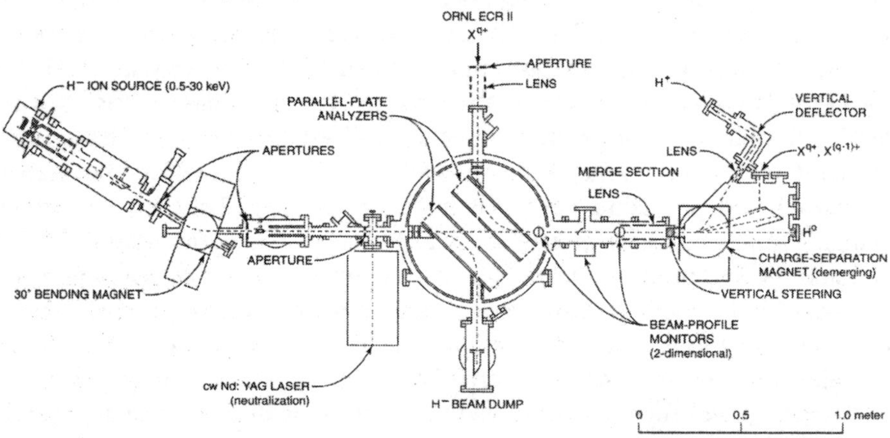

Figure 6.1. Schematic of the current ORNL ion-atom merged-beams apparatus.

Fig. 6.1 is a schematic of the current ORNL ion-atom merged-beams apparatus [14] which illustrates the merged-beams technique for the generic electron capture reaction

$$X^{q+} + H \rightarrow X^{(q-1)+} + H^{+} \tag{6.2}$$

for some multicharged ion, X^{q+} colliding with atomic H. An intense multicharged ion beam (1-10 μA) with energies up to $q \times 25$ keV is provided by the

ORNL 2ω CAPRICE ECR ion source [31, 32] and merged electrostatically with a neutral H (or D) beam. A ground state H beam of up to 100 nA is produced by photodetachment of a 2-3μA beam of H^- in a 100~W~cw~Nd:YAG laser cavity (λ= 1.06 μm) where up to a kW of power circulates. The 6-10 keV H^- beam is extracted from a duoplasmatron, analyzed by a 30^0 bending magnet, focused and steered horizontally and vertically so that subsequent to neutralization, a nearly parallel neutral beam will travel down the merge path, interact with the multicharged ion beam, and then be collected in the neutral detector. Production of a neutral beam by photodetachment has several advantages over the more common technique of using a charge exchange or detachment gas cell. These advantages, as described by Van Zyl *et al.* [33], include compatibility with ultrahigh vacuum, generation of a ground state beam, and the ability to perform *in situ* absolute calibration of the neutral beam detector. The neutral beam passes through an ionizer where an electric field of up to 30 kV/cm (see fig. 6.1) removes by ionization a significant fraction of the small excited state component (currently < 0.01%) present in the neutral H beam. This excited state component is produced by collisional detachment of H^- with background gas.

The merged beams interact in a field-free region for a distance of 47.1 cm, after which the primary beams are magnetically separated from each other and the "signal" H^+ ions (see fig. 6.1). The merge path is kept at ultrahigh vacuum conditions (1 x 10^{-10} Torr) to minimize the H^+ background generated by stripping of the H beam as it travels through the background gas. To maximize the angular acceptance of the apparatus for collecting the H^+products, a cylindrical Einzel lens is placed toward the end of the merge path. Due to a limited magnetic dispersion, the $X^{(q-1)+}$ product of the reaction is not measured separately but is collected with the X^{q+} beam in a large Faraday cup. Since only the H^+signal is measured, the apparatus actually measures electron loss, a sum of electron capture and ionization. Ionization of H at the low energies of interest here, though, is negligible compared to the electron capture process [1]. The neutral-beam intensity is measured by secondary-electron emission from a stainless-steel plate [22], and the signal H^+ is recorded by a 1" diameter channel electron multiplier (CEM) operated in pulse counting mode. The H^+ undergo a second (vertical) dispersion out of the plane of magnetic dispersion in order to attenuate transmission of photons to the CEM. The photons are produced by ion impact on the Faraday cup.

The following discussion is meant to illustrate several aspects of merged-beams and is not meant to completely describe the experimental procedure. Such details can be found elsewhere [14].

Figure 6.2. Schematic of two beams of particles with masses m_1 and m_2, traveling at velocities $\mathbf{v}_1$ and $\mathbf{v}_2$, intersecting at angle θ.

2.1 Kinematics of merged beams

Fig. 6.2 shows two intersecting beams with energy E_1, E_2, mass m_1, m_2 and traveling at velocities $\mathbf{v}_1$, $\mathbf{v}_2$ intersecting at angle θ. The interaction energy in the center-of-mass frame W is defined [17] to be

$$W = \frac{\mu}{2}\left|\mathbf{v_1} - \mathbf{v_2}\right|^2 \tag{6.3}$$

where μ is the reduced mass given by $\frac{m_1 m_2}{m_1+m_2}$. The relative collision energy in the center-of-mass frame, E_{rel}, is defined by

$$E_{rel} = \frac{W}{\mu} = \frac{1}{2}\left(\mathbf{v_1^2} - \mathbf{v_2^2} - 2\mathbf{v_1}\mathbf{v_2}\cos\left(\theta\right)\right) \tag{6.4}$$

which can be written in terms of the lab frame energies of the two beams, E_1, E_2, as

$$E_{rel} = \frac{E_1}{m_1} + \frac{E_2}{m_2} - 2\sqrt{\frac{E_1 E_2}{m_1 m_2}}\cos\left(\theta\right) \tag{6.5}$$

The relative collision energy, which has been defined as the interaction energy divided by the reduced mass, is only dependent on the square of the relative collision velocity. E_{rel} can be directly compared to the $E(q \times V)/m$ collision energy in beam-stationary target experiments (e.g., beam-gas cell), where the collision energy is provided by the ion of mass m, charge q, and accelerated to the source potential V.

An important advantage of the merged-beams technique is the ability to perform collisions over a wide range of energy (meV/amu – keV/amu) with good energy resolution even at the low collision energies. A particular collision energy can be obtained in different ways, with different velocity primary beams, with one beam faster or slower than the other, or with the use of isotopes (e.g., H replaced with D). For $E_1/m_1 = E_2/m_2$ and for perfectly merged beams with θ=0, eq. (6.5) gives the collision energy in the center-of-mass frame as zero. However, due to the fact that θ can not be made identically zero for all interacting particles and that the energy spreads of each beam are non-zero, there is an energy spread in the center-of-mass frame and hence a practical limit to the lowest energy achievable. By taking the partial derivative of E_{rel} with respect to E_1 (assuming θ=0) in eq. (6.5) one obtains an expression for $\Delta E_{rel}^{E_1}$ in terms of the energy spread of beam 1 (ΔE_1) given by

$$\Delta E_{rel}^{E_1} = \frac{m_2}{m_1 + m_2} \left(1 - \sqrt{\frac{m_1}{m_2} \frac{E_2}{E_1}} \right) \Delta E_1. \tag{6.6}$$

A similar expression for $\Delta E_{rel}^{E_2}$ in terms of the spread in the energy of beam 2 (ΔE_2) may be obtained by exchanging subscripts 1 and 2. The expression multiplying ΔE_1 is the energy deamplification factor. This factor is always less than unity, and decreases for lower center-of-mass collision energies E_{rel}.

The energy spread in the center-of-mass frame also depends on the angular spreads of the interacting beams which is given by [25]

$$\Delta E_{rel}^{\theta} \approx \sqrt{\frac{E_1 E_2}{m_1 m_2}} \theta^2 \quad . \tag{6.7}$$

The resulting energy spread in the center-of-mass frame is proportional to the product of the beam speeds in the laboratory frame and the square of the relative beam divergence angle. For optimal energy resolution the beams should be merged as parallel as possible at moderate speeds. As an illustration, the uncertainties in E_{rel} for O^{5+} + D measurements [22] are presented in table (6.1). Different collision energies are obtained by varying the O^{5+} beam energy while the D beam energy is held constant at 8 keV. The total uncertainty in collision energy is obtained by taking a quadrature sum of the uncertainties ΔE_{rel} due to ΔE_1, ΔE_2, and an appropriate $\Delta\theta$. Note that as the collision energy decreases, the uncertainties due to ΔE_1, ΔE_2 are increasingly deamplified. At the lowest energies, the spread in angle is the major source of collision energy uncertainty and determines the low energy limit of the apparatus.

Another important aspect of the merged-beams technique is the potentially large angular collection of the reaction products. The low-energy electron capture collisions under study are typically exoergic, and since both products are positively charged, significant angular scattering can occur [2] in the center-of-mass frame. Therefore, experiments must be designed with as large an angular collection of products as possible. In this respect the merged-beams technique is ideal since the angular scattering in the center-of-mass frame becomes significantly compressed in the laboratory frame, the frame where the product ions

Table 6.1. Uncertainties in E_{rel} due to uncertainties (spreads) in the beam energies and merge angles.

E_{rel} (eV/amu)	D E_1 (keV)	O^{5+} E_2 (keV)	ΔE_{rel} (eV/amu)			
			$\Delta E_1 =$ 6eV	$\Delta E_2 =$ 46eV	$\Delta\theta =$ 0.1^0	$\Delta\theta =$ 0.35^0
100	8.00	45.00	0.47	0.55	0.011	0.13
1	8.00	61.50	0.05	0.047	0.012	0.15
.01	8.00	63.33	0.005	0.004	0.012	0.15

are collected. This angular compression can be evaluated by examining the kinematic transformation in more detail.

After the collision the exoergicity (endoergicity) of the reaction, Q, is added to the relative energy of the colliding particles. Hence, the final energy in the center-of-mass frame is

$$W^f = \frac{1}{2}\mu\, {v_r^f}^2 = \frac{1}{2}\mu\, {v_r^i}^2 + Q\,. \tag{6.8}$$

This additional energy is shared between the two product ions in accordance with the conservation of linear momentum. After the collision, the final velocities v_1^f and v_2^f are given by

$$v_i^f = V_{cm} + \frac{\mu}{\mathrm{m}_i} v_r^f \tag{6.9}$$

Fig. 6.3 shows the vector (Newton) diagram for a collision where the lighter projectile m_1 is travelling faster than m_2. The faster of the two collision partners gains in velocity for an exoergic reaction, while the slower loses. The angle θ_{cm} and θ_{lab} is shown for the lighter projectile m_1.

It is evident that the larger the center-of-mass velocity V_{cm}, the greater the angular compression (θ_{cm}/θ_{lab}) in the laboratory frame. As can be seen from fig. 6.3, the laboratory scattering angle θ_{lab} is related to the angle of scattering in the center of mass, θ_{cm}, by

$$\tan(\theta_{lab}) = \frac{(\mu/\mathrm{m}_1)v_r^f \sin(\theta_{cm})}{V_{cm} + (\mu/\mathrm{m}_1)v_r^f \cos(\theta_{cm})} \tag{6.10}$$

Using eq. (6.5)-(6.7) and an estimate [34] of 2.3^0 for the angular acceptance of the apparatus in the laboratory frame, the angular acceptance in the center-of-mass frame was calculated for the collision system O^{5+} + D and shown in

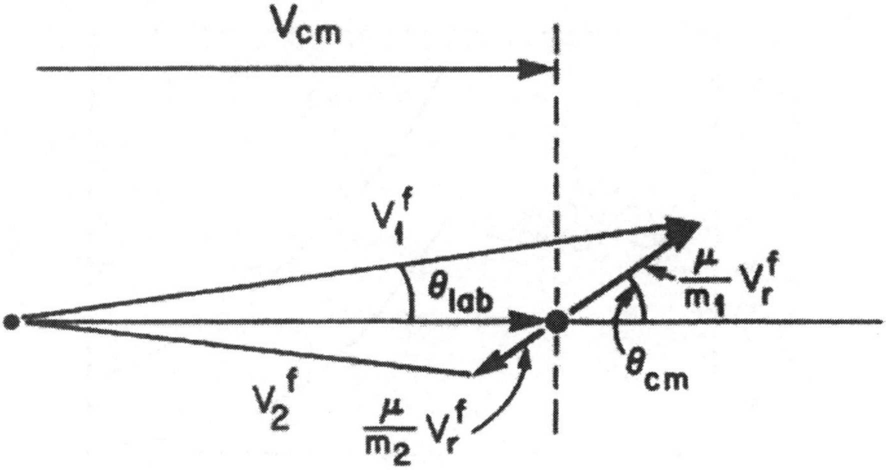

Figure 6.3. Vector (Newton) diagram after the collision. The symbols v_1^f and v_2^f correspond to the final velocities of the projectiles with masses m_1 and m_2, respectively, in the laboratory frame. v_r^f is their final relative velocity.

fig. 6.4. At these energies, capture takes place predominantly into the n=4 shell of O^{4+}, which, if one assumes a statistical population of the n=4 sublevels, has an average exoergicity of 8.2 eV. As can be seen in the figure for measurements with D at 8 keV, up to 90^0 scattering of the D^+ products in the center-of-mass frame can be collected at collision energies below 2 eV/amu. When the angular acceptance in the center-of-mass frame is $< 90^0$, the effect of angular scattering must be considered. Usually the only estimate readily available is that given by "half-Coulomb" Rutherford minimum scattering [2] which is shown in fig. 6.4 for the O^{5+} collision system. For this system, though, detailed differential cross-section calculations [35] were performed. They predict significantly larger scattering than the Rutherford "half-Coulomb" estimate, requiring an angular collection in the lab of at least 2.3^0 for collision energy between 1 and 10 eV/amu.

Another aspect of the merged-beams technique is that the exoergicity (endoergicity) Q of the reaction is amplified when transformed from the center-of-mass to the laboratory frame. This can also be illustrated using the O^{5+} + D collision system. Fig. 6.4 shows the maximum shift in energy for an 8-keV D beam as a function of E_{rel}, due to the 8.2 eV exoergicity. The maximum shift is calculated using eq. 6.9 and assuming that scattering is restricted to the forward angle ($\theta = 0^0$) for D faster than O^{5+} (upper curve) and to the backward angle ($\theta = 180^0$) for D slower (lower curve). Especially at low relative ener-

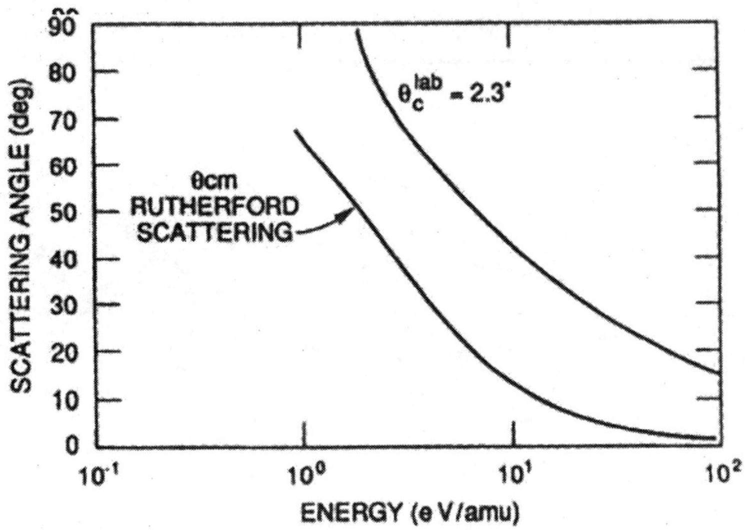

Figure 6.4. Estimated angular collection of D^+ in the center-of-mass frame for the O^{5+} + D collision system for an angular acceptance of the apparatus in the lab frame of $\theta_{cm} = 2.3^0$. Also shown is the "half-coulomb" Rutherford estimate (see text).

gies, these shifts can be large and, as was experimentally observed [22], can be correlated to the exoergicity of the reaction. This effect has been used not only to separate the signal from backgrounds [36] but also to extract detailed information about the reaction process [19, 37, 38].

This "amplification" effect was used at ORNL to measure state-selective cross sections and angular distributions for electron capture between Si^{4+} + D at eV/amu collision energies [39]. A multi-grid energy analyzer was installed after the signal was de-merged from the primary beam (see fig. 6.1) and was used to energy analyze the product D^+ from the reaction. Electron capture to the Si^{3+}(3d) $Q = 11.66$ and Si^{3+}(4s) $Q = 7.49$ were identified and estimates for the state-selective cross section were obtained. Unfortunately the large angular scattering observed complicated the analysis and prevented accurate determination of the state-selective cross sections.

2.2 Cross-section measurement

Absolute cross sections are determined by measuring the signal produced by the beam-beam interaction over the merged-length L. For ion-atom merged beams the electron capture cross section is determined from the measured quantities using

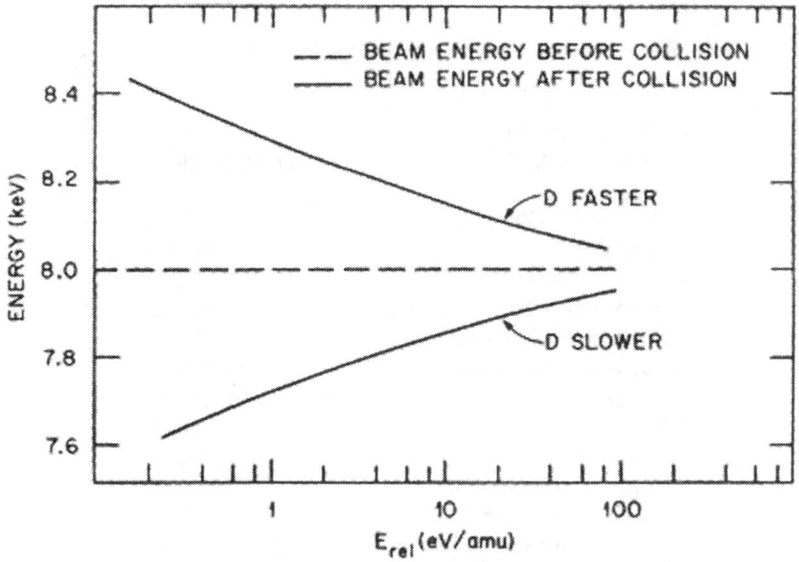

Figure 6.5. Resultant energy in the laboratory frame of the D^+ beam before and after the collision with O^{5+}. 8.2 eV is used as the exoergicity of the reaction.

$$\sigma = \frac{R}{\varepsilon} \frac{\gamma \, qe^2}{I_1 I_2} \frac{v_1 v_2}{v_r} \frac{1}{L\Omega} \tag{6.11}$$

where R, I_1, and I_2 correspond to the measured H^+ signal count rate, the effective current produced by neutral particles, and the ion current, respectively. The true signal rate is given by R/ε, with ε being the efficiency of the CEM and electronics for detecting H^+ (or D^+) at velocity v_1, and the true neutral "current" is I_1/γ, with γ being the measured effective secondary-electron emission coefficient from a simple stainless steel disc [22]. The quantity q is the charge of the ion, and e is the electronic charge. v_r is the relative velocity of the two beams. Ω is the overlap integral (effective form factor) [17] and is a measure of the spatial overlap of the two beams along the merged path L. Using ion-trajectory calculations [22], the merge-path length L is determined by considering positions along the merge path from which collision products (signal) can be transmitted to the detector. Table 6.2 gives typical operating parameters for the current ORNL ion-atom merged-beams apparatus.

The merged-beams technique is difficult, mainly due to the low target densities of the beams. For a merged-beams experiment the target thickness that one beam generates for the other beam can be defined [14] by

$$\Pi_{b-b} = \frac{I}{qe} \frac{v_r}{\mathrm{v}_1 \mathrm{v}_2} L\Omega \tag{6.12}$$

Table 6.2. Operating parameters for N^{2+} + D measurements [40] at $E_{rel} = 21.8$ eV/amu.

D^- @ 7.266 keV	1 μA
D @ 7.266 keV	47 nA (particle)
γ	1.11
N^{2+} @ 42.990 keV	7 μA
Signal	588 Hz
ε	0.97
Background from D stripping	16 kHz
Background from photons	400 Hz
Signal-to-noise ratio	3.7 x 10^{-2}
$v_1 v_2/v_r$	9.912 x 10^8 cm/sec
Ω	9.52 cm^{-2}
σ_{21}	2.3 x 10^{-16} cm^2
Time for 10% statistics	20 seconds
Merge path length, L	47.1 cm
Vacuum in merge section	1 - 2 x 10^{-10} Torr
Angular collection (lab frame)	2.3^0

For the N^{2+} + D operating parameters presented in Table 6.2, Π_{b-b} = 1 x 10^7 cm^{-2}. As one can see from the equation, Π_{b-b} is proportional to v_r, so as the collision energy decreases, the signal decreases as $\sqrt{E_{rel}}$. One can compare to a common beam-gas cell experiment where the target thickness Π is given by product of the target density and the effective length of the gas cell. For a typical 1 cm gas cell maintained at 1 mTorr, $\Pi = 3.3 \times 10^{13}$ cm $^{-2}$, over six orders-of-magnitude larger than the target thickness of the ion-atom merged-beams experiment.

There are background processes which further complicate the measurement of the beam-beam signal. When the neutral H or D beam strips on residual gas in the merge section ($H + H_2 \rightarrow H^+$), the H^+ is also detected by the CEM detector. To minimize this background, the merge path is kept at ultrahigh vacuum (10^{-10} Torr). Another source of background is photons created by the neutralization of multicharged ions as they are collected in the Faraday cup. Only after double dispersion of the H^+ products and gold blacking of possible reflecting surfaces could the photons reaching the CEM be reduced to an acceptable level ($<$ 100 Hz/μA). The signal/noise ratio generally decreases for lower energies, due mainly to the fact that the Π_{b-b} is proportional to v_r while the background rates essentially remain the same.

To separate the signal from background, a two-beam modulation technique [22] is used. Both beams are switched on and off at rates of 500 Hz - 2 kHz by applying 100-200 V between parallel deflecting plates placed near where the ion beams are extracted from their respective sources. No dependence on chopping frequency is observed except possibly at the very lowest frequency where background pressure modulation by the beams may have an effect on the signal and at the very highest frequencies where spurious effects due to switching the beams on and off may be significant. Spurious signals due to spatial beam intermodulation can be both serious and subtle [41]. Merged-beams experiments are particularly susceptible to such effects due to the extended interaction length. Local electric fields produced by beam space charge are enhanced for multiply charged ion beams and can affect the trajectories of the background and signal H^+. Any loss of background due to spatial displacement has serious consequences on the determination of the beam-beam signal. The demerging scheme used has, for this reason, a low dispersion to maximize the H^+ transmission to the detector. The chopping scheme incorporates various refinements [22] and is capable of separating a signal from background nearly 10^5 larger.

2.3 Overlap determination

In order to determine an accurate cross section from the measured signal, it is necessary to accurately measure the overlap of the two beams. This is certainly one of the technical challenges of the merged-beams technique. Rigorously, the overlap of the beams, Ω should be determined by integrating the product of the current densities j_1 and j_2 of the two beams at every point along the merged-path. Properly normalized Ω is defined as

$$\Omega = \frac{\iiint j_1(x,y,z) j_2(x,y,z) dxdydz}{\iiint j_1(x,y,z) dxdydz \iiint j_2(x,y,z) dxdydz} \tag{6.13}$$

where z is the direction of propagation of the beams and xy defines a plane perpendicular to the beams. Since Ω is nonzero only where beams overlap, Ω is called the overlap integral [17]. The particular method chosen to measure Ω is often selected on the basis of the spatial distribution and stability of the beams, suitability in vacuum, and whether beams can be selectively turned on/off. An early scheme [42] to determine the full three-dimensional profiles used square apertures formed by a pair of crossed slits scanned over the entire interaction region. Faster, two-dimensional profile monitoring [43] using a fluorescent screen and digitized video techniques at several (adjustable) positions along the merge path is currently in operation on an electron-ion merged-beams apparatus [29].

A simpler method is used on the ORNL apparatus where the heavy beams are not expected to significantly change trajectories along the merge path. Two orthogonal one-dimensional scans of each beam, I_i, give at $z = z_j$

$$I_i(y) = \int j_i(x, y, z = z_j) dx, \tag{6.14}$$

$$I_i(x) \quad = \quad \int j_i(x, y, z = z_j)dy \tag{6.15}$$

Assuming that the numerator of eq. (6.13) is separable in x and y and substituting what is actually measured (eqs. (6.14,6.15)) Ω(z) is given by

$$\Omega(z = z_j) = \frac{\int I_1(y)I_2(y)dy \int I_1(x)I_2(x)dx}{\int I_1(y)dy \int I_1(x)dx \int I_2(y)dy \int I_2(x)dx}. \tag{6.16}$$

The integrals in the denominator are just the total integrated beam currents. Currently the ORNL merged-beams apparatus uses both a transmission measurement as slits are translated into the beam (see fig. 6.6a) and a commercial rotating wire beam profile monitor (see fig. 6.6b). The beam overlap integral is then determined using a finite-sum approximation to eq. (6.16).

Fig. 6.6c shows two-dimensional beam profiles taken at three positions along the merge path for Si^{4+} + D measurements [44]. At each position there are horizontal and vertical profiles of the ion and neutral beams, the latter being the narrowest, with a FWHM of 1.5 to 2 mm. The Si^{4+} is generally larger, with FWHM of 2 to 3 mm, making the overlap of the two beams stable. The first and third profiles, labeled M1 and M2, are taken with respect to a 45^0 rotated frame of reference, thereby indicating the elliptical nature of the beam, if any. Fig. 6.6d shows the values of the three beam overlaps calculated from the measured profiles plotted as a function of position along the merged path. Values for the remainder of the merge path are estimated by linear interpolation or extrapolation as shown in the figure. An Einzel lens is located 51 cm into the merged path to aid in collection of the products. Any products created inside the lens will be born at a different potential and will not be transmitted through the magnetic and electrostatic dispersion elements to the detector (see fig. 6.1).

3. Merged-Beams Measurements

The ORNL ion-atom merged-beams experiment has been successful in providing benchmark total electron capture cross sections for a variety of multicharged ions (see table 6.3) in collisions with H or D. References for most measurements can be found at the web site www-cfadc.phy.ornl.gov or in the merged-beams reviews [14, 25].

Merged-beams measurements provide new insight into the low energy electron capture process. Resolving discrepancies between experiment and theory, the experimental and theoretical study [45] for N^{4+} + H (D) found that theoretical predictions of the cross section can be very sensitive to the quasi-molecular potentials used in the calculations. Similar results were found when merged-beams measurements [46] for B^{4+} + H (D) were compared to theory.

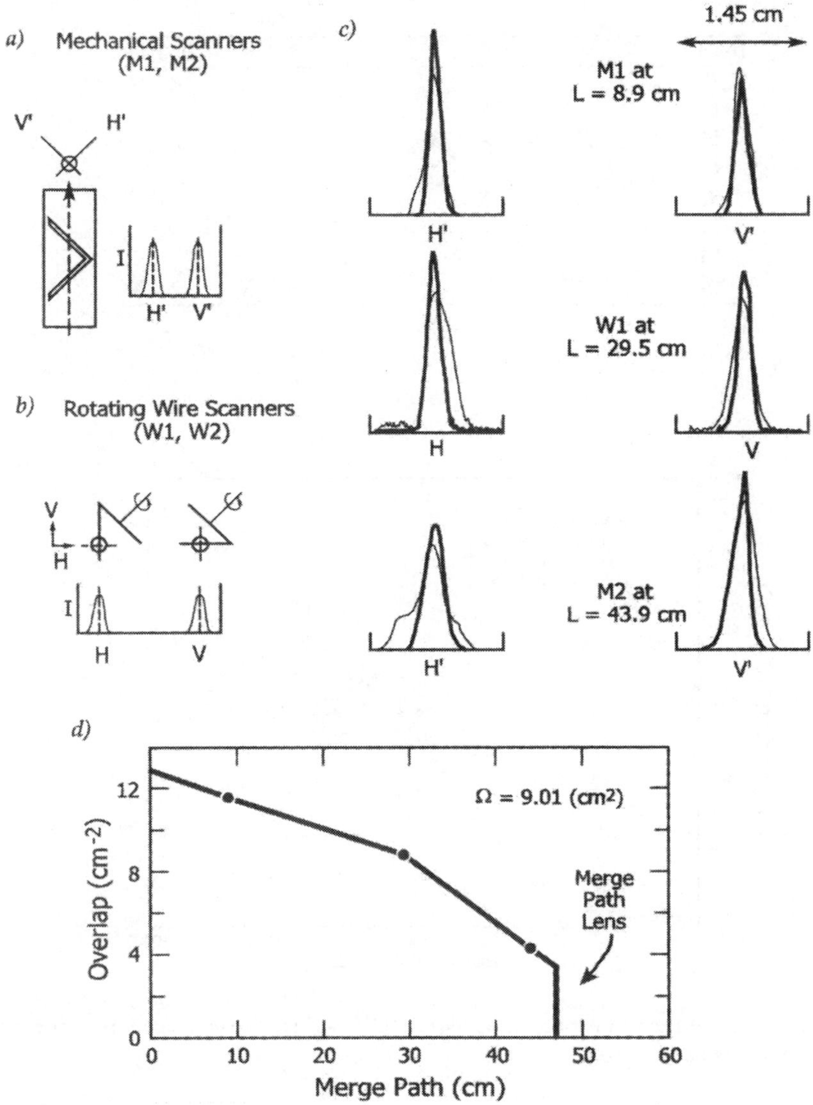

Figure 6.6. Two-dimensional beam profiles taken at the three positions along the merge path for the measurements for Si^{4+} + D [44]. In (c) the thick lines denote the D profiles while the thin lines denote the Si^{4+} profiles (see text).

Table 6.3. A list of the multicharged ions used in ORNL ion-atom merged-beams measurements. Total electron capture cross-sections have been measured for the ions with H or D.

Ion q =	1	2	3	4	5	6	7	8	9	10	11
B				X							
C	X		X	X							
N		X	X	X	X						
O			X	X	X						
Ne		X	X	X							
Si				X							
Cl							X				
Mo				X	X		X		X		X

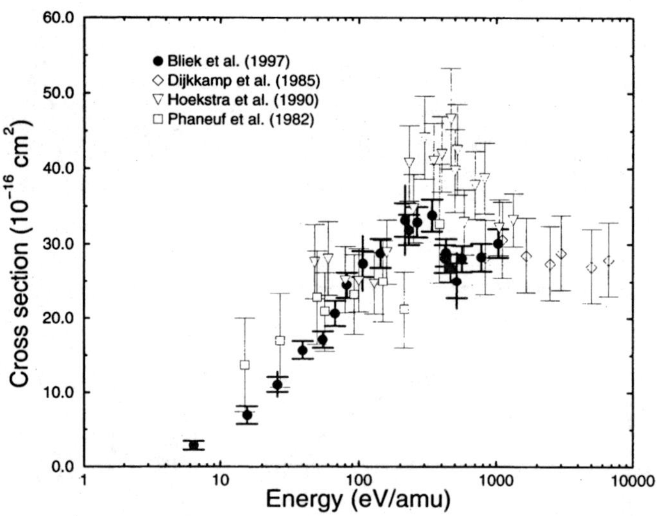

Figure 6.7. Merged-beams measurements [47] for C^{4+} + H (D) are compared to other experimental measurements (see text).

Merged-beams measurements [47] for C^{4+} + H (D) are shown in fig. 6.7. As can be seen in the figure, the merged-beams measurements have relatively low uncertainties compared to other (beam-gas) measurements in this energy range. For the merged-beams data error bars with caps denote the relative error at a 90% confidence level. The absolute error (a quadrature sum of the relative error and systematic error of 12% (see [22]) is denoted at a few energies by error bars

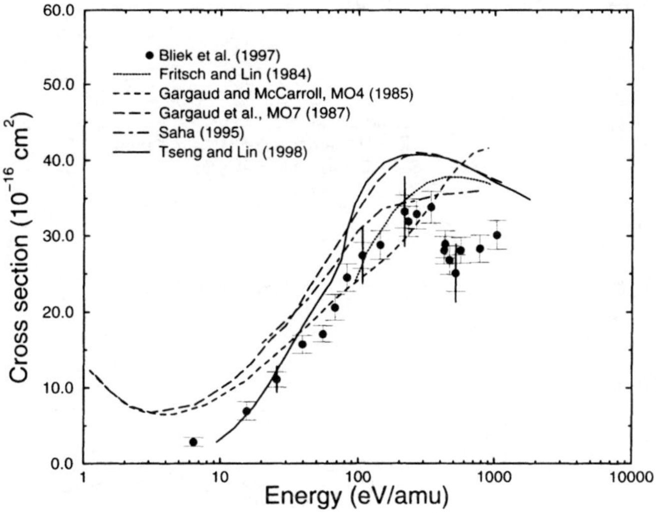

Figure 6.8. Merged-beams measurements for C^{4+} + H (D) [47] are compared with various theories (see text).

that extend beyond the caps. Excellent agreement is found at 1 keV/amu with the total-capture measurements of Dijkkamp *et al.* [4]. At in-between energies good agreement is found with Phaneuf *et al.* [49] and with Hoekstra *et al.* [48]. The total-electron-capture measurements of Hoekstra *et al.* were obtained by summing over the cross sections for individual ℓ-subshells. The merged-beams measurements are made with sufficient precision that structure was observed near the peak in the cross section. The state-selective measurements by Hoekstra *et al.* [48] support the position of the structure observed in the total electron capture cross section: It is in this energy range that the cross section for capture to the dominant Si^{3+}(3p) level was found to be sharply decreasing, while capture to the Si^{3+}(3s) was found to be increasing. By 1-keV/amu, capture to the 3s level becomes comparable to capture to 3p. Comparison of the merged-beams measurements for C^{4+} with various theories is shown in fig. 6.8. A full range of calculations have been applied to this system. They include fully quantal molecular orbital calculations of Gargaud and McCarroll [50] and Gargaud et al. [51], a semi-classical impact-parameter coupled state method by Saha [52] and atomic-orbital coupled-channel calculations by Fritsch and Lin [53]. The observation of the structure in the merged-beams measurements has led to a reevaluation [54] of theory but, as yet, no calculation of C^{4+} has been able to reproduce the structure. Slight structure was observed in the merged-beams measurements for C^{3+} + H (D) between 1000 eV/amu and 2000 eV/amu and was reproduced by theory (see [55]). More dramatic structure has been observed in measurements for N^{2+} + H (D) [40].

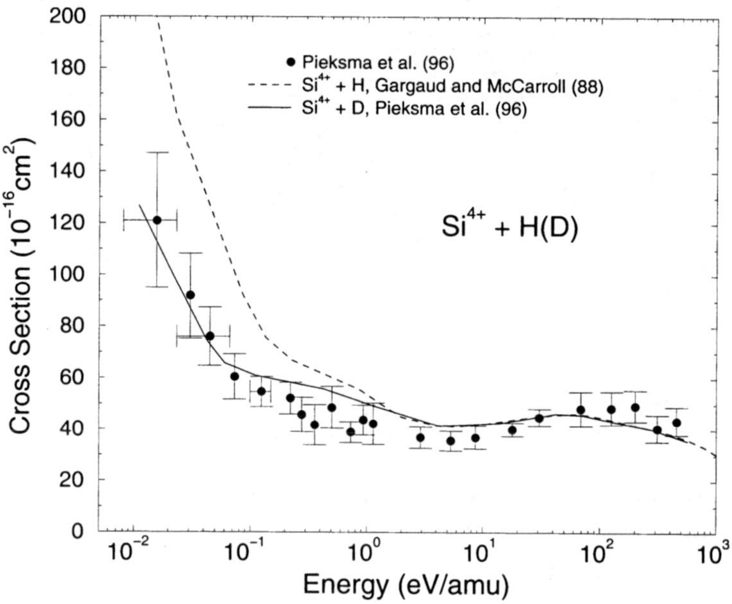

Figure 6.9. Merged-beams measurements for Si^{4+} + D [44] are compared to calculations for H and D (see text).

Merged-beams measurements for Si^{4+} + H (D) [44] are presented in fig. 6.9. These represent the first cross-section measurements in the thermal energy regime for collisions of multicharged ions with D atoms. Measurements with H could not be performed below 1 eV/amu due to the limited range of the acceleration voltage of the ECR ion source. The measurements are compared in fig. 6.9 to state-of-the-art molecular-orbital coupled-channel calculations [44, 56]. Excellent agreement is found above 1 eV/amu, where strong rotational coupling between capture to the Si^{3+}(3d) and Si^{3+}(4s) final state configurations exist. At energies below 1 eV/amu, the ion-induced dipole attraction between reactants is relatively strong and leads to trajectory effects [34, 57]. These trajectory effects result in the cross section increasing as $1/v$ toward lower energies and a strong isotope effect. As shown in the figure, the cross section for H is calculated to be a factor of 1.9 larger than that for D at an energy of 0.01 eV/amu. The factor of 1.9 is larger than the factor of 1.4 (the square root of the ratio of the reduced masses) predicted for the isotope effect by the Langevin model [58]. The failure of this classical model is not surprising since the model assumes the probability for capture to be unity, neglecting the quantum mechanical transition probability, which depends on, among other things, the radial velocity.

In fig. 6.10 the merged-beams measurements [59] for C^+ + D are shown. Note that the recommended data for fusion (solid line, [60]) is too high at low energies and is based on previous measurements using a thermal-dissociation atomic hydrogen target [61]. These measurements used a C^+ beam which was probably contaminated with metastables (long-lived excited state configurations). The merged-beams cross section measurements are over an order of magnitude lower at eV/amu energies and were taken with a C^+ beam essentially free of metastables. The metastable content of ion beams at ORNL can be estimated by observing electron-impact ionization below threshold. The ionization measurements were performed using the ORNL electron-ion crossed beams apparatus [62]. It is important to note that the electron capture process by ground state C^+ ion from H(D) is endothermic by 2.33 eV [59]. For endothermic reactions, the cross section decreases with decreasing energy, becoming exponentially small near threshold. ORNL molecular-orbital coupled-channel calculations [59] verify this lower cross section.

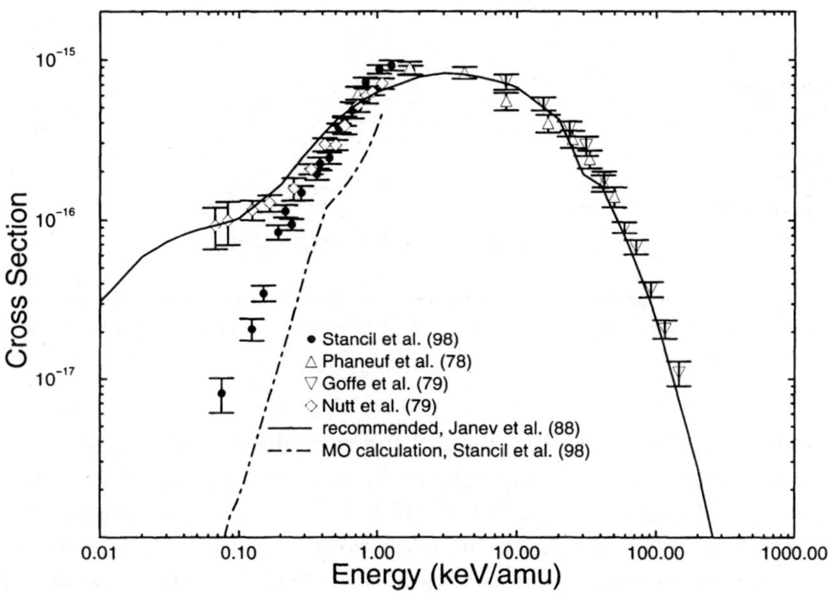

Figure 6.10. ORNL merged-beams measurements [59] for C^+ + D compared to other measurements and theory.

Recent measurements with the merged-beams apparatus involve heavier and higher charge state ions. The measurements [63] with Cl^{7+} + D are shown in fig. 6.11. The measured cross section shows a decrease toward eV/amu energies.

This is in contradiction to previous measurements for other 7+ ions, i.e., Ar^{7+} [64] and Fe^{7+} [65] (also shown if figure 6.11) and to a simple multichannel Landau-Zener (MCLZ) analysis [63] which predicts that the cross section is slightly increasing toward lower energies. The cross section energy behavior predicted by MCLZ theory reflects the expectation that for multielectron highly charged ions the cross section will be flat or increasing [57] with decreasing collision energy.

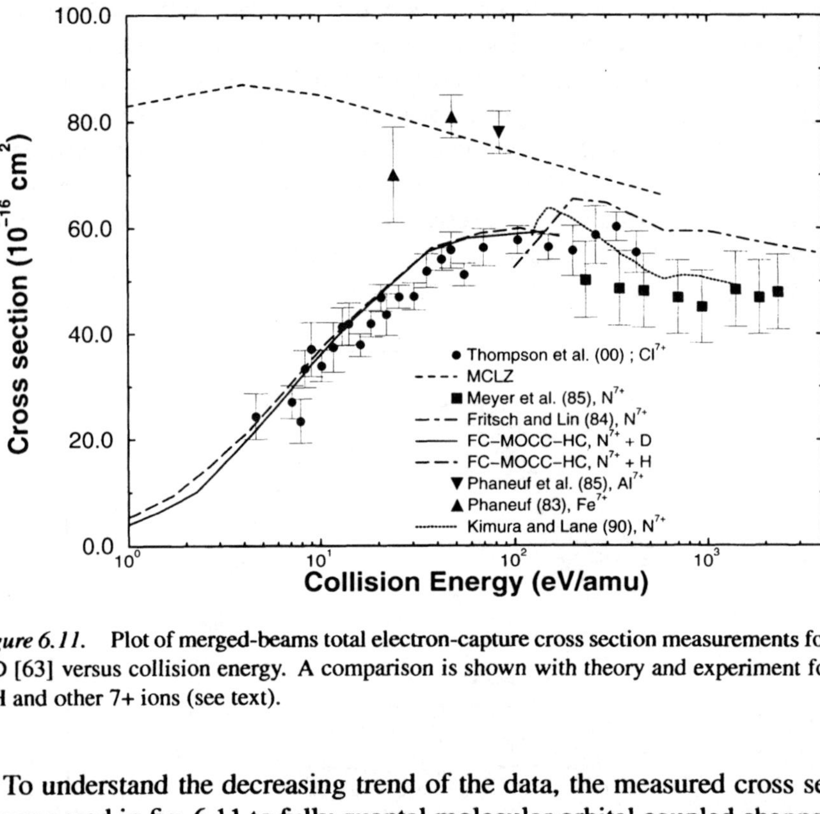

Figure 6.11. Plot of merged-beams total electron-capture cross section measurements for Cl^{7+} + D [63] versus collision energy. A comparison is shown with theory and experiment for N^{7+} + H and other 7+ ions (see text).

To understand the decreasing trend of the data, the measured cross section is compared in fig. 6.11 to fully quantal molecular orbital coupled channel hidden crossing calculations (FQ-MOCC-HQ) [63] for N^{7+} + D. Agreement is excellent, which seems to validate the assumption that at low energies electron capture occurs at large internuclear separations which does not probe the Cl^{7+} [Ne] core. The core may have an effect at larger energies, though, where the electron transfer process accesses smaller internuclear separations. The atomic orbital coupled-channel (AOCC) calculations of Fritsch and Lin [53] and the MOCC calculation of Kimura and Lane [66] for the N^{7+} + H system are in fair agreement with the plateau values of the current measurements of the Cl^{7+} + D system as well as previous experimental values of Meyer *et al.* [23] for N^{7+}. Although the calculations did not extend below 100 eV/amu, they suggest a decreasing cross section toward lower energies. The FQ-MOCC-HC calculation was also performed for N^{7+} + H (see fig. 6.11) which shows a slightly larger cross section for H due to trajectory effects. These trajectory effects were more pronounced at lower energies. Although not shown in the figure, the present calculation predicts a cross section that rises at energies below 1 eV/amu. The fact that the cross section does not remain flat toward decreasing energies shows that the actual quasi-molecular structure and associated dynamics remain important even for high charge state ions with multielectron cores.

4. Summary and Future of Merged Beams

With the intense multicharged ion beams made available by current ECR ion source technology, the ORNL ion-atom merged-beams apparatus is able to investigate total electron capture at energies not accessible by other techniques. The merged-beams measurements are absolute and are of sufficient precision to benchmark theory. The low energy behavior of the cross section has been observed to range from exponentially decreasing for endoergic reactions to increasing as $1/v$ for collision systems where the dynamics is characterized by the ion-induced dipole potential. Experimental measurements have resolved discrepancies between calculations whose difference is attributed to the quality of molecular potentials used. Structure observed in the total cross section can not always be reproduced by theory suggesting the need for continued experimental and theoretical investigations.

An upgrade in progress at ORNL will place the ECR ion source on a 250 kV high voltage platform. This will provide much higher energy multicharged ion beams and allow, e.g., merged-beams measurements to be performed with heavier atomic and molecular ions. Near-thermal collisions will be able to be performed with both H and D to allow a direct observation of the isotope effect. In addition, the negative ion source on the ion-atom merged-beams apparatus has been upgraded to include a new Cs negative ion sputter source which allows measurements with a wide variety of neutral atomic and molecular beams. Such neutral beams include Li, B, Na, Al, P, K, Ca, Cr, Fe, O_2, and CH_2. The merged-beams technique is the only technique available to explore collisions at eV/amu energies and below for these targets. Preliminary measurements for Ar^{2+} + Li have been reported [69]. For collisions of He^{2+} + Li, the large polarizability of Li should allow observations of predicted shape resonances [68] at eV/amu energies. Observation of these resonances would provide a new benchmark for theory. For vapor targets like Fe that can only be produced at high temperatures, the electron capture process is unexplored. For symmetric collisions like Fe^{q+} + Fe, the merged-beams technique can be used to explore the relative contribution of single to multiple resonant electron capture.

Acknowledgments

This work has been supported by the U.S. Department of Energy, Office of Fusion Energy Sciences and Office of Basic Energy Sciences, at Oak Ridge National Laboratory, which is managed by UT-Battelle, LLC, under Contract No. DE-AC05-00OR22725, and by the NASA SARA program under Work Order No. 10,060 with UT-Battelle, LLC.

References

[1] R. K. Janev, L. P. Presnyakov and V. P. Shevelko, *Physics of Highly Charged Ions*, Springer-Verlag, Berlin (1985)

[2] R. E. Olson and M. Kimura, J. Phys. B:At. Mol. Phys. **15** 4231 (1982)

[3] R. A. Phaneuf, Phys. Rev. A **24** 1138 (1981)

[4] D. Dijkkamp, D. Ciric, E. Vlieg, A. de Boer and F. J. de Heer, J. Phys. B: At. Mol. Phys. **18** 4763 (1985)

[5] F. G. Wilkie, R. W. McCullough and H. B. Gilbody, J. Phys. B: Atom. Mol. Phys. B **19** 239 (1986)

[6] P. Leputsch, D. Dumitriu, F. Aumayr and H. P. Winter, J. Phys. B: At. Mol. Opt. Phys. **30** 5009 (1997)

[7] J. Wang and V. H. S. Kwong, Rev. Sci. Instrum. **68** 3812 (1997)

[8] S. Yaltkaya, E. Y. Kamber and S. M. Ferguson, Phys. Rev. A **48** 382 (1993)

[9] V. H. S. Kwong and Z. Fang, Phys. Rev. Lett. **71** 4127 (1993)

[10] K. Okuno, J. Phys. Soc. Jpn. **55** 1504 (1986)

[11] K. Okuno, Jpn. J. Appl. Phys. **28** 1124 (1989)

[12] K. Okuno, K. Soejima and Y. Kaneko, Nuc. Instrum. Meth. In Phys. Res. **B53** 387 (1991)

[13] G. Lubinski, Z. Juhasz, R. Morgenstern and R. Hoekstra, Phys. B: At. Mol. Opt. Phys. **33** 5275 (2000)

[14] C. C. Havener, p.117-145 in *Accelerator-Based Atomic and Molecular Physics*, ed. Stephen M. Shafroth and James C. Austin,AIP Press, Woodbury, New York, (1997)

[15] S. M. Trujillo, R. H. Neynaber, L. Marino and E. W. Rothe, p.260 in *Proceedings of the IVth International Conference on the Physics of Electronic and Atomic Collisions* Science Bookcrafters, Inc., New York (1965)

[16] V. A. Belyaev, B. G. Brezhnev and E. M. Erastov, JETP Letters **3** 207 (1966)

[17] S. M. Trujillo, R. H. Neynaber and E. W. Rothe, Rev. Sci. Instrum. **37** 1655 (1966)

[18] V. A. Belyaev, B. G. Brezhnev and E. M. Erastov, JETP **25** 777 (1967)

[19] M. H. Cherkani, S. Szucs, H. Hus and F. Brouillard, J. Phys. B: At. Mol. Opt. Phys. **24** 2367 (1991)

[20] K. Olamba, S. Szucs, J. P. Chenu, N. El Arbi, and F. Brouillard, J. Phys. B: At. Mol. Phys. **29** 2837 (1996)

[21] J. Arianer and R. Geller Ann. Rev. Nucl. Part.Sci. **31** 19 (1981)

[22] C. C. Havener, M. S. Huq, H. F. Krause, P. A. Schultz and R. A. Phaneuf, Phys. Rev. A **39** 1725 (1989)

[23] F. W. Meyer, Nucl. Instrum. Methods Phys. Res., Sect. B **9** 532 (1985)

[24] R. H. Neynaber, p.476 in *Merging beams methods of experimental physics Atomic and Electron Physics. Part A. Atomic Interactions vol. 7* ed. B. Bederson and W L Fite Academic Press, New York (1968)

[25] R. A. Phaneuf, C. C. Havener, G. H. Dunn and A. Muller, Rep. Prog. Phys. **62** 1143 (1999)

[26] D. Auerbach *et al.*, At. Mol. Phys. B **10** 3797 (1977)

[27] S. Datz and P. F. Dittner, Z. Phys. D **10** 187 (1988)

[28] S. J. Smith, A. Chutjian, J. Mitroy, S. S. Tayal, R. J. W. Henry, K-F. Man, R. J. Mawhorter and I. D. Williams, Phys. Rev. A **48** 292 (1993)

[29] E. W. Bell *et al.*, Phys. Rev. A **49** 4585 (1994)

[30] J. R. Mowat *et al.*, Phys. Rev. Lett. **74** 50 (1995)

[31] B. Jacquot and M. Pontonnier, Nucl. Instrum. Methods **A295** 5 (1990)

[32] F. W. Meyer, M. E. Bannister, J. W. Hale, C. C. Havener, O. Woite, and Q. Yan, p.102 in *Proceedings of the 13th International Workshop on ECR Ion Sources*, ed. D.P. May and J. E. Ramirez Texas A&M University, College Station, TX (1997)

[33] B. Van Zyl, N. G. Utterback and R. C. Amme, Rev. Sci. Instrum. **47** 814 (1976)

[34] C. C. Havener, F. W. Meyer, and R. A. Phaneuf, p.381 in *Proceedings XVII Int. Conf. on the Physics of Atomic and Electronic Collisions, Brisbane* ed. W. R. MacGillivray *et al.* IOP Publishing Ltd. (1992)

[35] L. R. Andersson, M. Gargaud and R. McCarroll, J. Phys. B: At. Mol. Opt. Phys. **24** 2073 (1991)

[36] R. H. Neynaber and S. M. Trujillo, Phys. Rev. **167** 63 (1968)

[37] P. K. Rol, E. A. Entemann and K. L. Wendell, J. Chem. Phys. **61** 2050 (1974)

[38] W. R. Gentry, D. J. McClure, and C. H. Douglass, Rev. Sci. Instrum. **46** 367 (1975)

[39] W. Wu and C. C. Havener, J. Phys. B: At. Mol. Opt. Phys. **30** L213 (1997)

[40] , M. Pieksma, M. E. Bannister, W. Wu and C. C. Havener, Phys. Rev. A **55** 3526 (1997)

[41] P. O. Taylor, R. A. Phaneuf and G. H. Dunn, Phys. Rev. A **22** 435 (1980)

[42] S. K. Sethuraman, J. R. Gibson and J. L. Moruzzi, J. Phys. E: Sci. Instrum. **10** 14 (1977)

[43] J. L. Forand *et al.*, Rev. Sci. Instrum. **61** 3372 (1990)

[44] , M. Pieksma, M. Gargaud, R. McCarroll and C. C. Havener, Phys. Rev. A **54** R13 (1996)

[45] L. Folkerts, M. A. Haque and C. C. Havener, Phys. Rev. A **51** 3685 (1995)

[46] M. Pieksma and C. C. Havener, Phys. Rev. A **57** 1892 (1998)

[47] F. W. Bliek, R. Hoekstra, M. E. Bannister and C. C. Havener, Phys. Rev. A **56** 426 (1997)

[48] R. Hoekstra, J. P. M. Beijers, A. R. Schlatmann, R. Morgenstern and F. J. de Heer, Phys. Rev. A **41** 4800 (1990)

[49] R. A. Phaneuf, I. Alvares, F. W. Meyer and D. H. Crandall, Phys. Rev. A **26** 1892 (1982)

[50] M. Gargaud and R. McCarroll, J. Phys. B: At. Mol. Opt. Phys. **18** 463 (1985)

[51] M. Gargaud, R. McCarroll and P. Valiron, J. Phys. B: **20** 1555 (1987)

[52] B. C. Saha Phys. Rev. A **51** 5021 (1995)

[53] W. Fritsch and C. D. Lin, Phys. Rev. A **29** 3039 (1984)

[54] H. C. Tseng and C. D. Lin, Phys. Rev. A **58** 1966 (1998)

[55] C. C. Havener, A. Muller, P. A. Zeijlmans van Emmichover and R. A. Phaneuf, Phys. Rev. A **52** 2982 (1995)

[56] M. Gargaud and R. McCarroll, J. Phys. B: At. Mol. Opt. Phys. **21** 513 (1988)

[57] P. C. Stancil and B. Zygelman, Phys. Rev. Lett. **75** 1495 (1995)

[58] G. Giomousis and D. P. Stevenson, Phys. Rev. A **17** 534 (1958)

[59] , P. C. Stancil, J-P. Gu, C. C. Havener, P. S. Krstic, D. R. Schultz, M. Kimura, B. Zygelman, G. Hirsch, R. J. Buenker and M. E. Bannister, J. Phys. B: At. Mol. Opt. Phys. **31** 3647 (1998)

[60] R. K. Janev, R. A. Phaneuf, and H. T. Hunter, At. Data Nucl. Data Tables **40**, 240 (1988)

[61] W. L. Nutt, R. W. McCullough and H. B. Gilbody, J. Phys. B: At. Mol. Phys. **12** L157 (1979)

[62] M. E. Bannister, Phys. Rev. A **54** 1435 (1996)

[63] J. S. Thompson, A. M. Covington, P. S. Krstic, M. Pieksma, J. L. Shinpaugh, P. C. Stancil and C. C. Havener, Phys. Rev. A **63** 012717 (2000)

[64] R. A. Phaneuf, M. Kimura, S. Sato and R. E. Olson, Phys. Rev. A **31** 2914 (1985)

[65] R. A. Phaneuf, Phys. Rev. A **28** 1310 (1983)

[66] M. Kimura and N. F. Lane, Adv. At. Mol. Opt. Phys. **26** 79 (1990)

[67] F. W. Meyer, A. M. Howald, C. C. Havener and R. A. Phaneuf, Phys. Rev. A **32** 3310 (1985)

[68] M. Rittby, N. Elander, E. Brandas and A. Barany, J. Phys. B: At. Mol. Phys. **17** L677 (1984)

[69] R. Rejob and C. C. Havener, *to be published in the Bulletin of the American Physical Society*, DAMOP 2003

Chapter 7

APPLICATION OF THE BEAM GUIDE TECHNIQUE TO LOW ENERGY COLLISION EXPERIMENTS

K. Okuno

Department of Physics, Tokyo Metropolitan University, Minami-ohsawa1-1, Hachioji-shi, Tokyo 192-0397, Japan

please supply e-mail address

Abstract A combination of the ion beam guide technique pioneered by Teloy & Gerlich [2] and the Mini-EBIS developed by the author are described with particular reference to the application to low energy collision experiments of highly charged ions with atoms and molecules. In conventional beam experiments at low energies, the preparation of an intense ion beam can be interrupted by space charge, stray fields etc. Furthermore the perfect collection of product ions becomes difficult due to the dispersion of scattering angles after collisions. The ion beam guide is an effective tool to overcome such difficulties. In this chapter, the principal of the ion-beam guide technique and the analysis of the ionic motion in a multi-pole ion beam guide are given in section 7.2. A case study of development of a tandem octopole ion-beam guide is introduced in section 7.3 to illustrate its nature and performance. In section 7.4, the combination of the OPIG and the Mini-EBIS as applied to low energy cross section measurement is described and some features of charge transfer cross sections of highly charged ions in collision with atoms and molecules at low energies below 1 keV, where data are scarce, are introduced.

Keywords: octopole ion beam guide (OPIG), electron beam ion source (EBIS), highly charged ions, single electron transfer, double electron transfer, cross section, orbiting effects

1. Introduction

In the last two decades the advance of ion source devices has contributed to studies on charge transfer of highly charged ions in collision with atoms and molecules. Cross sections for such reactions are not only fundamentally

F.J. Currell (ed.), The Physics of Multiply and Highly Charged Ions, Vol. 2, 219-236.

important in atomic physics but also quite valuable for researches of fusion plasmas and astrophysical plasmas. At present, however, there is a scarcity of experimental and theoretical data in the low energy region below 1 keV/amu. Probably this is due to the experimental difficulties, especially in preparation of a stable beam of slow highly charged ions with a narrow energy spread.

In conventional beam experiments at low energies, preparation of an intense ion beam can be interrupted by space charge, stray fields etc. Also at low energies, perfect collection of product ions becomes difficult due to the dispersion of scattering angles after collisions. Historically, the flow tube technique and the drift tube technique have proved very effective for low energy collision experiments up to nearly 10eV. However, there is no non-reactive buffer gas for highly charged ions. The merging-beam technique developed by Trujillo et al [1] is excellent in principle for low energy collisions involving multiply charged ions, but it is technically difficult. On the other hands, Teloy and Gerlich [2] developed a beam-guided technique for the collision experiments in the thermal energy region. The technique has also been shown to be useful in studies at energies higher than the thermal region [3, 4]. The octopole ion-beam guide (OPIG) is an ideal tool for low energy collision experiments. Its structure is simple, its operation easy and its ion-confinement characteristics are good. It can be used as a linear ion-trap, a beam guide in a differentially pumped region and in a collision cell, a time-of-flight tube, and so on. Recently the beam guide technique has started to be used in various fields. Its remarkable performance, related to its ability to transport charged particles from a high pressure region to a high vacuum region without any loss of intensity, leads to an improvement in the detection efficiency of the analysis by more than one order of magnitude. For example it is being used in the fields of medicine, pollutants in the environment and unstable nuclear isotopes in high energy and nuclear physics [5].

The principal of the ion-beam guide technique and the analysis of the ionic motion in a multi-pole ion beam guide are given in section 7.2. To provide understanding of the nature and performance of the OPIG, a case study of a tandem OPIG is described in section 7.3. The results given in section 7.3 led to development of a combination technique of the OPIG and the Mini-EBIS as described in section 7.4. The use of this combination to measure charge transfer cross sections of highly charged ions at low energies below 1 keV is described.

2. The Ion-Beam Guide Technique and Ionic Motion in a Multi-Pole Ion-beam Guide

An ion-beam guide is composed of parallel 2N multi-poles (N$\geq$2) equally spaced around an axis. When oscillatory voltages $\pm V_{rf}\cos(\omega t)$ are supplied to poles alternatively in opposite phase, the potential created near the axis is represented by

$$U = (\frac{r}{a})^N \; V_{rf} \; \cos(N\theta)\cos(\omega t), \tag{7.1}$$

where r and θ are position parameters in cylindrical coordinates, a is the nearest distance from the axis to poles, V_{rf} is the amplitude of the R-F voltage with a frequency of $\omega/2\pi$ and t is time. For the case of N=4, a contour map of the oscillatory potential of eq. 7.1 at t=0 is shown at 20% intervals in fig. 7.1.

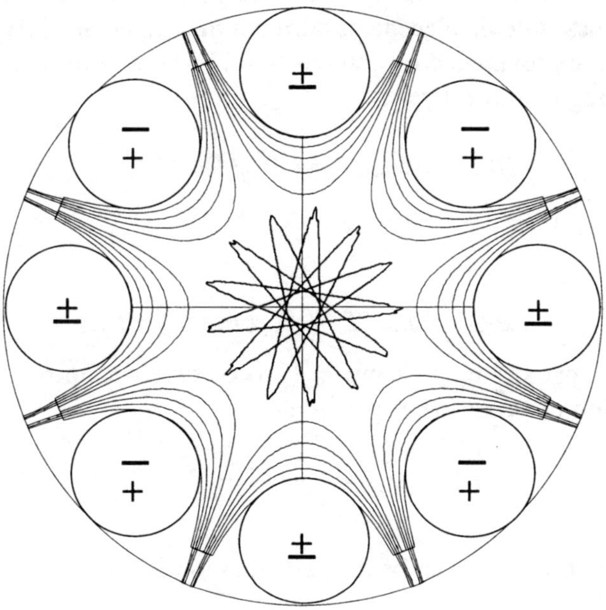

Figure 7.1. Oscillatory potential and ion trajectory in the OPIG.

In a multi-pole ion beam guide with $N \geq 2$, the motion of an ion with mass m and charge q is given by the following differential equations in the cylindrical coordinates normalized with $T = \omega t$, $R = r/a$ and $A = (NqV_{rf})/m(a\omega)^2$.

$$\frac{d^2R}{dT^2} - R(\frac{d\theta}{dT})^2 = -AR^{N-1}\cos(N\theta)\cos(T) \tag{7.2}$$

$$R\frac{d^2\theta}{dT^2} + 2\frac{dR}{dT}\frac{d\theta}{dT} = AR^{N-1}\sin(N\theta)\cos(T) \tag{7.3}$$

$$\frac{d^2z}{dT^2} = 0 \tag{7.4}$$

In the coordinate system (R, θ, z, T), velocity and energy are scaled by $a\omega$ and $ma^2\omega^2$ respectively. The oscillatory electric field created in the multi-pole ion beam guide modulates only the radial motion of charge particles on the $R - \theta$ plane and never affects the drift motion along the z-axis. The radial motion on the $R - \theta$ plane is decided by initial conditions of R_0, θ_0, $\left(\frac{dR}{dT}\right)_0$ and $\left(\frac{d\theta}{dT}\right)_0$ at $T = 0$ and the A value. The drift motion along the z-axis keeps the initial velocity component $\left(\frac{dz}{dT}\right)_0$. At a sufficiently high frequency ω where the ionic motion has little displacement during a time interval of $2\pi/\omega$, the radial motion can be approximated by composition of a slow periodical motion and a small oscillating part as follows:

$$R(T) = R_{av}(T) + R_{os}(R_{av}, \theta_{av}; T) \tag{7.5}$$

and

$$\theta(T) = \theta_{av}(T) + \theta_{os}(R_{av}, \theta_{av}; T). \tag{7.6}$$

The average parts of $R_{av}(T)$ and $\theta_{av}(T)$ coincide with those for a conserved periodic motion in a centripetal potential

$$U_{eff} = \frac{1}{4}A^2R^{2(N-1)}. \tag{7.7}$$

The average trajectory can be derived from the following relations,

$$\frac{1}{2}\left(\frac{dR_{av}}{dT}\right)^2 + \frac{1}{2}\left(\frac{L_{r\theta}}{R_{av}}\right)^2 + \frac{1}{4}A^2R_{av}^{2(N-1)} = E_{r\theta} \tag{7.8}$$

and

$$R_{av}^2\frac{d\theta_{av}}{dT} = L_{r\theta}, \tag{7.9}$$

where $E_{r\theta}$ and $L_{r\theta}$ are the energy and angular momentum at T=0 in the scaled coordinate system. Both of these quantities are conserved during the average motion.

The oscillatory parts of R_{os} and θ_{os} can be approximated by

$$R_{os} = AR_{av}^{(N-1)}\cos(N\theta_{av})\cos(T) \tag{7.10}$$

and

$$\theta_{os} = -AR_{av}^{(N-2)} \sin(N\theta_{av}) \cos(T). \tag{7.11}$$

As the frequency ω increases, amplitudes of the oscillatory parts of eqs. (7.10) and (7.11) rapidly reduce so the exact trajectory becomes close to the average one. Therefore, the maximum radius, R_{max}, of the trajectory on the R-θ plane can be roughly estimated from the relationship $\frac{1}{4}A^2 R_{\max}^{2(N-1)} \approx E_{r\theta}$ at a sufficiently high frequency. As the number N is increased, the region of ionic motion enlarges with expansion of the week field area as is seen in fig. 7.2. The radial ionic trajectory simulated for an octopole ion beam guide (OPIG) with $N = 4$, which is calculated by means of a step-wise integration, is illustrated in fig. 7.1.

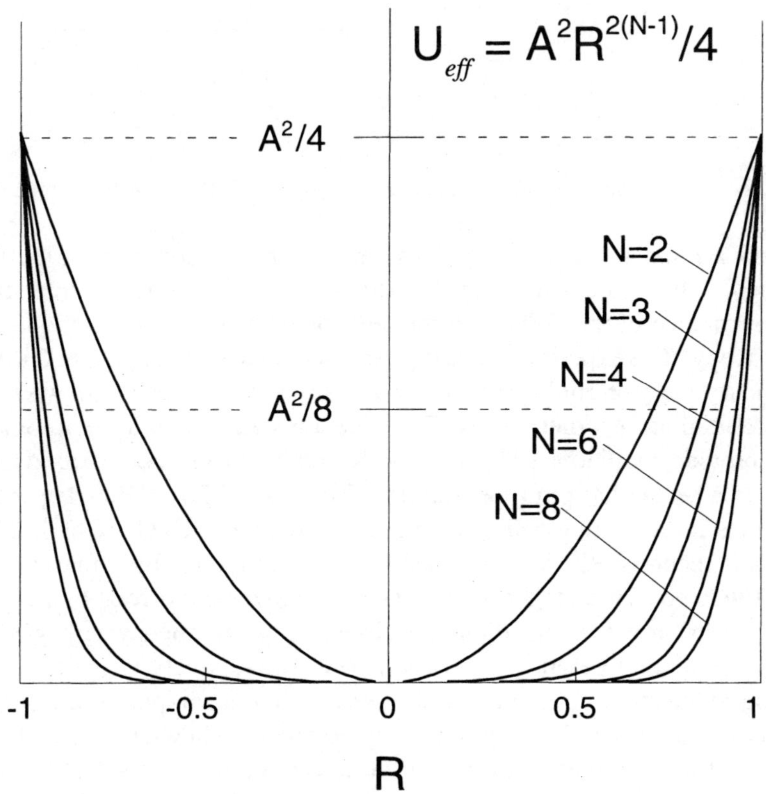

Figure 7.2. Effective potential for the radial motion.

Thus, this multi-pole beam guide, in which the oscillating R-F field modulates and confines only the radial motion of charge particles and never affects the drift motion along the axis, is of great advantage for transportation and confinement of charged particles. It is applicable for various purposes such as to make a collision cell, a TOF drift tube, a liner trap, an ion guide through a differential-pumping region and so on.

3. Application of a Tandem OPIG to Ion Spectroscopy

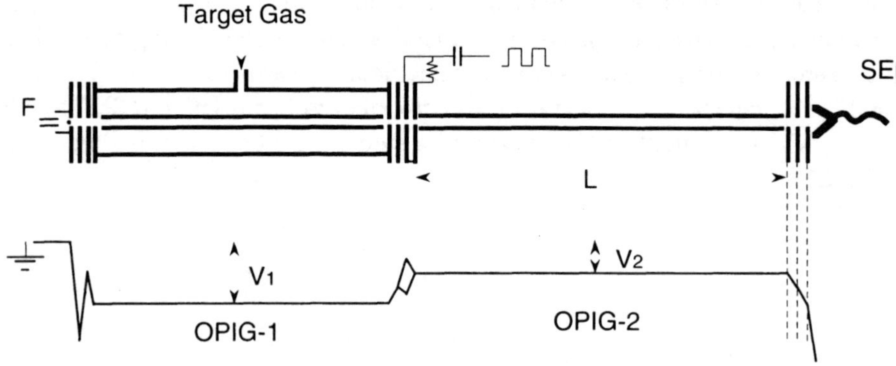

Figure 7.3. Setup of a tandem OPIG system and arrangements of the electric potentials.

In this section a tandem OPIG system, which was constructed in 1984, is described. This system was constructed with the two aims of understanding the performance of the OPIG and the development of a new technique for ion spectroscopy. The experimental setup consists of an ion source, two OPIGs and a secondary electron multiplier, which are aligned in series as shown together with potential arrangement in fig. 7.3. The ion source is of a thermionic type, able to provide alkali ions Li^+, Na^+ and K^+ emitted from a compound material pasted on a heated hairpin type filament. The first OPIG , OPIG-1, is 150 mm long and is installed in a collision cell. The second OPIG, OPIG-2, is 250 mm long. This second OPIG is used as a drift tube for time-of-flight measurements. Both OPIGs consist of eight equally spaced molybdenum rods of 1.5 mm diameter, inscribing a 7 mm diameter circle. High frequency voltages of 6.4 MHz are supplied to eight rods of each OPIG alternatively in opposite phases by using an electronic circuit usually used with a quadrapole mass filter. An interface circuit to supply high frequency voltages as shown in fig. 7.4 has been developed for the operation of the OPIG. Ions injected into the OPIG-1 collide with target gases. After collisions, the ion beam is chopped by a rectangular pulse supplied to a deflecting electrode at front of the OPIG-2 and their mass and energy-loss are analyzed by means of time-of-flight measurements. The collision energy of ions with target gases in OPIG-1 and the flight energy of ions in OPIG-2 are decided by potential differences of V_1 and V_2 measured with respect to the potential at the ion source.

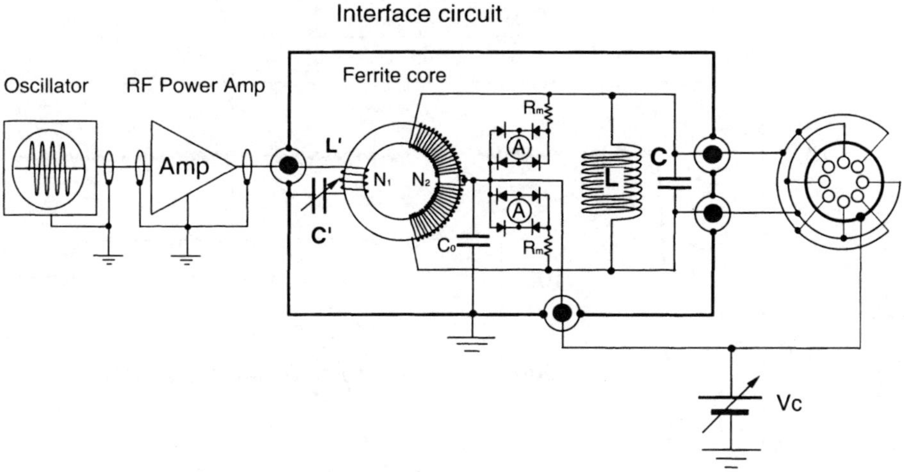

Figure 7.4. An interface circuit for operation of the OPIG.

In the preliminary test of ion-injection into the OPIG-1, an unexpected great benefit of using the OPIG was found when the ion intensity was measured as a function of V_1 by the detector placed at the rear of the OPIG-1 as shown at the top of fig. 7.5. As seen in fig. 7.5, the measured ion intensity curve is almost flat until it rapidly falls down at $V_1 \approx 0$. This means that the OPIG prevents ions from diverging not only in the OPIG but also in the retardation region in front of the OPIG perfectly. The energy distribution of injected ions can be roughly estimated from the differential curve around the onset as to be 0.24eV (FWHM). Thus, the OPIG makes it possible to provide a stable low-energy beam with a narrow energy spread.

In this experimental setup, it is easy to control the mass- and energy-resolution of TOF spectra by changing the fight velocity of ions in the OPIG-2. Since the fight time is given by $T = L \times (2E/M)^{-1/2}$, the mass resolution in the time spectrum of slow ions,

$$\frac{\Delta M}{M} = \left|\frac{\Delta E}{E}\right| + 2\left|\frac{\Delta L}{L}\right| + 2\left|\frac{\Delta T}{T}\right|, \qquad (7.12)$$

is mainly dominated by the initial energy spread of ions rather than uncertainty of the path length and the time resolution of electronic circuits used. As seen in fig. 7.6, $^6Li^+$ ions make a hump on the left side tail of $^7Li^+$ peak at the flight

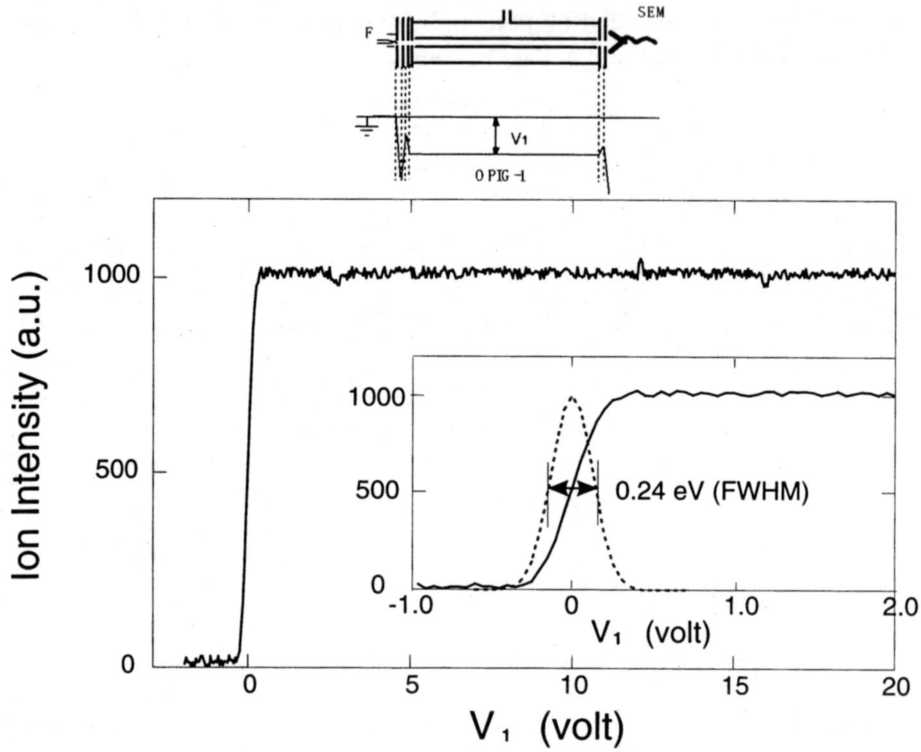

Figure 7.5. Ion intensity curve as a function of V_1.

energy of E_2=2.28 eV and the $^6Li^+$ peak perfectly separates from the $^7Li^+$ peak at E_2=36.5eV. On the other hand, the slower ions can be expected to give the higher energy resolution,

$$\frac{\Delta E}{E} = 2\left|\frac{\Delta L}{L}\right| + 2\left|\frac{\Delta T}{T}\right|. \tag{7.13}$$

The width of peaks observed in the TOF spectra corresponds to an energy width of about 0.3eV (FWHM), almost the same as the initial ion energy spread of 0.24eV (FWHM) estimated from the differential curve around the onset of ion intensity curve.

Using the tandem OPIG system, energy loss spectra of Li^+ and Na^+ ions in collisions with H_2 and CO introduced into the OPIG-1 were investigated. In fig. 7.7, the energy loss spectra of Li^+ ions with H_2 at the collision energy of E_1=2.09eV are shown, as measured at several gas pressures, with the potential of the OPIG-2 is fixed at V_2=2.09 volts. The energy position of shoulders observed in the right side of the Li^+ peak seems to coincide with expected

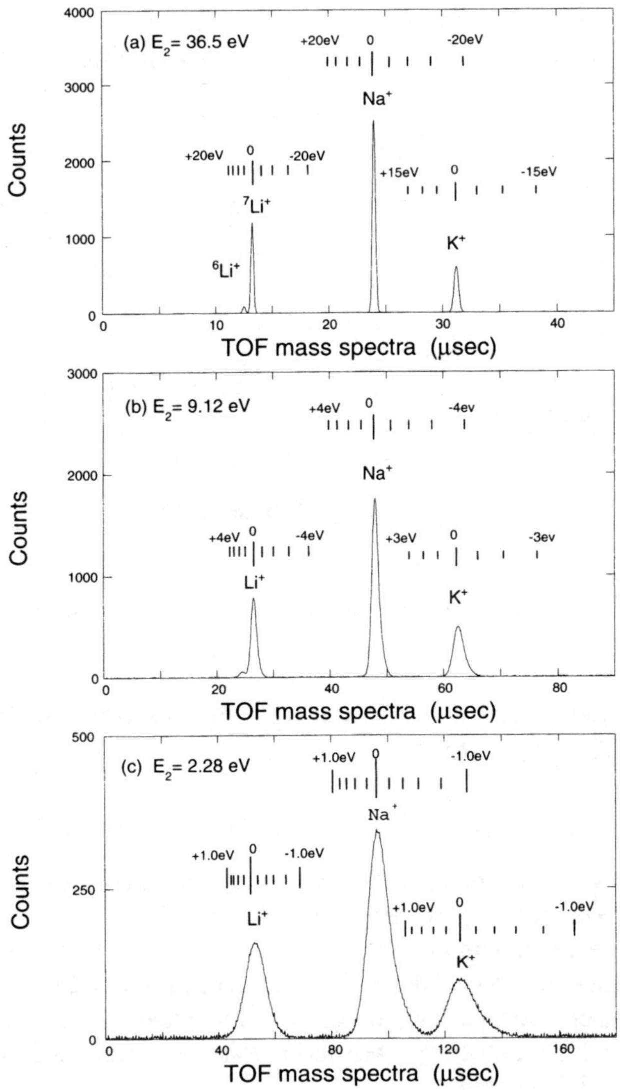

Figure 7.6. Flight energy dependence of the TOF spectra: Flight energies of ions in the OPIG-2 are 36.5, 9.12 and 2.28 eV for (a), (b) and (c), respectively.

energy losses (ΔE=0.54eV for H_2(v=0$\rightarrow$1)) due to the vibrational excitation of H_2 from v=0 to v=1 and 2. If this is true, it is very interest that the Li^+-H_2 collisions at a low energy of 0.46eV in the center-of-mass system is dominated by inelastic processes accompanied with a relatively large energy conversion into the internal energy. Furthermore, the cross section of such processes seems

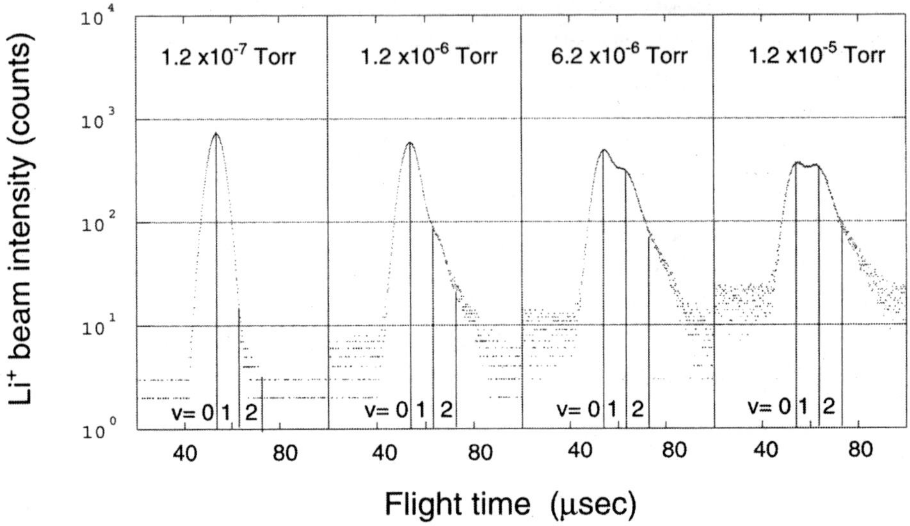

Figure 7.7. Energy loss spectra of Li^+ ions in collisions with H_2 at the collision energy of 2.09 eV.

to be unexpectedly large at very low energy contrary to the previous results at energies of 100-1200eV [6] .

$$\mathrm{Li}^+ + \mathrm{H}_2(\mathrm{v}=0) \rightarrow \mathrm{Li}^+ + \mathrm{H}_2(\mathrm{v}=1,2) - \Delta\mathrm{E} \tag{7.14}$$

In studies using the tandem OPIG system, it is found that the R-F field penetrated from the entrance aperture of the OPIG probably plays an important role to keep a stable beam with an almost constant intensity. This is an unexpected great advantage of using the OPIG. Thus, it has been proved that the beam guide technique is useful for the low energy collision experiments and even for ion spectroscopy with a high energy-resolution if an ion beam with enough narrow energy spreads is prepared.

4. Application of the OPIG to Low Energy Cross Section Measurements

In order to measure charge transfer cross sections of highly charged ions, the next issue is how to provide a stable beam of slow highly charged ions with a narrow energy spread.

An experimental method of cross section measurements using the OPIG technique has been established [3, 4]. Furthermore, a small electron beam ion source (Mini-EBIS) was constructed based on a new idea of cooling magnetic solenoid coils with liquid nitrogen [7, 8]. This Mini-EBIS can provide a stable beam of highly charged ions with a narrow energy spread when used in the DC mode of operation.

The principle and structure of the Mini-EBIS are as follows. Electrons emitted from a hot cathode are collimated to form a thin beam under a magnetic field. The ions produced by successive electron impact are trapped within the electron beam in the radial direction by its space charge potential, and in the axial direction by an electric potential applied from outside. Once ions have been trapped in the electron beam, successive ionization by electron impact takes place and the charge state of the ions increases unless the space charge of the electron beam is neutralized by ions. In the DC ion extraction mode, a suitable potential barrier of a few eV is set at the exit of the ionization region. The setting of such potential barrier is most effective to increase the relative population of higher charge state ions and to extract the ion beam with a narrower energy spread. Usually the contamination of ions in metastable states, which can cause significant and dramatic effects on the observed cross sections, is a very troublesome problem for cross section measurements. However, the DC mode operation of the Mini-EBIS is very effective to suppress production of excited ions with a long life. The Mini-EBIS is operated at very low pressure (lower than 10^{-10} torr). At such low densities, there is little probability of metastable production via electron capture collisions and, even if excited ions with a long life are produced, they will be quenched during long ion-confinement times in the DC mode of operation. Practically, no existence of long-lived excited ions has been found in multiply charged ion beams except for low charged ion beams of C^{2+} and O^{2+} only.

The cross section measurements for charge transfer of multiply charged ions with atoms and atmospheric molecules have been performed systematically by combining both techniques of the OPIG and the Mini-EBIS [8]. The apparatus used in these cross section measurements is basically similar to a tandem mass spectrometer that consists of an ion source of the Mini-EBIS type, a mass selector, a collision cell, a mass analyzer and an ion detector in sequence. In fig. 7.8, a schematic diagram of the experimental setup is shown together with the arrangement of electrostatic potentials. Only a brief description is given here since details of experimental setup and procedure have been reported previously [3, 8].

Ions extracted from the ion source are accelerated and mass analyzed by the first electromagnetic analyzer, MS1. Ions with the selected m/q are decelerated just before injection into a collision cell through an entrance aperture of 0.5 mm diameter. In the collision cell an OPIG system is situated, consisting of eight molybdenum poles with a diameter of 1.5 mm and 160 mm length inscribing an outlet ceramic pipe of 7 mm inner diameter. High frequency voltages are supplied to eight poles of the OPIG alternatively in an opposite phase using the interface circuit shown in fig. 7.4. The ions leaving the OPIG are accelerated

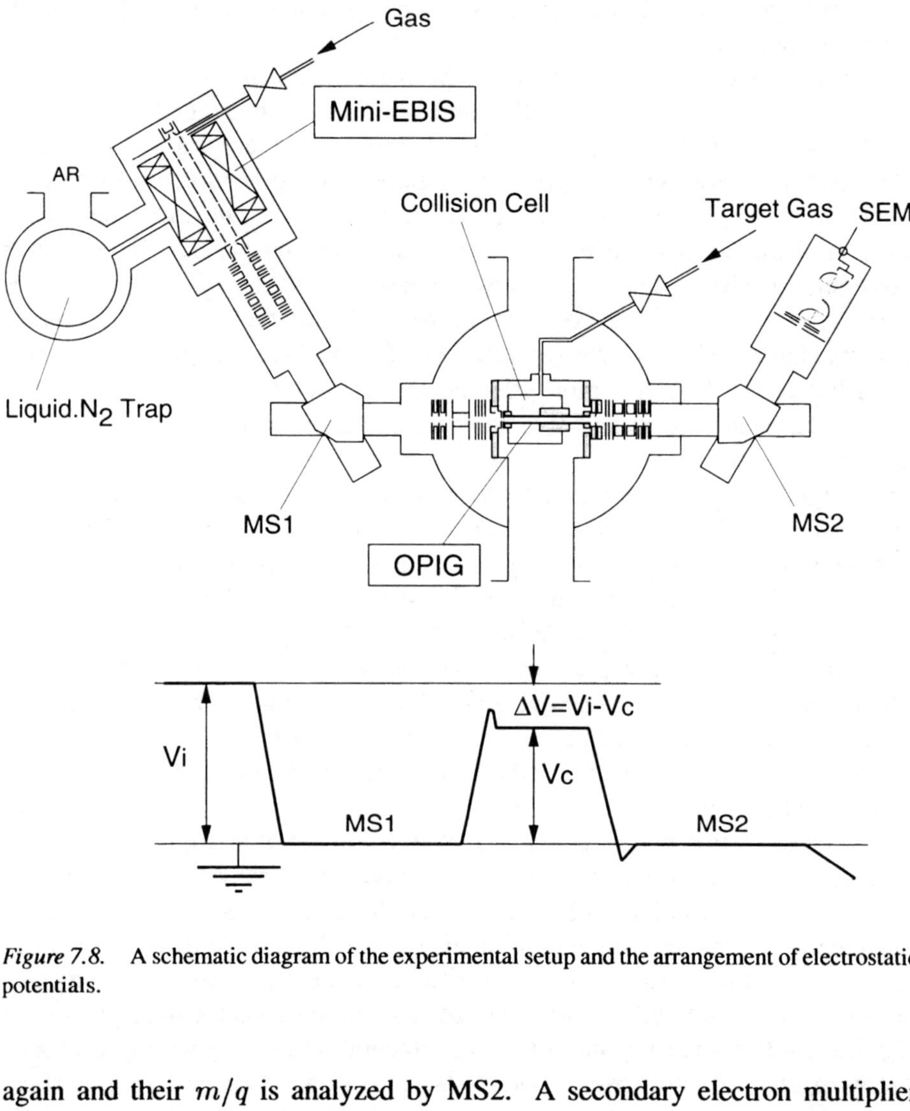

Figure 7.8. A schematic diagram of the experimental setup and the arrangement of electrostatic potentials.

again and their m/q is analyzed by MS2. A secondary electron multiplier counts the primary ion and product ions. A MKS Baratron gauge measures the pressure of target gas indirectly.

The collision energy is determined by the charge q times the potential difference ΔV between V_i at the ion source and V_c at the collision cell, so that $E_{lab} = q \times \Delta V$. When beam intensity is measured as a function of $\Delta V = V_i - V_c$ by changing V_c, it is almost constant until it falls off rapidly near the zero energy as seen in fig. 7.9. This is a great advantage of using the OPIG for cross section measurements at low energies. Energy spreads of ions extracted from the Mini-EBIS are estimated from a differential curve of the beam intensity, as typically $q\times(0.4\sim0.7)$eV (FWHM).

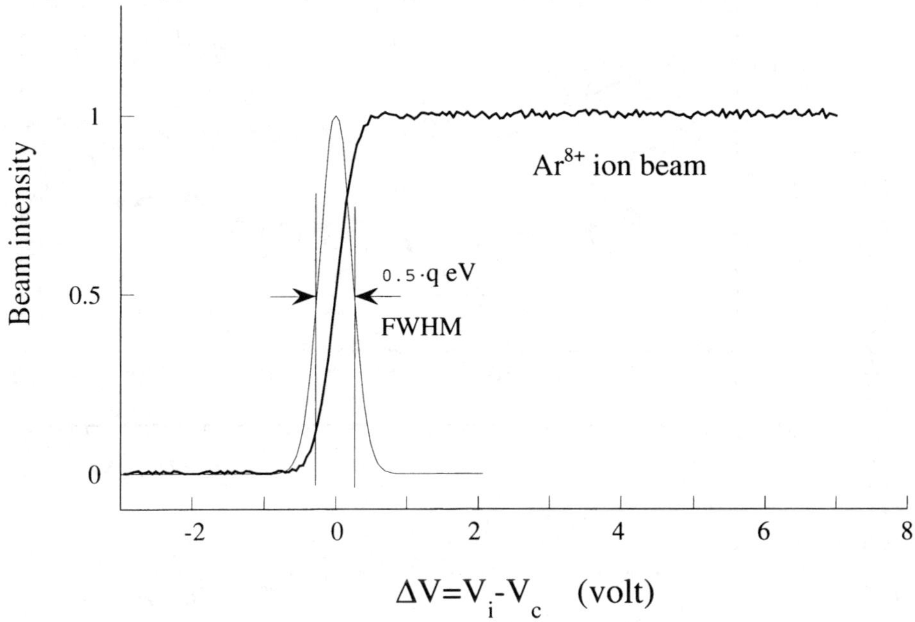

Figure 7.9. Ion intensity curve as a function of V_1.

Cross sections for single- and multiple-charge transfer reactions of multiply charged ions are determined from the initial growth of product ion intensity with increasing the target gas pressure. In the present measurements, since the product ions are identified by q'/m for the final charge state of q', measured cross sections of $\sigma_{q-q'}$ include contributions from the transfer ionization processes of $\sigma^{TI}_{q'',q'}$ in addition to the pure electron capture cross sections of $\sigma_{q,q'}$,

$$\sigma_{q-q'} = \sigma_{q,q'} + \sum_{q''} \sigma^{TI}_{q'',q'} \qquad (q'' < q' < q). \tag{7.15}$$

Here, q, q' and q'' are the initial, final and intermediate charge states of projectile ions, respectively.

Using the combination of the OPIG and the Mini-EBIS, single- and multiple-charge transfer cross sections of multiply charged ions in collisions with atoms and atmospheric molecules have been measured systematically in the low energy range of 0.5 to 2000eV per ion charge. Here, representative single- and double-electron transfer cross sections of Ar^{q+} ($q = 6$ and 8) in collision with the two electron targets He and H_2 [9, 10] are shown together with previous experimental and theoretical data in figs. 7.10 and 7.11. The single electron transfer cross sections, σ_1, are compared to three simple models of the Langevin cross sections, σ_L [11], the scaling law, σ_{MS} [12], and the absorbing sphere

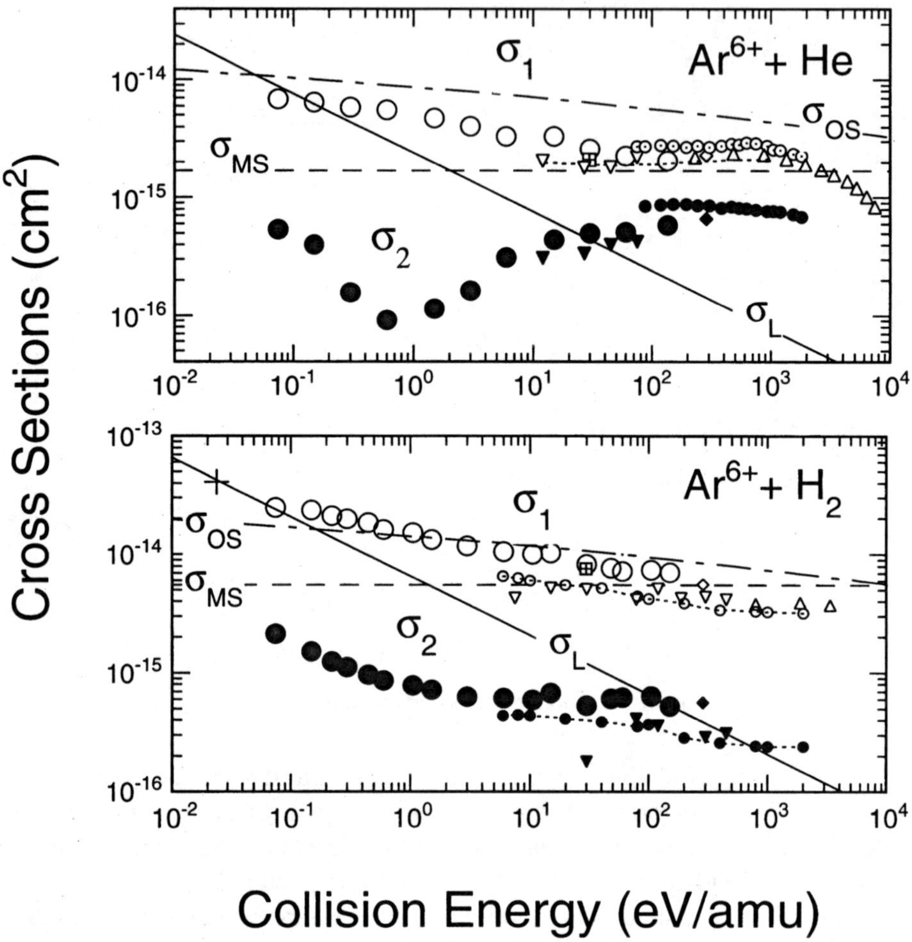

Figure 7.10. Single- and double-electron transfer cross sections for Ar^{6+} - He and H_2 collision systems [9, 10]. Large open and full circles denote present σ_1 and σ_2, respectively. Legend; σ_L for Langevin [11], σ_{MS} for Müller and Salzborn [12], σ_{OS} for absorbing sphere model [13], small open and full circles for molecular-orbital expansion of σ_1 and σ_2 [9].

model, σ_{OS} [13]. Cross sections σ_1 and σ_2 calculated using a molecular orbital expansion method for Ar^{6+} and Ar^{8+} + H_2 systems in the energy range from 6 to 2000eV/amu are in good agreement with the present data [9]. As seen in fig. 7.10, the measured σ_1 cross sections are all between σ_{OS} and σ_{MS} above about 1eV/amu and converge to σ_L at the low energy end. These simple models are useful to know the general behavior of the single electron transfer cross sections. For the Ar^{6+} + H_2 system, one measurement of the total cross section $\sigma_1 + \sigma_2$ measured at 50meV using Penning ion trap [14] is just on a

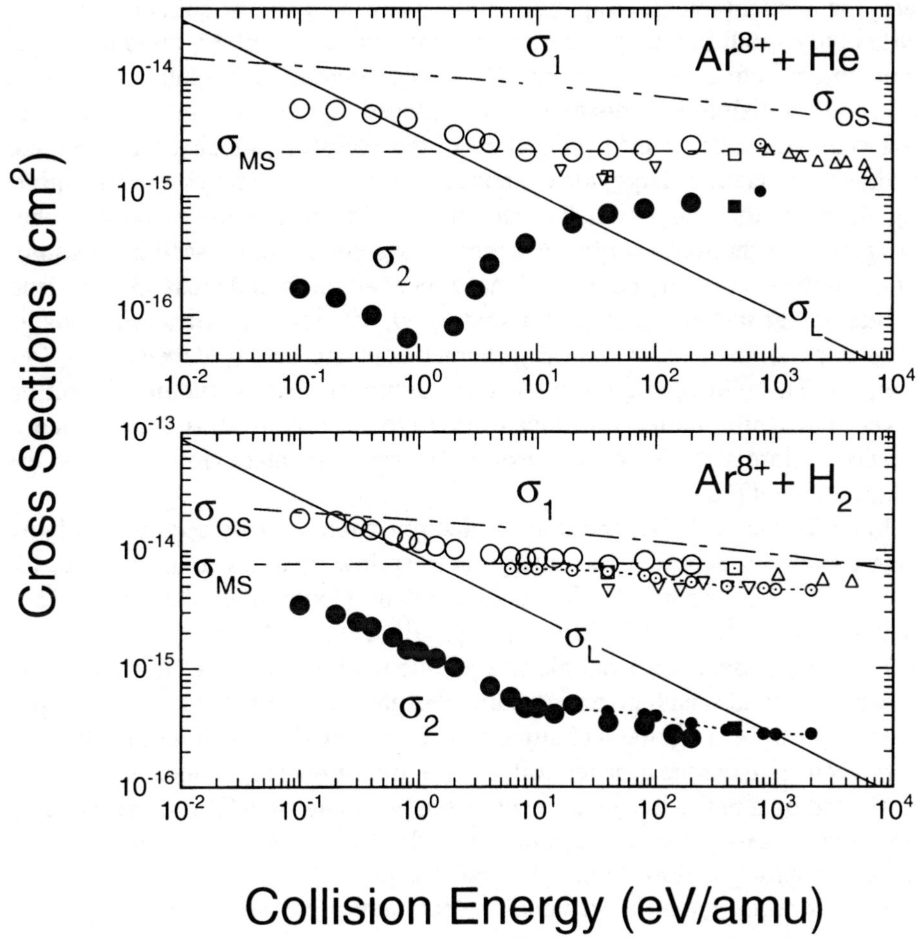

Figure 7.11. Single- and double-electron transfer cross sections for Ar^{8+} - He and H_2 collision systems [9, 10]. Labelling is the same as that of fig. 7.10.

line extrapolated from the present data converging to σ_L. The cross sections of σ_1 for the H_2 target are almost three times as large as those for the He target and the ratio of $\sigma_1(He)/\sigma_1(H_2)$ is close to $[I(He)/I(H_2)]^{-2.72}$ predicted by the scaling law. On the other hand, the double-electron transfer cross sections of σ_2 are smaller than one tenth of σ_1, and strongly depend on the collision energy in contrast to σ_1 and also increase at low energies.

In the low energy collisions, the collision dynamics drastically changes giving rise to various new effects and interesting phenomena. The Coulomb field created by the ionic charge polarizes the neutral target and an ion-induced dipole leads to a mutual attraction between collision partners. The induced dipole captures the incident ion into a spiraling trajectory closing to the collision center at sufficiently small impact parameters and into a circular orbit around the target at an appropriate impact parameter [11]. At sufficiently low collision energies where the collision is dominated by the induced dipole, the orbiting cross section is enhanced in inverse proportion to the collision velocity. As the ionic charge state becomes large, the enhancements of cross section due to the orbiting effects on the charge transfer reactions should become observable at higher energies than thermal energy. A series of systematic cross section measurements have successfully confirmed that almost all measured cross sections tend to increase at the low energy end studied [9, 10, 15–20]. Generally, charge transfer cross sections of multiply charged ions are energy dependent at low energies. Those involving multi-electron capture collisions with many electron targets have a minimum in the range from 0.1 to 10eV/amu, and this dependence should be closely related to competition between the interaction range and the orbiting radius [16–19].

In collisions with H_2, the double electron capture cross section of highly charged ions is expected to be large at very low energies as is the case for Ar^{q+} ions. The product $(H_2)^{2+}$ dissociates by Coulomb explosion and each of two protons produced will carry approximately 9eV. Such process should be great importance, for example as a production mechanism of fast protons in interstellar clouds and high temperature plasmas. Recently, the variety of applications of the beam guide technique has been expanding. Coulomb explosion of molecules in charge transfer collisions of multiply charged ions has been investigated by a new triple coincident technique using twin OPIG systems [21]. Groningen's group has also applied the OPIG to the study of state-selective electron capture of slow highly charged ions [22, 23].

Furthermore, quantum mechanical calculations predict that some sharp peaks should appear on the electron capture cross sections of multiply charged ion in very low energy collisions with atoms due to the orbiting resonance, which is a kind of a shape resonance due to the formation of the transient quasibound molecule [24]. The experimental verification of the orbiting resonance needs still more advanced experimental techniques. However, it should be performed in due course.

References

[1] S. M. Trujillo, R. H. Neynaber and E. W. Rothe, Rev. Sci. Instrum, **37**, 1655 (1966)

[2] E. Teloy and D. Gerlich, Chem. Phys. **4**, 417 (1974)

[3] K. Okuno, J. Phys. Soc. Jpn., **55**, 1504 (1986)

[4] K. Okuno and Y. Kaneko, Mass Spectrom. Jpn., **34**, 351 (1986)

[5] H. J. Xu, M. Wada, J. Tanaka, H. Kawakami, I. Katayama and S. Ohtani, Nucl. Instr.& Meth., **A333**, 274(1993); M. Wada, H. J. Xu, J. Tanaka, H. Kawakami, I. Katayama, K. Okada and S. Ohtani, Hyp. Int., **81**, 161 (1993)

[6] N. Kobayashi, Y. Itoh and Y. Kaneko, J. Phys. Soc. Jpn., **46**, 208 (1979)

[7] K. Okuno, AIP Conf. Proc. **188**, 33 (1988); Jpn. J. Appl. Phys., **28**, 1124 (1989)

[8] K. Okuno, K. Soejima and Y. Kaneko, Nucl. Instr. Meth. Phys. Res., **B53**, 387 (1991)

[9] S. Kravis, H. Saitoh, K. Okuno, K. Soejima, M. Kimura, I. Shimamura, Y. Awaya, Y. Kaneko, M. Oura, and N. Shimakura, Phys. Rev., **A52**, 1206 (1995)

[10] K. Okuno, H. Saitoh, K. Soejima, S. Kravis and N. Kobayashi, AIP Conf. Proc., **360**, 867 (1995)

[11] G. Gioumousis and D. P. Stevenson, J. Chem.. Phys., **29**, 294 (1958)

[12] A. Müller and E. Saltzborn, Phys. Lett., **62A**, 391 (1977)

[13] R. E. Olson and A. Salop, Phys. Rev., **A14**, 579 (1976)

[14] S. D. Kravis, D. A. Church, B. M. Johnson, M. Meron, K. W. Jones, J. C. Levin, I. A. Sellin, Y. Azuma, N. Berrah-Mansour, and H. G. Berry, Phys. Rev., **A45**, 6379 (1992)

[15] K. Okuno, K. Soejima and N. Kaneko, J. Phys. B: At. Mol. Opt. Phys., **25**, L105 (1992)

[16] K. Ishii, K. Okuno and N. Kobayashi, Physica Scripta, **T80**, 176 (1999)

[17] K. Soejima, K. Okuno and N. Kaneko, Org. Mass Spectrometry, **28**, 344 (1993)

[18] K. Soejima, C. J. Latimer, K. Okuno, N. Kobayashi and Y. Kaneko, J. Phys. B: At. Mol. Opt. Phys., **25**, 3009 (1992)

[19] K. Suzuki, K. Okuno and N. Kobayashi, Physica Scripta, **T73**, 172 (1997)

[20] K. Ishii T. Tanabe R. Lomsdze and K. Okuno, Physica Scripta, **T92**, 332 (2001)

[21] K. Okuno, T. Kaneyasu, K. Ishii, M. Yoshino and N. Kobayashi, Physica Scripta, **T80**, 173 (1999); T. Kaneyasu, K. Matsuda, M. Ehrich, M. Yoshino and K. Okuno, Physica Scripta, **T92**, 341 (2001)

[22] N. Shimakura and M. Kimura, Phys. Rev., **A44**, 1659 (1991); S. Suzuki, N. Shimakura and M. Kimura, J. Phys. B: At. Mol. Opt. Phys., **31**, 1741 (1998)

[23] P. Stancil, B. Zygelman, N. Clarke, and D. Cooper, J. Phys. B: At. Mol. Phys., **30**, 1013 (1997)

[24] F. W. Bliek, R. Hoekstra and R. Morgenstern, Phys. Rev. **A56**, 426 (1997); F. W. Bliek, G. R. Woestenek, R. Hoekstra and R. Morgenstern, Phys. Rev., **A57**, 221 (1998)

Chapter 8

THEORETICAL DESCRIPTION OF LOW ENERGY COLLISIONS.

Close-coupling Semiclassical Treatments.

L. F. Errea, A. Macías * , L. Méndez and A. Riera

Laboratorio Asociado al CIEMAT de Física Atómica y Molecular en Plasmas de Fusión Nuclear, Departamento de Química C-9, Universidad Autónoma de Madrid, Cantoblanco, 28049-Madrid, Spain

l.mendez@uam.es

Abstract Semiclassical close-coupling methods to treat ion-atom collisions are described. Computational techniques for single- and many-electron systems are considered, as well as the extension of these methods to ion-molecule collisions, which is illustrated with numerical examples of charge transfer in collisions of multicharged ions with H_2

Keywords: ion-atom collisions, ion-molecule collisions, close-coupling methods

1. Introduction

Collisions of ions with atoms and molecules at impact energies above 50 eV/a.m.u. are usually described by semiclassical methods. In these methods, the relative ion-atom (molecule) motion is treated by means of a classical trajectory and the remaining degrees of freedom are described quantally. There are two main reasons to introduce this approximation: Firstly, at these energies the full quantal treatment is very cumbersome because a large number of partial waves must be included. Secondly, it allows following the time evolution of the system, characteristic of classical descriptions, which helps to understand the collisional mechanisms. A complete explanation of the basis of the semiclassi-

* Also at: Instituto de Estructura de la Materia CSIC, Serrano 113 bis, 28006 Madrid, Spain.

F.J. Currell (ed.), The Physics of Multiply and Highly Charged Ions, Vol. 2, 237-274.

cal approximation can be found in the text of Bransden and McDowell [1] and the references cited in it and will not be repeated here. The basic equations of this approach are outlined in section 8.2.

In this chapter we shall consider close-coupling (CC) methods to describe ion-atom and ion-molecule collisions. In the CC treatments the collisional wavefunction is expanded in terms of a complete set of functions. Therefore, the method is formally convergent, which in practice means that the accuracy of the results can be improved by adding more terms to the expansion, but the speed of convergence of a CC expansion depends on the basis set chosen and on the collision energy.

At low impact energies, charge transfer and excitation in ion-atom collisions are usually described by means of a CC expansion in terms of molecular functions, which is described for one-electron systems in section 8.3. When applying the CC molecular expansion it is required to include the so-called electron translation factors (ETF) (a review on this topic can be found in ref. [2]) and we summarize this point in section 8.4, using again a single-electron system to introduce it. For projectile velocities (v) larger than the target electron velocity (v_0), ionization starts to be sizeable, and this is the main limitation on the application of CC methods [3], which is also discussed in section 8.4. The application of CC methods at high v is particularly relevant because of the very detailed experiments, like those based on ion recoil momentum spectroscopy [4], are carried out in this energy range.

In section 8.5 we consider the application of the molecular expansion to many-electron systems. The main difficulty of this extension is the accurate calculation of energies and non-adiabatic couplings, which requires the application of quantum chemistry techniques (see [5]). We also describe in this section the use of some quantum chemistry packages (MRCI) and effective potentials, which are fundamental tools for calculations in ion-molecule collisions (see e.g. [6], [7])

In section 8.6 we describe briefly the use of CC expansions in terms of atomic functions (see e.g.[8], [1]), which are commonly applied at higher energies ($v_0/2 \lesssim v \lesssim 2v_0$) than the molecular expansion. Several works have treated ionization in the framework of this atomic method by adding the so-called pseudostates to the basis (recent applications of this kind of formalism can be found in refs. [9–13]), and we shall describe several ways to define these pseudostates.

At even higher projectile velocities ($v > v_0$), non-CC methods, which will not be considered in this chapter, are often used. For example:

1 Perturbative methods: the Born series and modified treatments like those based on the distorted wave approximation. An explanation of them can be found in ref. [14].

2 Statistical Monte Carlo treatments, where the motion of all particles is treated classically (see [15]).

An illustration of the application of different methods (molecular and atomic CC, classical Monte Carlo and distorted wave approximation) to multiply-charged ions collisions with H can be found in the recent work of ref. [16],

Ion-molecule collisions give rise to new physical processes like electron capture and excitation into particular vibrorotational states. Also charge transfer or excitation can lead to molecular dissociation by transitions to either vibrational or electronic continuum states. Multiple electron capture can also yield molecular dissociation; for example, double electron capture in collisions of multicharged ions with H_2 gives rise to the Coulomb explosion process, where two protons are formed. Although the basic theoretical methods to treat ion-molecule collisions, based on the CC method, are well established (see the reviews of refs. [17] [18]), the complexity of these systems, in particular the description of the molecular internal motion, has limited the number of calculations; as an example, only a few applications to ion-H_2 collisions were reported in ref. [19], published in 1998, and in practice, additional approximations are employed. In section 8.7, we consider, as a benchmark case, the CC treatment of ion-H_2 collisions, and we describe the application of the sudden approximation [20] for molecular rotation and vibration that we have employed in recent works (see e.g. [21]) to evaluate vibrationally resolved cross sections.

Atomic units are used throughout except where otherwise stated.

2. The Semiclassical Approximation

We consider a collision system A + B, where B is a fully stripped ion and A is a single-electron ion or atom. In the semiclassical method, the nuclei follow a classical trajectory $\boldsymbol{R}(t)$, where $\boldsymbol{R}$ is the internuclear distance (see fig. 8.1). In almost all applications the *eikonal* method is used, where the trajectory is rectilinear:

$$\boldsymbol{R} = \boldsymbol{b} + \boldsymbol{v}t \tag{8.1}$$

with $\boldsymbol{b}$, the impact parameter and $\boldsymbol{v}$ the relative velocity. The electronic motion is described by the wavefunction $\Psi(\boldsymbol{r}, t)$, which fulfills the equation (see e.g. [1]):

$$\left[H_{\mathrm{el}} - \mathrm{i} \left. \frac{\partial}{\partial t} \right|_{\mathbf{r}} \right] \Psi = 0 \tag{8.2}$$

where $\boldsymbol{r}$ is the electron position vector with respect to the origin O (see fig. 8.1) and H_{el} is the clamped nuclei Hamiltonian in the Born-Oppenheimer approximation:

$$H_{\mathrm{el}} = -\frac{1}{2}\nabla_r^2 - \frac{Z_A}{r_A} - \frac{Z_B}{r_B} + \frac{Z_A Z_B}{R} \tag{8.3}$$

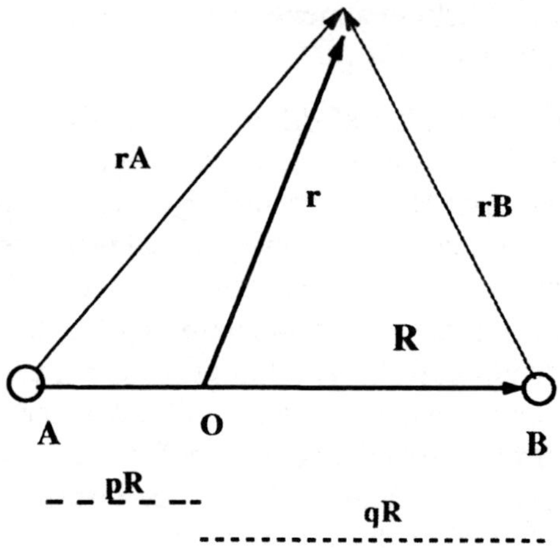

Figure 8.1. Definition of variables used in ion-atom collisions.

with $Z_{A,B}$ the nuclear charges, and $r_{A,B}$ the electron distances to the nuclei (see fig. 8.1). The symbol $\frac{\partial}{\partial t}\big|_{\boldsymbol{r}}$ indicates that the derivative is carried out by keeping constant the electronic coordinates.

Equation (8.2) must be solved with the appropriate initial condition; this can be obtained by looking for the solutions of (8.2) in the limit when $t \to -\infty$. In particular, for a single-electron system, the semiclassical wavefunction $\Psi(\boldsymbol{r}, t)$ fulfills:

$$\begin{aligned} \Psi(\boldsymbol{r}, t) \quad & \underset{t \to -\infty}{\sim} \quad \Phi_i^A(\boldsymbol{r}_A) \exp\left[-\mathrm{i}\left(p\boldsymbol{v} \cdot \boldsymbol{r} + \frac{1}{2}p^2v^2t + E_it + \alpha_A\right)\right] \\ & = \quad \Phi_i^A(\boldsymbol{r}_A) D^A \exp\left[-\mathrm{i}(E_it + \alpha_A)\right] \end{aligned} \tag{8.4}$$

which can be easily checked by substituting in (8.2) and taking into account the identity:

$$\left.\frac{\partial}{\partial t}\right|_{\boldsymbol{r}} = \left.\frac{\partial}{\partial t}\right|_{\boldsymbol{r}_A} + p\boldsymbol{v} \cdot \nabla_{\boldsymbol{r}} \tag{8.5}$$

where pR is the distance from the origin O to the nucleus A (see fig. 8.1), and $\frac{\partial}{\partial t}\big|_{\boldsymbol{r}_A}$ indicates that the derivative is carried out by keeping constant the electronic coordinates referred to the nucleus A.

The function (8.4) includes:

1 The atomic orbital (AO) $\Phi_i^A(\boldsymbol{r}_A)$, which describes the electronic orbital motion and is solution of

$$H^A(\boldsymbol{r}_A)\Phi_i^A(\boldsymbol{r}_A) = E_i\Phi_i^A(\boldsymbol{r}_A) \tag{8.6}$$

with

$$H^A(\boldsymbol{r}_A) = -\frac{1}{2}\nabla_r^2 - \frac{Z_A}{r_A} \tag{8.7}$$

2 The ETF [22] $D^A = \exp\left[-\mathrm{i}(p\boldsymbol{v}\cdot\boldsymbol{r} + p^2v^2t/2\right]$, which is a plane-wave that describes the translation of the electron attached to the nucleus A, with respect to the origin O.

3 The phase $\alpha_A = Z_B(Z_A - 1)v^{-1}\log(R - vt)$ that comes from the Coulomb interaction when the system is initially formed by two ions with charges Z_B and $Z_A - 1$.

Analogously, for elastic or excitation processes, the electron is attached to nucleus A in the limit $t \to \infty$, and a final state has the form

$$\begin{aligned}\Phi_j^A(\boldsymbol{r}_A)\exp\left[-\mathrm{i}\left(p\boldsymbol{v}\cdot\boldsymbol{r} + \frac{1}{2}p^2v^2t + E_jt + \alpha_A\right)\right] \\ = \Phi_j^A(\boldsymbol{r}_A)D^A\exp\left[-\mathrm{i}\left(E_jt + \alpha_A\right)\right]\end{aligned} \tag{8.8}$$

while for electron capture processes, one has:

$$\begin{aligned}\Phi_k^B(\boldsymbol{r}_B)\exp\left[-\mathrm{i}\left(-q\boldsymbol{v}\cdot\boldsymbol{r} + \frac{1}{2}q^2v^2t + E_kt + \alpha_B\right)\right] \\ = \Phi_k^B(\boldsymbol{r}_B)D^B\exp\left[-\mathrm{i}\left(E_kt + \alpha_B\right)\right]\end{aligned} \tag{8.9}$$

where qR is the distance from the origin O to the nucleus B. The ETF D^B describes the translational motion of the electron joined to nucleus B. Φ_k^B is solution of:

$$H^B(\boldsymbol{r}_B)\Phi_k^B(\boldsymbol{r}_B) = E_k\Phi_k^B(\boldsymbol{r}_B) \tag{8.10}$$

with

$$H^B(\boldsymbol{r}_B) = -\frac{1}{2}\nabla_r^2 - \frac{Z_B}{r_B} \tag{8.11}$$

and $\alpha_B = Z_A(Z_B - 1)v^{-1}\log(R - vt)$.

If $\Psi(\boldsymbol{r}, t)$ is solution of (8.2) with the initial condition (8.4), the probabilities for transitions to specific excitation and capture states are given respectively by:

$$P_{ij}^{\mathrm{exc}}(b,v) = \lim_{t\to\infty}\left|\langle\Phi_j^A D^A|\Psi\rangle - \delta_{ij}\right|^2 \;;\; P_{ik}^{\mathrm{cap}}(b,v) = \lim_{t\to\infty}\left|\langle\Phi_k^B D^B|\Psi\rangle\right|^2 \tag{8.12}$$

and the corresponding total cross sections can be obtained by applying the semiclassical approximation on the quantal boundary conditions leading to (see e.g. [1]):

$$\sigma_{if}(v) = 2\pi\int_0^\infty bP_{if}(b)\mathrm{d}b \tag{8.13}$$

3. The Molecular Expansion.

3.1 Basic equations.

At low collision energies, the interaction of the electron of the initial AO with the incoming ion takes place slowly, and this AO delocalizes forming a wavefunction of the (quasi)molecule. According to this physical picture, a CC expansion in terms of molecular orbitals (MOs) is employed for $\Psi(\mathbf{r}, t)$:

$$\Psi^{\mathrm{MO}}(\mathbf{r}, t) = \sum_j a_j(t)\psi_j(\mathbf{r}; R) \exp\left(-\mathrm{i}\int_0^t \varepsilon_j \mathrm{d}t'\right) \tag{8.14}$$

where the MOs, $\psi_j(\mathbf{r}; R)$, are eigenfunctions of the clamped nuclei electronic Hamiltonian H_{el} of eq. (8.3):

$$H_{\mathrm{el}}\psi_j(\mathbf{r}; R) = \varepsilon_j(R)\psi_j(\mathbf{r}; R) \tag{8.15}$$

Substitution of eq. (8.14) in (8.2) leads to the system of differential equations:

$$\mathrm{i}\dot{a}_k = \sum_{j\neq k} \left\langle \psi_k \left| -\mathrm{i}\frac{\partial}{\partial t} \right| \psi_j \right\rangle a_j \exp\left[-\mathrm{i}\int_0^t (\varepsilon_j - \varepsilon_k)\mathrm{d}t'\right] \tag{8.16}$$

with $\dot{a}_k = \mathrm{d}a_k/\mathrm{d}t$. Initially, the system is described by the AO Φ_i^A, which corresponds to the asymptotic form of the MO ψ_i. At very low velocities, the system behaves *adiabatically* and it can be described by the MO, ψ_i. As v increases, *non-adiabatic* transitions take place between the MOs, i.e., the coefficients a_k with $k \neq i$ are not zero and Ψ^{MO} becomes a superposition of several MOs, with $|a_k(t)|^2$ the population of a given MO. Since the MOs ψ_k can dissociate into excited AO of the target or with the electron attached to the projectile, excitation and charge transfer reactions take place through transitions between the MOs. From eq. (8.16), the importance of these transitions is determined by:

1 The values of the integrals $\langle\psi_k \left| \frac{\partial}{\partial t} \right| \psi_j\rangle$, which are called *dynamical* or *non-adiabatic couplings*.

2 The energy differences $\varepsilon_j - \varepsilon_k$: When these differences are large, the phase factors $\exp\left[-\mathrm{i}\int_0^t(\varepsilon_j - \varepsilon_k)\mathrm{d}t'\right]$ oscillate quickly, and the effect of the dynamical couplings is canceled out when the system (8.16) is integrated along the trajectory.

Since $t = Z/v$, with Z the nuclear coordinate ($Z = \widehat{Z} \cdot \boldsymbol{R}$), when v increases, the dynamical couplings increase and the phases $\int_0^t(\varepsilon_j - \varepsilon_k)\mathrm{d}t'$ decrease. Consequently, non-adiabatic transitions are more significant at high v. In practice, this requires to enlarge the CC expansion (8.14) as v increases.

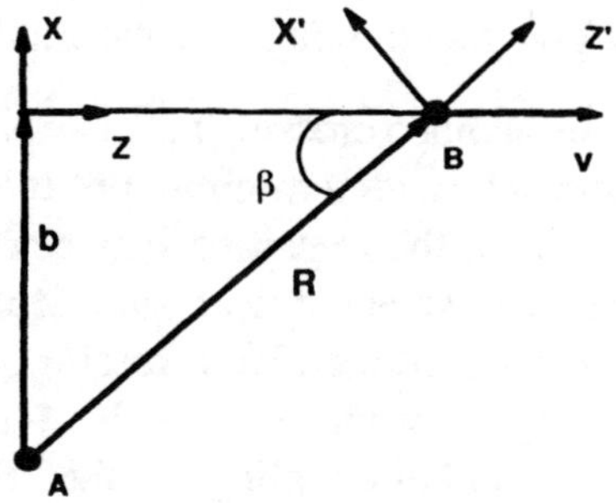

Figure 8.2. Definition of a molecule-fixed reference frame.

While eqs. (8.2),(8.4), (8.8) and (8.9) are expressed in a laboratory reference frame XYZ, with the Z axis in the direction of $\hat{\boldsymbol{v}}$ and the X axis along the direction of $\hat{\boldsymbol{b}}$, ï the MOs are calculated in a molecule-fixed rotating frame $X'Y'Z'$, with the Z' axis along the internuclear axis direction $\hat{\boldsymbol{R}}$ (see fig. 8.2). When the dynamical couplings are expressed in the same frame, one obtains:

$$\left\langle \psi_k \left| \frac{\partial}{\partial t} \right| \psi_j \right\rangle = \frac{v^2 t}{R} \left\langle \psi_k \left| \frac{\partial}{\partial R} \right| \psi_j \right\rangle + \frac{bv}{R^2} \left\langle \psi_k \left| \mathrm{i} L_{y'} \right| \psi_j \right\rangle \tag{8.17}$$

where $L_{y'}$ is the Y' component of the electronic angular momentum:

$$\mathrm{i} L_{y'} = z' \frac{\partial}{\partial x'} - x' \frac{\partial}{\partial z'} \tag{8.18}$$

The dynamical couplings in (8.17) have two components:

1 *Radial couplings*: $\langle \psi_k \left| \frac{\partial}{\partial R} \right| \psi_j \rangle$, which are due to the change of the MOs with the internuclear distance.

2 *Rotational couplings*: $\langle \psi_k \left| \mathrm{i} L_{y'} \right| \psi_j \rangle$, which are due to the rotation of the internuclear axis during the collision.

The first step in the application of the molecular expansion is the solution of eq. (8.15) to obtain the MOs and their energies. For one-electron systems, efficient techniques has been developed [23] to solve (8.15) and also to evaluate the dynamical couplings ([24]). Electronic energies and couplings are then substituted in the system (8.16), which is solved to obtain the coefficients $a_k(t)$ of eq. (8.14).

3.2 Mechanism of non-adiabatic transitions.

Some physical effects that give rise to non-adiabatic transitions are:

- **Avoided crossings.** In the avoided crossing between two potential energy curves, ε_j and ε_{j+1} approach, with a minimum of the energy difference $\varepsilon_{j+1} - \varepsilon_j$ at $R = R_0$. Then, the transition between the corresponding MOs, ψ_j and ψ_{j+1} becomes sizeable near R_0. Moreover, these MOs interchange their "characters" in the avoided crossing. Calling δ the width of the avoided crossing region, we have $\psi_{j+1}(R_0 + \delta/2) \simeq \pm\psi_j(R_0 - \delta/2)$, which gives rise to a radial coupling. In fact, for small δ, we can approximate the radial coupling between these MOs in the form:

$$\left\langle \psi_j \mid \frac{\partial \psi_{j+1}}{\partial R} \right\rangle \simeq \pm\frac{1}{\delta} \langle \psi_j(R_0 - \delta/2)|\psi_{j+1}(R_0 + \delta/2)\rangle \simeq \pm\frac{1}{\delta}, \tag{8.19}$$

  Radial couplings exhibit Lorentzian peaks in the neighborhood of avoided crossings, and transitions there are often described by means of the linear or Landau-Zener model [25].

- **Rotational transitions.** As an example of rotational transitions, we consider two MOs verifying:

$$\psi_j \underset{R\to 0}{\sim} p_{z'}\,; \quad \psi_{j+1} \underset{R\to 0}{\sim} p_{x'} \tag{8.20}$$

  where $p_{x'}$, $p_{z'}$ are orbitals of the united atom, defined in the molecule fixed frame, which fufil

$$< p_{z'}|\mathrm{i}L_{y'}|p_{x'} >= 1$$

  A rotation of a small angle θ about the axis Y' is carried out by the rotation operator

$$\exp(-\mathrm{i}\theta L_{y'}) \simeq 1 - \mathrm{i}\theta L_{y'}, \tag{8.21}$$

  which yields in the limit $R \to 0$:

$$< p_{z'}|\mathrm{i}L_{y'}|p_{x'} >\simeq -(\theta)^{-1} < p_{z'}|\exp(-\mathrm{i}\theta L_{y'})|p_{x'} > \tag{8.22}$$

  Physically, the effect of the quick rotation of the internuclear axis near $R = b$ is to leave the electron in an orbital which is a superposition of the two orbitals $p_{x'}$, $p_{z'}$, defined in the rotating frame.

- **Delocalization** We consider two orbitals with asymptotic behaviours:

$$\psi_j \underset{R\to\infty}{\sim} \Phi_j^A(\boldsymbol{r}_A)\,; \quad \psi_{j+1} \underset{R\to\infty}{\sim} \Phi_{j+1}^B(\boldsymbol{r}_B) \tag{8.23}$$

As R decreases the MOs delocalize and can be represented by combinations of the form:

$$\psi_j = N_j \left(\Phi_j^A + \alpha(R)\Phi_{j+1}^B\right) \; ; \; \psi_{j+1} = N_{j+1} \left(\Phi_{j+1}^B + \beta(R)\Phi_j^A\right) \tag{8.24}$$

At large internuclear distances, $< \Phi_j^A | \Phi_{j+1}^B > \simeq 0$, and then:

$$\left\langle \psi_j \mid \frac{\partial \psi_{j+1}}{\partial R} \right\rangle \simeq N_j N_{j+1} \frac{\mathrm{d}\beta}{\mathrm{d}R} \tag{8.25}$$

The mechanism of a given collision involves several radial and rotational transitions, whose relative importance is determined by the electronic energies and couplings of the particular system considered. An useful illustration of this mechanism is the "collision history"; i.e. the plot of the populations of the MOs, $|a_j(t)|^2$ as functions of t, which indicates how transitions occur during the collision.

In practice, one has to study the electronic energies and dynamical couplings to decide which MOs must be included in the CC expansion; this selection procedure involves in many cases a trial-and-error process, where more MOs are added to the initial expansion after checking that substantial transitions take place to them. Given that energies and dynamical couplings are different for each system, an independent calculation is required for each collision, which is not generally the case in the high v methods where calculation for series of systems are carried out by changing the input parameters.

3.3 Diabatic bases.

In some cases, radial couplings are very large (for example in narrow avoided crossings), which can cause numerical difficulties when solving (8.16). Then, it can be useful to employ a *diabatic* basis $\{\psi_j^d\}$, formed by linear combinations of the MOs $\{\psi_j\}$ (the adiabatic states), and therefore, the diabatic states are no longer eigefunctions of H_{el}. The CC expansion has then the form:

$$\Psi^d(\mathbf{r}, t) = \sum_j a_j^d(t) \psi_j^d(\mathbf{r}; R) \exp\left(-\mathrm{i} \int_0^t H_{jj}^d \mathrm{d}t'\right) \tag{8.26}$$

where

$$H_{jj}^d = \langle \psi_j^d | H_{\text{el}} | \psi_j^d \rangle \,, \tag{8.27}$$

and the system of differential (8.16) must be replaced by:

$$\mathrm{i}\dot{a}_k^d = \sum_j \left\langle \psi_k^d \middle| H_{\text{el}} - \mathrm{i}\frac{\partial}{\partial t} \middle| \psi_j^d \right\rangle a_j^d \exp\left[-\mathrm{i} \int_0^t (H_{jj}^d - H_{kk}^d) \mathrm{d}t'\right] \tag{8.28}$$

It must be noted that the electronic couplings $\langle\psi_k^d|H_{\rm el}|\psi_j^d\rangle$ are independent of v, while the dynamical couplings are proportional to v (see eq. (8.17)). So that, if the diabatic states are built up in such a way that the dynamical couplings are small, the transitions at high v between them will decrease with respect to those between the original adiabatic states.

In the particular case of a narrow avoided crossing of two energy curves, $\varepsilon_j, \varepsilon_{j+1}$, a diabatic basis $\{\psi_j^d, \psi_{j+1}^d\}$ can be defined by imposing that the peak of the radial coupling $\left\langle \psi_j^d|\frac{\partial}{\partial R}|\psi_{j+1}^d\right\rangle$ vanishes, which leads to a crossing of the diabatic Hamiltonian diagonal matrix elements H_{jj}^d, H_{j+1j+1}^d. At high v, this crossing is traversed *diabatically*; i.e. the transitions between the diabatic states are small because the remaining electronic coupling is uneffective. Hence, the diabatic basis is also useful to simplify the collision mechanism. Several definitions of diabatic states can be found in [5].

4. Translation Factors. The CTF Method.

The so-called *momentum transfer* or *electron translation problem* has been treated in many papers since it was pointed out in 1958 [22]. This difficulty is related to the presence of the ETFs in the initial condition (8.4). However, when the expansion (8.14) is employed, each basis function fulfils:

$$\psi_k(\boldsymbol{r}; R) \underset{R\to\infty}{\sim} \Phi_k^{A,B}(\boldsymbol{r}_{A,B}), \tag{8.29}$$

which, unlike the correct initial and final conditions (eqs. (8.4), (8.8), (8.9)), does not contain any translation factor. At first sight, this does not seem to be a serious difficulty, because the molecular set is complete, and the linear combination (8.14) is able to describe any initial condition. However, in practice, truncated expansions are used that do not include continuum MOs, and then they cannot reproduce the ETFs which are continuum wavefunctions. It must be remarked that this is not only a formal problem, but neglecting the ETFs produces important practical difficulties such as non-vanishing asymptotic couplings and non-Galilean invariant cross sections.

The generally accepted solution to the ETF problem, is to modify the expansion (8.14) by multiplying each MO by a ETF to ensure that each basis function verifies asympotically the semiclassical equation. The most usual way to modify the molecular expansion is the common translation factor (CTF) method, proposed in ref. [26], and applied in many works (see [2]). In this method, all molecular functions are multiplied by *the same* ETF $D(\boldsymbol{r}, t)$, so that the CC expansion has the form:

$$\Psi^{\rm CTF}(\mathbf{r}, t) = D(\boldsymbol{r}, t) \sum_k a_k(t)\psi_k(\mathbf{r}; R) \exp\left(-{\rm i}\int_0^t \varepsilon_k {\rm d}t'\right) \tag{8.30}$$

Where the coefficients $a_k(t)$ are determined by substituting Ψ^{CTF} in the semi-classical equation (8.2), which leads to the system of differential equations:

$$\mathrm{i}\dot{a}_k = \sum_j \left\langle \psi_k D \left| H_{\mathrm{el}} - \varepsilon_j - \mathrm{i}\frac{\partial}{\partial t} \right| \psi_j D \right\rangle a_j \exp\left[-\mathrm{i}\int_0^t (\varepsilon_j - \varepsilon_k)\mathrm{d}t'\right] \tag{8.31}$$

The usual form of a CTF is:

$$D(\boldsymbol{r}, t) = \exp\left[\mathrm{i}U(\boldsymbol{r}, t)\right] \tag{8.32}$$

where the function U verifies:

$$\begin{aligned} \lim_{R\to\infty, \boldsymbol{r}_A \text{ finite}} U &= -p\boldsymbol{v}\cdot\boldsymbol{r} - p^2v^2t/2 \\ \lim_{R\to\infty, \boldsymbol{r}_B \text{ finite}} U &= q\boldsymbol{v}\cdot\boldsymbol{r} - q^2v^2t/2 \end{aligned} \tag{8.33}$$

By comparing the definition of the CTF (eq. (8.32)) with its asymptotic forms (8.33) , a simple interpretation of the CTF emerges: $\nabla U(\boldsymbol{r}, t)$ is an electronic velocity field, which is defined by means of a switching function, $f(\boldsymbol{r}, R)$:

$$U(\boldsymbol{r}, t) = f(\boldsymbol{r}, R)\boldsymbol{v}\cdot\boldsymbol{r} - f^2(\boldsymbol{r}, R)v^2t/2 \tag{8.34}$$

where

$$\lim_{R\to\infty, \boldsymbol{r}_A \text{ finite}} f(\boldsymbol{r}, R) = -p; \qquad \lim_{R\to\infty, \boldsymbol{r}_B \text{ finite}} f(\boldsymbol{r}, R) = q \tag{8.35}$$

Several switching functions have been used (see e.g., [26], [27], [28], [29]) and many other can be proposed. An example is [28]:

$$f(\boldsymbol{r}, R) = \frac{R}{R^2 + \beta^2}\boldsymbol{r}\cdot\widehat{\boldsymbol{R}} \tag{8.36}$$

where β is a parameter, which determines the extension of the region $R < \beta$ where U is small.

With the definitions of eqs. (8.34) and (8.35), each term of the CTF expansion (8.30) has the correct asymptotic behaviour, and the initial condition has the form:

$$\lim_{t\to-\infty} a_k(t) = \delta_{ik} \exp\left[-\mathrm{i}\int_0^t [E_i - \varepsilon_i(R)]\mathrm{d}t\right], \tag{8.37}$$

while the transition probabilites are:

$$P_{if}(b, v) = \lim_{t\to\infty} |\langle \psi_f D | \Psi \rangle - \delta_{if}|^2 = \lim_{t\to\infty} |a_f(t) - \delta_{if}|^2 \tag{8.38}$$

The main advantages of the CTF method are:

1 Since all MOs are multiplied by the same ETF, and since the MOs $\{\psi_k\}$ form a complete set, the set of modified functions $\{\psi_k D\}$ is also complete, and therefore the method is formally convergent.

2 The modified dynamical couplings are calculated with a relatively modest computational effort when compared with not-common treatments.

3 The method is easily generalized to many-electron systems and diabatic bases.

When a CTF is used, a kinematic effect is described: the nuclei drag the electron, but the CTF also introduces a dynamic effect: the dynamical couplings, and therefore the transitions between the MOs, are modified (see eq. (8.31)). However, the velocity field ∇U is irrotational [30] and cannot describe rotational transitions. Then, an appropriate CTF verifies $\nabla U = 0$ at small R, which permits that the rotational transitions be completely described by the unmodified rotational couplings. For example, the switching function of eq. (8.36) contains the cut-off factor $\frac{R}{R^2+\beta^2}$ in (8.36) yielding $\lim_{R\to 0} f = 0$, and accordingly $\lim_{R\to 0}(\nabla U) = 0$.

As mentioned above, the CTF expansion is formally convergent independently of the switching function employed, but different functions will lead to different convergence speeds. In fact, the choice of CTF is a problem similar to the choice of a basis set to evaluate atomic and molecular wavefuctions. This point of view has been used as a criterion, first suggested in ref. [31], to optimize CTFs; it requires that the choice of the CTF minimizes the couplings with the MOs not included in the basis. This idea is the basis of the so-called norm-method of ref. [32]. There is however a natural boundary of the method [33]: When v increases, ionization starts to be sizeable, and one finds that the CTF with the asymptotic behaviour of eq. (8.35) cannot describe the motion of the ionized electron and is no longer appropriate.

The behaviour of the CTF molecular method at high v has been studied in [2], [34] [35], [36] and [3], where several collisions of fully stripped ions with H were considered. There are two motivations for this study: First, the suggestion that the ionization process can be described by bound MOs, as a consequence of the "hidden crossing" mechanism (see [37]). Second, the results of several CC calculations that showed good agreement between calculated electron-loss (charge transfer + ionization) cross sections and the experimental ones, pointing out to a common molecular mechanism for charge transfer and ionization processes. We can summarize the main conclusions of these works as follows:

1 the CTF method overestimates charge transfer total cross sections calculated for impact energies beyond their maxima.

2 Calculated charge transfer cross sections at high v show good agreement with experimental target electron-loss cross sections.

3 At short R, some modified functions $\chi_j D$ have positive energy with respect to both moving nuclei and then they can describe ionization, which explains the good agreement with experimental electron-loss cross sections. Nevertheless, this description is neccessarily incomplete, since the functions $\chi_j D$ are asympotically charge transfer or excitation channels, and the electronic flux that should yield ionization in the limit $t \to \infty$ overpopulates the charge transfer (and in some cases also the excitation) channels.

4 From the comparison with accurate ionizing wavefunctions [12], we have found that the CTF expansion attemps to represent the ionizing wavefunction by means of a very complicate combination of the modified functions $\chi_j D$, with a CTF that does not describe the motion of an ionized electron; this unphysical behaviour of the CTF does not allow to correctly represent the time evolution of the ionizing wavefunction, and the population remains "trapped" in some charge transfer states. The consequence of this trapping effect is that the wavefunction can be very different from the exact one and therefore, the expansion (8.30) is not useful to evaluate parameters such as momentum components of the ionized electron, which are more sensitive than total cross sections to the quality of the wavefunction.

5 The apparent interlocking of charge transfer and ionization mechanisms at high v is mainly a consequence of the limitation of the CC expansion to describe ionization. The method attempts to represent the ionizing wavefunction by combining the states available, but this process is largely artificial.

5. Many-electron Systems

In order to generalize the CC molecular expansion to many-electron systems, one must require that the collisional wavefunction be antisymmetric with respect to the interchange of the coordinates of two electrons. This requirement is easily fulfilled by the CTF wavefunction by taking:

$$U(\boldsymbol{r}_1, \boldsymbol{r}_2, ..., \boldsymbol{r}_N, t) = \sum_{i=1}^{N} \left[f(\boldsymbol{r}_i, t)\boldsymbol{v} \cdot \boldsymbol{r}_i - f^2(\boldsymbol{r}_i, t)v^2 t/2 \right] \tag{8.39}$$

which is symmetric with respect to the electrons exchange, and with

$$\Psi^{\rm CTF}(\boldsymbol{r}_1, \boldsymbol{r}_2, ..., \boldsymbol{r}_N, t) = \exp\left[{\rm i}U(\boldsymbol{r}_1, \boldsymbol{r}_2, ..., \boldsymbol{r}_N, t) \right] \times$$

$$\times \sum_k a_k(t)\psi_k(\boldsymbol{r}_1, \boldsymbol{r}_2, ..., \boldsymbol{r}_N; R) \exp\left(-\mathrm{i}\int_0^t \varepsilon_k \mathrm{d}t'\right) \qquad (8.40)$$

where the many-electron molecular functions $\psi_k(\boldsymbol{r}_1, \boldsymbol{r}_2, ..., \boldsymbol{r}_N; R)$ are antisymmetric.

The main difference with the single-electron case is the difficulty of evaluating the many-electron molecular wavefunctions. In fact, because of the presence of the electron-electron repulsion potential, analytical molecular wavefunctions cannot be obtained, and the functions ψ_k of eq. (8.40) are approximate solutions of the Born-Oppenheimer electronic equation (8.15), which are evaluated by employing quantum chemistry techniques. A detailed explanation of these techniques can be found in quantum chemistry textbooks (see e.g. [38]) and their application in collisions has been considered in ref. [5]. In practice, the approximate molecular functions are expressed in terms of suitable basis functions, and the most usual are:

- Slater type orbitals (STO):

$$\xi^{STO}(\boldsymbol{r}; n, l, m, \alpha) = N_{\alpha,n,l} r^{n-1} Y_{lm}(\widehat{\boldsymbol{r}}) \exp(-\alpha r) \qquad (8.41)$$

  where $N_{\alpha,n,l}$ is a normalization constant, n, l, m are integers, Y_{lm} is a spherical harmonic and α is a non-linear parameter.

- Gaussian type orbitals (GTO):

$$\xi^{GTO}(\boldsymbol{r}; n, l, m, \alpha) = N_{\alpha,n,l,m} x^n y^l z^m \exp(-\alpha r^2) \qquad (8.42)$$

  where $N_{\alpha,n,l,m}$ is a normalization constant, n, l and m are integers; x, y, z, the electron cartesian coordinates and α is a non-linear parameter.

5.1 Configuration interaction.

In the Configuration Interaction (CI) method, we look for an approximate solution of the electronic Schrödinger equation:

$$H_{\mathrm{el}}(\boldsymbol{r}_1, \boldsymbol{r}_2, ..., \boldsymbol{r}_N; R)\psi_j(\boldsymbol{r}_1, \boldsymbol{r}_2, ..., \boldsymbol{r}_N; R) = \varepsilon_j(R)\psi_j(\boldsymbol{r}_1, \boldsymbol{r}_2, ..., \boldsymbol{r}_N; R) \qquad (8.43)$$

where

$$\begin{aligned} H_{\mathrm{el}}(\boldsymbol{r}_1, \boldsymbol{r}_2, ..., \boldsymbol{r}_N; R) &= \sum_{i=1}^{N}\left[-\frac{1}{2}\nabla_i^2 - \frac{Z_A}{r_{iA}} - \frac{Z_B}{r_{iB}}\right] + \\ &\quad \sum_{i<j}\frac{1}{r_{ij}} + \frac{Z_A Z_B}{R} \\ &= H_{\mathrm{core}} + \sum_{i<j}\frac{1}{r_{ij}} \end{aligned} \qquad (8.44)$$

where ψ_j takes the form of a linear combination of N-electron wavefunctions ϕ_m called *configurations*:

$$\psi_j = \sum_m c_{mj} \phi_m \tag{8.45}$$

The configurations are built up from products of one-electron functions (*spin-orbitals*) $\chi_p(\boldsymbol{r}_i, \sigma_i)$. The spin-orbitals are products of orbitals $\varphi(\boldsymbol{r}_i)$ (e.g., STOs, GTOs, etc.), and spin wavefunctions α, β, with spin quantum number $m_s = +1/2$ and for $m_s = -1/2$ respectively. A configuration is a linear combination of a few products of spin-orbitals of the form:

$$\phi_m = \sum_\lambda d_{m\lambda} \prod_p \chi_p^\lambda(\boldsymbol{r}_p, \sigma_p) \tag{8.46}$$

where the constant coefficients $d_{m\lambda}$ are chosen so as the configuration ϕ_m verifies the following conditions:

- It is antisymmetric with respect to the interchage of two electrons.
- It is an eigenfuction of the spin operators S^2 (it has a given multiplicity) and S_z.
- It is symmetry adapted: it has the symmetry of an irreducible representation of the molecular point group. For example, for atomic calculations, a configuration must be an eigenfunction of the orbital angular momentum operators L^2 and L_z.

To illustrate these definitions, we consider the well-known example of He 1s2s configurations (see e.g. [38]) One considers two orbitals (1s and 2s) and then there are four spin-orbitals: 1sα, 1sβ, 2sα, 2sβ. The possible spin-orbital products are: 1sα(1) 2sα(2), 1sα(1) 2sβ(2), 2sα(1) 1sα(2), 2sα(1) 1sβ(2), 1sβ(1) 2sα(2), 1sβ(1) 2sβ(2), 2sβ(1) 1sα(2), 2sβ(1) 1sβ(2), and a ^{1}S configuration is:

$$\frac{1}{\sqrt{2}} [\, \| 1\mathrm{s}\alpha(1) 2\mathrm{s}\beta(2) \| - \| 1\mathrm{s}\beta(1) 2\mathrm{s}\alpha(2) \| \,]$$

where the symbol $\|\ \|$ indicates a (normalized) Slater determinant; this is equivalent to

$$\frac{1}{2} [(1\mathrm{s}(1)2\mathrm{s}(2) + 2\mathrm{s}(1)1\mathrm{s}(2)) \, (\alpha(1)\beta(2) - \beta(1)\alpha(2))]$$

This function is obviously antisymmetric, the combination of spin wavefunctions is eigenfunction of S^2 with $S = 0$, and the spatial wavefunction is eigenfunction of L^2 with $L = 0$, then it is a ^1S configuration. Analogously, in the molecular case, configurations are built from products of spin-molecular-orbitals. For example, a configuration $^1\Sigma_g^+$ of H_2 is of the form:

$$\|\sigma_g\alpha(1)\sigma_g\beta(2)\| = \frac{1}{\sqrt{2}}\sigma_g(1)\sigma_g(2)\left[\alpha(1)\beta(2) - \beta(1)\alpha(2)\right]$$

where σ_g is a MO of this symmetry.

To determine the coefficients c_{jm} of eq. 8.45 the variational method is applied (see e.g. [38]). In this method, one looks for the coefficients that minimize the energy expectation value:

$$\overline{\varepsilon_j} = \frac{\langle\psi_j|H_{\text{el}}|\psi_j\rangle}{\langle\psi_j|\psi_j\rangle} \tag{8.47}$$

The minimum condition $\delta\overline{\varepsilon_j} = 0$ leads to the secular equation:

$$(\boldsymbol{H} - \overline{\varepsilon_j}\boldsymbol{S})\boldsymbol{c_j} = 0 \tag{8.48}$$

where $\boldsymbol{H}$, and $\boldsymbol{S}$ are respectively the Hamiltonian and overlap matrices, whose elements are:

$$\begin{aligned} H_{mn} &= \langle\phi_m|H_{\text{el}}|\phi_n\rangle \\ S_{mn} &= \langle\phi_m|\phi_n\rangle\,, \end{aligned} \tag{8.49}$$

and the solution of eq. (8.48) is the column vector $\boldsymbol{c_j}$ whose elements are the coefficients c_{mj}.

The application of the molecular expansion with CI wavefunctions involves the calculation of the dynamical couplings $\langle\psi_j|\partial/\partial t|\psi_k\rangle$, which will be considered in section 8.5.5, and the solution of the system of differential equations (8.31). An important difficulty of the use of the variational method comes from the fact that the phase of the approximate solutions ψ_j of eq. (8.44) is arbitrary. In practice, since the configurations ϕ_m are usually real functions, this results in an arbitrary sign of the solutions of (8.48), (and consequently, arbitrary signs of the dynamical couplings) which in actual calculations changes randomly as a function of R. Therefore, the method involves a painstaking first step (also required for SCF and MRCI wavefunctions) of removing random signs before substituting the corresponding couplings in (8.31). The CI method has been applied in our laboratory to obtain molecular wavefunction for several two-electron collision systems; e.g. He^+-H [39] H^+-H^- [40], Li^{2+}-H [41].

An alternative to the CI method is to expand directly the collisional wavefunction in a basis set of configurations [42]. For a two-electron system, this expansion has the form:

$$\Psi_{\text{cf}}^{\text{CTF}}(\boldsymbol{r}_1, \boldsymbol{r}_2, t) = D(\boldsymbol{r}_1, \boldsymbol{r}_2, t)\sum_k a_k(t)\phi_k(\boldsymbol{r}_1, \boldsymbol{r}_2; R)\exp\left(-\mathrm{i}\int_0^t H_{kk}\mathrm{d}t'\right) \tag{8.50}$$

where the configurations ϕ_k are products of MOs of a one-electron system, and then they are not eigenfunctions of H_{el}. The advantage of this method is that the calculation of dynamical coupling is easier than between CI wavefunctions and the sign of the configurations do not change randomly. The disadvantage is that larger basis sets than for the CI expansion are required. Expansion (8.50) has been applied in refs. [29], [43] to study single and double capture in collisions of multiply charged ions (C^{6+}, N^{7+}, O^{8+}, Ne^{8+}, Ne^{10+}) with He, including double capture into autoionizing states.

5.2 The SCF method.

In the Self Consistent Field (SCF) method, a single configuration approach is employed, and the variational method is applied to determine the "best" orbitals, which define this configuration. The simplest case is a closed-shell configuration, in which all orbitals are doubly occupied and this configuration has the form:

$$\psi_j = \left\| \varphi_1\alpha(1)\varphi_1\beta(2)...\varphi_{N/2}\alpha(N-1)\varphi_{N/2}\beta(N) \right\| , \qquad (8.51)$$

which is called the Restricted Hartree-Fock (RHF) method. In the variational method, one requires the condition: $\delta[\langle\psi_j|H_{\text{el}}|\psi_j\rangle] = 0$ for small variations $\delta\varphi_p$ with the restrictions $\langle\varphi_p|\varphi_q\rangle = \delta_{pq}$.

Usually, the orbitals φ_p are expressed as linear combinations of a set of n ($n > N$) basis orbitals, $\{\xi_s\}$, (for example STOs or GTOs):

$$\varphi_p = \sum_{s=1}^{n} b_{sp}\xi_s \, , \qquad (8.52)$$

and the application of the variational method yields a set of equations with the form of a secular equation for the coefficients b_{ps}, which define the so-called canonical orbitals: [44]:

$$(\boldsymbol{\mathcal{F}} - \boldsymbol{\mathcal{S}}\varepsilon_p)\, \boldsymbol{b}_p = 0 \, , \qquad (8.53)$$

where $\boldsymbol{b}_p$ is a column vector with components b_{sp}, $\boldsymbol{\mathcal{S}}$ is the overlap matrix in the $\{\xi_s\}$ basis and $\boldsymbol{\mathcal{F}}$ is the matrix of the Fock operator:

$$\begin{aligned} \mathcal{F}_{st} &= \langle\xi_s|H_{\text{core}}|\xi_t\rangle + \qquad (8.54) \\ &+ \sum_{p=1}^{N/2}\sum_{q,r}^{n} b_{qp}b_{rp}\left[2\left\langle \xi_s\xi_q\left|\frac{1}{r_{12}}\right|\xi_t\xi_r\right\rangle - \left\langle \xi_s\xi_q\left|\frac{1}{r_{12}}\right|\xi_r\xi_t\right\rangle\right] \end{aligned}$$

The eigenvalue ε_p in (8.53) is interpreted as the energy of the electron decribed by the orbital φ_p in the average field created by the nuclei and the remainder $N-1$ electrons, which occupy the orbitals that define the configuration

(8.51). Since the Fock matrix elements (8.54) depend on the coefficients b_{qp}, which are the solutions of (8.53), this latter equation must be solved iteratively.

For open-shell configurations, the unrestricted Hartree-Fock (UHF) method is usually employed, where a treatment similar to that of eqs. (8.52)- (8.54) is applied, but the restriction of the same orbital part for two spin-orbitals (with α and β spin functions) of eq. (8.51) is removed. Therefore, a configuration of the form:

$$\psi_j = \|\varphi_1\alpha(1)...\varphi_m\alpha(m)\varphi_{m+1}\beta(m+1)\varphi_N\beta(N)\| \ , \tag{8.55}$$

is used. Although, SCF methods have been widely employed in static problems, their usefulness in dynamics is limited by the fact that they are not appropriate to evaluate excited states wavefunctions, and are normally used as the starting point of a CI calculation.

5.3 MRCI methods.

These methods start from the SCF treatment, and although the same arguments apply to the UHF method, we take, as an example, the RHF-SCF case; this provides a set of n canonical orbitals $\{\varphi_s\}$, solutions of (8.53), with n the size of the basis set (see eq. (8.52)). These orbitals can be classified into two sets: 1) *Occupied* orbitals: the $N/2$ orbitals of lowest energy, which are those defining the ground state configuration (8.51). 2) *Virtual* or unoccupied orbitals: the remainder $n - N/2$ orbitals.

In order to obtain wavefunctions for excited states, the SCF orbitals can be used to construct a set of configurations by promoting electrons from the occupied orbitals to the virtual ones. As an example, the excitation of two electrons from the last occupied orbital $\varphi_{N/2}$ in eq. (8.51) to the first virtual $\varphi_{N/2+1}$ gives rise to the antisymmetrized product of spin-orbitals:

$$\psi_k = \left\|\varphi_1\alpha(1)\varphi_1\beta(2)...\varphi_{N/2+1}\alpha(N-1)\varphi_{N/2+1}\beta(N)\right\| \ , \tag{8.56}$$

This set of configurations is then employed as a basis for a CI calculation. As long as all excitations are permitted, the difference between occupied and virtual orbitals is irrevelant. However, the size of the resulting configurations set can be very large, and some simplifications are often introduced in many-electron calculations; for example, only single and double excitations are allowed. This approximation provides an accurate energy for the ground state, for which the occupied orbitals were optimized in the SCF step, but less accurate results for the excited states. In dynamical calculations, the transitions are determined by the energy differences (see eq. (8.16) of section 8.2), and it is desirable to evaluate with similar accuracy all states included in the basis, and a better alternative to this treatment are the multirreference CI (MRCI) methods. The scheme of a MRCI calculation is as follows:

1 SCF calculation to obtain a set of canonical orbitals $\{\varphi_p\}$.

2 Definition of a reference configurations set: This set is formed by a few configurations build up from products of the spin-orbitals obtained in the SCF step, and which usually includes the SCF configuration. These configurations are assumed to provide a reasonable description of the states required.

3 Construction of a CI space by allowing excitations (normally single and double excitations) from the reference configurations.

4 Solution of the secular equation (8.48) with $\boldsymbol{S} \equiv \mathbf{1}$.

In practice, a first MRCI calculation is carried out at infinite internuclear separation for the atomic states involved in the collision, to compare the calculated energies with the corresponding spectroscopic values. The reference configurations are chosen in order to achieve a given precision in the atomic energies. For finite R, where experimental data are not in general available, the reference set is changed with the criterion of keeping an approximate constant weight (the sum of the squares of the CI coefficients) (of about 95%) of the reference configurations in the final wavefunctions. The application of MRCI methods in atomic collision is carried out by employing the quantum chemistry packages in which this technique has been implemented. Some examples of the application of MRCI packages to collisions involving doubly charged ion are the following:

- Ne^{2+} + H collisions have been studied [45] by using the package GAMESS [46].

- C^{2+} + H collisions have been studied in refs. [47] and [48], by using the packages MRDCI [49] and MELD [50], respectively

- O^{2+} + H collisions have been considered in ref. [51] . The MRCI calculation employed a CI method based on the the package CIPSI [52].

- C^{2+} + He collisions have been studied in [53] by using MELD.

5.4 Effective potentials.

In some collision processes there is a clear distinction between the *active* electron, which is exchanged or excited during the collision, and the remainder (*core*) electrons. As an example, in the collision C^{4+} + H, at low energies ($E <$ 10keV/a.m.u.), only the electron capture process:

$$C^{4+}(1s^2\ {}^1S) + H(1s) \rightarrow C^{3+}(1s^2nl) + H^+ \tag{8.57}$$

is relevant. In this process, the active electron interacts with a core formed by the nuclei and the $1s^2$ closed-shell, which does not significantly change in this process. To treat this problem approximately, the electronic Hamiltonian $H_{\rm el}$ is replaced by the one-electron effective Hamiltonian:

$$h^e = -\frac{1}{2}\nabla^2 + V^e(r_C) - \frac{1}{r_H} \tag{8.58}$$

where r_C and r_H are respectively the electron distances to the C and H nuclei, and V^e is an effective potential that describes the interaction of of the active electron with the core $C^{4+}(1s^2)$.

A widely employed type of effective potentials are model potentials, and a simple example is that of ref. [54]:

$$V^e = -\frac{Z_C - N_c}{r} - \frac{N_c}{r_C}(1 + \alpha r_C)\exp(-\alpha' r_C) \tag{8.59}$$

where Z_C(=6) is the charge of the C nucleus, N_c the number of core electrons (2 in the present example) and α, α' two parameters, which are determined by fitting the $C^{3+}(1s^2 2s)$ and $C^{3+}(1s^2 2p)$ ionization potentials to the experimental values.

Some works have also treated systems with two active electrons, where the model potential describes the interaction of the active electrons with the atomic cores and the wavefunction for the active electrons is obtained by means of a CI calculation (see [55]). This method has been applied for example to $C^{3+}(1s^2 2s)$+ H(1s) collisions [56], where the CI calculation for the active electrons was carried out as explained in section 8.5.1.

5.5 Calculation of dynamical couplings.

In general, the evaluation of rotational couplings is carried out analytically and it does not involve a special difficulty from the computational point of view (see [5]). Also, the switching function f that defines the CTF in eq. (8.39) is chosen in order to simplify the calculation of the new couplings introduced by the CTF. The main difficulty lies on the calculation of the radial couplings $\langle\psi_i|\frac{\partial}{\partial R}|\psi_j\rangle$, and there are two kinds of methods to evaluate them:

1 *Analytical methods.* This includes methods based on the Hellman-Feynman theorem (see ref. [57]), which assume that the electronic wavefunctions ψ_j are exact eigenfunctions of $H_{\rm el}$, and the methods of refs. [58, 59], which are not limited by this assumption. In particular, the method of ref. [58] has been employed for two-electron CI wavefunctions (see section 8.5.1 and 8.5.5) in terms of GTOs, where it is very useful because the radial couplings are expressed in terms of the derivatives with respect to R of the matrix elements H_{mn}, S_{mn} of eq. (8.49) that are analytical functions of R.

2 *Numerical methods.*

Although analytical methods, like that of ref. [58], are in principle the most appropriate for wavefunctions built from GTOs, their implementation in quantum chemistry packages involves a considerable programming effort, and numerical methods are usually employed. (see [60], [6]). These techniques are based on the finite difference differentiation approximation:

$$\left\langle \psi_i \left| \frac{\partial}{\partial R} \right| \psi_j \right\rangle \simeq \delta^{-1} \langle \psi_i(\boldsymbol{r}, R) | \psi_j(\boldsymbol{r}, R+\delta) \rangle \tag{8.60}$$

In general, the use of these numerical methods requires to check the stability with respect to the differentiation step δ. A practical disadvantage is the increase of the computational effort because the need of evaluating the wavefunctions in two nearby points R and $R+\delta$ and it is necessary to ensure that the sign of the wavefunctions is the same in both points.

6. Atomic expansions.

6.1 Basic equations.

At relatively high collision energies ($E \gtrsim$10keV/a.m.u, for ion-H collisions), an atomic expansion becomes useful. A complete description of the method, together with references to many applications can be found in the book of ref. [1]. To describe the main characteristics of the atomic expansion, we consider the one-electron semiclassical equation (8.2), and we expand its solutions in the set of "travelling" AOs $\{\Phi_j^A D^A,\ \Phi_k^B D^B\}$ (see eqs. (8.4)-(8.9)):

$$\begin{aligned} \Psi^{AO}(\boldsymbol{r}, t) &= \sum_j a_j(t) \Phi_j^A(\boldsymbol{r}_A) D^A(\boldsymbol{r}, t) \exp(-\mathrm{i} E_j^A t) + \\ &+ \sum_k a_k(t) \Phi_k^B(\boldsymbol{r}_B) D^B(\boldsymbol{r}, t) \exp(-\mathrm{i} E_k^B t) \end{aligned} \tag{8.61}$$

where the AOs $\Phi_j^{A,B}$ are eigenfunctions of the atomic Hamiltonians $H^{A,B}$ of eqs. (8.6) and (8.10)

As in the molecular method, substitution of the expansion (8.61) yields a system of differential equations for the coefficients $a_j(t)$. However, the transitions between the AOs are mainly caused by electrostatic couplings of the form:

$$M_{ij}^{AB} = \int \Phi_i^{A*} H_{\mathrm{el}} \Phi_j^B \exp(-\mathrm{i}\boldsymbol{v} \cdot \boldsymbol{r}) \mathrm{d}\boldsymbol{r} \tag{8.62}$$

From the computational point of view, the presence of the quickly oscillating phase factor $(D^A)^* D^B = \exp(-\mathrm{i}\boldsymbol{v} \cdot \boldsymbol{r})$ in (8.62), makes difficult the evaluation of these integrals. Moreover, because of these phase factors, the integrals depend on b and v and must be calculated for each nuclear trajectory.

Atomic expansions have been applied in general to one-electron systems with some applications to two-electron systems (an example is the calculation for C^{3+}-H of ref. [61]). For many-electron systems, the calculations are based on the use of the independent electron approximation [62], and a detailed example of this kind of calculations can be found in ref. [11].

6.2 Pseudostates.

The atomic expansion is usually employed in a region of impact energies where ionization starts to be competitive with charge exchange and excitation processes. Therefore, many works have studied the extension of expansion (8.61) to describe the ionization continuum; this is often carried out by adding to this expansion the so-called *pseudostates*. To introduce these pseudostates, we consider a two-centre basis set (e.g. STOs) $\{\xi_s^{A,B}\}$, similar to that of eq. (8.52), which is used to obtain approximations to the AO $\Phi_k^{A,B}$. We write

$$\Phi_k^{A,B} = \sum_s b_{sk}^{A,B} \xi_s^{A,B} \tag{8.63}$$

The coefficients $b_{sk}^{A,B}$ are then evaluated variationally; namely, they are solutions of the secular equations :

$$(\boldsymbol{H}^{\boldsymbol{A},\boldsymbol{B}} - \overline{E_j^{A,B}}\boldsymbol{S}^{\boldsymbol{A},\boldsymbol{B}})\boldsymbol{b}_j^{\boldsymbol{A},\boldsymbol{B}} = 0 \; , \tag{8.64}$$

which are analogous to the equation (8.48) of section 8.5.1 but for a one-electron atomic case. The elements of the matrices $\boldsymbol{H}^{\boldsymbol{A},\boldsymbol{B}}$ and $\boldsymbol{S}^{\boldsymbol{A},\boldsymbol{B}}$ are

$$H_{st}^{A,B} = \left\langle \xi_s^{A,B} | H^{A,B} | \xi_t^{A,B} \right\rangle \quad ; \quad S_{st}^{A,B} = \left\langle \xi_s^{A,B} | \xi_t^{A,B} \right\rangle \tag{8.65}$$

The solutions of (8.64) with $\overline{E_j^{A,B}} < 0$ are approximations to the AOs, while those with $\overline{E_j^{A,B}} > 0$ are approximations to the continuum states, called pseudostates. Their inclusion in the expansion, together with a set of bound states, has the following practical consequences:

1 Since the pseudostates complete the basis set, delocalization effects are better represented in the the augmented basis. Therefore their inclusion makes it possible to extend the treatment to low energies.

2 The pseudostates furnish a representation of the ionizing flux through transitions from the bound AO to the pseudostates. This implies that the populations of the bound AOs are also improved, and therefore the values of charge transfer and excitation cross sections.

3 One can calculate ionization cross sections; these are approximated by the sum of the cross sections for transitions to the pseudostates.

Atomic calculations with pseudostates have been carried out with one-centre basis, formed for example by orbitals of the target atom $\{\Phi_A\}$. If the origin of electronic coordinates is on the nucleus A then $p = 0$ and $D^A \equiv 1$ (see fig. 8.1 and eq. (8.4)), and the expansion has the form:

$$\Psi^{OC}(\boldsymbol{r},t) = \sum_j aj(t)\Phi_j^A(\boldsymbol{r}_A)\exp(-\mathrm{i}E_j^A t) , \qquad (8.66)$$

which simplifies the calculation of the required matrix elements. This kind of expansion can include a set of pseudostates, which represent continuum states and also charge transfer (with the corresponding ETF). A systematic method to build up these pseudostates, applied in several papers of the Frankfurt group, is the basis generator method, (see [63, 64, 11]) where a set of pseudostates $\{\Phi_j^\mu\}$ is created by repeated application of the operator $[H_{\rm el} - \mathrm{i}\partial/\partial t]$ [65] on a set of one-centre AOs $\{\Phi_j^A\}$:

$$\Phi_j^\mu = [H_{\rm el} - \mathrm{i}\partial/\partial t]^\mu \Phi_j^A \qquad (8.67)$$

In this way, the electronic flux leaving the zeroth order functions (the AO, Φ_j^A) is absorbed by the first order pseudostates Φ_j^1. The second order pseudostates Φ_j^2 absorb the flux that leaks the first-order ones and does not come back to the original functions, and so on (see [36]). The decreasing populations of increasing hierarchy absorber states provides a check of the convergence of this procedure. An obvious practical difficulty in the application of this scheme is the possible redundances between absorber states, which leads, after orthogonalization of the set, to vanishing functions. Another difficulty is related to the Coulomb singularity in the potential Z_B/r_B, which is solved by using regularized potentials (see [66]). Finally, the solutions must be projected on the physical charge transfer states of (8.8) to obtain charge transfer cross sections, and it is in general difficult to separate charge transfer from ionization.

Two-center AO calculations employ expansion (8.61) with pseudostates centered in one or both nuclei. An illustration of state-of-the-art atomic ([67], [68]) and molecular [69] calculations is shown in fig. 8.3 for the particular case of C^{4+}+ H collisions (see section 8.5.4). The excelent agreement between calculations using different methods indicates that CC calculations for (quasi)one-electron systems can be more accurate than the experimental results.

Three-center bases have been also used to define pseudostates [70, 71]. A CC three-center expansion has the form (see e.g. [72]):

$$\Psi^{TC}(\boldsymbol{r},t) \quad = \quad D^A \sum_j aj(t)\Phi_j^A(\boldsymbol{r}_A)\exp(-\mathrm{i}E_j^A t)$$

$$+ \quad D^B \sum_k a_k(t)\Phi_k^B(\boldsymbol{r}_B)\exp(-\mathrm{i}E_k^B t)$$
$$+ \quad D^C \sum_l a_l(t)\Phi_l^C(\boldsymbol{r}_B)\exp(-\mathrm{i}E_l^C t) \qquad (8.68)$$

where Φ_l^C are orbitals centred in an intermediate point C of the internuclear axis and the ETF $D^C(\boldsymbol{r}, t)$ represents the translational motion of an electron with the velocity of the point C with respect to the coordinates origin. The inclusion of the third center basis functions is based on the physical picture of ionization at low v (see [73]) in which the electron is not able to follow the motion of the fast nuclei and remains between them, which is also the basis of the definition of the molecular pseudostates of ref. [35].

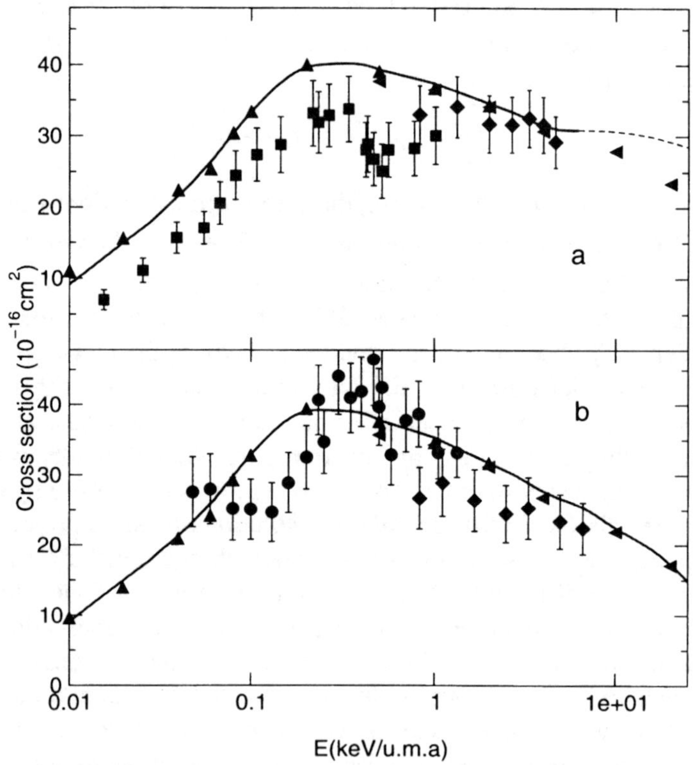

Figure 8.3. (a) Total electron capture cross section in C^{4+}+ H(1s) collisions: Full curve, [69]; ▲,[68] with rectilinear trajectories; ◁, [67]. Experimental results: ♦, [74]; ■, [75]. (b) Cross section for electron capture into C^{3+}(n=3): Full curve, [69]; ▲,[68] with rectilinear trajectories; ♦, [74]; ● [76].

7. Extension to Ion-H_2 Collisions.

The methods explained in the preceeding sections can be extended to treat ion-molecule collisions. In particular, at low impact energies, CC molecular expansions are used and the rotational motion is described by means of the sudden approximation; this is the basis of the so-called infinite order sudden approximation (IOSA), where a full quantal treatment is employed (see [77]). For $E \gtrsim 50$eV/a.m.u., the semiclassical method with three-centre molecular expansion is employed, which is described in this section. Only a few works have tried to extend the atomic CC expansion ([78], [79], [80]), where the target molecule is approximated by means of a spherically symmetric effective potential. Classical trajectory Monte Carlo methods have been also applied (see e.g. [81] and references therein). At high energies the calculations use the sudden approximation for the ro-vibrational motion, which is explained in section 8.7.2 and in many cases the Franck-Condon approximation, which can be obtained as an additional approximation to the sudden vibrational approach.

7.1 Close-coupling treatment.

To describe the main features of the CC treatment of ion-H_2 collisions, we consider the electron capture:

$$X^{q+} + H_2 \rightarrow X^{(q-1)+} + H_2^+ \tag{8.69}$$

and excitation

$$X^{q+} + H_2 \rightarrow X^{q+} + (H_2)^* \tag{8.70}$$

processes, where the products of these reactions can be found in electronic or vibrational excited states. Here we do not consider processes leading to either electronic or vibrational dissociative states.

In the semiclassical approach, the ion-diatomic relative motion is described with straight-line trajectories $\boldsymbol{R} = \boldsymbol{b} + \boldsymbol{v}t$, where $\boldsymbol{R}$ is the ion position vector with respect to the centre of the H_2 internuclear axis (see fig. 8.4). The remaining degrees of freedom are treated quantally, and the corresponding wavefunction $\Psi(\boldsymbol{r}, \boldsymbol{\rho}, t)$ is solution of the semiclassical equation:

$$\left(H_{\rm i} - {\rm i} \left. \frac{\partial}{\partial t} \right|_{\boldsymbol{r},\boldsymbol{\rho}} \right) \Psi(\boldsymbol{r}, \boldsymbol{\rho}, t) = 0 \tag{8.71}$$

where $\boldsymbol{\rho}$ is the target internuclear vector in the laboratory frame (see fig. 8.5), and $\boldsymbol{r}$ denotes the coordinates of all electrons. The equation (8.71) is the equivalent of (8.2) for ion-diatomic collisions. In this equation, $H_{\rm i}$ is the total Hamiltonian of the three-centre system without the kinetic energy term asociated to the nuclear motion along the coordinate $\boldsymbol{R}$ that is treated classicaly. Explicitly:

$$H_{\rm i}(\boldsymbol{r}, \boldsymbol{R}, \boldsymbol{\rho}) = -\frac{1}{2\mu}\nabla^2_\rho + H_{\rm el} \tag{8.72}$$

where μ is the H_2 reduced nuclear mass and

$$H_{\rm el}(\boldsymbol{r}_1, \boldsymbol{r}_2, ..., \boldsymbol{r}_N, R, \rho, \gamma) = \sum_{i=1}^{N}(T_i + V_i) + \sum_{j<k}^{N} \frac{1}{r_{jk}} + V_{\rm nuc}\ , \tag{8.73}$$

being γ the angle between vectors $\boldsymbol{R}$ and $\boldsymbol{\rho}$, T_i the electronic kinetic energy operators, V_i the attraction potential of electron i by the three nuclei, and $V_{\rm nuc}$ the nuclear repulsion potential.

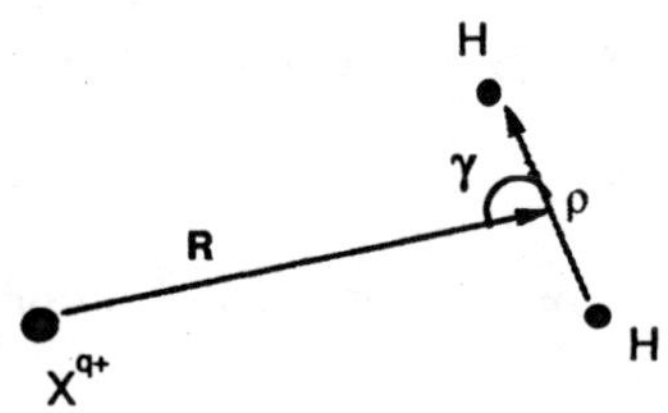

Figure 8.4. Nuclear coordinates for ion-H_2 collisions.

The equation (8.71) must be solved with the initial condition:

$$\Psi \underset{t\to-\infty}{\sim} \Psi^{JM}_{i\mu} = \rho^{-1} Y_{JM}(\widehat{\boldsymbol{\rho}})\psi_{i\mu}(\boldsymbol{r}, \rho, t) \tag{8.74}$$

with

$$\psi_{i\mu} = \Phi_i(\boldsymbol{r}; \rho)\chi_{i\mu}(\rho)D^i(\boldsymbol{r}, t)\exp(-{\rm i}E_i t) \tag{8.75}$$

where Y_{JM} is the spherical harmonic that corresponds to the H_2 initial rotational state, D^i is a plane-wave translation factor, $\chi_{i\mu}$ is the wavefunction for the H_2 vibrational initial state and Φ_i is the electronic initial state, which is the product of the initial electronic states of the ion X^{q+} and the H_2 molecule:

$$\Phi_i(\boldsymbol{r}, \rho) = \Phi_i^{\rm H_2}(\boldsymbol{r}_1, \boldsymbol{r}_2; \rho)\Phi_i^{\rm X}(\boldsymbol{r}_3, ..., \boldsymbol{r}_N) \tag{8.76}$$

Analogously, a reaction exit channel has the form:

$$\Psi^{J'M'}_{n\nu} = \rho^{-1} Y_{J'M'}(\widehat{\boldsymbol{\rho}})\psi_{n\nu} \tag{8.77}$$

with

$$\psi_{n\nu} = \Phi_n(\boldsymbol{r};\rho)\chi_{n\nu}(\rho)D^n(\boldsymbol{r},t)\exp(-\mathrm{i}E_f t) \tag{8.78}$$

For electronic and vibrational excitation processes , $\chi_{n\nu}$ is a H_2 vibrational wavefunction and

$$\Phi_n(\boldsymbol{r},\rho) = \Phi_n^{H_2}(\boldsymbol{r}_1,\boldsymbol{r}_2;\rho)\Phi_n^{X}(\boldsymbol{r}_3,...,\boldsymbol{r}_N)\,, \tag{8.79}$$

while for electron capture, $\chi_{n\nu}$ is a vibrational function of H_2^+ and

$$\Phi_n(\boldsymbol{r},\rho) = \Phi_n^{H_2^+}(\boldsymbol{r}_1;\rho)\Phi_n^{X}(\boldsymbol{r}_2,...,\boldsymbol{r}_N)\,, \tag{8.80}$$

In the range of impact energies E >50eV/a.m.u., we can consider that the characteristic collision period is short compared to the rotation period of the diatomic molecule and use the *sudden approximation* for rotation. In this approximation, we assume that the rotational component of the the collisional wavefunction does not change during the collision, and accordingly we write:

$$\Psi = \rho^{-1}Y_{JM}(\widehat{\boldsymbol{\rho}})\psi(\boldsymbol{r},\boldsymbol{\rho},t) \tag{8.81}$$

It can be shown (see ref. [82]) that assuming that all the M values are equally probable in the initial condition, the probability for transition to all rotational states of a given vibronic state $\psi_{n\mu}$, can be expressed as an orientation averaged magnitude:

$$\mathcal{P}_{n\nu}(\boldsymbol{b},v) \;=\; (4\pi)^{-1}\int \mathrm{d}\widehat{\boldsymbol{\rho}}\, P_{n\nu}(\widehat{\boldsymbol{\rho}},\boldsymbol{b},v) \tag{8.82}$$

where

$$P_{n\nu}(\widehat{\boldsymbol{\rho}},\boldsymbol{b},v) = \lim_{t\to\infty}\left|\langle\psi_{n\nu}|\psi\rangle_{\boldsymbol{r},\rho}\right|^2 \tag{8.83}$$

is the transition probability for a fixed orientation of the H-H internuclear axis with respect to the direction of $\boldsymbol{v}$. The corresponding orientation averaged cross section for transition to the vibronic state $\psi_{n\nu}$ is

$$\sigma_{n\nu}(v) = \int \mathrm{d}\boldsymbol{b}\mathcal{P}_{n\nu}(\boldsymbol{b},v) = (4\pi)^{-1}\int \mathrm{d}\widehat{\boldsymbol{\rho}}\int P_{n\nu}(\widehat{\boldsymbol{\rho}},\boldsymbol{b},v)\mathrm{d}\boldsymbol{b} \tag{8.84}$$

The CC method can be applied to the function ψ, which is expanded in the vibronic basis $\{\phi_j(\boldsymbol{r};\boldsymbol{R},\boldsymbol{\rho})\chi_{j\nu}(\rho)\}$. Explicitly:

$$\begin{aligned}\psi \;&=\; D(\boldsymbol{r},t)\sum_j\sum_\nu a_{j\nu}(t;\widehat{\boldsymbol{\rho}},\boldsymbol{b},v)\phi_j\chi_{j\nu} \\ &\times\; \exp\left[-\mathrm{i}\int_0^t(\varepsilon_j(R,\gamma)+\epsilon_{j\nu})\mathrm{d}t'\right]\end{aligned} \tag{8.85}$$

where D is a CTF, $\phi_j(\boldsymbol{r};\ \rho, R, \gamma)$ are the electronic wavefunctions, which are (approximate) solutions of the fixed nuclei Schrödinger equation (the analogous to eq. (8.15) for the three-centre problem):

$$H_{\mathrm{el}}(\boldsymbol{r};\ \rho, R, \gamma)\phi_j(\boldsymbol{r};\ \rho, R, \gamma) = \varepsilon_j(\rho, R, \gamma)\phi_j(\boldsymbol{r};\ \rho, R, \gamma) \tag{8.86}$$

As for ion-atom collisions, diabatic basis sets can be employed (see section 8.3.3), where functions ϕ_j are not solutions of (8.86) (see [77]). The vibrational wavefunctions $\chi_{j\nu}(\rho)$ can be defined in several ways, but the simplest choice, which fulfills the asymptotic conditions (8.75) and (8.78), is to multiply each electronic state ϕ_j by the vibrational functions $\chi_{j\nu}$ that correspond to either H_2 or H_2^+ molecules depending on the asymptotic structure of this electronic state. They are solutions of the equations:

$$\left[-\frac{1}{2\mu}\nabla_\rho^2 + \varepsilon_j(\rho, R=\infty) - \varepsilon_j(\rho_{0j}, R=\infty)\right]\chi_{j\nu}(\rho) = \epsilon_{j\nu}\chi_{j\nu}(\rho) \tag{8.87}$$

with ρ_{0j} the equilibrium internuclear separation of the diatomic molecule in the electronic state j.

Substitution of eq. (8.85) in (8.71) leads to a system of differential equations for the coefficients $a_{j\nu}$, from which one can calculate the transition probabilities and cross sections of eqs. (8.83) and (8.84) in a similar way to the ion-atom case. However, the problem is very complex from the computational point of view. It involves:

1 Evaluation of potential energy surfaces ε_j and dynamical couplings, which are now functions of three variables R, ρ, γ (they only depend on R in the ion-atom case).

2 Calculation of a set of vibrational functions and evaluation of the dynamical couplings in the vibronic basis:

$$\left\langle \phi_j\chi_{j\nu}D \left| H_{\mathrm{i}} - \mathrm{i}\frac{\partial}{\partial t} \right| \phi_k\chi_{k\mu}D \right\rangle_{\boldsymbol{r},\rho}$$

A large number of coupled channels can be involved when using the vibronic expansion (8.85). In fact, including only bound vibrational states of H_2 and H_2^+ and six electronic states, the vibronic basis includes about 100 functions.

3 Solution of the system of differential equations for different orientations of the diatom and integration (eq. (8.84)) of the transition probabilities to obtain the corresponding cross sections

A reason for the complexity of ion-molecule collisions is the anisotropy of the interaction potential. In particular, when using the expansion (8.85), potential energy surfaces and dynamical couplings depend on the angle γ (see fig. 8.4), which changes along the trajectory. To simplify the calculation, an additional *isotropic* approximation can be introduced by substituting energies and couplings calculated at a fixed molecular orientation, γ_0, which leads to the isotropic transition probabilities $P^{\rm iso}_{n\nu}(\boldsymbol{b}, v; \gamma_0)$ and cross sections, $\sigma^{\rm iso}_{n\nu}(v; \gamma_0)$. The ensuing average over γ_0 yields the corresponding orientation averaged cross sections:

$$\sigma_{n\nu}(v) \simeq \int_0^{\pi/2} \mathrm{d}\gamma_0 \sin\gamma_0 \sigma^{\rm iso}_{n\nu}(v; \gamma_0) \tag{8.88}$$

This is the approach generally employed in IOSA calculations [77]. In a drastic approximation the orientation average cross section $\sigma_{n\nu}(v)$ is replaced by the corresponding isotropic result for a given γ_0, $\sigma^{\rm iso}_{n\nu}(v; \gamma_0)$.

Until now, only a few works [83] have considered the validity of the isotropic approximation. In general, charge transfer in collisions of multiply-charged ions with H_2 (see e.g. results for C^{4+}-H_2 of ref. [84]) take place in the neighborhood of avoided crossings of the potential energy surfaces at relatively large internuclear distances where anisotropy effects are unimportant, and $\sigma^{\rm iso}_{n\nu}(v; \gamma_0)$ is practically independent on γ_0. On the other hand, collisions of singly charged ions (Li^+, H^+) with H_2 take place at shorter R and the dependence of $\sigma^{\rm iso}_{n\nu}(v; \gamma_0)$ on γ_0 is noticeable for energies $E \lesssim 250$eV/a.m.u.. However, results of ref. [83] show that this dependence cancels out when the cross sections are averaged over the molecular orientation with good agreement between cross sections calculated using eq. (8.88) and the more exact values from eq. (8.84), and that even good agreement is found with $\sigma^{\rm iso}_{n\nu}(v; \gamma_0)$ an intermediate angle $\gamma_0 \simeq 60^0$.

7.2 The sudden vibrational approximation

Because of the complexity of ion-molecule collisions, some additional approximations have been proposed, which are particularly useful when v increases, because of the very large vibronic basis set required. In particular, when the vibrational period is larger than the time interval in which electronic transitions take place, one can use the sudden approximation for vibration (see ref. [17]) where it is assumed that the initial vibrational wavefunction $\chi_{i\mu}$ remains unchanged during the collision. So that, the CC wavefunction is substituted by:

$$\psi = D(\boldsymbol{r}, t)\chi_{i\mu} \sum_j a_j(t; \boldsymbol{\rho}, \boldsymbol{b}, v)\phi_j \exp\left[-\mathrm{i}\int_0^t \varepsilon_j(R, \gamma)\mathrm{d}t'\right] \tag{8.89}$$

This is the basis of the SEIKON method of ref.[82], where it was shown that the transition probabilities are:

$$P_{n\nu}(\widehat{\boldsymbol{\rho}}, \boldsymbol{b}, v) = \left| \int \mathrm{d}\rho \chi_{i\mu} \chi_{n\nu} \exp\left[\int_0^\infty \mathrm{d}t(\varepsilon_n - E_n)\right] a_n(t = \infty; \boldsymbol{\rho}, \boldsymbol{b}, v) \right|^2 \tag{8.90}$$

Although the SEIKON approach notably simplifies the calculation with respect to the vibronic CC method, it still involves a considerable computational effort. In particular, for a given trajectory and molecular orientation $\hat{\rho}$, it requires:

- Calculation of potential energy surfaces and dynamical couplings for a grid of values of R, ρ and γ, and 2D interpolation of these molecular data as functions of R and γ
- Integration for each value of ρ of a system of differential equation to obtain the coefficients a_n.
- Integration of eq. (8.90) to evaluate the transition probabilities $P_{n\nu}$.

The SEIKON method has been successfully applied to several collisions, and, as an example, we show in fig. 8.5 the results [21] for electron capture into individual vibrational states in the process

$$C^{2+}(2s2p;{}^3P) + H_2(X^1\Sigma_g^+, \nu = 0) \rightarrow C^+(2s2p^2;{}^2D) + H_2^+(X^2\Sigma_g^+, \nu) \tag{8.91}$$

compared to the experimental results of [85], which shows the accuracy of this method.

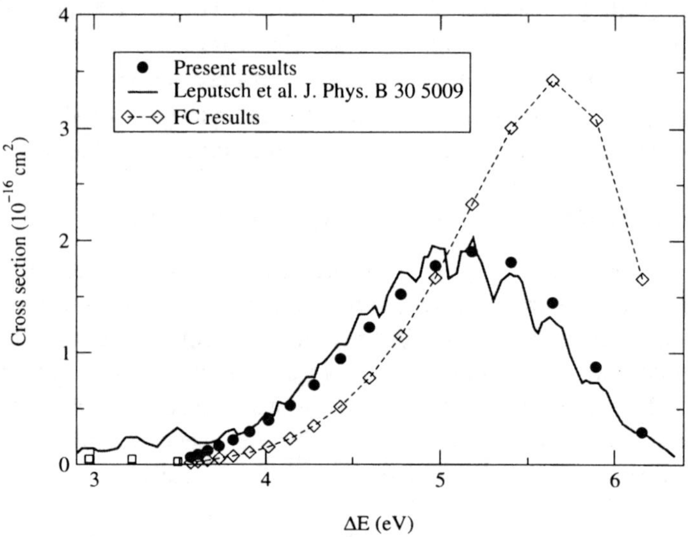

Figure 8.5. Comparison of electron capture cross sections into individual H_2^+ vibrational levels for the reaction (8.91): • Calculations from ref. [21] using SEIKON ◇ FC results. Line, experimental results from ref.[85]

An additional approximation is obtained by assuming that the coefficients a_n can be taken as constants in the interval of ρ where the vibrational functions

$\chi_{i\mu}$, $\chi_{n\nu}$ have appreciable values. If we take the values of the coefficients at the equilibrium internuclear distance of the initial electronic state ρ_{0i}, we obtain:

$$P_{n\nu} = \left| \int \mathrm{d}\rho \chi_{i\mu} \chi_{n\nu} \right|^2 |a_n(t = \infty; \boldsymbol{\rho_{0i}}, \boldsymbol{b}, v)|^2 \ , \tag{8.92}$$

which is the familiar Franck-Condon (FC) approximation. The FC approximation only requires to calculate the coefficients a_n for a single value of ρ and is then much simpler than the vibrational sudden approximation from the computational point of view.

In general, the FC approximation becomes more accurate as v increases, but there is a reduced experience in comparing FC results with the more accurate ones obtained by applying (8.89) and (8.90) (see [21] and references therein). These results indicate that the FC approximation is appropriate to evaluate total cross sections at E larger than about 500 eV/a.m.u., but important disagreements have been found for partial cross sections, in particular for vibrational distributions, as can be observed in the illustration of fig. 8.5.

In a recent work, [86] we have studied the low v limit of the SEIKON method by comparing total cross sections calculated by employing this approximation with those obtained from the vibronic CC expansion (eq. (8.85)) for the benchmark collisions:

$$\mathrm{H}^+ + \mathrm{H}_2(X^1\Sigma_g^+, \nu = 0) \rightarrow \mathrm{H}(1\mathrm{s} + \mathrm{H}_2(X^2\Sigma_g^+, \nu') \tag{8.93}$$

and

$$\mathrm{H}^+ + \mathrm{H}_2(X^1\Sigma_g^+, \nu = 0) \rightarrow \mathrm{H}^+ + \mathrm{H}_2(X^1\Sigma_g^+, \nu') \tag{8.94}$$

The calculations show (see fig. 8.6) an important disagreement between the total charge transfer cross sections calculated with both methods for E <250 eV/amu, indicating a limitation of the SEIKON approach. The analysis of ref. [86] has pointed out that there are two mechanisms leading to the charge transfer process:

- At $E \gtrsim$ 250eV, the charge transfer process takes place through direct transitions from the the entrance channel to the charge transfer ones.

- At $E \leq$ 250eV, the reaction (8.93) takes place through a two-stage mechanism first suggested in ref. [87]. It involves the vibrational excitation to H_2 levels with $\nu \simeq 4$ followed by transitions from those states to the quasidegenerate charge transfer state with $\nu' = 0$ in the outgoing part of the collision.

The low-v mechanism cannot be reproduced by employing the sudden vibrational approximation, which explains the limitation of the SEIKON calculation.

However, since this limitation is mainly a consequence of the quasidegeneracy of the vibronic levels in H^+- H_2, further work is needed to know the limitations of the method for other systems.

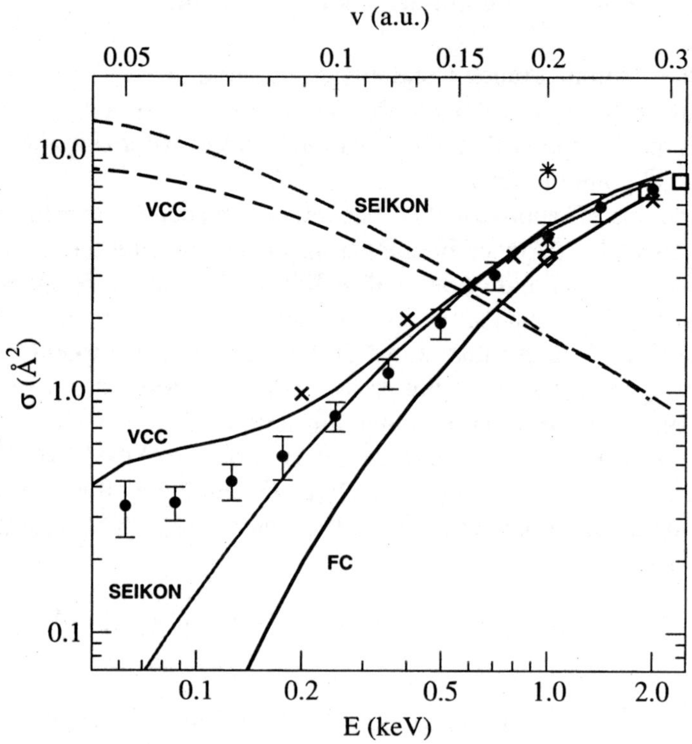

Figure 8.6. Cross sections for $H^+ + H_2$ collisions from ref. [86]. Lines, results of ref. [86] (VCC, vibronic close-coupling; SEIKON: vibrational sudden; FC, Franck-Condon). Solid lines and symbols: charge transfer (Eq. 8.93); dashed lines: vibrational excitation (Eq. 8.94). ×, [88] multiplied by 2 following author's suggestion. Experiments: •, [89]; ◇, [90]; ○, [91]; ⋆, [92].

8. Concluding Remarks

In this chapter we have described the main characteristics of CC treatments of ion-atom and ion-molecule collisions. We can summarize the present situation with regards to the practical application of these methods as follows:

1. Low energy ion-atom collisions are generally described using molecular CC expansions. These methods are well established for single-electron systems at low velocities ($v \lesssim v_0$) where ionization is not important. In particular, the computing facilities allow to employ relatively large expansions, for which the choice of ETFs is not critical. As an example,

electron capture cross sections for collisions between fully stripped ions and hydrogen atoms were calculated in ref. [93], using basis sets that include about 100 molecular functions multiplied by a common translation factor. At higher v new techniques are needed to describe ionization.

2. In the last ten years, several calculations have considered many electron ion-atom systems by using quantum chemistry ab initio packages. This work has been motivated by state selective measurements [94] which in many cases involve the contamination of the initial beam by unknown amounts of metastable ions. Although the theoretical methods are similar to that employed in single electron systems, the application is not a routine work, which explains the necessity of calculations for these systems.

3. Because of the complexity of ion-molecule collisions, relatively few calculations have been performed for these systems. In practice, methods based on the use of the sudden approximation are applied. Although a first study has been recently presented for the benchmark system $H^+ + H_2$ [86], further work is required to know the energy limits of these approximations. The study of anisotropy effects is also needed in particular in relation to recent experiments [95], where the dissociation of molecules after collisions with ions has been studied as function of the molecular orientation. Finally, data for collisions with more complex molecules are required in view of possible applications in fusion research and astrophysics.

Acknowledgments

This work has been partially supported by DGICYT projects BFM2000-0025 and FTN2000-0911.

References

[1] B. H. Bransden and M. H C. McDowell. *Charge Exchange and the Theory of Ion-Atom Collisions.* Clarendon, Oxford, (1992)

[2] L. F. Errea, C. Harel, H. Jouin, L. Méndez, B. Pons, and A. Riera. J. Phys. B: At. Mol. Opt. Phys. **27**, 3603, (1994)

[3] L. F. Errea, C. Harel, H. Jouin, L. Méndez, B. Pons, A. Riera, and I.Sevila. Phys. Rev. A **65** 022711, (2002)

[4] J. Ullrich, R. Moshamer, R. Dorner, O. Jagutski, V. Mergel, H. Schmidt-Bocking, and L. Spielberger. J. Phys. B: At. Mol. Opt. Phys. **30**, 2917 (1997)

[5] A. Macías and A. Riera. Phys, Rep **81**, 299 (1982)

[6] J. F. Castillo, L. F. Errea, A. Macías, L. Méndez, and A. Riera. J. Chem. Phys. **103**, 2113 (1995)

[7] M. Kimura, Y. Li, G. Hirsch, and R. J. Buenker. Phys. Rev. A **54**, 5019 (1996)

[8] W. Fritsch and C. D. Lin. Phys. Rep. **202**, 1 (1991)

[9] J. Kuang and C. D. Lin. J. Phys. B: At. Mol. Opt. Phys. **30**, 101 (1997)

[10] N. Toshima. Phys. Rev. A **59**, 1981 (1999)

[11] T. Kirchner, H. J. Lüdde, M. Horbatsch, and R. M. Dreizler. Phys. Rev. A **61**, 052710 (2000)

[12] B. Pons. Phys. Rev. Lett. **84**, 4569 (2000)

[13] T.G. Winter and J. R. Winter. Phys. Rev. A **61**, 052709 (2000)

[14] Dz Belkič, R. Gayet, and A. Salin. Phys, Rep **56**, 279 (1979)

[15] R. Abrines and I. C. Percival. Proc. Phys. Soc. **88**, 861 (1966)
R.E. Olson and A. Salop. Phys. Rev. A **16**, 631 (1977)

[16] P. C. Stancil, A. R. Turner, D. L. Cooper, D. R. Schultz, M. J. Raković, W. Fritsch, and B. Zygelman. J. Phys. B: At. Mol. Opt. Phys. **34**, 2481 (2001)

[17] V. Sidis. Adv. At. Mol. Phys **26**, 161 (1990)

[18] E. A. Gislason, G. Parlant, and M. Sizun. *in State-Selected and State to State Ion-Molecule Reaction Dynamics. Advances in Chemical Physics Vol LXXXII*, ed. Michael Baer and Cheuk-Yiu Ng John Wiley & Sons, Inc., (1992)

[19] P. C. Stancil, B. Zygelman, and K. Kirby. in *Photonic, Electronic and Atomic Collisions* ed. F. Aumayr and H. P. Winter. Clarendon, Oxford, (1998)

[20] W. Pauli. Handb. Phys. **24**, 164 (1933)

[21] L. F. Errea, A. Macías, L. Méndez, I. Rabadán, and A. Riera. J. Phys. B: At. Mol. Opt. Phys. **33**, L615 (2000)

[22] D. R. Bates and R. McCarroll. Proc. Roy. Soc. A **245**, 175 (1958)

[23] J. D. Power. Phil. Soc. Trans. R. Soc **274**, 663 (1973)

[24] A. Salin. Comput. Phys. Commun **14**, 121 (1978)
A. Salin. Comput. Phys. Commun **15**, 443 (1978)
A. Salin. Comput. Phys. Commun **20**, 462 (1980)

[25] L. D. Landau. Pys. Z. Sowetjetunion **2**, 46 (1932)
C. Zener. Proc. R. Soc. London Ser. A **137**, 696 (1932)

[26] S. B. Schneiderman and A. Russek. Phys. Rev. **181**, 311 (1969)

[27] H. Levy II and W. R. Thorson. Phys. Rev. **182**, 256 (1969)
K. Taulbjerg, J. Vaaben, and B. Fastrup. Phys. Rev. A **12**, 2325 (1975)

N. Shimakura, H. Inouye, F. Koike, and T. Watanabe. J. Phys. B: At. Mol. Phys. **14**, 2203 (1981)

[28] L. F. Errea, L. Méndez, and A. Riera. J. Phys. B: At. Mol. Opt. Phys. **15**, 101 (1982)

[29] C. Harel and H Jouin. J. Phys. B: At. Mol. Opt. Phys. **25**, 221 (1992)

[30] V. H. Ponce. J. Phys. B: At. Mol. Phys. **12**, 3731 (1979)

[31] J. Rankin and W. R. Thorson. Phys. Rev. A **16**, 1990 (1978)

[32] A. Riera. Phys. Rev. A **30**, 2304 (1984)

[33] C. Illescas and A. Riera. Phys. Rev. Lett. **80**, 3029 (1998)

[34] C. Harel, H. Jouin, B. Pons, L. F. Errea, L. Méndez, and A. Riera. Phys. Rev. A **55**, 287 (1997)

[35] L. F. Errea, C. Harel, C. Illescas H. Jouin, L. Méndez, B. Pons, and A. Riera. J. Phys. B: At. Mol. Opt. Phys. **31**, 3199 (1998)

[36] A. Riera. Comments Mod. Phys. **1D**, 131 (1999)

[37] J. P. Grozdanov and E. A. Solov'ev. Phys. Rev. A **42**, 2703 (1990.

[38] R. McWeeny. *Methods in Molecular Quantum Mechanics.* Academic Press, London, (1989)

[39] A. Macías, A. Riera, and M. Yáñez. Phys. Rev. A **23**, 2941 (1981)
A. Macías, A. Riera, and M. Yáñez. Phys. Rev. A **27**, 206 (1983)
A. Macías, A. Riera, and M. Yáñez. Phys. Rev. A **27**, 213 (1983)
L. F. Errea, L. Méndez, and A. Riera. Z. Phys. D **14**, 229 (1989)

[40] F. Borondo, A. Macías, and A. Riera. Phys. Rev. Lett. **46**, 420 (1981)
F. Borondo, A. Macías, and A. Riera. J. Chem. Phys. **74**, 6126 (1981)

[41] L. F. Errea, L. Méndez, A. Riera, M. Yáñez, J. Hanssen, C. Harel, and A. Salin. J. Physique **46**, 709 (1985)
L. F. Errea, L. Méndez, A. Riera, M. Yáñez, J. Hanssen, C. Harel, and A. Salin. J. Physique **46**, 719 (1985)

[42] C. Harel and A. Salin. J. Phys. B: At. Mol. Phys. **13**, 785 (1980)

[43] C. Harel and H Jouin. Europhys. Lett. **11**, 2121 (1990)
C. Harel, H Jouin, and B. Pons. J. Phys. B: At. Mol. Opt. Phys. **24**, L425 (1991)
X. Fléchard, C. Harel, H. Jouin, B. Pons, L. Adoui, F. Freémont, A. Cassimi, and D. Hennecart. J. Phys. B: At. Mol. Opt. Phys. **34**, 2759 (2001)

[44] C. C. J. Roothaan. Rev. Mod. Phys. **23**, 69 (1951)

[45] C. Forster, I. L. Cooper, A. S. Dickinson, D. R. Flower, and L. Méndez. J. Phys. B: At. Mol. Opt. Phys. **24**, 3433 (1991)

[46] M. F. Guest. *GAMESS User's Guide and Reference Manual, Revision A Daresbury Laboratory* (1989)

[47] J. P. Gu, G. Hirsch, R. J. Buenker, M. Kimura, C. M. Dutta, and P. Norlander. Phys. Rev. A **57**, 4483 (1998)

[48] L. F. Errea, A. Macías, L. Méndez, and A. Riera. J. Phys. B: At. Mol. Opt. Phys. **33**, 1369 (2000)

[49] R. J. Buenker, S. D. Peyerimhoff, and P. J. Bruna. *Computational Theoretical Organic Chemistry*, ed. I. G. Csizmadia and R. Daudel Reidel, Dordrecht, (1981)

[50] E. Davidson. In E. Clementi, editor, *MOTECC, Modern Techniques in Computational Chemistry*. ESCOM Publishers B. V., Leiden (1990)

[51] P. Honvault, M.C. Bacchus-Montabonel, and R. M. McCarroll. J. Phys. B: At. Mol. Opt. Phys. **27**, 3115 (1994)

[52] B. Huron, J. P. Malrieu, and P. Rancurel. J. Chem. Phys. **58**, 5745 (1973)

[53] J. F. Castillo, I. L. Cooper, L. F. Errea, L. Méndez, and A. Riera. J. Phys. B: At. Mol. Opt. Phys. **27**, 5011 (1994)

[54] M. Gargaud, J. Hansen, R. McCarroll, and P. Valiron. J. Phys. B: At. Mol. Phys. **14**, 2259 (1981)

[55] O. Mó, A. Riera, and M. Yáñez. Phys. Rev. A **31**, 3977 (1985)

[56] L. F. Errea, B. Herrero, L. Méndez, O. Mó, and A. Riera. J. Phys. B: At. Mol. Opt. Phys. **24**, 4049 (1991)

[57] P. Habitz and C. Votava. J. Chem. Phys. **72**, 5532 (1980)

[58] A. Macías and A. Riera. J. Phys. B: At. Mol. Phys. **10**, 861 (1977)
A. Macías and A. Riera. J. Phys. B: At. Mol. Phys. **11**, 1077 (1978)

[59] B. H. Lengsfield III, P. Saxe, and D. R. Yarkony. J. Chem. Phys. **81**, 4549 (1984)
P. Saxe, B. H. Lengsfield III, and D. R. Yarkony. Chem. Phys. Lett. **113**, 159 (1985)

[60] G. Hirsch, P. J. Bruna, R. J. Buenker, and S. D. Peyerimhoff. Chem. Phys. **45**, 335 (1980) I. D. Petsalakis G. Theodorakopoulos and R. J. Buenker. Chem. Phys. Lett. **148**, 285 (1988)
M.C. Bacchus-Montabonel, C. Courbin, and R. McCarroll. J. Phys. B: At. Mol. Opt. Phys. **24**, 4409 (1991)
I. L. Cooper. J. Phys. B: At. Mol. Opt. Phys. **24**, 1517 (1991)

[61] H. C. Tseng and C. D Lin. J. Phys. B: At. Mol. Opt. Phys. **32**, 5271 (1999)

[62] J. H. McGuire and L. Weaver. Phys. Rev. A **16**, 41 (1977)
V.A. Sidorovich. J. Phys. B: At. Mol. Phys. **14**, 4805 (1981)
H. J. Lüdde and R. M. Dreizler. J. Phys. B: At. Mol. Opt. Phys. **18**, 107 (1985)

[63] H. J. Lüdde, A. Henne, T. Kirchner, and R. M. Dreizler. J. Phys. B: At. Mol. Opt. Phys. **29**, 4423 (1996)

[64] A. Henne, H. J. Lüdde, and R. M. Dreizler. J. Phys. B: At. Mol. Opt. Phys. **30**, L565 (1997)

[65] L. F. Errea, L. Méndez, and A. Riera. Chem. Phys. Lett. **164**, 261 (1989)

[66] R. Dreizler, L. F. Errea, A. Henne, H. J. Lüdde, and A. Riera. Phys. Rev. A **47**, 3852 (1993)

[67] W. Fritsch and C. D. Lin. J. Phys. B: At. Mol. Phys. **17**, 3271 (1984)

[68] H. C. Tseng and C. D Lin. Phys. Rev. A **58**, 1966 (1998)

[69] L. F. Errea, J. D. Gorfinkiel, C. Harel, H. Jouin, A. Macías, L. Méndez, B. Pons, and A. Riera. J. Phys. B: At. Mol. Opt. Phys. **32**, L673 (1999)

[70] D. G. M. Anderson, M. J. Antal, and M. B. McElroy. J. Phys. B: At. Mol. Phys. **7**, L118 (1974)

[71] W. Fritsch and C. D. Lin. J. Phys. B: At. Mol. Phys. **15**, 1255 (1982)

[72] N. Toshima. Phys. Scr. **73**, 144 (1997)

[73] R.E. Olson. Phys. Rev. A **27**, 1871 (1983)

[74] D. Dijkkamp, D. Ciric, E. Vlieg, A. de Boer, and J. de Heer. J. Phys. B: At. Mol. Opt. Phys. **18**, 4763 (1985)

[75] F. W. Bliek, R. Hoekstra, M. E. Bannister, and C. C. Havener. Phys. Rev. A **56**, 426 (1997)

[76] R. Hoekstra, J. P. M. Beijers, A. R. Schalatmann, R. Morgenstern, and F. J. de Heer. Phys. Rev. A **41**, 4800 (1990)

[77] M. Baer. *in State-Selected and State to State Ion-Molecule Reaction Dynamics. Advances in Chemical Physics Vol LXXXII*, ed. Michael Baer and Cheuk-Yiu Ng John Wiley & Sons, Inc., (1992)

[78] R. Shingal and C. D. Lin. Phys. Rev. A **40**, 1302 (1989)

[79] W. Fritsch. Phys. Rev. A **46**, 3910 (1992)

[80] D. Elizaga, L. F. Errea, J. D. Gorfinkiel, L. Méndez, A. Macías, A. Riera, A. Rojas, O. J. Kroneisen, T. Kirchner, H. J. Lüdde, A. Henne, and R. M. Dreizler. J. Phys. B: At. Mol. Opt. Phys. **32**, 857 (1999)

[81] C. Illescas and A. Riera. Phys. Rev. A **A60**, 4546 (1999)
C.J. Wood and R. E. Olson. Phys. Rev. A **59**, 1317 (1999)

[82] L. F. Errea, J. D. Gorfinkiel, A. Macías, L. Méndez, and A. Riera. J. Phys. B: At. Mol. Opt. Phys. **30**, 3855 (1997)

[83] D. Elizaga, L. F. Errea, J. D. Gorfinkiel, A. Macías, L. Méndez, A. Riera, and A. Rojas. J. Phys. B: At. Mol. Opt. Phys. **33**, 2037 (2000)
L. F. Errea, A. Macías, L. Méndez, I. Rabadán, and A. Riera. Int. J. Mol. Sci. **3**, 142 (2002)

[84] L. F. Errea, J. D. Gorfinkiel, A. Macías, L. Méndez, and A. Riera. J. Phys. B: At. Mol. Opt. Phys. **32**, 1705 (1999)

[85] P. Leputsch, D. Dumitriu, F. Aumayr, and H. P. Winter. J. Phys. B: At. Mol. Opt. Phys. **30**, 5009 (1997)

[86] L. F. Errea, A. Macías, L. Méndez, I. Rabadán, and A. Riera. Phys. Rev. A **65**, 010701(R) (2001)

[87] G. Niedner, M. Noll, J. P. Toennies, and Ch. Schlier. J. Chem. Phys. **87**, 2685 (1987)

[88] M. Kimura. Phys. Rev. A **32**, 802 (1985)

[89] M. W. Gealy and B. I. Van Zyl. Phys. Rev. A **36**, 3091 (1987)

[90] C. J. Abbe and P. Adolf. Bull. Soc. Chim. Fr. **6**, 11212 (1964)

[91] S. E. Chambers. *Report No 4 CRL-14214* (1965)

[92] Yu S. Gordeev and N. M. Pranov. Zh. Tekh. Fiz. **34**, 857 (1964)

[93] C. Harel, H. Jouin, and B. Pons. At. Data. Nucl. Data Tables **68**, 279 (1998)

[94] J. B. Greenwood, D. Burns, R. W. McCullough, J. Geddes, and H. B. Gilbody. J. Phys. B: At. Mol. Opt. Phys. **29**, 5867 (1996)
D. Burns, J. B. Greenwood, K.R. Bajajova, R. W. McCullough, J. Geddes, and H. B. Gilbody. J. Phys. B: At. Mol. Opt. Phys. **30**, 1531 (1997)
D. Voulot, D. R. Gillen, W. R. Thompson, H. B. Gilbody, R. W. McCullough, L. Errea, A. Macías, L. Méndez, and A. Riera. J. Phys. B: At. Mol. Opt. Phys. **33**, L189–L192 (2000)

[95] F. Frémont, C. Bedouet, M. Tarisien, L. Adoui, A. Cassimi, A. Dubois, J-Y Chesnel, and X. Husson. J. Phys. B: At. Mol. Opt. Phys. **33**, L249 (2000)
H. Bräuning, I. Reiser, A. Diehl, A. Theiß, E. Sidkey, C. L. Cocke, and E. Salzborn. J. Phys. B: At. Mol. Opt. Phys. **34**, L321 (2001)

Chapter 9

QUANTUM DYNAMICS OF ION-ATOM COLLISIONS

R. McCarroll

Laboratoire de Dynamique des Ions, Atomes et Molécules (UMR 7066 du CNRS)
Université Pierre et Marie Curie
4, place Jussieu, Paris, France
mccarrol@ccr.jussieu.fr

Abstract The dynamics of collisions involving charge transfer between multiply charged ions and neutral atoms at low energies (less than a few keV/amu) is treated within the framework of an adiabatic representation using a fully quantum mechanical formalism. Considerable attention is given to the practical problems of reducing the quantum mechanical equations to manageable proportions for numerical solution. In particular, the adiabatic-diabatic transformation to remove rapidly varying couplings at avoided crossings, the partial wave expansion used to reduce the Schrödinger equation to a set of coupled second differential equations and the numerical methods used to solve these equations are reviewed in detail.

The presence of non-vanishing asymptotic coupling terms which originate from the incapacity of a finite adiabatic basis to represent correctly the translation of the collision partners is discussed. It is pointed out that the problem can be overcome in principle by the introduction of appropriate reaction coordinates, but only those types which preserve the essential simplicity of the adiabatic representation would seem practical. A review is given of some of the more promising systems of reaction coordinates.

Keywords: quantum dynamics, slow collisions, multiply charged ions, electron capture, Jacobi coordinates, adiabatic representation, adiabatic/diabatic transformation, partial wave expansion, non-adiabatic coupling, S-matrix, log-derivative algorithms, reaction coordinates, translation factors

1. Introduction

The aim of this chapter is to present a fully quantum mechanical model capable of treating the dynamics of inelastic, rearrangement or ionization processes

F.J. Currell (ed.), The Physics of Multiply and Highly Charged Ions, Vol. 2, 275-308.

taking place in collisions involving multiply charged ions with atoms (or other ions). The first question is to determine under which conditions such a model is really necessary. The answer depends on the relative energy of the colliding particles. When the centre of mass (CM) energy is of the order of a few hundred eV/amu or greater, that is to say much greater than the electronic energy exchanged in the collision, the de Broglie wave-length of the nuclear motion is much less than a_0. So while, a quantum description of the internal electronic motion of the collision system is always required, a semi-classical description of the nuclear motion is adequate. On the other hand, when the CM energy is less than a few tens of eV/amu, electronic energy exchanged in the collision is comparable with the collision energy and a quantum mechanical description of the dynamics is required. In the intermediate energy range between 50-500 eV/amu, the specific choice of method depends to a large extent on the aims of the calculation. Conceptually, the semi-classical approach is simpler, but its implementation is practical only when small angle scattering is dominant (which is the case if we are concerned with total cross sections in the energy range beyond few tens of eV/amu). But if we are concerned with the contribution of large angle scattering processes, a quantum calculation often proves to be more convenient even for energies as high as 1keV/amu.

We may observe that throughout the entire energy range for which quantum methods are either required or convenient to use, the relative collision velocity is much smaller than the velocity of the bound electrons. Even for energies of the order of 1keV/amu, the velocity is only 1/5 of that of an electron in a Bohr orbit. Under these conditions, we may expect the collision system to evolve in a nearly adiabatic manner. So, just as in molecular spectroscopy, it is convenient to use the adiabatic electronic states as a basis for the representation of the system in collision. Of course, we may expect the influence of non-adiabatic effects to be much greater than they are for the bound spectroscopic levels of stable molecules, but the underlying physical model is very similar.

In the adiabatic model, the Schrödinger equation is first solved assuming the nuclei to be clamped at some fixed internuclear distance R. In this way a series of adiabatic eigenfunctions may be generated, whose eigenenergies $E_n(R)$ depend parametrically on R. Because of the axial symmetry of the interaction potential, the projection Λ_n of the electronic angular momentum on the internuclear axis is conserved. If the adiabatic energy curves are well separated from each other for all R, the system evolves adiabatically during a collision and energy exchange is unlikely to occur. The crossing of adiabatic energy curves corresponding to states of the same symmetry is forbidden by the *von Neumann-Wigner* non-crossing rule, but well defined avoided crossings may exist. Non-adiabatic coupling may then become important and electronic transitions between different adiabatic states can take place. When the adiabatic energies of states with different symmetry cross or (which is often the case) become quasi degenerate, they too can be strongly coupled by non-adiabatic Coriolis coupling.

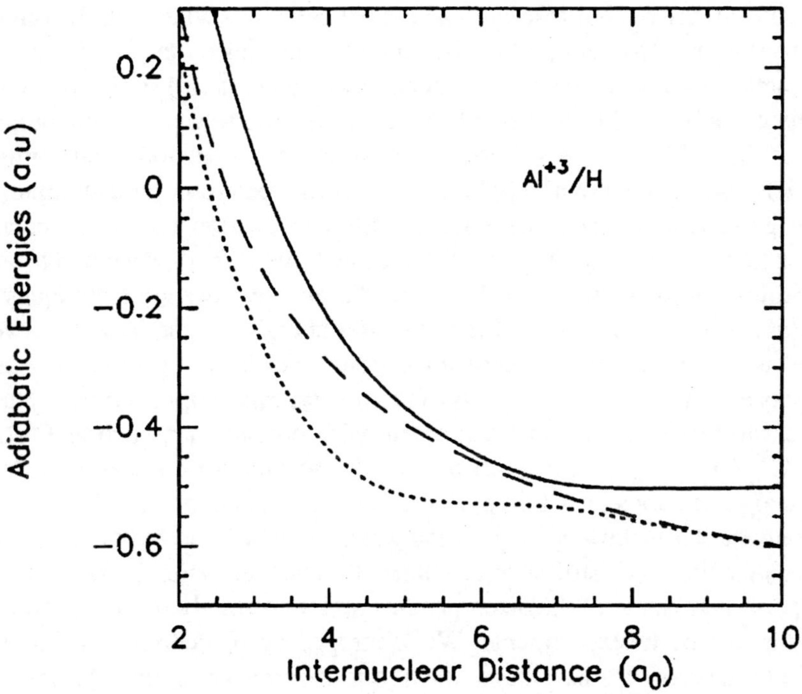

Figure 9.1. Adiabatic potential energies of the AlH^{3+} molecular ion. The full curve designates the $^2\Sigma$ state correlated to the $[Al^{3+}/H(Is)]$ entry channel. The dotted and dashed curves designate the $^2\Sigma$ and $^2\Pi$ states correlated to the $[Al^{3+}(3p)/H^{+}]$ electron capture channels.

Fortunately, for a wide variety of ion atom systems, such accidents, involving either avoided or real curve crossings, tend to be well localized, so that in practice, only a small number of non-adiabatic transitions occur for most systems. So, even if many non-adiabatic (excitation or charge transfer) channels are classically accessible, only those connected by strong non-adiabatic coupling are likely to be activated in low energy collisions. A typical example is provided by the AlH^{3+} system [1] where we observe a simple network of three states (two of $^2\Sigma g$ symmetry, one of $^2\Pi g$ symmetry), which interact strongly via the avoided $\Sigma - \Sigma$ crossing around $7a_0$ and the quasi-degeneracy of the Π and Σg states at large internuclear distances (fig. 9.1). The electron capture reaction of Al^{3+} ions with atomic hydrogen at low energies is well described in terms of the non-adiabatic coupling between these three states.

At the present time, there is a considerable wealth of experimental data, based either on photon emission spectroscopy [2] or on ion translation spectroscopy techniques [3], which support the predictions of the adiabatic model. Particularly striking is the success of the model in predicting the state selective nature of electron capture by multiply charged ions in collisions with neutral atomic or molecular targets. However, while the non-adiabatic selection rules for the primary reaction channels are now well understood, there is still some uncertainty as to the capacity of the theoretical model to determine accurate quantitative values of the collision cross sections. Absolute cross section measurements are difficult and still few in number. Translation spectroscopy techniques do provide good relative cross sections for different rearrangement or inelastic channels, but most existing experiments of this type were not designed to measure absolute cross sections. Besides, translation spectroscopy techniques are most efficient only at relatively high collision energies in the range of several hundred eV/amu. In spite of these limitations, translation spectroscopy measurements have provided severe tests of the theory, especially for those systems where several reactions channels compete with one another, such as O^{2+}/He [4, 5], C^{4+}/H [6, 7]. As a general rule, the results are consistent with the theoretical predictions to within the experimental error bars.

In contrast, it is in the low eV/amu energy range, where theory should be most reliable, that there are still several unresolved discrepancies between theory and experiment. Some of the discrepancies almost certainly originate from the difficult nature of the experiments. While merged beam techniques [8] make it possible to measure cross sections down to low eV energies, the interpretation of the results often depends on unknown factors (such as the metastable fraction of the ion beam, possible contamination from ions with the same charge/mass ratio). Sometimes the agreement between theory and experiment is excellent [9]. Sometimes, it is less so [10]. Many of the discrepancies have not yet been completely elucidated and some interesting theoretical predictions [11] still await experimental confirmation. Since the eV energy range is also important for many applications in astrophysical and laboratory plasmas, there is clearly still much interest in exploring the problem.

However, while an adiabatic representation of the system allows us to visualize how the collision process evolves, there are many conceptual difficulties [12–14] in actually carrying out the calculations (even leaving aside the problem of an accurate determination of the adiabatic eigenenergies and eigenfunctions). Electronic adiabatic states generated from the clamped nuclei approximation, can only match the correct asymptotic boundary conditions in an approximate way. Firstly, it is clear that beyond some sufficiently large internuclear distance, the translation energy of the bound electrons with respect to the centre of mass of the nuclei exceeds the adiabatic interaction potential. Secondly, the adiabatic basis is defined with respect to a rotating reference frame so that account must be taken of Coriolis coupling between degenerate states with Λ_n in the dissociation limit. As we shall see, the use of a finite adiabatic basis set to describe the collision is responsible for the existence of non-vanishing asymptotic couplings between certain reaction channels, which make it formally impossible to extract transition amplitudes.

But before trying to understand some of the subtler aspects of the problem, it is best to begin with the standard adiabatic model. The different contributions to non-adiabatic coupling will be analyzed in detail in order to clarify where difficulties may arise in practical applications. It will be argued that, despite its formal shortcomings, the standard adiabatic model does yield useful results in the eV collision energy range. Subsequently, a simple modification is proposed which eliminates the defects of the standard model but which preserves the essential features of an adiabatic representation.

Atomic units will be used throughout, except where stated otherwise.

2. Standard Adiabatic Model

For the sake of simplicity, let us consider two heavy particles A and B (typically two ionic cores) of masses m_a and m_b and one light particle *e* (typically an electron) with mass m_e(=1 in atomic units). The many particle kinetic energy operator takes its simplest form in Jacobi coordinates (see figure 9.2)

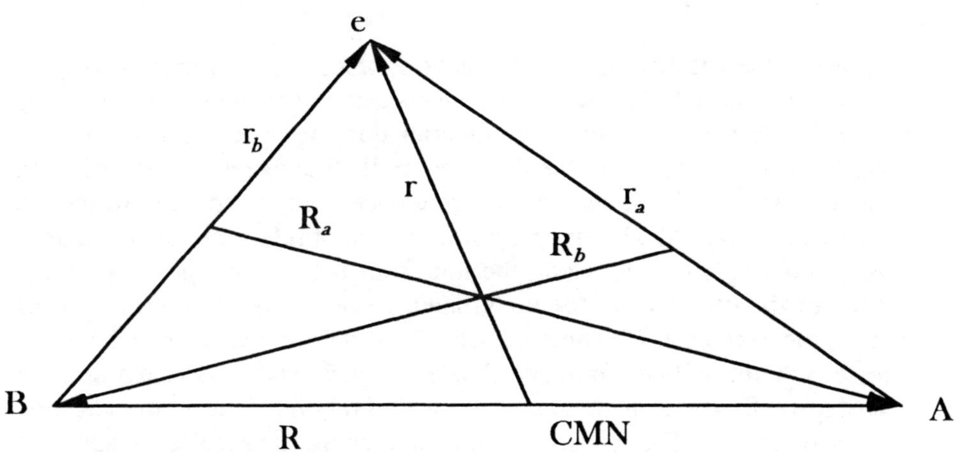

Figure 9.2. Schematic diagram of the Jacobi coordinates of a three particle system.

For three particles, there exist three equivalent sets of Jacobi coordinates $(\mathbf{r}, \mathbf{R})$, $(\mathbf{r}_a, \mathbf{R}_a)$, $(\mathbf{r}_b, \mathbf{R}_b)$ where $\mathbf{r}_a, \mathbf{r}_b, \mathbf{r}$ are respectively the position vectors of the electron relative to A, B and CMN (the centre of mass of nuclear cores

A and B) while $\mathbf{R}_a, \mathbf{R}_b$ are respectively the position vectors of A relative to the centre of mass of (B+e) and of B relative to the centre of mass of (A+e). Since the adiabatic states are defined by a clamped nuclei approximation, the most natural choice of Jacobi coordinates would appear to be $(\mathbf{r}, \mathbf{R})$.

After separation of the kinetic operator of the centre of mass of the total system, the Hamiltonian for the entire system may be written as

$$H = -\frac{1}{2m_{a,b}}\Delta_{\mathbf{R}} - \frac{1}{2m_{ab,e}}\Delta_{\mathbf{r}} + V_{int}(\mathbf{r}, \mathbf{R}) \tag{9.1}$$

where the operator $V_{int}(\mathbf{r}, \mathbf{R})$ contains all interactions between the particles, $m_{a,b} = m_a m_b/(m_a + m_b)$ is the reduced nuclear mass and $m_{ab,e} = m_e(m_a + m_b)/(m_e + m_a + m_b) \approx m_e = 1$ is the reduced electron mass. In the clamped nuclei limit, the motion of the electrons is governed by the electronic Hamiltonian

$$H_e = -\frac{1}{2}\Delta_{\mathbf{r}} + V_{int}(\mathbf{r}, \mathbf{R}) \tag{9.2}$$

whose eigenfunctions $\chi_j(\mathbf{r}; \mathbf{R})$ and eigenvalues $E_j(R)$ are defined by

$$H_e \chi_j(\mathbf{r}; \mathbf{R}) = E_j(R)\chi_j(\mathbf{r}; \mathbf{R}) \tag{9.3}$$

It may be observed that since the interaction potential is invariant with respect to the orientation of $\mathbf{R}$ in space, the eigenvalues $E_n(R)$ depend only on the distance R. On the other hand, the eigenfunctions $\chi_j(\mathbf{r}; \mathbf{R})$ depend not only on the distance R but also on the direction of $\mathbf{R}$. It is therefore convenient to define the adiabatic eigenfunctions in a reference frame fixed with respect to the internuclear axis AB. In this *body-fixed* frame of reference, the O$\bar{z}$ axis is taken, by convention, to lie along the line from B to A: that is to say in the direction of the vector $\mathbf{R}$. In the body-fixed frame, the position vector of the electron is denoted by $\bar{\mathbf{r}}$. Geometrically, the vectors $\bar{\mathbf{r}}$ and $\mathbf{r}$ are identical, but $\bar{\mathbf{r}}$ is expressed in terms of body fixed coordinates $(\bar{x}, \bar{y}, \bar{z})$ whereas $\mathbf{r}$ is expressed in terms of space-fixed coordinates (x, y, z). In the body-fixed frame, the adiabatic eigenfunctions $\chi_j(\bar{\mathbf{r}}; R)$, depend only on $\bar{\mathbf{r}}$ and on the distance R. In addition, they can also be chosen as eigenfunctions of the component $L_{\bar{z}}$ of the electronic angular momentum along the O$\bar{z}$ axis

$$L_{\bar{z}}\chi_j(\bar{\mathbf{r}}, R) = \pm\Lambda_j\chi_j(\bar{\mathbf{r}}, R) \tag{9.4}$$

where $\Lambda_j = 0, 1, 2, \ldots$, corresponding to Σ, Π, Δ states.

The wave function of the whole system is then written in the form

$$\psi(\mathbf{r},\mathbf{R}) = \sum_{j}^{N} F_j(\mathbf{R})\chi_j(\bar{\mathbf{r}},R) \tag{9.5}$$

where N is the number of adiabatic states included in the basis. Substituting eq. 9.5 in the Schrödinger equation

$$\left[-\frac{1}{2m_{a,b}}\nabla^2_{\mathbf{R}} - \frac{1}{2}\nabla^2_{\mathbf{r}} + V_{int}(\mathbf{r},\mathbf{R})\right]\psi(\mathbf{r},\mathbf{R}) = E_T\psi(\mathbf{r},\mathbf{R}) \tag{9.6}$$

we obtain a system of coupled partial differential equations, which in matrix form can be written as

$$\left[\nabla^2_{\mathbf{R}} + 2m_{a,b}\left\{E_T\mathbf{I} - \mathbf{E}(R)\right\}\right]\mathbf{F}(\mathbf{R}) = \left[\mathbf{Q}(\mathbf{R}) + \mathbf{P}(\mathbf{R}).\nabla_{\mathbf{R}}\right]\mathbf{F}(\mathbf{R}) \tag{9.7}$$

where $\mathbf{E}$ is a diagonal matrix, whose elements are the adiabatic energies

$$E_{ij}(R) = E_j(R)\delta_{ij} \tag{9.8}$$

The matrix elements of $\mathbf{Q}$ and $\mathbf{P}$ involve the non-adiabatic coupling terms

$$Q_{ij}(R) = -\int \chi_i^*(\bar{\mathbf{r}},R)\nabla^2_{\mathbf{R}}\chi_j(\bar{\mathbf{r}},R)d\bar{\mathbf{r}} \tag{9.9}$$

$$P_{ij}(R) = -2\int \chi_i^*(\bar{\mathbf{r}},R)\nabla_{\mathbf{R}}\chi_j(\bar{\mathbf{r}},R)d\bar{\mathbf{r}} \tag{9.10}$$

It should be recalled that the differential operators $\nabla^2_{\mathbf{R}}$ and $\nabla_{\mathbf{R}}$ in eq. 9.9 and eq. 9.10, are defined with respect to the space fixed reference frame (x,y,z), whereas the adiabatic basis functions are expressed in terms of coordinates in the body-fixed system $(\bar{x},\bar{y},\bar{z})$. In calculating the matrix elements of **P** and **Q**, we may choose to express the adiabatic basis function either in terms of **r** and **R** or to transform the operators with respect to **r** fixed to operators with respect to $\bar{\mathbf{r}}$ fixed. As a general rule the second alternative is simpler. In other words we first transform the partial differential operators $\partial/\partial R, \partial/\partial\Theta$, $\partial/\partial\Phi$ taken with respect to fixed (x,y,z) to the partial differential operators $\partial'/\partial R$, $\partial'/\partial\Theta$, $\partial'/\partial\Phi$ taken with respect to $(\bar{x},\bar{y},\bar{z})$ fixed, where (Θ,Φ) designate the polar angles of **R** in the space-fixed frame.

The transformation from the space fixed axis (x,y,z) to the body fixed axis $(\bar{x},\bar{y},\bar{z})$ (where O$\bar{z}$ is in the direction of **R**) is specified by a series of two rotations of the reference frame: firstly a rotation by Φ about Oz, to yield a new set of axes $(x', \bar{y}, z)$, followed by a second rotation by Θ about O$\bar{y}$ to yield $(\bar{x}, \bar{y}, \bar{z})$. The O$\bar{z}$ axis is now in the direction of **R** and O$\bar{y}$ is in the space-fixed Oxy plane. The components of the position vector in the body fixed $\bar{x}, \bar{y}, \bar{z}$ frame are related to those in the space-fixed frame (x, y, z) by the matrix relation

$$\begin{pmatrix} \bar{x} \\ \bar{y} \\ \bar{z} \end{pmatrix} = \begin{pmatrix} \cos\Theta & 0 & -\sin\Theta \\ 0 & 1 & 0 \\ \sin\Theta & 0 & \cos\Theta \end{pmatrix} \begin{pmatrix} \cos\Phi & -\sin\Phi & 0 \\ \sin\Phi & \cos\Phi & 0 \\ 0 & 0 & 1 \end{pmatrix} \begin{pmatrix} x \\ y \\ z \end{pmatrix} \tag{9.11}$$

or explicitly

$$\bar{x} = x\cos\Theta\cos\Phi + y\cos\Theta\sin\Phi - z\sin\Theta \tag{9.12}$$

$$\bar{y} = -x\sin\Phi + y\sin\Phi \tag{9.13}$$

$$\bar{z} = x\sin\Theta\cos\Phi + \sin\Theta\sin\Phi + z\cos\Theta \tag{9.14}$$

Using these transformations (9.12-9.14), one can easily verify that

$$\begin{array}{lll} \frac{\partial\bar{x}}{\partial\Theta} = -\bar{z} & \frac{\partial\bar{y}}{\partial\Theta} = 0 & \frac{\partial\bar{z}}{\partial\Theta} = \bar{y} \\ \frac{\partial\bar{x}}{\partial\Phi} = \bar{y}\cos\Theta & \frac{\partial\bar{y}}{\partial\Phi} = -\bar{x}\cos\Theta - \bar{z}\sin\Theta & \frac{\partial\bar{z}}{\partial\Phi} = \bar{y}\sin\theta \end{array} \tag{9.15}$$

from which we may deduce

$$\frac{\partial}{\partial R} = \frac{\partial'}{\partial R} \tag{9.16}$$

$$\begin{aligned} \frac{\partial}{\partial\Theta} &= \frac{\partial'}{\partial\Theta} + \frac{\partial x}{\partial\Theta}\frac{\partial}{\partial\bar{x}} + \frac{\partial y}{\partial\Theta}\frac{\partial}{\partial\bar{y}} + \frac{\partial z}{\partial\Theta}\frac{\partial}{\partial\bar{z}} \\ &= \frac{\partial'}{\partial\Theta} - \frac{i}{\hbar}L_{\bar{y}} \end{aligned} \tag{9.17}$$

$$\begin{aligned} \frac{\partial}{\partial\Phi} &= \frac{\partial'}{\partial\Phi} + \frac{\partial x}{\partial\Phi}\frac{\partial}{\partial\bar{x}} + \frac{\partial y}{\partial\Phi}\frac{\partial}{\partial\bar{y}} + \frac{\partial z}{\partial\Phi}\frac{\partial}{\partial\bar{z}} \\ &= \frac{\partial'}{\partial\Phi} - \cos\theta\frac{i}{\hbar}L_{\bar{z}} + \sin\Theta\frac{i}{\hbar}L_{\bar{x}} \end{aligned} \tag{9.18}$$

where $L_{\bar{x}}, L_{\bar{y}}, L_{\bar{z}}$ are the components of the electronic angular momentum in the body fixed system. Writing $\nabla_{\mathbf{R}}$ and $\nabla^2_{\mathbf{R}}$, in spherical polar coordinates

$$\nabla_{\mathbf{R}} = \mathbf{u}_R \frac{\partial}{\partial R} + \frac{1}{R}\mathbf{u}_\Theta \frac{\partial}{\partial \Theta} + \frac{1}{R\sin\Theta}\mathbf{u}_\Phi \frac{\partial}{\partial \Theta} \tag{9.19}$$

where $\mathbf{u}_R$, is the unit vector in the direction of $\mathbf{R}$, $\mathbf{u}_\Theta$ the unit vector perpendicular to $\mathbf{R}$ in the plane containing $\mathbf{R}$ and Oz, $\mathbf{u}_\Phi = \mathbf{u}_R \times \mathbf{u}_\Theta$ the unit vector perpendicular to $\mathbf{R}$ and Oz

$$\nabla^2_{\mathbf{R}} = \frac{\partial^2}{\partial R^2} + \frac{2}{R}\frac{\partial}{\partial R} + \frac{1}{R^2}\left[\frac{\partial^2}{\partial \Theta^2} \cot\Theta \frac{\partial}{\partial \Theta}\frac{1}{\sin^2\Theta}\frac{\partial^2}{\partial \Theta^2}\right] \tag{9.20}$$

Using the fact that the adiabatic basis functions depend neither on Θ nor on Φ, we find

$$\begin{aligned}
\mathbf{P}_{ij}.\nabla_{\mathbf{R}} &= -\langle i|\frac{\partial}{\partial R}|j\rangle \frac{\partial}{\partial R} + \frac{2i}{R^2}\langle i| L_{\bar{y}}|j\rangle \frac{\partial}{\partial \Theta} + \\
&\quad \frac{2i\Lambda_j}{R^2 \sin\Theta}\left[\langle i| L_{\bar{x}}|j\rangle + \Lambda_j \delta_{ij}\cot\Theta\right]\frac{\partial}{\partial \Phi}
\end{aligned} \tag{9.21}$$

$$\begin{aligned}
\mathbf{Q}_{ij} &= -\langle i|\frac{\partial^2}{\partial R^2} + \frac{2}{R}\frac{\partial}{\partial R}|j\rangle + \frac{1}{R^2}\langle i| L^2_{\bar{x}} + L^2_{\bar{y}}|j\rangle + \\
&\quad \frac{2\cot\Theta}{R^2}\Lambda_j \langle i| L_{\bar{x}}|j\rangle + \frac{\Lambda^2_j \cot^2\Theta}{R^2}\delta_{ij}
\end{aligned} \tag{9.22}$$

We observe that $\langle i| L_{\bar{x}}{}^2 + L_{\bar{y}}{}^2|j\rangle = 0$ if $\Lambda_i \neq \Lambda_j$ and $\langle i| L_{\bar{x}}|j\rangle = \langle i| L_{\bar{x}}|j\rangle = 0$ if $\Lambda_i \neq \Lambda_j \pm 1$

Using eqs. 9.21 and 9.22), the coupled equations (9.7) can now be written in the form

$$\begin{aligned}
&\left[\tfrac{\partial^2}{\partial R^2} - \tfrac{\mathbf{J}^2 - \Lambda^2}{R} + 2m_{a,b}\left\{E_T \mathbf{I} - \mathbf{E}(R)\right\}\right] R\mathbf{F}(\mathbf{R}) = \\
&\left\{\mathbf{V}^{(r)}(R) + \mathbf{V}^{(c)}(R)\right\} R\mathbf{F}(\mathbf{R})
\end{aligned} \tag{9.23}$$

where the matrix $\mathbf{V}^{(r)}$ is designated as the radial coupling matrix

$$\mathbf{V}^{(r)}_{ij} = \left[\langle i| - \frac{\partial^2}{\partial R^2} + \frac{L_{\bar{x}}{}^2 + L_{\bar{y}}{}^2}{R^2}|j\rangle - 2\langle i| - \frac{\partial}{\partial R}|j\rangle \frac{\partial}{\partial R}\right]\delta(\Lambda_i, \Lambda_j) \tag{9.24}$$

and the matrix $\mathbf{V}^{(c)}$ as the rotational (or Coriolis) coupling matrix

$$\begin{aligned}\mathbf{V}_{ij}^{(c)} &= \tfrac{1}{R^2}\left[2i\left<i\right| L_{\bar{y}} - \tfrac{1}{\sin\Theta} L_{\bar{x}} \left|j\right> \tfrac{\partial}{\partial\Theta} - 2\Lambda_j \left<i\right| L_{\bar{x}} \left|j\right>\right] \times \\ &\quad \delta(\Lambda_i, \Lambda_j \pm 1))\end{aligned} \tag{9.25}$$

The matrices Λ and $\mathbf{J}^2$ are both diagonal. The elements of Λ are Λ_j while those of $\mathbf{J}^2$ are differential operators defined as

$$\mathbf{J}^2 = \left[-\frac{\partial^2}{\partial\Theta^2} + \cot\Theta\frac{\partial}{\partial\Theta} + \frac{1}{\sin^2\Theta}\left(\frac{\partial}{\partial\Phi} - i\Lambda\cos\Theta\right)^2 - \Lambda^2\right] \tag{9.26}$$

It will now be shown that $\mathbf{J}$ is none other than the total angular momentum, that is to say the sum of $\mathbf{N}$ the nuclear angular momentum and $\mathbf{L}$ the electronic angular momentum. It is recalled that while neither $\mathbf{N}$ nor $\mathbf{L}$ are conserved during a collision, the total angular momentum is a constant of the motion.

To demonstrate this interpretation of $\mathbf{J}^2$, we first make use of the relations (9.16, 9.16) to obtain explicit forms of the components of the angular momentum operators $\mathbf{N}$ and $\mathbf{L}$. It is straightforward to verify that

$$\begin{aligned}J_x + iJ_y &= (N_x + iN_y) + (L_x + iL_y) \\ &= e^{i\Phi}\left[\left(\frac{\partial'}{\partial\Theta} + i\cot\Theta\frac{\partial'}{\partial\Phi}\right) + \frac{L_{\bar{z}}}{\sin\Theta}\right]\end{aligned} \tag{9.27}$$

$$\begin{aligned}J_x - iJ_y &= (N_x - iN_y) + (L_x - iL_y) \\ &= e^{-i\Phi}\left[\left(-\frac{\partial'}{\partial\Theta} + i\cot\Theta\frac{\partial'}{\partial\Phi}\right) + \frac{L_{\bar{z}}}{\sin\Theta}\right]\end{aligned} \tag{9.28}$$

$$J_z = -i\frac{\partial'}{\partial\Phi} \tag{9.29}$$

Using the fact that the adiabatic basis functions $\chi_j(\bar{\mathbf{r}}, R)$ depend neither on Θ or on Φ, it can be shown without difficulty that

$$\begin{aligned}&\mathbf{J}^2\chi(\bar{\mathbf{r}}, R)F_j(\mathbf{R}) = \\ &-\chi(\bar{\mathbf{r}}, R)\left[\tfrac{\partial'^2}{\partial\Theta^2} + \cot\Theta\tfrac{\partial'}{\partial\Theta} + \tfrac{1}{\sin^2\Theta}\left(\tfrac{\partial'}{\partial\Phi} - i\Lambda_j\cos\Theta\right)^2 - \Lambda_j^2\right] F_j(\mathbf{R})\end{aligned} \tag{9.30}$$

Recalling that when operating on a function only of $\mathbf{R}$, there is no distinction between the primed and unprimed partial differentials: in other words in eqs. 9.30, $\partial'/\partial\Theta = \partial/\partial\Theta$, $\partial'/\partial\Phi = \partial/\partial\Phi$. This justifies the interpretation of $\mathbf{J}^2$ as the square of the total angular momentum.

3. Partial Wave Expansion

In order to reduce the system of coupled partial differential equations to a system of coupled ordinary differential equations which can be solved numerically, it is necessary to make a partial wave expansion of the functions F_n. The most natural choice is to use as expansion basis the simultaneous eigenfunctions of $\mathbf{J}^2$ and J_z, These are designated by $U^J_{M\Lambda}(\Theta, \Phi)$

$$\mathbf{J}^2 U^J_{M\Lambda} = J(J+1)U^J_{M\Lambda} \qquad J_z U^J_{M\Lambda} = M U^J_{M\Lambda} \tag{9.31}$$

The explicit form of the functions $U^J_{M\Lambda}(\Theta, \Phi)$ can be expressed as

$$U^J_{M\Lambda}(\Theta, \Phi) = (-)^{\Lambda+M} d^J_{M\Lambda}(\Theta) \exp iM\Phi \tag{9.32}$$

where $d^J_{M\Lambda}(\Theta)$ is the reduced rotation matrix [15]

$$d^J_{M\Lambda}(\Theta) = N^J_{M\Lambda} \frac{\sqrt{1-\mu}(\Lambda - M)}{\sqrt{1+\mu}(\Lambda + M)} \left[\left(\frac{\partial}{\partial\mu} \right)^{J-M} (1-\mu)^{J-\Lambda}(1+\mu)^{J+\Lambda} \right] \tag{9.33}$$

with $\mu \equiv \cos\Theta$ and $N^J_{M\Lambda}$ is a normalization constant, chosen such that

$$\int_{-1}^{1} d\mu \int_{0}^{2\pi} d\Phi U^{J*}_{M\Lambda}(\Theta, \Phi) U^{J'}_{M'\Lambda}(\Theta, \Phi) = \frac{4\pi}{2J+1} \delta(J, J')\delta(M, M') \tag{9.34}$$

We may observe that in the case $M = 0$,

$$U^J_{0\Lambda}(\Theta, \Phi) = (-)^{\Lambda} \sqrt{\frac{4\pi}{2J+1}} Y_{J,\Lambda}(\Theta, \Phi = 0) \tag{9.35}$$

Expanding each channel function in the form

$$F_j(\mathbf{R}) = \frac{1}{R} \sum_{J=0}^{\infty} f^J_j(R) U^J_{M\Lambda_j}(\Theta, \Phi) \tag{9.36}$$

and making use of the orthogonality property eq. 9.34 and the following relation (see[16])

$$\left[\mp\frac{\partial}{\partial\Theta} + \frac{i}{\sin\Theta}\frac{\partial}{\partial\Phi} + \Lambda\cot\Theta\right] U^J_{M\Lambda}(\Theta,\Phi) = \sqrt{(J\pm\Lambda+1)(J-\Lambda)}U^J_{M\Lambda\pm1}(\Theta,\Phi) \tag{9.37}$$

it is be easily demonstrated that the radial functions are solutions of the

$$\left[\frac{d^2}{dR^2} + \frac{\Lambda_j^2 - J(J+1)}{R^2} + 2m_{a,b}\left\{E_T - E_j(R)\right\}\right] f_j^J(R) = \sum_j \left(V_{ij}^{(r)} + V_{ijJ}^{(c)}\right) f_j^J(R) \tag{9.38}$$

or in matrix form

$$\left[\frac{d^2}{dR^2} + \frac{\Lambda^2 - J(J+1)}{R^2} + 2m_{a,b}\left\{E_T\mathbf{I} - \mathbf{E}(R)\right\}\right] \mathbf{f}^J(R) = \left[\mathbf{V}^{(r)}(R) + \mathbf{V}_J^{(c)}(R)\right] \mathbf{f}^J(R) \tag{9.39}$$

where the radial coupling matrix (independent of J) is given by eq. 9.24 and the Coriolis coupling (which depends on J) can be reduced to the form

$$\begin{aligned} V_{ijJ}^{(c)}(R) &= \frac{2}{R^2}\langle i|\, iL_{\bar{y}}\,|j\rangle \quad \Big[\delta(\Lambda_i,\Lambda_j-1)\sqrt{(J+\Lambda_i)(J-\Lambda_j+1)} - \\ &\qquad \delta(\Lambda_i,\Lambda_j+1)\sqrt{(J-\Lambda_i)(J+\Lambda_j+1)}\,\Big] \end{aligned} \tag{9.40}$$

We may observe that while the radial matrix $\mathbf{V}^{(r)}$ only connects channels i, j if $\Lambda_i = \Lambda_j$, the Coriolis part $\mathbf{V}^{(c)}$ connects channels i, j if $\Lambda_i = \Lambda_j \pm 1$.

4. Non-adiabatic Coupling Terms

In order to acquire some intuition of the standard adiabatic model, it is desirable to have some quantitative idea of the nature of the radial and rotation coupling terms $V_{ij}^{(r)}$ and $V_{ij}^{(c)}$. Let us begin with the term $V_{ij}^{(r)}$.

$$\mathbf{V}_{ij}^{(r)} = \left[\langle i| - \frac{\partial^2}{\partial R^2} + \frac{{L_{\bar{x}}}^2 + {L_{\bar{y}}}^2}{R^2}|j\rangle - 2\langle i| - \frac{\partial}{\partial R}|j\rangle\frac{\partial}{\partial R}\right]\delta(\Lambda_i,\Lambda_j) \tag{9.41}$$

It should be stressed that this term operates on the radial function $F_j(R)$. Except for the case of extremely low energies, $F_j(R)$ is quite an oscillatory function since its associated wave number $k_j >> 1$. For that reason, the second term of eq. 9.41 dominates (roughly by a factor proportional to $\sqrt{m_{a,b}}$). The radial coupling is therefore controlled by the matrix element $\langle i| \frac{\partial}{\partial R} |j\rangle$, which we designate by A_{ij}. A precise determination of A_{ij} involves a numerical differentiation of the adiabatic eigenfunctions with respect to the internuclear distance. But it is possible to gain some qualitative indication of its size from elementary considerations.

But, first of all, it is important to establish a phase convention for the adiabatic wave functions $\chi_j(\bar{\mathbf{r}}; R)$. Introducing explicitly the coordinates of $\bar{\mathbf{r}}$ as (r, θ, ϕ) where ϕ is the azimuthal coordinate and (r, θ) the ensemble of non-azimuthal coordinates, the axial symmetry of the system makes it possible to express the eigenfunction $\chi_j(\bar{\mathbf{r}}; R)$ in the form

$$\chi_j(\bar{\mathbf{r}}; R) = \zeta(r, \theta; R)\Phi(\phi) \tag{9.42}$$

The phase of the azimuthal part, being independent of R, is well defined for a given Λ_j and has no effect on the radial coupling matrix element. In contrast, the phase of the non-azimuthal part is not well defined for a given R. It is therefore customary to adopt the convention that the non-azimuthal part be real. This still leaves an arbitrary overall factor of ± 1. The convention is to assume that the sign does not change with respect to small changes in R. In practice this can be ensured by calculating numerically the overlap of the adiabatic functions at R and $R + \delta R$, where $\delta R << R$. If the overlap is close to +1 the phases are correct, if the overlap is close to -1, the phases are adjusted accordingly. The problem of ensuring a systematic phase convention is far from trivial in numerical calculations, when there are more than 2 coupled states. Errors arising from a failure to respect the phase conventions are known to be present in some published calculations.

With the above choice of phase convention, it is easy to show that the diagonal elements of matrix $\mathbf{A}$ vanish. Since the eigenfunctions are normalized for all R

$$\frac{d}{dR}\langle i| i\rangle = \left\langle \frac{\partial}{\partial R} i \,\middle|\, i \right\rangle + \left\langle i \,\middle|\, \frac{\partial}{\partial R} i \right\rangle = 2\langle i| \frac{\partial}{\partial R} |i\rangle = 0 \tag{9.43}$$

As for the off-diagonal elements, we use the fact that since the matrix element $\langle i| H_e |j\rangle = 0$

$$\frac{d}{dR}\langle i| H_e |j\rangle \;=\; \left\langle \frac{\partial}{\partial R} i \,\middle|\, H_e |j\rangle + \langle i| \frac{\partial H_e}{\partial R} |j\rangle + \langle i| H_e \,\middle|\, \frac{\partial}{\partial R} j \right\rangle$$

$$\begin{aligned} &= E_j(R)\left\langle \frac{\partial}{\partial R} i \,\middle|\, j \right\rangle + \langle i | \frac{\partial H_e}{\partial R} | j \rangle + E_i(R)\left\langle i \,\middle|\, \frac{\partial}{\partial R} j \right\rangle \\ &= + \langle i | \frac{\partial H_e}{\partial R} | j \rangle + (E_i(R) - E_j(R))\left\langle i \,\middle|\, \frac{\partial}{\partial R} j \right\rangle \\ &= 0 \end{aligned} \qquad (9.44)$$

It therefore follows that

$$A_{ij} = \left\langle i \,\middle|\, \frac{\partial}{\partial R} j \right\rangle = \frac{-1}{(E_i(R) - E_j(R))} \langle i | \frac{\partial H_e}{\partial R} | j \rangle \quad \text{if} \quad i \neq j \qquad (9.45)$$

We thus see from eq. 9.45 that the radial coupling can become large if the energy separation of the adiabatic states concerned becomes small. Of course, the energy separation between adiabatic states of the same symmetry (Wigner-von Neumann no crossing rule) can never vanish, but it can become quite small. Typical examples are shown in figures 9.3 - 9.6.

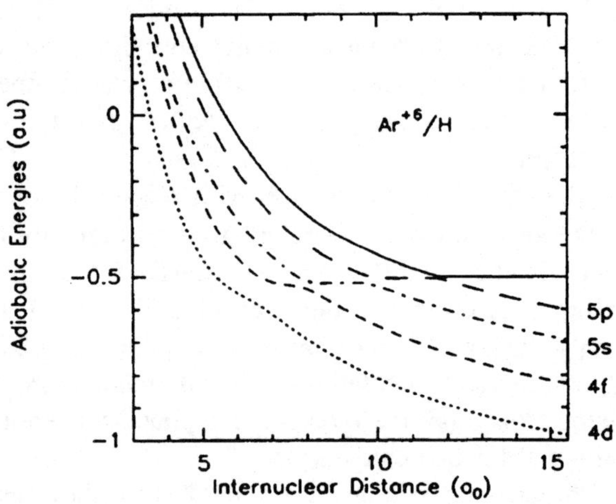

Figure 9.3. Adiabatic potential energies of the ArH^{6+} molecular ion:
solid curve 1 [$Ar^{6+}(3s^2)$+H(1s)], long dashed curve 2 [$Ar^{5+}(3s^2 5p)+H^+$], dot dashed curve 3 [$Ar^{5+}(3s^2 5s)+H^+$], short dashed curve 4 [$Ar^{5+}(3s^2 4f)+H^+$], dotted curve 5 [$Ar^{5+}(3s^2 4d)+H^+$].

for the systems ArH^{6+} and $ArHe^{6+}$, where several well defined avoided crossings are observed [17]. The corresponding radial coupling matrix elements are strongly peaked at the internuclear distances corresponding to the avoided crossings. As a consequence, transitions between the adiabatic states tend to be localized.

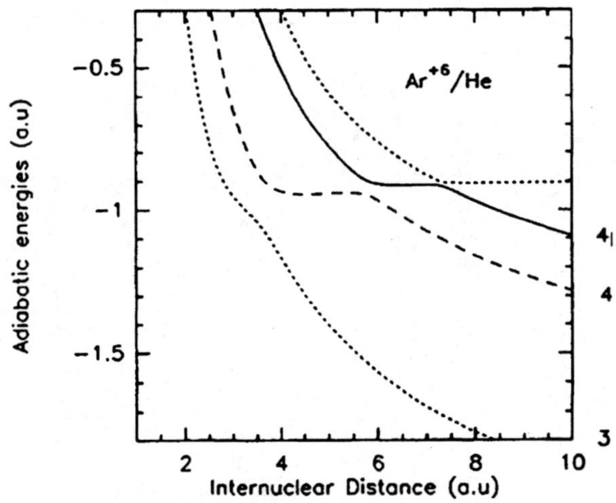

Figure 9.4. Adiabatic potential energies of the $ArHe^{6+}$ molecular ion:
top dotted curve 1 [$Ar^{6+}(3s^2)$+$He(1s^2)$], solid curve 2 [$Ar^{5+}(3s^2 4p)$+He^+], dashed curve 3 [$Ar^{5+}(3s^2 4s)$+He^+], bottom dotted curve 4 [$Ar^{5+}(3s^2 4f)$+He^+].

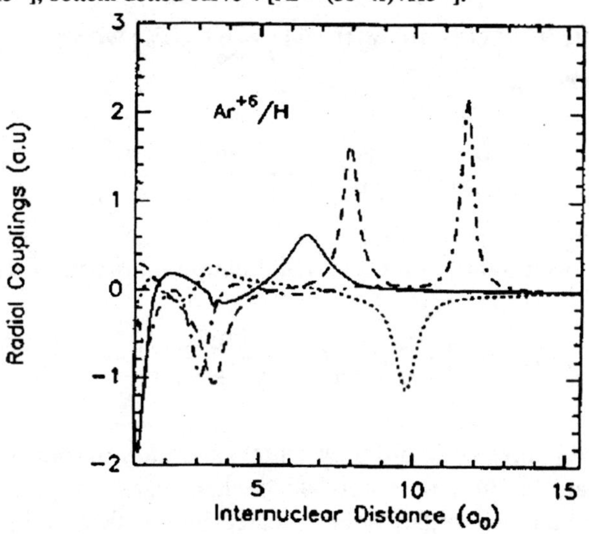

Figure 9.5. ArH^{6+} radial coupling matrix elements:
dot-dashed curve A_{12}, dotted curve A_{23}, dashed curve A_{34}, solid curve A_{45}.

It is also of interest to examine the radial coupling matrix elements in the asymptotic region as $R \to \infty$. The derivative $\partial/\partial R$ is calculated with respect to $\bar{\mathbf{r}}$. Let us assume that as $R \to \infty$, the adiabatic eigenfunction $\chi_j(\bar{\mathbf{r}}; R)$ tends to

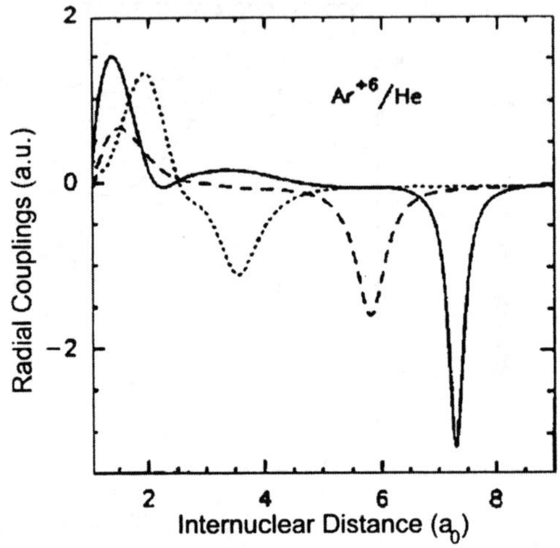

Figure 9.6. $ArHe^{6+}$ radial coupling matrix elements: solid curve A_{12}, dashed curve A_{23}, dotted curve A_{34}.

state j of $(A+e)$. In other words $\chi_j(\bar{\mathbf{r}}; R) \underset{R\to\infty}{\approx} \phi_j(\bar{\mathbf{r}}_a)$ where $\bar{\mathbf{r}} = \bar{\mathbf{r}}_a + \frac{m_b}{m_a m_b}\mathbf{R}$. It is easy to show that

$$\left.\frac{\partial}{\partial R}\right|_{\bar{\mathbf{r}}} = \left.\frac{\partial}{\partial R}\right|_{\bar{\mathbf{r}}_a} + \frac{m_b}{m_a + m_b}\left.\frac{\partial}{\partial \bar{z}}\right|_{\mathbf{R}} \tag{9.46}$$

and that, in the asymptotic limit, the radial coupling matrix is given by

$$A_{ij} = \langle i|\,\frac{\partial}{\partial R}\,|j\rangle \underset{R\leftarrow\infty}{\approx} \frac{m_b}{m_a + m_b}\,\langle i|\,\frac{\partial}{\partial \bar{z}_a}\,|j\rangle \tag{9.47}$$

If states i and j correlate to different centres in the asymptotic limit, the matrix element will vanish. But it does not vanish when states i and j are correlated to the same centre and when they are connected by an allowed dipole transition. The existence of such non-vanishing off-diagonal couplings makes it formally impossible to express the asymptotic solution of the radial equations in terms of the S-matrix. In practice, the real effect of these non-vanishing coupling terms is fairly weak in the low energy regime below 10eV/amu but with increasing energy they can and do introduce large uncertainties in the calculations. We shall return to this point later.

We have already remarked that the second term of eq. 9.41 is dominant because of the fact that it involves both A_{ij} and the derivative of the radial function $\frac{df_j^J}{dR}$. To within an order of magnitude $\frac{df_j^J}{dR} \approx k_j f_j^J$ so that using eq. 9.45 we may write

$$2\langle i| - \frac{\partial}{\partial R}|j\rangle \frac{\partial}{\partial R} \approx \frac{2k_j}{a|\Delta_{ij}|}|\langle i| V |j\rangle| \tag{9.48}$$

where Δ_{ij} is the minimum value of $E_i(R) - E_j(R)$ at the avoided crossing and a is a length characteristic of the crossing region. Let us now examine the first term of eq. 9.41. The matrix **B** designates the radial coupling matrix involving second derivative terms, defined as follows

$$B_{ij} = \langle i| \frac{\partial^2}{\partial R^2}|j\rangle \tag{9.49}$$

The matrix elements of **B** are related to those of **A** by the relation

$$\frac{d}{dR}A_{ij}(R) = \frac{d}{dR}\langle i| \frac{\partial}{\partial R}|j\rangle = \langle i| \frac{\partial^2}{\partial R^2}|j\rangle + \left\langle \frac{\partial}{\partial R}i \middle| \frac{\partial}{\partial R}j \right\rangle \tag{9.50}$$

If the adiabatic basis set were complete, the second term on the right hand side of eq. 9.50 may be written as

$$\left\langle \frac{\partial}{\partial R}i \middle| \frac{\partial}{\partial R}j \right\rangle = \sum_k^{\infty} \langle i| \frac{\partial}{\partial R}|k\rangle \langle k| \frac{\partial}{\partial R}|j\rangle = \sum_k^{\infty} A_{ik}A_{kj} = (\mathbf{A}^2)_{ij} \tag{9.51}$$

If the sum over k is restricted to a finite basis, the relation (9.51) is not valid. However, it is commonly assumed in most practical calculations that to a good approximation we may write

$$\left\langle \frac{\partial}{\partial R}i \middle| \frac{\partial}{\partial R}|j\right\rangle \cong \sum_{k=1}^{N} A_{ik}A_{kj} = (\mathbf{A}^2)_{ij} \tag{9.52}$$

in which case, we may write

$$\mathbf{B} \cong \mathbf{A}^2 + \frac{d}{dR}\mathbf{A} \tag{9.53}$$

(Additional discussion of this point will be given in section 9.5). The formula (9.53) gives us some idea of the order of magnitude of the terms B_{ij}. First of all we note that the diagonal elements of **B** depend only on $\mathbf{A}^2$. Secondly, the non-diagonal term B_{ij} is dominated by the term $d\mathbf{A}/dR$. So to a crude approximation, the off-diagonal terms of **B** are given by

$$|B_{ij}| \cong \frac{1}{a_2 |\Delta_{ij}|} |\langle i| V |j\rangle| \tag{9.54}$$

This term is much smaller than A_{ij} provided $k_j a >> 1$. As for the matrix elements involving the operator $(L_{\bar{x}^2} + L_{\bar{y}^2})/R^2$ in eq. 9.41, their magnitude is much less affected by the presence of an avoided crossing than are the matrix elements involving radial derivatives. So while they may be of comparable magnitude to the elements B_{ij} away from an avoided crossing, they are much smaller than these elements in the vicinity of the crossing. As a general rule, they are of interest only in the investigation of small isotope effects. They may be safely neglected in collision problems.

5. Diabatic Representation

The radial coupling matrix elements $V_{ij}^{(r)}(R)$ may undergo a very rapid variation in the vicinity of an avoided crossing. For this reason, great care must be exercised in the numerical integration of the radial eqs. 9.37 and 9.38 and it is often desirable to make a unitary transformation of the adiabatic basis functions to what is termed a *diabatic* basis, in which the coupling elements vary smoothly with R. The *diabatic* basis functions, designated by $\chi_j^d(\bar{\mathbf{r}}; R)$, have the same symmetry properties as the adiabatic basis functions from which they are derived and are defined in terms of these latter by means of a unitary transformation matrix $\mathbf{C}(R)$ according to

$$\chi_j^d(\bar{\mathbf{r}}; R) = \sum_k^N \chi_k(\bar{\mathbf{r}}, R) C_{kj}(R) \tag{9.55}$$

Introducing the column vector $\mathbf{g}^J(R)$, related to $\mathbf{f}^J(R)$ by the transformation

$$\mathbf{f}^J(R) = \mathbf{C}(R)\mathbf{g}^J(R) \tag{9.56}$$

the differential equation (9.39) transforms to

$$\left[\frac{d^2}{dR^2} + \frac{\Lambda^2 - J(J+1)}{R^2} + 2m_{a,b}\mathbf{C}^{-1}\left\{E_T\mathbf{I} - \mathbf{E}(R)\right\}\mathbf{C}\right]\mathbf{g}^J(R) =$$
$$\mathbf{C}^{-1}\left[\mathbf{V}^{(r)}(R) + \mathbf{V}_J^{(c)}(R)\right]\mathbf{C}\mathbf{g}^J(R) \tag{9.57}$$

Posing

$$\begin{aligned} \mathbf{U}^d &= \mathbf{C}^{-1}\,\mathbf{E}(R)\,\mathbf{C} \\ {}^d\mathbf{V}^{(r)} &= \mathbf{C}^{-1}\,\mathbf{V}^{(r)}\,\mathbf{C} \\ {}^d\mathbf{V}_J^{(c)} &= \mathbf{C}^{-1}\,\mathbf{V}^{(c)}\,\mathbf{C} \end{aligned} \tag{9.58}$$

where $\mathbf{E}$ is a diagonal matrix, whose elements are the adiabatic energies, eq. 9.57 may be written in the form

$$\begin{aligned} &\left[\frac{d^2}{dR^2} + \frac{\Lambda^2 - J(J+1)}{R^2} - 2m_{a,b}\mathbf{U}(R) + 2m_{a,b}E_T\right]\mathbf{g}^J(R) = \\ &\left[{}^d\mathbf{V}^{(r)}(R) + {}^d\,\mathbf{V}_J^{(c)}(R)\right]\mathbf{g}^J(R) \end{aligned} \tag{9.59}$$

We may therefore consider the matrix $\mathbf{U}$ as the electronic Hamiltonian in the *diabatic* basis. Of course, since the *diabatic* basis functions are not eigenfunctions of the electronic Hamiltonian, $\mathbf{U}$ is non-diagonal, except in the asymptotic limit where it can be made diagonal by imposing the condition

$$\mathbf{C}(R) \underset{R\to\infty}{\approx} \mathbf{I} \tag{9.60}$$

As we have already remarked, the principal contribution to $\mathbf{V}^{(r)}$ comes from the matrix $\mathbf{A}$. Let us therefore consider the contribution of $\mathbf{A}$ to ${}^d\mathbf{V}^{(r)}$. In the *diabatic* representation, $\mathbf{A}$ transforms to $\mathbf{A}^d = \mathbf{C}^{-1}\mathbf{A}\mathbf{C}$, so that

$$\begin{aligned} A_{ij}^d &= \sum_{k,l} \langle\chi_k|C_{ki}^*(R)\frac{\partial}{\partial R}C_{lj}(R)\,|\chi_l\rangle \\ &= \sum_{k,l} C_{ki}^*(R)\,\langle\chi_k|\frac{\partial}{\partial R}\,|\chi_l\rangle\,C_{lj}(R) + \sum_k C_{ki}^*(R)\frac{d}{dR}C_{kj}(R) \end{aligned} \tag{9.61}$$

where we have used the orthogonality relation $\langle\chi_k \mid \chi_l\rangle = \delta_{kl}$. In matrix notation eq. 9.61 can be written as

$$\mathbf{A}^d = \mathbf{C}^{-1}\mathbf{A}\mathbf{C} + \mathbf{C}^{-1}\frac{d}{dR}\mathbf{C} \tag{9.62}$$

The usual *diabatic* transformation is to require that $\mathbf{A}^d$ vanishes. This can be achieved by requiring the transformation matrix to be a solution of the equation

$$\mathbf{AC} + \frac{d\mathbf{C}}{dR} = 0 \tag{9.63}$$

with eq. 9.60 as the boundary condition.

Let us now consider how the matrix **B** transforms under a diabatic representation. The matrix elements may be written in the form

$$\begin{aligned} B^d_{ij} &= \sum_{k,l} \langle \chi_k | C^*_{ki}(R) \tfrac{\partial^2}{\partial R^2} C_{lj}(R) | \chi_l \rangle = \sum_{k,l} C^*_{ki}(R) \langle \chi_k | \tfrac{\partial^2}{\partial R^2} | \chi_l \rangle C_{lj}(R) \\ &+ 2 \sum_{k,l} C^*_{ki}(R) \langle \chi_k | \tfrac{\partial}{\partial R} | \chi_l \rangle \tfrac{d}{dR} C_{lj}(R) + \sum_k C^*_{ki}(R) \tfrac{d^2}{dR^2} C_{kj}(R) \end{aligned} \tag{9.64}$$

or, in matrix notation

$$\mathbf{B}^d = \mathbf{C}^{-1}\mathbf{BC} + 2\mathbf{C}^{-1}\mathbf{A}\frac{d}{dR}\mathbf{C} + \mathbf{C}^{-1}\frac{d^2}{dR^2}\mathbf{C} \tag{9.65}$$

Taking the derivative of eq. 9.63, we obtain

$$\frac{d^2}{dR^2}\mathbf{C} = -\mathbf{A}\frac{d\mathbf{C}}{dR} - \frac{d\mathbf{A}}{dR}\mathbf{C} \tag{9.66}$$

from which we may deduce, using eq. 9.65

$$\begin{aligned} \mathbf{B}^d &= \mathbf{C}^{-1}\mathbf{BC} + \mathbf{C}^{-1}\mathbf{A}\tfrac{d}{dR}\mathbf{C} - \mathbf{C}^{-1}\tfrac{d\mathbf{A}}{dR}\mathbf{C} \\ &= \mathbf{C}^{-1}\mathbf{BC} - \mathbf{C}^{-1}\mathbf{A}^2\mathbf{C} - \mathbf{C}^{-1}\tfrac{d\mathbf{A}}{dR}\mathbf{C} \\ &= \mathbf{C}^{-1}\left(\mathbf{B} - \mathbf{A}^2 - \tfrac{d\mathbf{A}}{dR}\right)\mathbf{C} \end{aligned} \tag{9.67}$$

We thus see that it is only in the case of a complete basis set ($N \to \infty$), when eq. 9.53 is satisfied, that the matrix $\mathbf{B}^d$ also vanishes. In the case of a finite basis set, $\mathbf{B}^d$ does not vanish exactly, even though it may be reasonable approximation to assume it to be small.

Recently [18, 19], a quantitative test of this point has been carried out on the two-state system describing electron capture in Si^{3+}/He, for which matrix elements

$$D_{ij} = \left\langle \frac{\partial \chi_i}{\partial R} \middle| \frac{\partial \chi_j}{\partial R} \right\rangle \tag{9.68}$$

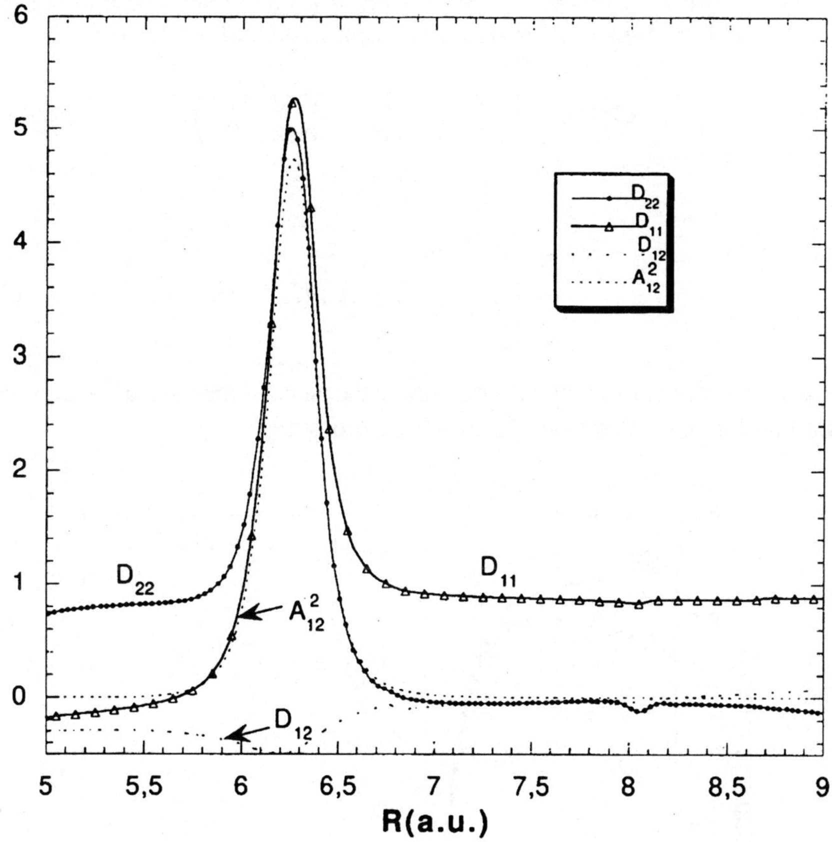

Figure 9.7. A comparison of the radial coupling matrix elements of **D** and **A** for the two $^2\Sigma$ states of $SiHe^{3+}$ responsible for electron capture by Si^{3+} from He. The radial derivatives have been calculated with the origin of electron coordinates on the Si nucleus. We may note that in this case D_{11} does not vanish in the asymptotic limit. Had the origin of electronic coordinates been taken on the He nucleus, D_{11} would vanish at the asymptotic limit but then D_{22} would be non-zero.

have been calculated directly by numerical integration and compared with the matrix elements of $\mathbf{A}^2$ (fig. 9.7). We note that in the vicinity of the avoided crossing **D** is indeed approximately equal to $\mathbf{A}^2$. But away from the crossing, D_{11} and D_{22} differ significantly from the diagonal elements of $\mathbf{A}^2$. The off diagonal elements of **D**, however remain small. Fortunately, in most calculations of collision dynamics, it is the coupling terms in the vicinity of the avoided crossing that are the most important. So in practice, it is probably reasonable to neglect $\mathbf{B}^d$. However, it can be important for the study of non-adiabatic effects on pre-dissociating vibrational states in molecular systems [20].

In the case of a two-state approximation, it is easily verified that the transformation matrix is none other than a rotation matrix of the form

$$\mathbf{C}(R) = \begin{pmatrix} \cos\omega(R) & \sin\omega(R) \\ -\sin\omega(R) & \cos\omega(R) \end{pmatrix} \tag{9.69}$$

where

$$\omega(R) = \int_R^\infty A_{12}(R')dR'. \tag{9.70}$$

Such a transformation matrix constitutes the basis of the Landau-Zener model of a non-adiabatic transition at an avoided crossing.

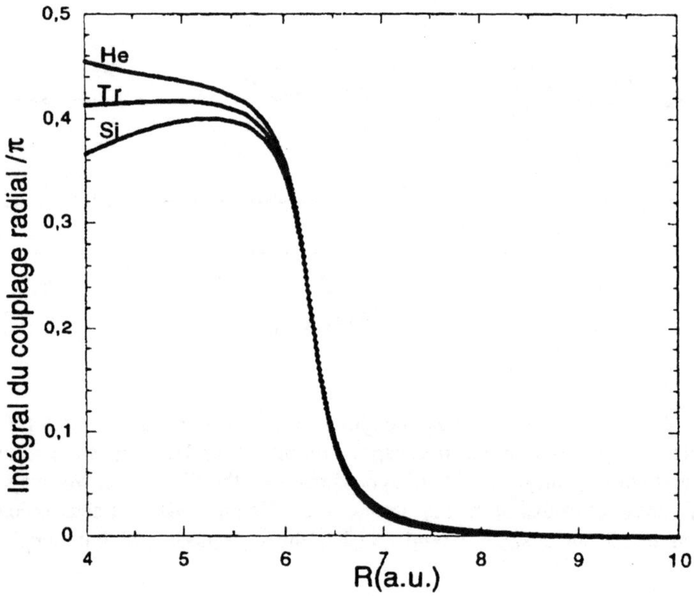

Figure 9.8. Diagram to illustrate how $\omega(R)$ varies as a function of R for the radial coupling between the two $^2\Sigma$ states of $SiHe^{3+}$ in the vicinity of their avoided crossing. The upper and lower curves refer to calculations where the radial derivative is taken with respect to the He nucleus and to the Si nucleus. The intermediate curve is a modified coupling to take account of translation effects (see section 9.6). Considerable deviation from $\pi/2$ is observed at short distances.

If the system is well described by a two-state approximation, we may expect that the rotation angle $\omega(R)$ to vary from zero in the asymptotic region $R \to \infty$

to about $\pi/4$ in the vicinity of the crossing, reaching a limit of $\pi/2$ at small R. In practice, this is often found to be the case in many well defined two state systems. But there is no rigorous proof of this result and many cases have been found where $\omega(R)$ departs considerably from $\pi/2$ at small R (see figure 9.8). It would therefore is a little unwise to give a purely physical interpretation of the *diabatic* energy matrix. The problem is related to the defects of the adiabatic model to represent the translation energy of the collision in the asymptotic region (just like the non-vanishing asymptotic radial couplings.

In the more general case (three or more coupled states) the determination of the transformation matrix requires a numerical solution of the eq. 9.63. The specific techniques [21] used to solve eq. 9.63 are fairly straightforward but great care must be exercised in the vicinity of very long distance avoided crossings where the radial coupling matrix elements of **A** can become very highly peaked. In that case, a very fine mesh of grid points is necessary to obtain an accurate solution. But it should be remarked that since the *diabatic* transformation is independent of the collision energy, only one calculation for each system is needed. The main practical advantages of a *diabatic* representation are that both the *diabatic* energy and the transformed Coriolis coupling matrices vary smoothly with R. In addition, the absence of first order derivatives in the radial equations makes them much easier to solve numerically.

6. Solution of the Radial Equations and Extraction of the S-matrix

The radial eqs. 9.59 are amenable to a numerical solution provided the number of coupled channels is reasonably small (less than 20 or so) and provided the wave number is not too large ($< 10^3\ a_0^{-1}$). As the wave number increases, so does the number of partial waves contributing to the cross sections. This does not necessarily increase the amount of computation since the partial wave contributions to the cross sections vary smoothly with J and it is not necessary to carry out the calculations for every J. The limiting feature is the oscillatory nature of the radial functions with large wave number. As a general rule, quantum calculations are feasible in the energy range up to about 1keV/amu or so. For higher energies, semi-classical methods are to be preferred.

The transition amplitudes are expressed in terms of the scattering S-matrix, which is defined by the solutions of eq. 9.59 which vanish at R=0 and which have the asymptotic form

$$f_j^J(R) \underset{R\to\infty}{\approx} g_j^J(R) \underset{R\to\infty}{\approx} \frac{1}{\sqrt{k_j}} \Big\{ \exp[-i\,(k_jR + \vartheta_{J,j}(R) - J\pi/2)] \\ - S_{ji}^J \exp[i\,(k_jR + \vartheta_{J,j}(R) - J\pi/2)] \Big\} \tag{9.71}$$

where $\vartheta_{J,j}(R)$, the Coulomb phase associated with channel j, is given by

$$\vartheta_{J,j}(R) = \arg\Gamma(J+1+i\gamma_j) - \frac{m_{a,b}q_j^a q_j^b}{k_j}\log 2k_jR \tag{9.72}$$

q_j^A, q_j^B are the residual charges centred on A and B in the asymptotic channel j and $\gamma_j = q_j^A q_j^B / k_j$. If one of the residual charges, either q_j^A or q_j^B, becomes zero (as is the case for the elastic and excitation channels in ion-neutral collisions), the Coulomb phase for that channel vanishes.

It is not practical to generate directly the solutions of eqs. 9.59 which satisfy simultaneously the condition at the origin and the asymptotic condition (9.71). But, since the eqs. 9.59 are linear, it is possible to construct any desired solution from a complete set of linearly independent solutions of the form

$$f_j^J(R) \underset{R\to\infty}{\approx} \frac{1}{\sqrt{k_j}} \Big\{ X_{jl}^J \sin(k_jR + \vartheta_{J,j}(R) - J\pi/2) \\ - Y_{jl}^J \cos(k_jR + \vartheta_{J,j}(R) - J\pi/2) \Big\} \tag{9.73}$$

satisfying the required boundary condition at the origin. Clearly, there is an infinite number of such solutions but we only need N arbitrary solutions which are linearly independent. It is then simple to derive the **K**-matrix defined by the solution of eq. 9.59 with the asymptotic conditions

$$f_j^J(R) \underset{R\to\infty}{\approx} \frac{1}{\sqrt{k_j}} \Big\{ \sin(k_jR + \vartheta_{J,j}(R) - J\pi/2) \\ - K_{ji}^J \cos(k_jR + \vartheta_{J,j}(R) - J\pi/2) \Big\} \tag{9.74}$$

The elements of **K**, are real and they are related to the matrix elements X_{jl}^J, Y_{jl}^J by the relation

$$\mathbf{K} = \mathbf{X}^{-1}\mathbf{Y} \tag{9.75}$$

It is easily verified if the N solutions are linearly independent, for only then does $\mathbf{X}^{-1}$ exist. The **S**-matrix can then be extracted using the simple relation

$$\mathbf{S} = i\frac{\mathbf{I} + i\mathbf{K}}{\mathbf{I} - i\mathbf{K}} \tag{9.76}$$

Since the elements of **K** are real, the elements of the **S**-matrix are complex. It is recalled that the **S**-matrix is unitary and symmetric.

Although standard algorithms for integrating the radial eqs. 9.59 are stable, there are problems in generating linearly independent solutions when the integration is started deep inside the classically forbidden region, that is to say for small values of R. So conventional propagator methods cannot be used directly to integrate from the origin and special starting procedures are required. An alternative approach, which avoids the need for any special starting procedure in the classically forbidden region is use the log-derivative method [22] proposed by Johnson, modified to treat the case of Coulomb interactions [23]. It is of some interest to explain the method.

Let us designate the N independent column vectors $\mathbf{g}^J$ in the form of a matrix $\mathbf{G}^{J\cdot}$. Then we may write the coupled equations in the form

$$\left[\frac{d^2}{dR^2}\right]\mathbf{G}^J(R) = \mathbf{W}(R)\mathbf{G}^J(R) \tag{9.77}$$

where

$$\mathbf{W}(R) = \frac{J(J+1) - \Lambda^2}{R^2} - 2m_{a,b}\left\{E_T\mathbf{I} - \mathbf{E}(R)\right\} + \mathbf{V}^{(r)}(R) + \mathbf{V}_J^{(c)}(R) \tag{9.78}$$

The log-derivative matrix is defined by

$$\mathbf{Z}(R) = \frac{\mathbf{F}'(R)}{\mathbf{F}(R)} \tag{9.79}$$

Differentiating eq. 9.79 and using eq. 9.77, we obtain the matrix Ricatti equation

$$\frac{d\mathbf{Z}}{dR} = \mathbf{W}(R) - \mathbf{Z}^2(R) \tag{9.80}$$

The log-derivative matrix becomes undefined whenever the determinant of the wave function vanishes. These singularities prohibit the use of standard numerical integration techniques to solve eq. 9.80 directly. However, it was shown by Johnson [22] that the log-derivative matrix may be propagated in the interval $[R', R'']$ using what is termed an imbedding type propagator according to, which

$$\mathbf{Z}(R'')) = \mathbf{z}_4(R', R'') - \mathbf{z}_3(R', R'') \left[\mathbf{Z}(R') + \mathbf{z}_1(R', R'')\right]^{-1} \mathbf{z}_3(R', R'') \tag{9.81}$$

The elements of the propagator matrices $\mathbf{z}_1, \mathbf{z}_2, \mathbf{z}_3, \mathbf{z}_4$, take on quite a simple analytic form. However, there are several variants [24] of the exact form of the propagators and the reader requiring more details should consult the original references.

The log-derivative method has proved to be a remarkably convenient and stable method to solve the radial equations and extract the **S**-matrix. It is not the most rapid integrator, but by avoiding the need for any special starting procedure in the classically inaccessible region it is easy to implement and can be applied to a wide variety of problems of involving collisions between heavy atomic particles without any instability problems.

The transition amplitude for scattering into state j from an initial state i can be immediately deduced from the **S**-matrix. For example if the initial state is a spherically symmetric state with $\Lambda_j = 0$

$$f_{ji}(\Theta) = \frac{i}{2\sqrt{k_i k_j}} \sum_J \left[\sqrt{4\pi(2J+1)}\right] U^J_{0\Lambda_j}(\Theta, \Phi) \left[\delta_{ji} - S^J_{ji}\right] \tag{9.82}$$

The differential cross section is given by

$$\frac{d\sigma_{ji}}{d\Omega} = \frac{k_j}{k_i} \left|f_{ji}(\Theta)\right|^2 \tag{9.83}$$

and the total cross section by

$$\sigma_{ji} = \frac{\pi}{k_i^2} \sum_J (2J+1) \left|S_{ji}\right|^2 \tag{9.84}$$

It is observed that the differential cross section depends on the phases of the elements of the **S**-matrix whereas the total cross section depends only on their amplitudes. This is the reason why differential cross sections may exhibit a highly periodic angular variation, whereas the total cross sections usually show a fairly slow variation with energy. Experimental measurements of differential and total cross sections provide complementary information on the collision process (provided that the angular resolution of the differential cross section is sufficiently well resolved).

In the case where there are non-vanishing non-diagonal matrix elements in the asymptotic region, it is strictly speaking not possible to define a unique transformation matrix **C** because the condition (9.59) cannot be satisfied. One way to obtain an approximate solution is to impose the condition (9.59) at some large value of R_m. The resulting **S**-matrix will then be a function of R_m. However, as R_m is varied, the elements of **S** generally oscillate about some mean value with small amplitude. In that way, it is possible to put some kind of error bar on the standard adiabatic model. At low eV energies the errors in the calculated cross sections are usually less than a few per cent if the cross sections are larger than about $10^{-16}cm^2$. So, the standard adiabatic method is reliable for the dominant channels. On the other hand for energies upwards of a few 100 eV the errors can become considerably greater, depending on the system under investigation. The standard adiabatic model then requires some modification.

7. Modified Adiabatic model: Reaction Coordinates

To overcome the inconsistencies of the standard adiabatic model, it was first proposed [10] to modify the adiabatic basis set by the inclusion of appropriate translation factors (ETF), which would allow for a matching to the correct asymptotic boundary equations. The concept of a translation factor is natural in a semi-classical (or impact parameter) approach, where it is implicitly assumed that the internuclear distance is a known function of time (particularly simple in the case of rectilinear trajectories). However, in a quantum mechanical calculation, the introduction of ETFs is not very practical. An alternative approach, first proposed by Mittleman [25] is to introduce a set of reaction coordinates that can automatically adjust to both the excitation and charge exchange channels. For example, if the electron is initially bound to centre A, the initial conditions can only be described correctly by the Jacobi coordinates $(\mathbf{r}_a, \mathbf{R}_a)$, whereas the final charge exchange channels are only described correctly by the Jacobi coordinates $(\mathbf{r}_b, \mathbf{R}_b)$. A suitable set of reaction coordinates would therefore be one, which for large R takes the form of the Jacobi cooordiantes $(\mathbf{r}_a, \mathbf{R}_a)$ or $(\mathbf{r}_b, \mathbf{R}_b)$, depending on whether $r_a << R$ or whether $r_b << R$. The idea was successfully implemented by Thorson and Delos [26]. A little later, Robert and Baudon [27] drew attention to the fact that Eckart [28] had already proposed a set of coordinates for the 3-body problem, which had the desired properties required for the problem. It is of interest to follow the Eckart approach.

Introducing the mass-scaled coordinates

$$\mathbf{R}_1 = \sqrt{M_{a,b}}\mathbf{R} \quad \mathbf{R}_2 = \sqrt{M_{ab,e}}\mathbf{r} \tag{9.85}$$

we may form the Gram matrix $\check{\mathbf{G}}$, defined as

$$\check{\mathbf{G}} = \begin{pmatrix} R_1^2 & \mathbf{R}_1.\mathbf{R}_2 \\ \mathbf{R}_1.\mathbf{R}_2 & R_2^2 \end{pmatrix} \tag{9.86}$$

whose two real positive eigenvalues r_1^2, r_2^2 with $r_1 > r_2$, may be expressed in the form

$$r_1^2 = R_1^2 \cos^2 q + R_2^2 \sin^2 q \qquad r_2^2 = R_1^2 \sin^2 q + R_2^2 \cos^2 q \tag{9.87}$$

where q, the angle of the rotation matrix which diagonalises $\check{\mathbf{G}}$, is given by

$$\tan 2q = \frac{2\mathbf{R}_1.\mathbf{R}_2}{R_1^2 - R_2^2} \tag{9.88}$$

The 3 unit vectors $\hat{\mathbf{u}}_i (i = 1, 2, 3)$ defined by the relations

$$r_1\hat{\mathbf{u}}_1 = \cos q\mathbf{R}_1 + \sin q\mathbf{R}_2 \quad r_2\hat{\mathbf{u}}_2 = -\sin q\mathbf{R}_1 + \cos q\mathbf{R}_2 \quad \hat{\mathbf{u}}_3 = \hat{\mathbf{u}}_1 \times \hat{\mathbf{u}}_2 \tag{9.89}$$

correspond to the principal axes of inertia of the instantaneous body formed by the 3 particles A, B and e. The coordinates r_1, r_2, q are designated as the Eckart coordinates.

In view of the fact that the masses m_a, m_b are much greater than the electron mass, the expression for the rotation angle can be simplified [29] to

$$\tan 2q = \frac{2\mathbf{R}_1.\mathbf{R}_2}{R_1^2 - R_2^2} \approx \frac{1}{\sqrt{m_{a,b}}}\frac{2\mathbf{r}.\mathbf{R}}{R^2} \tag{9.90}$$

The rotation angle is small (but non-zero) and is given by

$$q \approx \frac{1}{\sqrt{m_{a,b}}}\frac{\mathbf{r}.\mathbf{R}}{R} \tag{9.91}$$

and we have approximately

$$\begin{aligned} r_1\hat{\mathbf{u}}_1 &= \cos q\mathbf{R}_1 + \sin q\mathbf{R}_2 \\ &\approx \sqrt{m_{a,b}}\left(\mathbf{R} + \frac{\mathbf{r}.\mathbf{R}}{m_{a,b}R^2} - \frac{(\mathbf{r}.\mathbf{R})^2}{m_{a,b}R^4}\mathbf{R}\right) \end{aligned} \tag{9.92}$$

$$\begin{aligned} r_2\hat{\mathbf{u}}_2 &= -\sin q\mathbf{R}_1 + \cos q\mathbf{R}_2 \\ &\approx \mathbf{r} - \frac{\mathbf{r}.\mathbf{R}}{R^2}\mathbf{R} \end{aligned} \tag{9.93}$$

We may observe that in the limit as $r_b/R \to 0$

$$r_2\hat{\mathbf{u}}_2 \underset{R\to\infty}{\approx} \mathbf{r}_b \quad r_1\hat{\mathbf{u}}_1 \underset{R\to\infty}{\approx} \sqrt{m_{a,b}}\,\mathbf{R}_b \tag{9.94}$$

while in the limit as $r_a/R \to 0$

$$r_2\hat{\mathbf{u}}_2 \underset{R\to\infty}{\approx} \mathbf{r}_a \quad r_1\hat{\mathbf{u}}_1 \underset{R\to\infty}{\approx} \sqrt{m_{a,b}}\,\mathbf{R}_a \tag{9.95}$$

It is therefore seen that the Eckart coordinates allow for a correct description of the asymptotic states with no need for the introduction of any explicit translation factors if we choose r_1 as the adiabatic variable. However, there is a price to pay. The transformation of the Hamiltonian from the $(\mathbf{r}, \mathbf{R})$ Jacobi coordinates to the Eckart coordinates $(\mathbf{r}_1, \mathbf{r}_2, q)$ is quite complicated. The reader who is interested can work through all the details from the original Eckart paper. However, it can be easily shown, that the Eckart Hamiltonian with fixed $\mathbf{r}_1$ will only differ from the standard adiabatic Hamiltonian with fixed $\mathbf{R}$ by terms which are of the order of the electron to nuclear mass ratio. So, as in [26], it is legitimate to assume that $\xi_n(r_2, q; \mathbf{r}_1) \approx \chi_n(\bar{\mathbf{r}}; r_1)$. The only real difference between the standard adiabatic states and the Eckart states concerns the definition of the adiabatic variable, which leads to a modification of the non-adiabatic coupling terms. The total wave-function is therefore expanded in the form

$$\Psi(\mathbf{r}_2, \mathbf{r}_1) = \sum_n F_{eck,n}(\mathbf{r}_1)\xi_n(r_2, q; \mathbf{r}_1) \approx \sum_n \tilde{F}_n(\mathbf{r}_1)\chi_n(\bar{\mathbf{r}}; \mathbf{r}_1) \tag{9.96}$$

This is the identical to the expansion proposed in [26]. Compared with the standard adiabatic model, the equations have a similar form to eq. 9.7 but the radial and rotational coupling matrix elements are modified. A straightforward (but lengthy) calculation similar to that of section 9.2 yields a similar set of coupled equations to eq. 9.39 but with the radial and Coriolis coupling matrix elements modified according to

$$\langle\chi_i|\,\frac{\partial}{\partial R}\,|\chi_j\rangle \Rightarrow \langle\chi_i|\,\frac{\partial}{\partial R} + \frac{\bar{z}}{R}\frac{\partial}{\partial \bar{z}}\,|\chi_j\rangle \tag{9.97}$$

$$\langle\chi_i|\,iL_{\bar{y}}\,|\chi_j\rangle \Rightarrow \langle\chi_i|\,iL_{\bar{y}} + \bar{x}\frac{\partial}{\partial \bar{z}} + \bar{z}\frac{\partial}{\partial \bar{x}}\,|\chi_j\rangle \tag{9.98}$$

It is easy to verify that the matrix elements eqs. 9.97 and 9.98 will all vanish in the asymptotic limit. We may note that these modified matrix elements can be calculated using the same conventional tools of quantum chemistry.

A word of caution should be added. The simplifying assumption eq. 9.91, that makes for a practical implementation of the Eckart coordinates, may break down in close collisions when (r/R) and the rotation angle can become large. No simple way of going beyond the approximation (9.91) has yet been proposed. Fortunately, at the collision energies where quantum mechanical methods are required the influence of the close collision region is not important and this defect it is not expected to be serious.

8. Other Reaction Coordinates

Another way to circumvent the need for translation factors is to introduce hyperspherical coordinates. They differ from both the standard adiabatic coordinates and the Eckart coordinates in an important way. In hyperspherical coordinates, the Hamiltonian of the system is expressed in terms of a generalized operator [30] which includes all rotations (electronic and nuclear) whereas the standard adiabatic coordinates and the Eckart coordinates consider separately the electronic angular momentum relative to the body fixed axis. For that reason, the hyperspherical approach offers an interesting solution to the treatment of non-adiabatic effects in bound state systems for low angular momentum states. On the other hand it would appear to be less practical in most collision systems where large angular momentum effects are usually dominant.

To give the reader some idea of the hyperspherical approach, we draw attention to the approach of Solov'ev and Vinitsky [31], who assume from the outset that the angular momentum terms of the total Hamiltonian are unimportant. Making the scale transformation $\mathbf{r}' = \mathbf{r}/R$, and neglecting all angular momentum terms, the Schrödinger equation (9.6) may be written in the form

$$\begin{aligned}\Big[-\tfrac{1}{2M_{a,b}R^2}\tfrac{\partial}{\partial R}\left(R^2\tfrac{\partial}{\partial R}\right) \\ +\tfrac{1}{M_{a,b}R^2}\left(\mathbf{r}'.\nabla_{\mathbf{r}'}\right)\left(1+R\tfrac{\partial}{\partial R}\right)-\tfrac{\rho}{2R^2}\nabla^2_{\mathbf{r}'}+V\Big]\Psi &= E_T\Psi\end{aligned} \tag{9.99}$$

where

$$\rho = \left(1+\frac{1}{M_{a,b}}r'^2\right) \tag{9.100}$$

We may remark that the radial matrix elements

$$\langle \chi_i | \frac{\partial'}{\partial R} | \chi_j \rangle \underset{R \to \infty}{=} 0 \tag{9.101}$$

since the derivative is taken with $\mathbf{r}'$ constant and not $\mathbf{r}$. Of course, the scale transformation is not enough by itself, since the existence of cross derivatives in eq. 9.99 still give rise to potentially non-vanishing asymptotic matrix elements. But these can be removed by introducing the hyperspherical radius $\Re$ defined by

$$\Re = \sqrt{\rho} R \tag{9.102}$$

Making the transformation

$$\Psi = \Re^{-3/2} \sqrt{\rho} \psi_{hyp} \tag{9.103}$$

we obtain (after some tedious algebra)

$$\left[-\frac{1}{2M_{a,b}\Re^2} \frac{\partial}{\partial \Re} \left(\nabla \Re^2 \frac{\partial}{\partial \Re} \right) + -\frac{\rho}{2\Re^2} \nabla^2_{\mathbf{r}'} + \frac{3}{8M_{a,b}\Re^2} V \right] \Psi = E_T \Psi \tag{9.104}$$

Eq. 9.103 contains no cross derivatives and has the desirable properties, which allow for an adiabatic separation of the slow variable $\Re$ and the fast variable $\mathbf{r}'$. We then define the hyperspherical adiabatic states as eigenfunctions of

$$H_{hyp} = -\frac{\rho}{2\Re^2} \nabla^2_{\mathbf{r}'} + V \tag{9.105}$$

As in the case of the Eckart coordinates, the smallness of the mass ratio $1/M_{a,b}$ ensures that the hyperspherical adiabatic states do not differ significantly from the clamped nuclei states. We may also observe that when $r_a/R \to 0$

$$-\frac{\rho}{2\Re^2} \nabla^2_{\mathbf{r}'} + V \to \nabla^2_{\mathbf{r}_a} \tag{9.106}$$

and

$$-\frac{1}{2M_{a,b}\Re^2} \frac{\partial}{\partial \Re} \left(\nabla \Re^2 \frac{\partial}{\partial \Re} \right) \to -\frac{1}{2M_{ae,b} R_a^2} \frac{\partial}{\partial R_a} \left(R_a^2 \frac{\partial}{\partial R_a} \right) \tag{9.107}$$

In the case, when $r_b/R \to 0$

$$-\frac{\rho}{2\Re^2}\nabla^2_{\mathbf{r}'} + V \to \nabla^2_{\mathbf{r}_b} \tag{9.108}$$

$$-\frac{1}{2M_{a,b}\Re^2}\frac{\partial}{\partial\Re}\left(\nabla\Re^2\frac{\partial}{\partial\Re}\right) \to -\frac{1}{2M_{be,a}R_b^2}\frac{\partial}{\partial R_b}\left(R_b^2\frac{\partial}{\partial R_b}\right) \tag{9.109}$$

So we see that the hyperspherical coordinates, just like the Eckart coordinates make it possible to describe the initial and final states of the reaction correctly and to extract the S-matrix in a well-defined manner.

The evaluation of the radial coupling matrix with respect to $\mathbf{r}'$ constant is simple in spheroidal coordinates

$$\lambda = \frac{r_a + r_b}{R} \qquad \mu = \frac{r_a - r_b}{R} \qquad 1 \le \lambda < \infty \qquad -1 \le \mu < +1 \tag{9.110}$$

It suffices to calculate the radial derivative with respect to constant λ, μ.

When the neglect of Coriolis coupling is justified, the hyperspherical approach gives results in close agreement with the Eckart coordinate approach. But when Coriolis couplings are present, the simple introduction of the scaled coordinate $\mathbf{r}'$ does not solve the problems associated with slowly converging non-adiabatic Coriolis terms.

9. General Concluding Remarks

In spite of the formally attractive features of reaction coordinates, which avoid the use of explicit translation factors without the need of any additional ad-hoc assumptions, it must be stressed that they do not offer a unique solution to the problem. Both the Eckart and the hyperspherical coordinates are based on purely kinematical considerations. They are not optimized specifically to the interaction potentials involved in the collision process. So while the modification of the non-adiabatic coupling matrix elements ensures the correct boundary conditions, the problem has been displaced in some way from the asymptotic region into the molecular interaction region.

In our present state of knowledge, it is difficult to take the problem much further since experimental measurements are not yet sufficiently precise to establish clear guide-lines on the optimal choice of reaction coordinates (or translation factor). Fortunately, in the energy range less than 100eV or so, the limitations of the standard adiabatic model are not really too serious for most practical applications. Besides, it has been implicitly supposed throughout this chapter, that it is possible to determine the adiabatic energies and non-adiabatic couplings with high accuracy. But, it should be recognized that inaccurately determined couplings are another potential source of error.

On the other hand, for collision energies beyond a few 100eV/amu, defects of the standard adiabatic model become increasingly apparent and we need more detailed experimental measurements to guide the validity of the most appropriate theoretical model. However, in this energy range, it is simpler to analyse the problem in terms a semiclassical approach in which translation factors are explicitly introduced. (This topic is dealt with in chapter 8 on semiclassical methods.) It may be remarked however, that the Eckart coordinate system does provide a natural connection with the common translation factors of the semiclassical approach.

References

[1] M. Gargaud and R. McCarroll, J. Phys. B: At. Mol. Opt. Phys. **29**, 3673 (1996)

[2] G. Lubinski, Z. Juhαsz, R. Morgenstern and R. Hoekstra, J. Phys. B: At. Mol. Opt. Phys. **33**, 5275 (2000)

[3] D. Voulot, D.R. Gillen, W.R. Thomson, H.B. Gilbody, R.W. McCullough, L.Errea, A. Macias, L.Mendez, A. Riera, J. Phys. B: At. Mol. Opt. Phys. **33**, L187 (2000)

[4] M. Gargaud, M.C. Bacchus-Montabonel and R. McCarroll, J. Chem. Phys. **99**, 4495 (1993)

[5] T.K. McLaughlin, S.M. Wilson, R.W. McCullough and H.B. Gilbody, J. Phys. B: At. Mol. Opt. Phys. **23**, 737 (1990)

[6] M. Gargaud, R. McCarroll and P. Valiron, J. Phys. B: At. Mol. Opt. Phys. **20**, 1555 (1987)

[7] F.W. Bliek, R. Hoekstra, M.E. Bannister and C.C Havener, Phys. Rev.A **56**, 426 (1997)

[8] C.C Havener, A. Müller, P.A. Zeijlmans van Emmichoven and R.A. Phaneuf, Phys. Rev. A **51**, 2982 (1995)

[9] M. Pieksma, M. Gargaud, R. McCarroll and C.C Havener, Phys. Rev A **54**, R13 (1996)

[10] W. Wu and C.C Havener, J. Phys: At. Mol. Opt. Phys. **30**, L213 (1997)

[11] D. Rabli, M. Gargaud and R. McCarroll, Phys. Rev. A **64**, 22707 (2001)

[12] D.R. Bates and R. McCarroll, Proc. Roy. Soc. A **245**, 175 (1958)

[13] D.R. Bates and R. McCarroll, Phil. Mag. Suppl. **11**, 39 (1962)

[14] B.H. Bransden and M.R.C. McDowell, *Charge Exchange and the Theory of Ion-Atom Collisions*, Clarendon Press, Oxford (1992)

[15] A.R. Edmonds, *Angular Momentum in quantum mechanics*, Princeton University Press, Princeton (1960)

[16] J.B. Delos, Rev. Mod. Phys. **53**, 287 (1981)

[17] M. Gargaud, R. McCarroll and L. Benmeuraiem, Physica Scripta **51**, 752 (1995)

[18] D. Rabli, M. Gargaud and R. McCarroll, J. Phys: At. Mol. Opt. Phys. **34**, 1395 (2001)

[19] D. Rabli, *Thèse de doctorat*, Université Pierre et Marie Curie (2001)

[20] Yan Li, O Bludsky, G. Hirsch and R.J. Buenker, J. Chem. Phys. **107**, 3014 (1997)

[21] M. Gargaud, J. Hanssen, R. McCarroll and P. Valiron, J. Phys: At. Mol. Opt. Phys. **14**, 2259 (1981)

[22] B.R. Johnson, *J. Comput. Phys*. **13**, 445 (1973)

[23] R. McCarroll and P. Valiron, Astron. Astrophys. **44**, 465 (1975)

[24] D.E. Manolopoulos, J. Chem. Phys. **85**, 6425 (1986)

[25] M.H. Mittleman, Phys. Rev. **188**, 221 (1969)

[26] W.R. Thorson and J.B. Delos, Phys. Rev. A **18**, 135 (1978)

[27] J. Robert and J. Baudon, J. Phys. B: At. Mol. Phys. **19**, 171 (1986)

[28] C. Eckart, Phys. Rev. **46**, 383 (1934)

[29] R.McCarroll and D.S.F. Crothers, in: *Advances in Atomic, Molecular and Optical Physics* ed. D.R. Bates and B. Bederson, Academic Press, New York, **32**, 253 (1994)

[30] F.T. Smith, Phys. Rev. **120**, 1058 (1960)

[31] E.A. Solov'ev and S.I. Vinitsky, J. Phys. B: At. Mol. Phys. **18**, L557 (1985)

Chapter 10

STATE-SELECTIVE ELECTRON CAPTURE BY TRANSLATIONAL ENERGY SPECTROMETRY

R. W. McCullough, D. M. Kearns and H. B. Gilbody

Atomic and Molecular Physics Research Division, School of Mathematics and Physics, Queen's University Belfast, Belfast, U.K.

rw.mccullough@queens-belfast.ac.uk

Abstract Translational Energy Spectrometry is a powerful technique for the identification and determination of the relative importance of the excited states formed in collisions of multiply charged ions with atoms and molecules at energies less than 25keV u^{-1}. Double Translational Energy Spectrometry enables these studies to be carried out where the initial state of the ion is well defined and effects arising from metastable ions can be separately studied. Such studies are important to our understanding of solar and stellar winds, cometary X-ray emissions, planetary atmospheres and technological plasmas.

Keywords: **charge exchange, atomic hydrogen, multiply charged ions, state selective capture**

1. Introduction

Electron capture is the dominant inelastic process during the interaction of multiply charged ions with neutral atoms and molecules at energies less than 25 keV/amu. There is currently strong interest in these interactions because of their importance in many astrophysical and terrestrial environments. For example, electron capture by solar wind ions interacting with cometary gases [1], by auroral ions interacting with planetary atmospheres [2], and stellar wind ions with the interstellar medium [3], results in the formation of excited states of either or both collision partners with the resultant visible, UV and X-ray emissions. The thermal and ionisation balance of astrophysical plasmas are influenced by one-electron capture processes [4] involving a wide variety of partially ionised ions. The modeling of ion transport and radiative cooling

F.J. Currell (ed.), The Physics of Multiply and Highly Charged Ions, Vol. 2, 309-338.

processes in the divertor region of current large Tokamak fusion plasma devices requires detailed information on the relevant electron capture processes [5].

There have been a large number of experimental measurements of total one-electron capture cross sections by multiply charged ions in atomic and molecular targets (see for example [6, 7]). Experimental studies of state selective electron capture have been carried out primarily via Photon Emission Spectrometry (PES) or Translational Energy Spectrometry (TES), see for example [8–10]. Theoretical calculations of electron capture cross sections have been carried out using both semi-classical and quantal approaches. See for example [11, 12].

In many of the collision systems investigated, a comparison of theoretical and experimental data was complicated by the presence of unknown fractions of low-lying metastables in the primary ion beams used in the experiments. These metastable ions can have a significantly different capture probability than ground state ions [13]. Therefore, to obtain information on collision processes of ground state ions in atomic and molecular targets, it is necessary to either determine the metastable fraction of the ion beam or produce an ion beam free from metastable contamination. We have developed an experimental system that makes it possible to produce pure primary ion beams in either ground or metastable states using Double Translational Energy Spectrometry (DTES) [14]. This technique has been used to identify the significant capture channels and determine their relative importance, of a wide variety of multiply charged ions in both ground and metastable states in atomic and molecular targets (see [15, 16]). A particular feature of the DTES technique is its ability, in the case of molecular targets, to determine the relative importance of capture involving both dissociative and non-dissociative processes where the projectile is left in either a ground or metastable state. Information can also be obtained on the distribution of the final vibrational state population of the target product ion [17].

2. Principle of the Translational Energy Spectrometry Technique

In one-electron capture processes of the type symbolically characterised by equation (10.1), low velocity collisions can be thought of in terms of the so called 'quasi-molecular' model.

$$\mathrm{A}^{q+} + \mathrm{B} \rightarrow \mathrm{A}^{(q-1)+}(n,l) + \mathrm{B}^{+}(n',l') \qquad (10.1)$$

In this model the initial molecular state, $(\mathrm{A}^{q+} + \mathrm{B})$, is strongly coupled with a large number of states of the electronically rearranged system, ($\mathrm{A}^{(q-1)+}(n,l)$ + $\mathrm{B}^{+}(n,l)$).

These states may be represented by adiabatic potential energy curves. A schematic diagram of two such potential curves showing typical initial and final states for the reaction described by equation 10.1 is shown in figure 10.1

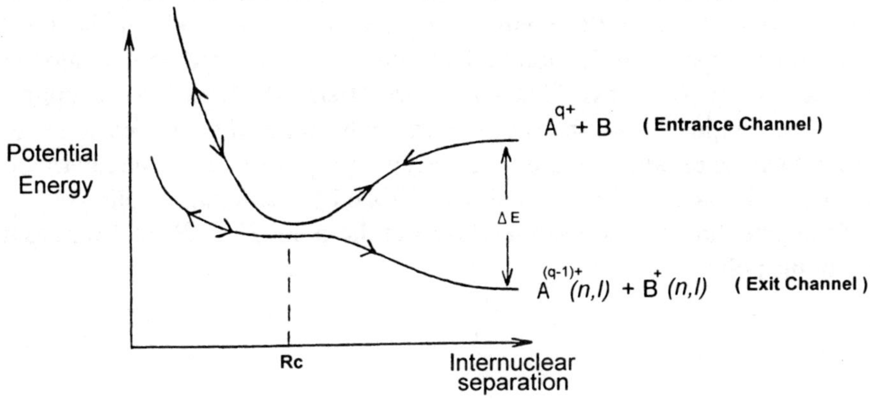

Figure 10.1. Schematic adiabatic potential energy curves for the system A^{q+}+B.

The entry channel (A^{q+} + B) is attractive at large internuclear separations due to polarisation of the target by the projectile charge q. The potential energy of the state is therefore given by:

$$V_i = -\frac{\alpha q^2 e}{2R^4} \tag{10.2}$$

where α is the target polarisability and R the internuclear separation. For the exit channel ($A^{(q-1)+}(n,l) + B^+(n',l')$) the potential is dominated by Coulombic repulsion between the projectile and target ions.

The potential has the form:

$$V_f \cong -\frac{(q-1)e^2}{R} - \Delta E \tag{10.3}$$

where ΔE is the internal energy difference between the initial and final channels.

As two atoms approach each other in a collision, a particular electronic state of the molecule is formed and a transition to another molecular state is possible if the potential curves of the molecular orbitals (MO's) describing these two states intersect. However, as shown in figure 10.1, for a typical system most of the crossings become avoided or 'pseudo'-crossings because of the Wigner non-crossing rule which states that adiabatic states with the same molecular symmetry cannot intersect. The molecular symmetry is defined by the projection of the electron orbital angular momentum along the internuclear axis and states are designated by Σ, Π etc. Transitions will then occur therefore when the MO's approach close enough in energy so that coupling can occur. The types of coupling that can occur are called *Radial* and *Rotational* couplings. Radial coupling is mainly due to residual electrostatic interactions between two electrons. This kind of coupling is effective between orbitals having the same parity and same projection of the electronic angular momentum along the internuclear axis (e.g. two Σ states or two Π states). Rotational coupling is due to the rotation of the internuclear axis with time. This mixes states with different values of orbital angular momentum projection along the internuclear axis (e.g. a Σ state can be coupled to a Π state by rotational coupling).

If the potential curves become closest at the point $R = R_c$ and we assume that at this point:

$$V_i \approx V_f \tag{10.4}$$

then:

$$\Delta E = V_i - V_f = \frac{(q-1)e^2}{R_c} + \frac{\alpha q^2 e}{2{R_c}^4} \tag{10.5}$$

If Coulombic repulsion in the exit channel is the dominant influence and the polarisation interaction between the reactants is ignored, then the psuedo-crossing distance R_c is given approximately by:

$$R_c \cong \frac{(q-1)}{\Delta E} \tag{10.6}$$

where the parameters are expressed in atomic units.

In figure 10.2 a sketch of the fundamental parameters is given by which an inelastic collision can be completely described by a classical treatment.

The projectile A^{q+} with mass m_p and initial kinetic energy T_1 collides with a stationary target B and undergoes electron capture. As a result the projectile is scattered at an angle ϑ with a final kinetic energy T_2 and the target recoils with energy ΔK at an angle φ.

By considering the conservation of energy, the energy change is:

$$\Delta T = T_2 - T_1 = \Delta E - \Delta K \tag{10.7}$$

In this expression ΔE is the energy defect corresponding to a particular reaction channel characterised by the initial and final states of both target and

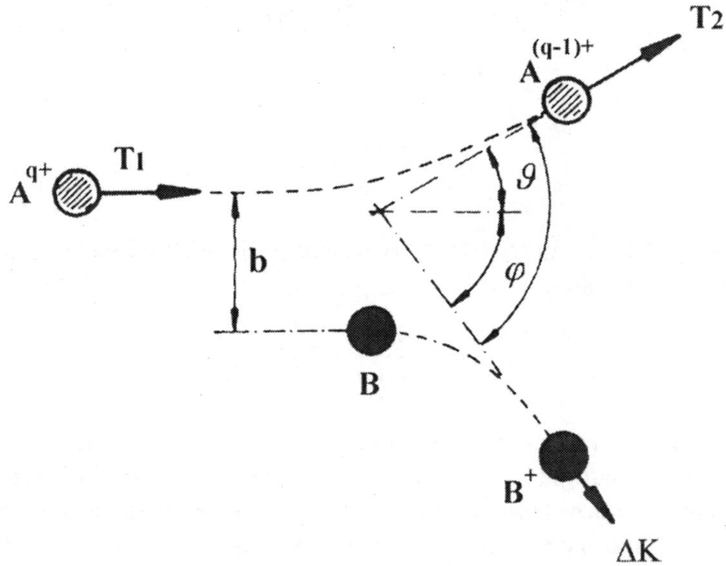

Figure 10.2. Collision geometry in one-electron capture collision.

projectile. From the conservation of energy and momentum the recoil energy of the target ΔK is given by:

$$\Delta K = \frac{1}{4}\frac{m_p}{m_t}\frac{(\Delta E)^2}{T_1}\cos\phi + \frac{m_p}{m_p + m_t}(1 - \cos\phi)\left(\frac{2m_t}{m_p + m_t}T_1 + \Delta E\right) \tag{10.8}$$

where m_p is the projectile mass, m_t the target mass. ϕ is the centre of mass scattering angle [18]:

$$\phi = \left(\frac{m_t + m_p}{m_t}\right)\vartheta \tag{10.9}$$

A rough estimate of the centre of mass scattering angle can be obtained using the assumptions of zero interaction in the entrance channel and Coulomb repulsion in the exit channel [19].

$$\phi = \frac{\pi}{2} - \cos^{-1}\left(\frac{\alpha}{(\alpha^2+4)^{\frac{1}{2}}}\right) \tag{10.10}$$

where

$$\alpha = \frac{Z_1 Z_2 e^2}{E_{cm} b} \tag{10.11}$$

Z_1 and Z_2 are the ion charges of the scattered projectile and target and E_{cm} is the centre of mass collision energy.

$$E_c m = \frac{m_t}{m_t + m_p} E_{lab} \tag{10.12}$$

It can be seen from equation (10.10) that the scattering angle increases as α gets larger. From equation (10.11) it can be seen that this means that scattering angles are greater for slow collisions at small impact parameters between heavy projectiles and light targets. Equation (10.10) was used to estimate scattering angles [20] in collisions between C^{3+} ions and atomic hydrogen at an impact energy of 900eV. For the major C^{2+} (3s) ^{3}S capture channel with R_c = 11.5 au, a value for ϕ was estimated to be 1.95° (0.15° in the laboratory frame). Other measurements [21] for collisions of C^{4+} ions in an Ar target predict laboratory scattering angles of around 0.5°.

From this analysis, an assumption of nearly forward scattering can therefore be justified i.e. $\phi \to 0$. Equation (10.8) therefore becomes:

$$\Delta K \cong \frac{1}{4}\frac{m_p}{m_t}\frac{(\Delta E)^2}{T_1} \tag{10.13}$$

In addition $\Delta E << T_1$, since energy defects are of the order of a few eV while T_1 is in the keV energy range. The recoil energy then becomes negligible and equation (10.7) can be approximated to:

$$\Delta T = T_2 - T_1 \cong \Delta E \tag{10.14}$$

The kinetic energy change as a result of the collision is equivalent to the internal energy change of the reactants. An accurate measurement of ΔT allows the electron capture reaction channel to be identified. The achievable state resolution of TES depends directly on the initial kinetic energy spread of the projectile particles. Further contributions to a finite energy resolution result from the finite angular resolution as well as the thermal motion of the target particles.

3. Experimental Approach

A schematic diagram of the DTES system used by our group at Queen's University Belfast (QUB) is shown in figure 10.3. In order to illustrate the technique, the production of a ground state beam of $N^{2+}(2s^22p)\ ^2P^o$ ions is described. A beam of ions is produced by an ECR ion source [22] is which the magnetic structure is provided by NdFeB permanent magnets and the microwave power by a 0 – 100 W frequency variable (9 - 10.5GHz) solid state supply. This source replaces the 5GHz ECR shown in figure 10.3. A characteristic of such sources is the existence of a positive plasma potential that must be accounted for in defining the final energy of the beam. In our case, the source plasma tube is held at a small negative potential approximately equal and opposite to the positive plasma potential. The plasma potential is measured using a retarding potential analyzer (RPA) at the rear of energy analyzer 1. The ion beam is then extracted from the ECR source via a 'floating beam-line accelerator' in which the beam-line is held at a potential of –4kV. N^{3+} ions are selected using a 90° double focusing analysing magnet. A switching magnet is used to deflect the beam into the DTES apparatus. After entering the DTES beamline, the N^{3+} ions are decelerated by a cylindrical lens system L1 into the first hemispherical energy analyser EA1. The beam passes through EA1 and then is accelerated by lens system L2 into the first target gas cell T1.

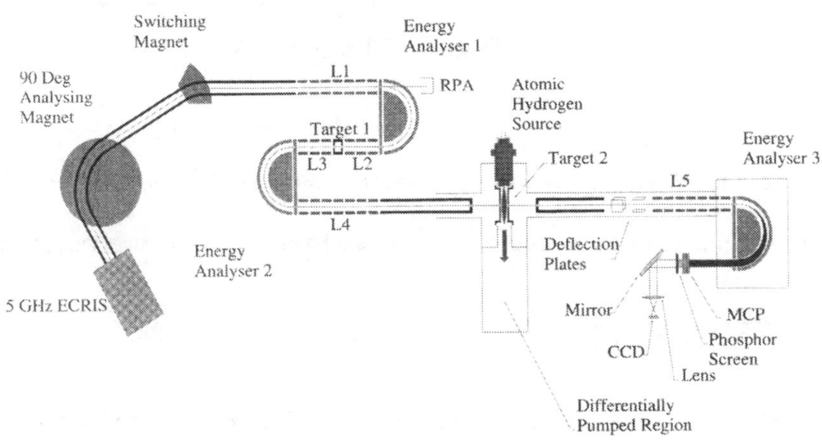

Figure 10.3. Schematic diagram of the QUB DTES apparatus.

The N^{3+} ions undergo one-electron capture collisions in this first gas target and the resulting beam, now containing N^{2+} and N^{3+} ions in both ground and metastable states, is accelerated by lens system L3 into a second energy analyser,

EA2. Applying a suitable retarding voltage on L3, L4, and EA2 ensures only the N^{2+}ions pass through. By scanning this voltage, a translational energy spectrum is obtained of the N^{2+} ions as shown in figure 10.4.

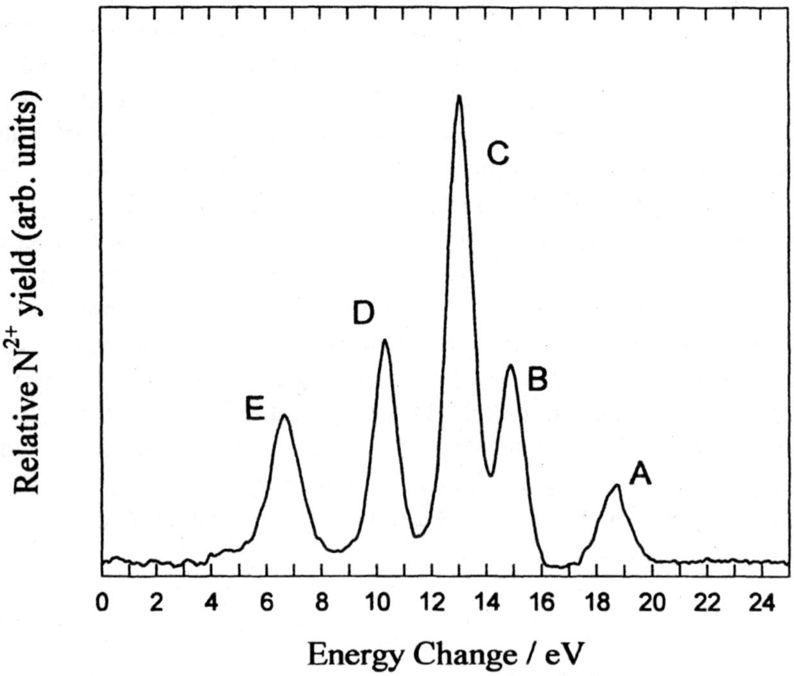

Figure 10.4. Energy change spectrum for one-electron capture by N^{3+} in He at 3 keV [28].

It is possible to select product N^{2+} (^{2}P) ions from the main observed channel:

$$N^{3+}(1s^2 2s2p)\ ^3\mathrm{P} + \mathrm{He} \rightarrow N^{2+}(2s2p^2)\ ^2\mathrm{P} + \mathrm{He}^+(2s) \tag{10.15}$$

corresponding to peak C. The excited N^{2+} (^{2}P) ions decay rapidly (1.4ns) to the ground state within the minimum flight time to the second target (10μs), and so produce a beam of ground state N^{2+} ions. This is the first stage of double translational energy spectrometry.

The ion beam, now consisting of only ground state N^{2+} ions, is accelerated by lens system L4 into the second target region (T2). T2 consists of an aluminium cell coupled to a microwave-driven atomic hydrogen source that can be used to provide beams of highly dissociated hydrogen [23] or other target species as required. A schematic diagram of this T2 region is shown in figure 10.5. It should be noted that at this stage the beam energy is uniquely defined and independent of any plasma potential changes. By applying a bias potential of 100V to the chamber in which the target cell is located it is possible in the case of measurements with an atomic hydrogen target to distinguish between collisions which occur outside the target cell from those occuring inside. This is important in measurements using an atomic hydrogen target where the outside region is composed mainly of molecular hydrogen formed by recombination of atomic hydrogen from the cell striking surfaces outside the cell. The N^{2+}ions undergo charge transfer collisions in the target gas and the product N^+ ions, forward scattered within an angular range of $\pm 3^o$, pass through a set of horizontal and vertical deflection plates into L5, where they are decelerated into the final energy analyser EA3. Again, by applying a suitable retarding voltage on L5 and EA3, only the passage of the product N^+ ions is allowed. The N^+ ions pass through EA3 and are detected by a position sensitive detector (PSD).

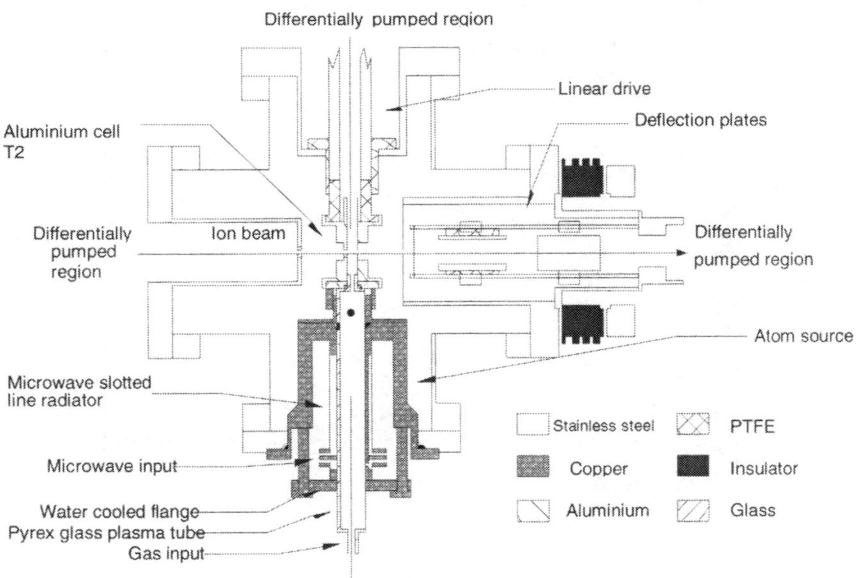

Figure 10.5. Schematic diagram of the target region and the microwave driven atomic hydrogen source.

By scanning the retarding voltage on L5 and EA3, a translational energy spectrum of the product N^+ ions can be obtained whilst maintaining a constant energy resolution of 0.5 qeV (q = primary ion charge). The translational energy change is equal to the energy defect of the particular capture channel with a small correction for non-zero degree scattering and target recoil.

The feasibility of DTES was first demonstrated in 1984 [24], but was not seriously applied until the series of measurements carried out in this laboratory. Using the DTES technique described above we have studied one-electron capture by C^{2+} ^1S ground state and C^{2+} ^3P metastable ions in He, Ne, Ar, H_2, N_2 and O_2 [25, 26] by N^{2+} ^2P ground state ions in He, Ne, Ar and H_2[27, 28] and by O^{2+} ^3P ground state ions in He, Ne and Ar [29]. We have also succeeded in carried out the first DTES measurements in atomic hydrogen for one-electron capture by C^{2+} ^1S ground state and C^{2+} ^3P metastable ions [30], by N^{2+} ^2P ground state ions [31] and by O^{2+} ^3P ground state ions [32].

Typical energy change spectra obtained at 3 keV for one-electron capture by pure N^{2+} $^2P^o$ ground-state ions are shown in figure 10.6.

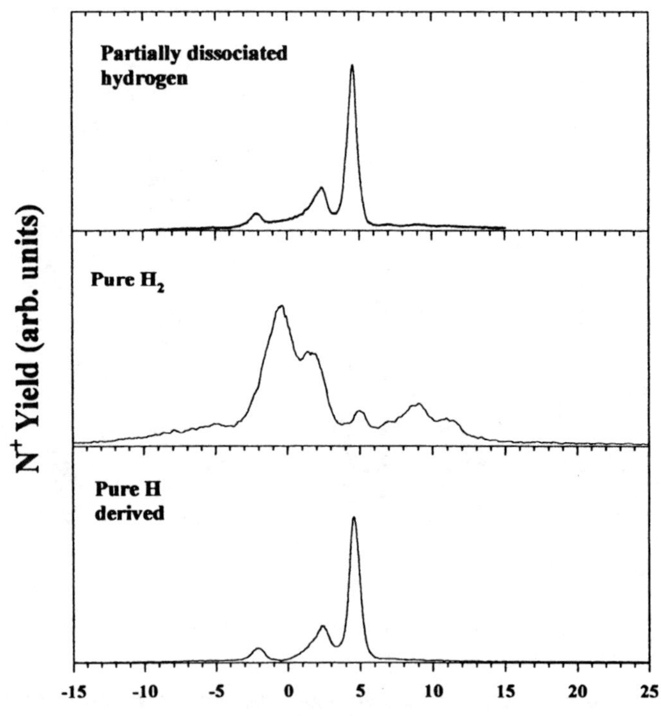

Figure 10.6. Energy change spectra at 3 keV measured for one-electron capture by pure N^{2+} $^2P^o$ ions in highly dissociated hydrogen, in pure H_2 and derived (see text) for pure H atoms.

Although highly dissociated hydrogen is injected into the target cell, H_2 molecules are formed by collisions with the cell walls so that the spectra obtained reflects the mixture of H and H_2 present in the cell. Energy change spectra for atomic hydrogen can be derived from the measured spectra by subtracting the molecular contribution from the spectra obtained in the 'mixed' target. The accurate subtraction process is facilitated by the ability of the hydrogen source to be rapidly switched from highly dissociated hydrogen to pure H_2. The spectra for highly dissociated (approximately 70 %) hydrogen and pure molecular hydrogen can be seen to be very different. The derived spectrum for a pure atomic hydrogen target exhibits clearly defined product channels suitable for quantitative analysis.

Whilst our DTES apparatus was not designed for absolute measurements of the total one-electron capture cross sections for state-prepared ions it has been possible, in a few measurements with C^{2+} ions, to use a simple beam attenuation technique to determine the ratios $\sigma(\,^3P\,)/\,\sigma(\,^1S\,)$ of the total cross sections for metastable and ground state species. Figure 10.7 shows the attenuation of a beam of pure metastable or ground state C^{2+} ions in helium as the target thickness μ is increased so that the total recorded beam intensity follows the simple relation $I = I_0\ e^{-\mu\sigma}$

Analysis of plots of this type in helium, neon and argon at 4 keV, provide ratios $\sigma\ (^3P)/\sigma\ (^1S) = 11.5 \pm 3$, 3.1 ± 0.2 and 3.5 ± 0.3 respectively. These observations indicate the need for cautious interpretation of any measurements of cross sections carried out with ion beams of unknown metastable content.

4. Some Examples of Measurements Using the QUB Translational Energy Spectrometer

4.1 Measurements of one-electron capture in molecular hydrogen

4.1.1 $C^{2+} + H_2$. Figure 10.8 shows the measured energy change spectrum obtained for C^{2+} ions in H_2 at 500 eV/amu. The main reaction channels are listed in table 10.1.

It can clearly be seen that one electron capture is dominated by capture from metastable C^{2+} ions. Figure 10.9 shows the energy dependence of both ground state and metastable reaction channels. It can be seen in the ground state spectra that the channel corresponding to peak G1 is decreasing in importance with decreasing energy, while peak G2 is increasing. In contrast, the metastable spectra show very little energy dependence. Leputsch et al [17] have carried out a TES study of this system with an ion kinetic energy resolution $\leq$ 200meV enabling the final vibrational state distribution of the H_2^+ target product ion to be identified.

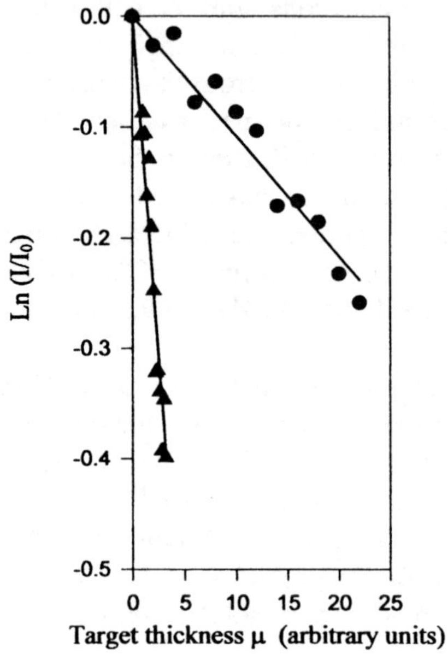

Figure 10.7. Attenuation by one-electron capture collisions in helium of pure ground state C^{2+} ^1S • and pure metastable C^{2+} ^3P ▲ ion beams at 4 keV as the target thickness μ is increased (from [14]).

Table 10.1. Main reaction channels for one electron capture by C^{2+} ions in H_2 $^1\Sigma_g^+$ [33]. DL indicates the dissociation limit.

	Peak	Product channels	Energy defect / eV
C^{2+} $(1s^22s^2)$ + H_2 $^1\Sigma_g^+$ →			
	G1	C^+ $(1s^22s^22p)$ + H_2^+ $^2\Sigma_g^+$	8.96 (v=0) – 6.31 (DL)
	G2	C^+ $(1s^22s2p^2)$ + H_2^+ $^2\Sigma_g^+$	-0.33 (v=0) – -2.98 (DL)
	G3	C^+ $(1s^22s2p^2)$ + H_2^+ $^2\Sigma_g^+$	-3.01 (v=0) – -5.66 (DL)
	G3	C^+ $(1s^22s2p^2)$ + H_2^+ $^2\Sigma_g^+$	-4.76 (v=0) – -7.41 (DL)
	G3	C^+ $(1s^22s^23s)$ + H_2^+ $^2\Sigma_g^+$	-5.49 (v=0) – -8.14 (DL)
C^{2+} $(1s^22s2p)$ + H_2 $^1\Sigma_g^+$ →			
		C^+ $(1s^22s^22p)$ + H_2^+ $^2\Sigma_g^+$	15.45 (v=0) – 12.80 (DL)
		C^+ $(1s^22s2p^2)$ + H_2^+ $^2\Sigma_g^+$	10.12 (v=0) – 7.47 (DL)
	M1	C^+ $(1s^22s2p^2)$ + H_2^+ $^2\Sigma_g^+$	6.16 (v=0) – 3.51 (DL)
		C^+ $(1s^22s2p^2)$ + H_2^+ $^2\Sigma_g^+$	3.49 (v=0) – 0.84 (DL)
		C^+ $(1s^22s2p^2)$ + H_2^+ $^2\Sigma_g^+$	1.73 (v=0) – -0.92(DL)

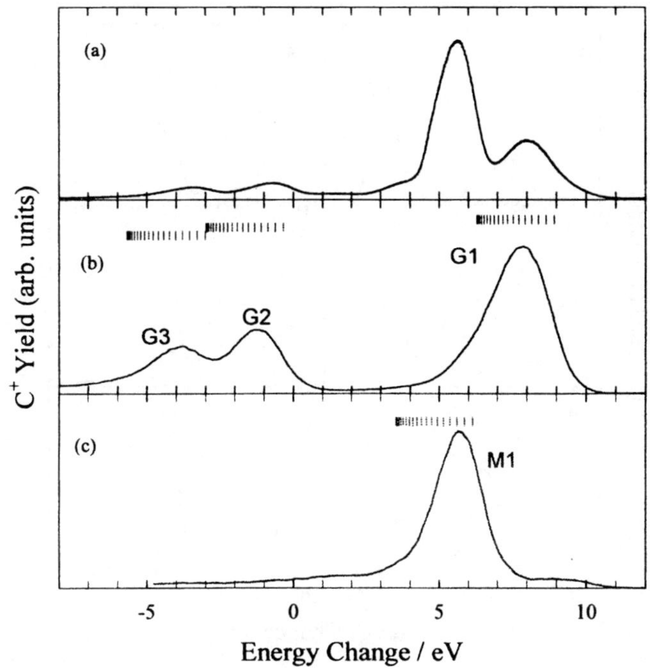

Figure 10.8. Energy change spectra for electron capture at 500eV/amu by (a) a mixed beam of ground state and metastable C^{2+} ions at, (b) a pure ground state beam of C^{2+} ions, and (c) a pure metastable C^{2+}beam taken from [33]. Lines above spectra show energy defects associated with individual reaction channels listed in table 10.1.

4.1.2 He^{2+} + H_2. Figure 10.10 shows the measured energy change spectra obtained for He^{2+} ions in H_2 at energies of 1 keV/amu and 500 eV/amu.

Table 10.2. Main product channels for one electron capture by He^{2+} ions in H_2 $^1\Sigma_g^+$

Product channels	Energy defects / eV
$He^+(n{=}2) + H_2^+$ $^2\Sigma_g$	-1.79 · · · -4.44
$He^+(1s) + H^+ + H(2s)$	4.30 · · · 18.36
$He^+(1s) + H^+ + H(2p)$	4.40 · · · 12.73

The peak positions were compared with calculated energy defects shown in table 10.2. The energy spectra contain 2 features. Firstly, a sharp endothermic

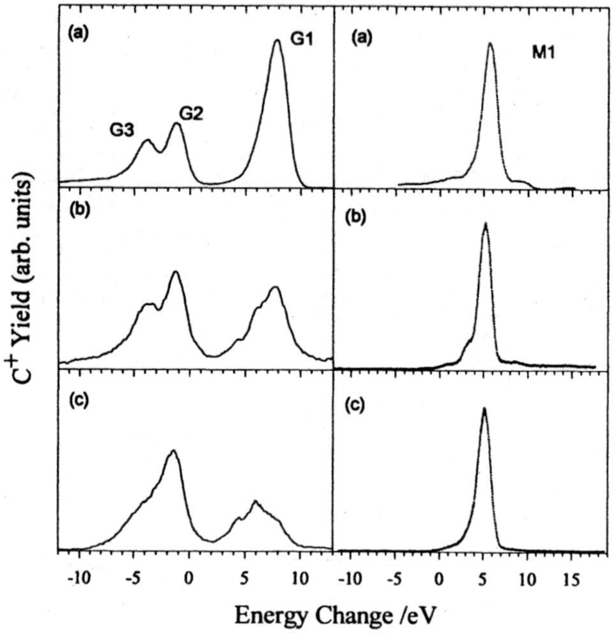

Figure 10.9. Energy change spectra obtained for C^{2+} ground and metastable ions in H_2 at (a) 500 eV/amu, (b) 250 eV/amu, and 125 eV/amu [33].

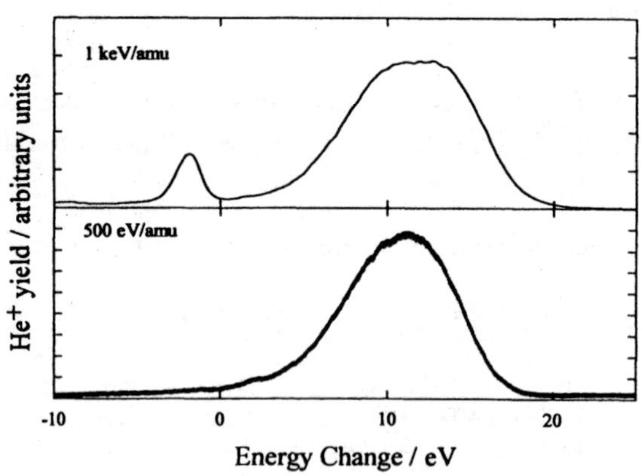

Figure 10.10. Translational energy spectra obtained for He^{2+} ions in H_2 at 1 keV/amu and 500 eV/amu.

peak corresponding to He^+(n=2) formation with H_2^+ in the $^2\Sigma_g$ state with vibrational excitation. Secondly, a broad distribution corresponding to dissociative capture processes:

$$He^{2+} + H_2 \rightarrow He^+(1s) + H^+ + H\ (n \geq 2) \tag{10.16}$$

It can be seen that at low energies only dissociative capture processes are important. Hoekstra et al [34] suggested that at energies > 5keV/amu, capture into excited states of He^+ is dominant, while at lower energies capture into the ground state of He^+ with dissociation and excitation of the target dominates. These results confirm this suggestion.

4.1.3 N^{5+} in H_2. Figure 10.11 shows the measured energy change spectra for N^{5+} ions in H_2 at energies between 857 eV amu^{-1} and 214 eV amu^{-1}. Energy defects corresponding to possible one-electron capture mechanisms are shown alongside the observed peaks in these spectra. As the impact energy is reduced, the structure of the spectra can be seen to change significantly. It is worth noting that, although the primary helium-like ions from the ECR ion source may have contained an unknown admixture of long-lived metastable as well as ground-state ions, our TES spectra contained no evidence of collision channels involving metastable primary ions. Indeed, theoretical considerations indicate there are no favourable channels for N^{5+}(1s2s) ^{3}S metastables to undergo charge transfer [35].

At the highest energy of 857 eV amu^{-1}, three main peaks centred on energy defects of about 8, 24 and 42 eV are apparent. The largest of these corresponds to the non-dissociative capture process

$$N^{5+}(1s^2)\ ^1\mathrm{S} + H_2(^1\Sigma_g^+) \ \rightarrow\ N^{4+}(1s^2 3s, 3p, 3d) \tag{10.17}$$

$$+\ H_2^+\ (v = 10 - 0) + (20.4 - 25.9)\mathrm{eV}$$

which involves capture into the three $N^{4+}(3s, 3p, 3d)$ sublevels together with some evidence of vibrational excitation of the H_2^+ product ion.

The second largest peak in the spectrum at 857 eV amu^{-1}can be identified with a similar process

$$N^{5+}(1s^2)\ ^1S + H_2(^1\Sigma_g^+) \ \rightarrow\ N^{4+}(1s^2 4s, 4p, 4d, 4f) \tag{10.18}$$

$$+\ H_2^+\ (v = 10 - 0) + (3.78 - 7.29)\mathrm{eV}$$

involving electron capture into the $N^{4+}(4s, 4p, 4d, 4f)$ sublevels.

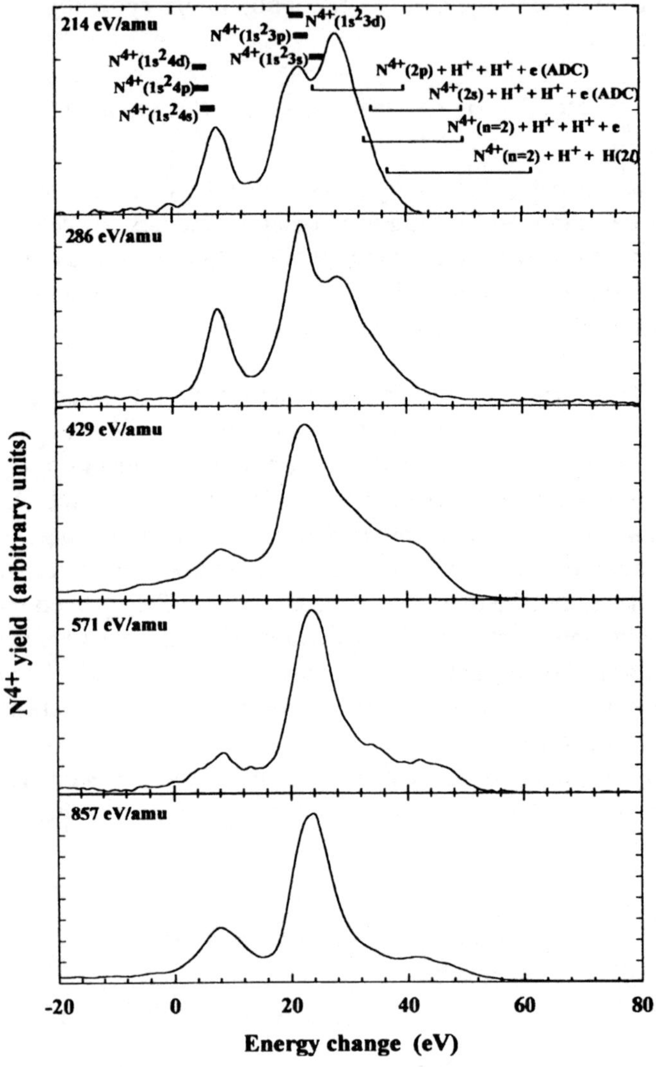

Figure 10.11. Energy change spectra obtained for one-electron capture by N^{5+} ions in H_2 at energies between 857eV amu^{-1} and 214 eV amu^{-1}. Energy defects corresponding to possible N^{4+} product channels are also shown.

The third peak in the spectrum at 857 eV amu^{-1}, which involves the largest energy defects, contains possible contributions from a number of incompletely resolved processes leading to $N^{4+}(n = 2)$ production. The main contribution appears to be due to the dissociative electron capture process

$$N^{5+}(1s^2)\ ^1S + H_2(^1\Sigma_g^+) \rightarrow N^{4+}(n=2) + H^+ + H(2l) + (36.87 - 61.61)\mathrm{eV} \qquad (10.19)$$

but contributions from

$$N^{5+}(1s^2)\ ^1S + H_2(^1\Sigma_g^+) \rightarrow N^{4+}(n=2) + H^+ + H^+ + e + (34.15 - 49.66)\mathrm{eV} \qquad (10.20)$$

also appear likely.

At impact energies below 857 eV amu^{-1}, the peaks corresponding to $N^{4+}(n = 3)$ and $N^{4+}(n = 4)$ formation remain as major features of the energy change spectra. However, additional collision mechanisms leading to $N^{4+}(n = 2)$ production become apparent. In particular, at our lowest energy of 214 eV amu^{-1}, the dominant peak (centred on an energy defect of about 28 eV) can be correlated with the autoionizing double capture channels

$$\begin{aligned} N^{5+}(1s^2)\ ^1S + H_2(^1\Sigma_g^+) &\rightarrow N^{3+(**)} + H^+ + H^+ \\ &\rightarrow N^{4+}(1s^2 2p) + H^+ + H^+ + e + (24.15 - 39.66)\mathrm{eV} \\ &\rightarrow N^{4+}(1s^2 2s) + H^+ + H^+ + e + (34.15 - 49.66)\mathrm{eV} \qquad (10.21) \end{aligned}$$

of which (10.20) is dominant. However, it will be noted that due to the overlap in the energy defects corresponding to (10.21) and the dissociative electron capture process (10.20), a clear quantitative assessment is impossible. The main observed product channels are summarised in table 10.3 while in table 10.4, we have estimated the fractions of N^{4+}ions formed in $n = 2$, $n = 3$ and $n = 4$ states through these collision mechanisms from an analysis of energy change spectra in the range 857 - 250 eV amu^{-1}.

From table 10.4 it will be seen that, over the present energy range, $N^{4+}(n = 3)$ formation through the non-dissociative electron capture channel decreases from about 66% at 857 eV amu^{-1} to 33% at 214 eV amu^{-1} while $N^{4+}(n = 4)$ formation through the same type of process (10.4) changes very little over the same energy range. An important feature highlighted by these TES data is

Table 10.3. Observed product channels for one-electron capture in $N^{5+}(1s^2)\ ^1S$ - $H_2(^1\Sigma_g^+)$ collisions. The energy defects shown were derived from potential energy curves [36] and energy level tabulations [37].

Product channels	Energy defects (eV)
Non-dissociative channels	
$N^{4+}(1s^2 3s) + H_2^+$	23.86 (v = 10) - 25.91 (v = 0)
$N^{4+}(1s^2 3p) + H_2^+$	21.17 (v = 10) - 23.22 (v = 0)
$N^{4+}(1s^2 3d) + H_2^+$	20.35 (v = 10) - 22.40 (v = 0)
$N^{4+}(1s^2 4s) + H_2^+$	5.23 (v = 10) - 7.29 (v = 0)
$N^{4+}(1s^2 4p) + H_2^+$	4.14 (v = 10) - 6.19 (v = 0)
$N^{4+}(1s^2 4d) + H_2^+$	3.80 (v = 10) - 5.85 (v = 0)
$N^{4+}(1s^2 4f) + H_2^+$	3.78 (v = 10) - 5.83 (v = 0)
Dissociative channels	
$N^{4+}(n = 2) + H^+ + H(2l)$	36.87 - 61.61
$N^{4+}(n = 2) + H^+ + H^+ + e$	32.87 - 49.86
Autoionizing two-electron capture channels	
$N^{3+(**)} + H^+ + H^+ \rightarrow$	
$N^{4+}(1s^2 2s) + H^+ + H^+ + e$	34.15 - 49.66
$N^{4+}(1s^2 2p) + H^+ + H^+ + e$	24.15 - 39.66

Table 10.4. Measured fractions of N^{4+} product ions formed in $n = 2$, $n = 3$ and $n = 4$ states through one-electron capture in $N^{5+}(1s^2)\ ^1S$ - $H_2(^1\Sigma_g^+)$ collisions.

Energy (eV/amu)	$n = 2$ (%)	$n = 3$ (%)	$n = 4$ (%)
857	16.9 ± 3.0	66.2 ±2.5	16.9 ± 4.0
786	19.0 ± 2.9	60.4 ± 2.4	20.6 ± 3.7
714	19.3 ± 3.0	61.5 ± 3.0	19.2 ± 4.1
643	25.8 ± 2.5	55.2 ± 3.0	19.0 ± 4.0
571	19.4 ± 2.7	67.3 ± 2.7	13.2 ± 5.4
500	26.6 ± 3.0	55.2 ± 3.0	18.3 ± 3.0
429	30.7 ± 3.0	53.8 ± 3.2	15.5 ± 2.7
357	37.9 ± 2.4	47.5 ± 3.7	14.6 ± 2.5
286	37.8 ± 3.2	43.1 ± 3.4	19.1 ± 2.5
250	59.0 ± 1.7	31.5 ± 3.8	9.5 ± 1.8
214	50.0 ± 1.8	33.0 ± 3.7	17.0 ± 2.6

the relative importance of $N^{4+}(n = 2)$ production which, through a number of possible collision mechanisms, accounts for 16.9% of the total at 857 eV amu^{-1} rising to 50% at 214 eV amu^{-1}. While dissociative electron capture leading to $N^{4+}(n = 2)$ production is clearly important at the higher energies, this is by no means a dominant process. It is interesting to note that this is contrary to the situation in He^{2+} - H_2 one-electron capture where our TES measurements [38] showed that dissociative excitation was the main electron capture mechanism at low energies. The most surprising feature in the present data is the major role at the lower energies of the two-electron capture autoionizing mechanism leading to $N^{4+}(n = 2)$ production. This is also the dominant one-electron capture process at our lowest energies.

In order to compare the present TES measurements with other relevant data, we have normalised our observed relative N^{4+} yields to our separately measured total cross sections to obtain the cross sections for N^{4+}(n = 2), $N^{4+}(n = 3)$ and $N^{4+}(n = 4)$ production. These values are shown in figure 10.12 together with experimental results of the KVI Groningen group [39–41] based on the PES approach.

The $N^{4+}(n = 3)$ and $N^{4+}(n = 4)$ cross sections [39] were obtained in the range 0.89 – 7.14 keV amu^{-1} from spectroscopic observations of the spontaneous decay of the relevant excited sublevels. The more recent work of [40] extended these measurements down to about 5 eV amu^{-1} but only cross sections for $N^{4+}(n = 3)$ formation were measured since other contributions to the total electron capture cross sections at these low energies were considered to be negligible. Following our present TES observations, the KVI group in a private communication [41] provided a reassessment of the $N^{4+}(n = 4)$ contributions in their PES data. These $N^{4+}(n = 4)$ cross sections (included in figure 10.13), while confirming that the n = 4 contribution at low energies is not negligible, can be seen to be generally considerably smaller than our TES values in the energy range. Our $N^{4+}(n = 3)$ formation cross sections can be seen to be considerably larger than the PES values of except at our lowest energy where, while our cross sections are continuing to fall, their values begin to increase again with decreasing energy.

Included in figure 10.12 are theoretical estimates of the $N^{4+}(n = 3)$ formation cross sections based on a semiclassical close coupling approach [42] and based on an *ab initio* method [43]. Both these calculations predict cross sections in accord with the downward trend of our measurements. However, the quantitative agreement with experiment can be seen to be generally poor. Unlike the PES measurements, our TES measurements have been able to record the substantial contribution of the $N^{4+}(n = 2)$ products to the total one-electron capture cross section. As already noted, the low energy dominance of the component arising from autoionizing two-electron capture is particularly noteworthy. We include in figure 10.12 recent theoretical estimates [43] of cross sections for autoionizing double capture leading to $N^{4+}(n = 2)$ formation. These values can be seen to be much smaller and in poor agreement with our measured cross sections for $N^{4+}(n = 2)$ formation even at our lowest energy where we estimate that the autoionizing double capture contribution is about 50%.

4.2 Measurements of one electron capture in atomic hydrogen

4.2.1 N^{2+} + H. We have carried out DTES measurements [31] of the process

$$N^{2+} + \mathrm{H}(1s) \rightarrow N^{+}(n, l) + \mathrm{H}^{+} \tag{10.22}$$

at energies within the range 0.8 – 6.0 keV using a beam of pure ground-state $N^{2+}(2s^22p)$ $^2P^o$ primary ions. These measurements allowed comparison with current theoretical predictions and a re-evaluation of previous TES measurements [44] of this process carried out in this laboratory with ion beams containing an unknown fraction of metastable $N^{2+}(2s2p^2)$ ^4P ions.

Figure 10.13 shows energy change spectra obtained for one-electron capture by 6 keV N^{2+} ions in atomic hydrogen. The 'mixed' beam spectrum for N^{2+} ions obtained directly from the ion source containing an unknown fraction of metastable ions may be compared with the spectrum obtained using a state-prepared pure N^{2+} ^2P ground-state beam. In this case, the 'mixed' beam and pure ground state spectra are very similar evidently because collision product channels with not greatly different energy defects provide the dominant contributions to both.

In the case of ground-state primary ions, the three main product channels can be identified as:

$$\begin{aligned}
N^{2+}\ {}^2\mathrm{P}^o + \mathrm{H}(1s) \quad &\rightarrow \quad N^{+}(2s2p^3)\ {}^3\mathrm{D}^o + \mathrm{H}^{+} + 4.57\mathrm{eV} \\
&\rightarrow \quad N^{+}(2s2p^3)\ {}^3\mathrm{P}^o + \mathrm{H}^{+} + 2.46\mathrm{eV} \\
&\rightarrow \quad N^{+}(2s^22p3s)\ {}^1\mathrm{P}^o + \mathrm{H}^{+} - 2.49\mathrm{eV} \\
&\rightarrow \quad N^{+}(2s^22p3s)\ {}^3\mathrm{P}^o + \mathrm{H}^{+} - 2.46\mathrm{eV} \\
&\rightarrow \quad N^{+}(2s2p^3)\ {}^1\mathrm{D}^o + \mathrm{H}^{+} - 1.87\mathrm{eV}
\end{aligned} \tag{10.23}$$

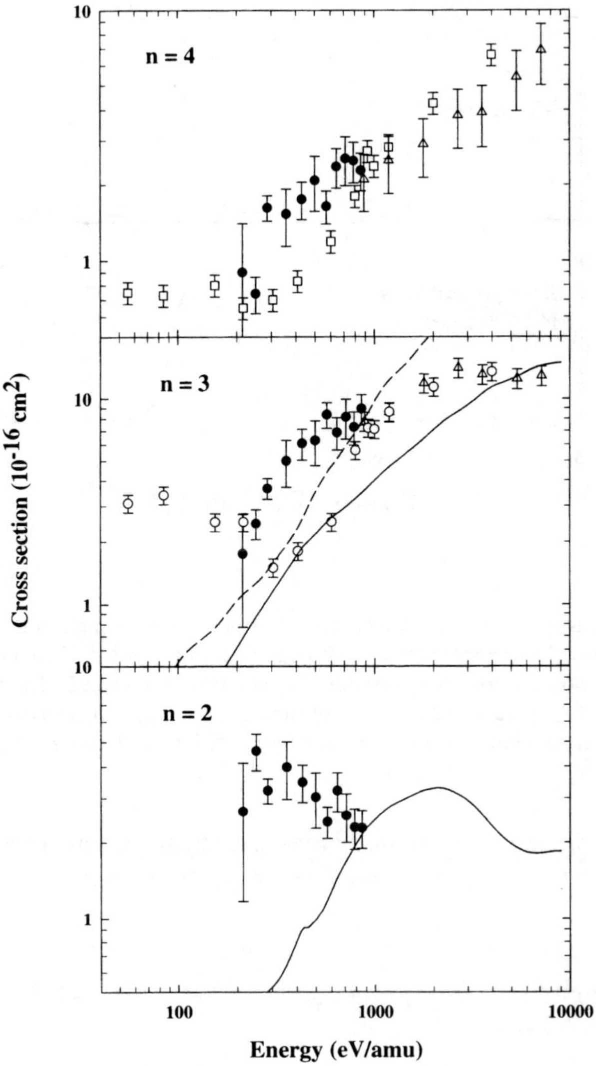

Figure 10.12. Cross sections for one-electron capture into $n = 2$, $n = 3$ and $n = 4$ states of N^{4+} by N^{5+} ions in H_2. Experiment: •, $n = 2$, $n = 3$, $n = 4$ states, present work. □, $n = 4$ states [41], Δ, $n = 3$, $n = 4$ states,[39], ∘, $n = 3$ states, [40]; Theory: long-dashed line, $n = 2$ states through autoionizing double capture, [43] solid line, $n = 3$ states, [43] short-dashed line, $n = 3$ states, [42]

of which the two exothermic channels (which involve core excitation) are dominant.

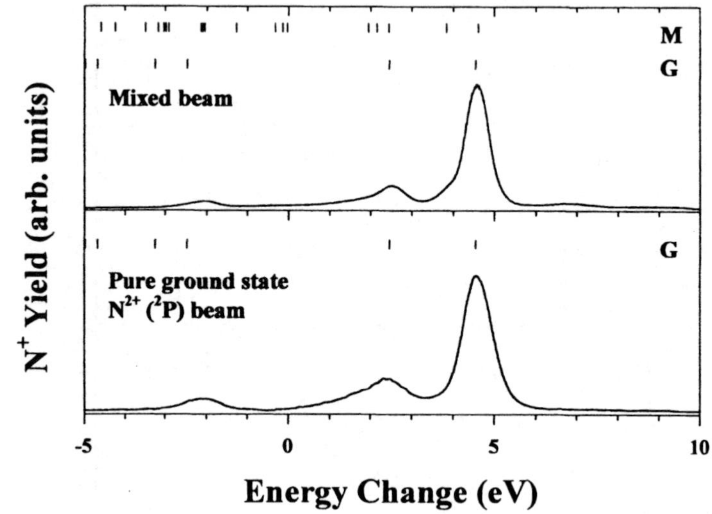

Figure 10.13. Energy change spectra (from [31]) obtained for one-electron capture by 6 keV N^{2+} ions in atomic hydrogen. The 'mixed' beam spectrum for N^{2+} ions obtained with ions containing an unknown fraction of metastable ions is compared with the DTES spectrum obtained using a pure $N^{2+}\ {}^2P^o$ ground-state beam. Positions of possible product channels corresponding to ground-state primary ions (G) and metastable primary ions (M) are also shown.

In the case of metastable primary ions, the channels which seem most likely to contribute to the observed 'mixed' beam spectrum are:

$$\begin{aligned}
\mathrm{N}^{2+}(2s2p^2)\ {}^4\mathrm{P} + \mathrm{H}(1s) \quad &\rightarrow \quad \mathrm{N}^+(2s^22p3s)\ {}^3\mathrm{P}^o + \mathrm{H}^+ + 4.63\mathrm{eV} \\
&\rightarrow \quad \mathrm{N}^+(2s2p^3)\ {}^3\mathrm{S}^o + \mathrm{H}^+ + 3.86\mathrm{eV} \\
&\rightarrow \quad \mathrm{N}^+(2^22p3p)\ {}^3\mathrm{D} + \mathrm{H}^+ + 2.45\mathrm{eV} \\
&\rightarrow \quad \mathrm{N}^+(2s^22p4p)\ {}^3\mathrm{D} + \mathrm{H}^+ - 2.04\mathrm{eV} \\
&\rightarrow \quad \mathrm{N}^+(2s^22p4p)\ {}^3\mathrm{P} + \mathrm{H}^+ - 2.10\mathrm{eV} \\
&\rightarrow \quad \mathrm{N}^+(2s^22p4p)\ {}^3\mathrm{S} + \mathrm{H}^+ - 2.14\mathrm{eV}
\end{aligned} \qquad (10.24)$$

The close similarity between the 'mixed' beam and pure ground-state energy change spectra using DTES confirms the results of our earlier TES measurements [44] in the range 0.6 – 8.0 keV using N^{2+} beams which contained an admixture of metastable ions. In those measurements, changes in the metastable fraction were found to have no detectable effect on the measured total electron capture cross sections leading us to conclude that the total cross sections for ground and metastable ions N^{2+} in atomic hydrogen were not greatly different. In view of the lack of dependence of these total cross sections on the metastable content of the primary beam, we have normalised the DTES data obtained at different energies to our previously measured total electron capture cross sections [44]. The latter values are in good agreement with previous measurements by [45, 46] in the energy ranges of overlap.

In figure 10.14, these cross sections for N^+ $^3D^o$, $^3P^o$ and $N^+(^1P^o+^3P^o+^1D^o)$ formation can be seen to be in good accord with the values obtained in our previous TES measurements [44]. However, it is important to note that, in the lower resolution measurements of [44], the small peak comprising the $N^+(^1P^o+^3P^o+^1D^o)$ contributions from endothermic channels were incorrectly identified as due to the $N^+(^1D^o)$ product channel alone. In addition, the measurements of [44] also recorded at and above 6 keV very small contributions from a $N^+(2s^22p3p)$ ^3P product channel corresponding to ΔE = - 5.17 eV; these are absent in the higher resolution DTES measurements. Figure 10.14 also includes cross sections for the main $^3D^o$ product channel calculated by Bienstock *et al* [47] using a molecular approach which extends earlier theoretical work by Heil et al [48]. These calculations predict the $^3D^o$ product channel is the only significant electron capture channel at the energies considered and are therefore at variance with experiment. More recent quantal calculations carried out by Herrero *et al* [49], in which molecular states were obtained using *ab initio* SCF-CI methods, correctly predict contributions from channels other than the dominant $^3D^o$ product channel. However, in the energy range shown in figure 10.14, their calculated cross sections for the $^3D^o$ and $^3P^o$ channels are not in very satisfactory agreement with the experimental values.

4.2.2 O^{2+} + H. The interpretation of previous total electron capture cross sections and TES measurements with O^{2+} ions has been complicated by the possible presence of O^{2+} $(2s^22p^2)$ ^1D and O^{2+} $(2s^22p^2)$ ^1S metastable ions in addition to O^{2+} $(2s^22p^2)$ ^3S ground state ions in the primary ion beams. For example, our previous TES studies [50] of one-electron capture in O^{2+}- H(1s) collisions at energies in the range 2 – 8 keV, revealed two peaks in the energy change spectra. The main peak, which accounts for about 70% of the total O^+ yield could be identified with the ground state primary ion channel

$$O^{2+}(2s^22p^2)\ ^3S + H(1s) \rightarrow O^+(2s2p^3)2p\ ^4P + H^+ + 6.64eV \qquad (10.25)$$

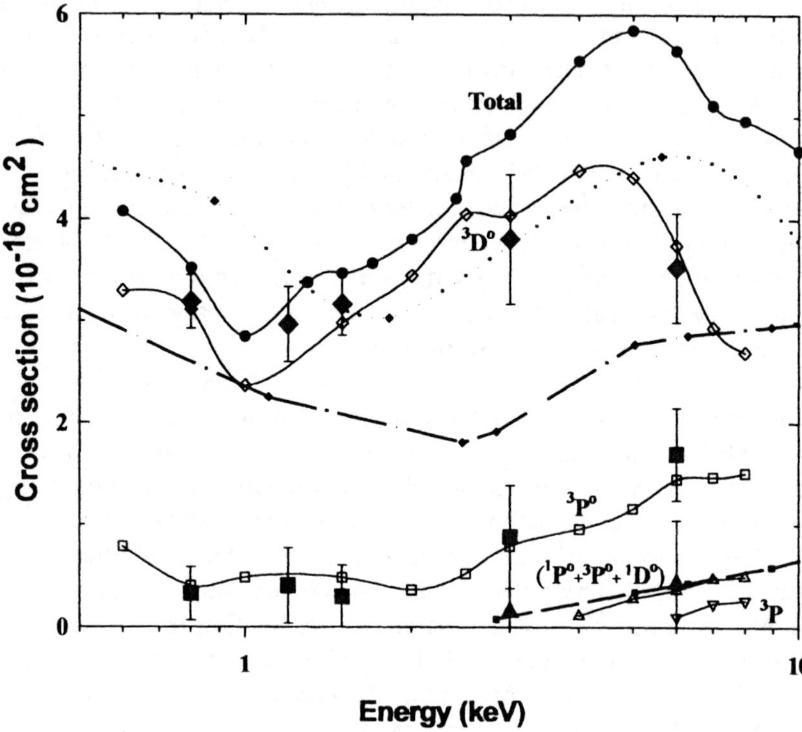

Figure 10.14. Cross sections for one-electron capture by N^{2+} ions in atomic hydrogen leading to specified N^+ (n, l) products. DTES measurements [31] with pure ground-state N^{2+} $^2P^o$ ions, ♦, $^3D^o$; ■, $^3P^o$; ▲, ($^1D^o$ + $^1P^o$+ $^3P^o$). Previous TES measurements [44] using an N^{2+} beam containing an unknown fraction of metastable ions, •, total; ◇, $^3D^o$; □, $^3P^o$; Δ, ($^3P^o$ + $^1P^o$ + $^1D^o$); ∇, ^{3}P. Theory [47], · · · · · · · ·, $^3D^o$. *Theory* [49], — · — ·, $^3D^o$; — — —, $^3P^o$.

but the available energy resolution could not exclude a possible small contribution from O^{2+} ^{1}S metastable primary ion channel

$$O^{2+}(2s^2 2p^2)\ ^1\mathrm{S} + \mathrm{H}(1s) \rightarrow O^+(2s2p^3)2p\ ^2\mathrm{D} + \mathrm{H}^+ + 6.29\mathrm{eV} \quad (10.26)$$

The second smaller peak in the observed spectra was correlated with the O^{2+} ^{1}D metastable primary ion channel

$$\mathrm{O}^{2+}(2s^2 2p^2)\ ^1\mathrm{D} + \mathrm{H}(1s) \rightarrow \mathrm{O}^+(2s2p^3)2p\ ^2\mathrm{D} + \mathrm{H}^+ + 3.45\mathrm{eV} \quad (10.27)$$

Figure 10.15 shows a DTES energy change spectrum [32] obtained at 6 keV for a beam of pure O^{2+} ^3S ground state ions compared with a 'mixed' beam spectrum measured for O^{2+} ions of unknown metastable content direct from the ion source. In the 'mixed' beam spectrum, the O^+ $(2s2p^3)2p$ ^2D contribution in the smaller of the two peaks arising from electron capture by O^{2+} ^1D metastable ions can be clearly seen while, in the pure O^{2+} ^3S spectrum, the single peak corresponds to the O^+ $(2s2p^3)2p$ ^4P product channel. These measurements confirm the low energy quantal calculations [51] which predict that only the O^+ $(2s2p^3)2p$ ^4P product channel should be significant for O^{2+} ^3S ground state ion impact.

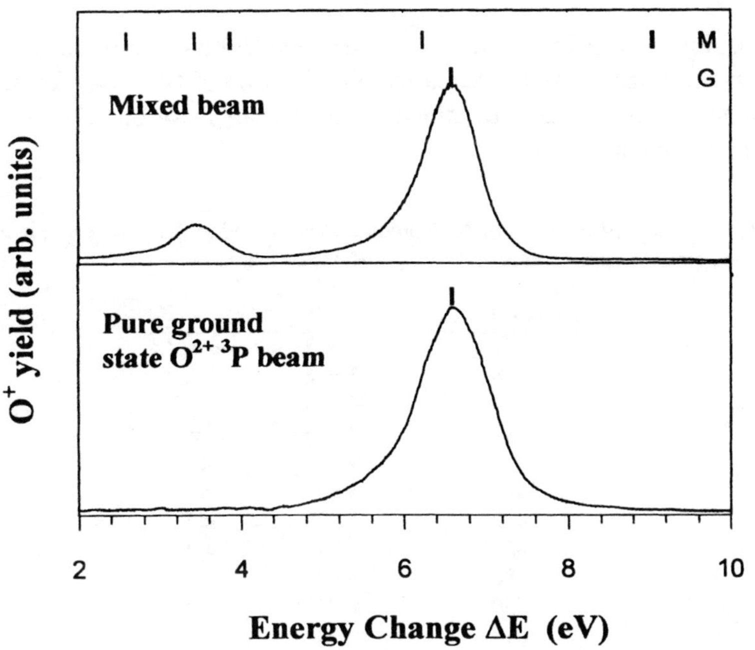

Figure 10.15. Energy change spectra (from [32]) for one-electron capture in O^{2+} - H(1s) collisions (a) by a beam of O^{2+} ions of unknown metastable content obtained directly from the ion source and (b) by pure O^{2+} ^3S ground state ions prepared by DTES. Possible product channels associated with O^{2+} ^1D metastable ions and O^{2+} ^3S ground state ions are indicated as M and G respectively.

4.2.3 N^{5+} + H. Figure 10.16 shows our measured energy change spectra for one-electron capture by N^{5+} ions in H at energies in the range from 857eV amu^{-1} down to 357eV amu^{-1}. These spectra show very clearly that capture into the n = 4 states of N^{4+}through

$$N^{5+}(1s^2)\ {}^1S + H(1s) \rightarrow \ N^{4+}(1s^2 4s, 4p, 4d, 4f) + H^+ \quad (10.28)$$
$$+9.09, 8.00, 7.66, 7.64eV$$

is dominant throughout the present energy range. A contribution from capture into the n = 3 states of N^{4+}through the more exothermic channels

$$N^{5+}(1s^2)\ {}^1S + H(1s) \rightarrow \ N^{4+}(1s^2 3s, 3p, 3d) + H^+ \quad (10.29)$$
$$+27.72, 25.03, 24.21eV$$

while negligible at 357eV amu^{-1}, can be seen to make an increasing contribution as the impact energy increases. The fraction of N^{4+} ions formed in the n= 4 and n = 3 states, as determined from an analysis of our TES spectra are summarised in table 10.5.

Table 10.5. Measured fractions of N^{4+} product ions formed in the $n = 3$ and $n = 4$ states through one-electron capture in $N^{5+}(1s^2)\ {}^1S - H(1s)$ collisions.

Energy (eV/amu)	$n = 4$ (%)	$n = 3$ (%)
857	84 ± 3	16 ± 3
786	87 ± 2	13 ± 2
714	94 ± 3	6 ± 3
643	97 ± 3	3 ± 3
571	> 99	< 1
500	> 99	< 1
429	> 99	< 1
357	> 99	< 1
286	> 99	< 1
250	> 99	< 1
214	> 99	< 1

Although there have been numerous experimental measurements of the total cross sections for electron capture in $N^{5+}(1s^2)\ {}^1S - H(1s)$ collisions using several different techniques over a wide energy range, the results from different laboratories exhibit large discrepancies and are subject to large uncertainties in the present energy range (cf. data in reference [52]). For this reason, and because the main purpose of the present measurements was to determine the relative importance of the different electron capture channels, we have made no attempt to normalise the present TES data to previously measured total cross sections.

In figure 10.16 we show our measured yields of N^{4+} product ions in the $n = 4$ and $n = 3$ states (expressed as a fraction of the total electron capture cross section) compared with the higher energy PES [39].

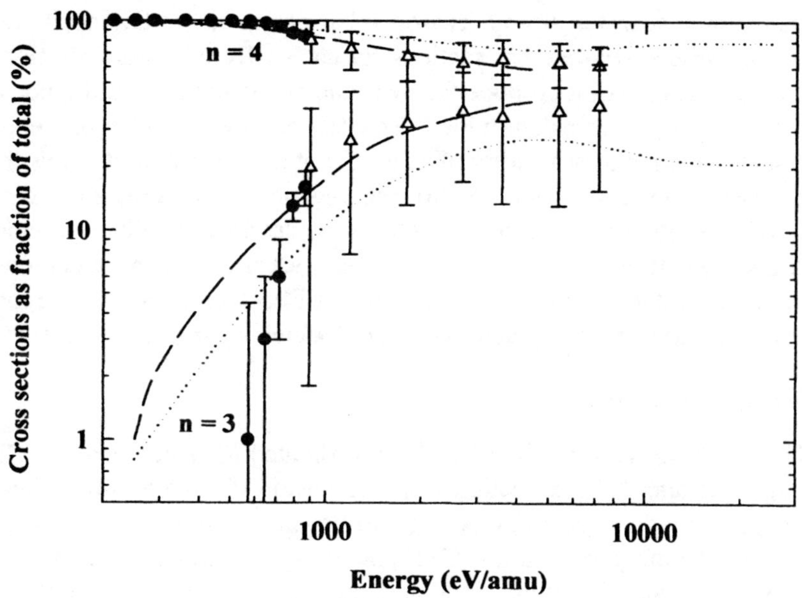

Figure 10.16. One-electron capture into the $n = 4$ and $n = 3$ states of N^{4+} by N^{5+} ions in atomic hydrogen shown as a fraction of the total electron capture cross section. Experiment:•, present work; Δ, [39]. Theory: ------, [54] ; ·······, [53]

There is good general accord between the two sets of experimental data. Theoretical estimates of the N^{5+} - H(1s) one-electron capture process based on a molecular orbital expansion method [53] and based on both quantum mechanical and semiclassical molecular-orbital expansion methods [54] extend down to the low energies considered here. Their predicted relative yields of both the $n = 4$ and $n = 3$ states are included in figure 10.16. The theoretical values of [53] can be seen to be in somewhat better accord with the low energy experimental values. For $n = 4$ capture. In the case of $n = 3$ capture, the yields predicted by [53] are significantly smaller than those calculated by [54]. However, neither theory satisfactorily describes the very rapid decrease in the $n = 3$ state yields observed at low energies in the present work.

5. Conclusion

Our measurements with state-prepared C^{2+}, N^{2+} and O^{2+} have demonstrated that DTES is a powerful technique capable of providing detailed information on state-selective electron capture without the ambiguities of interpretation associated with previous measurements carried out with ion beams of unknown metastable content. DTES measurements in several atomic and molecular gases of fusion interest and in atomic hydrogen have been made. Our apparatus does not yet provide absolute cross sections for each selected ground or metastable species but the collision product channels can be identified unambiguously and their relative importance determined. The data provide a much more reliable assessment of the range of validity of a number of theoretical predictions than has been possible previously and help to resolve some of the serious discrepancies in both TES measurements and in total cross sections obtained previously in different laboratories. In the case of N^{5+} our measurements have demonstrated the complementary nature of the DTES and PES techniques and highlighted differences between theory and experiment which need to be resolved.

Acknowledgments

This work forms part of the IAEA Co-ordinated Research Project (CRP) on 'Charge Exchange Cross Section Data for Fusion Plasma Studies'. The work was support by the U.K. Engineering and Physical Sciences Research Council and it also forms part of the E.C. Framework 5 Thematic Network on Low Energy Ion Beam Facilities. D.K acknowledges the award of a studentship from the Northern Ireland Department of Employment and Learning.

References

[1] T. E. Cravens, Geophysics Research Lett. **24** 105 (1997)

[2] W. Liu and D. R. Schultz, Astrophys. J. **526** 538 (1999)

[3] B. J. Wargelin and J. Drake, Astrophys. J. **546** L57 (2001)

[4] J. B. Kingdon and G. J. Ferland, Astrophys. J. **516** L107 (1999)

[5] C. F. Maggi et al, Plasma Physics and Controlled Fusion **42** 669 (2000)

[6] R. A. Phaneuf et al, Rep. Prog. Phys. **62** 1143 (1999)

[7] K. Ishii et al, Phys. Scr. **T92** 332 (2001)

[8] G. Lubinski et al, J. Phys. B: At. Mol. Opt. Phys. **33** 5275 (2000)

[9] E. Unterreiter.et al, J. Phys. B: At. Mol. Opt. Phys. **24** 1003 (1991)

[10] R. W. McCullough et al, .J. Phys. B: At. Mol. Opt. Phys. **17** 1373 (1984)

[11] L. F. Errea et al, J. Phys. B: At. Mol. Opt. Phys. **32** 4065 (1999)

[12] P. Honvault et al J. Phys. B: At. Mol. Opt. Phys. **27** 3115 (1994)

[13] HP. Winter, J. Phys. Chem. **99** 15448 (1995)
[14] J. B. Greenwood et al, J. Phys. B: At. Mol. Opt. Phys. **29** L599 (1996)
[15] D. Voulot et al, J. Phys. B: At. Mol. Opt. Phys. **33** L187 (2000)
[16] D.Voulot et al, J. Phys. B: At. Mol. Opt. Phys. **34** 1039 (2001)
[17] P. Leputsch et al, J. Phys. B: At. Mol. Opt. Phys **30** 5009 (1997)
[18] R. G. Cooks, *Collision Spectroscopy*, Plenum Press, (1978)
[19] R. E. Olson and M. Kimura, J. Phys. B. **15** 4231 (1982)
[20] R. W. McCullough et al, J. Phys. B. **17** 1373 (1984)
[21] H. Cederquist et al J. Phys. B. **18** 3951 (1985)
[22] F. Broetz et al, Physica Scripta. **T92** 278 (2001)
[23] R. W. McCullough et al, Meas. Sci. Technol. **4** 79 (1993)
[24] B.A. Huber et al, .J. Phys.B: At. Mol. Phys. **17** 2883 (1984)
[25] J. B.Greenwood et al, J.Phys.B: At. Mol. Opt. Phys. **29** 5867 (1996)
[26] D. Burns et al, J. Phys. B: At. Mol. Opt. Phys **30** 1531 (1997)
[27] D. Burns et al J. Phys. B: At. Mol. Opt. Phys. **30** 4559 (1997)
[28] D. Burns et al, J. Phys. B; At. Mol Opt. Phys. **30** L323 (1997)
[29] W. R. Thompson et al, Physica Scripta **T80** 362 (1999)
[30] D. Voulot et al, J. Phys.B: Atom.Mol.Opt. Phys. **33** L187 (2000)
[31] D. Voulot et al, J Phys.B: Atom.Mol.Opt.Phys. **34** 1039 (2001)
[32] D. M. Kearns et al, Physica Scripta vol **T92** 76 (2001)
[33] D. Voulot, PhD Thesis, Queen's University Belfast, United Kingdom (2000)
[34] R. Hoekstra et al, J Phys.B: Atom.Mol.Opt.Phys. **27** 2021 (1994).
[35] M. Gargaud and R. J. McCarroll, J. Phys.B: Atom.Mol.Opt.Phys. **18** 463 (1985).
[36] T. E. Sharp, Atomic Data **2** 119 (1971)
[37] S. Bashkin and J.O.Stoner Jr, *Atomic Energy Levels and Grotrian Diagrams* North Holland, Amsterdam (1978)
[38] J. M. Hodgkinson et al, J.Phys.B: At. Mol. Opt. Phys. **28** L395 (1995)
[39] D. Dijkkamp et al, J. Phys B: At. Mol. Phys **18** 763 (1985)
[40] G. Lubinski et al, J. Phys. B: At. Mol. Opt. Phys. **33** 5275 (2000)
[41] R. Hoekstra , private communication, (2002)
[42] A. Kumar and B. C. Saha Phys. Rev. A **59** 1273 (1999)
[43] L. Méndez, private communication, (2002)
[44] F. G. Wilkie et al, J. Phys B: At. Mol. Phys. **18** 479 (1985)

[45] W. Seim et al, J. Phys. B. At. Mol. Phys. **14** 3475 (1981)

[46] R. A. Phaneuf et al, Phys. Rev.A **17** 534 (1978)

[47] S. Bienstock et al, Phys. Rev. A **33** 2078 (1986)

[48] T. G. Heil et al, Phys. Rev. A **23** 1100 (1981)

[49] B. Herrero et al, J. Phys.B: At. Mol. Opt. Phys. **28** 711 (1995)

[50] T.K. McLaughlin et al , J.Phys. B: At. Mol. Opt. Phys. **23** 737 (1990)

[51] S. E. Butler, Astrophys. J. **241** 442 (1980)

[52] C. C. Havener et al, Phys. Rev A **39** 1725 (1989)

[53] M. Bendahman et al, J. Physique **46** 561 (1985)

[54] N. Shimakura and M. Kimura, Phys. Rev. A **44** 1659 (1991)

Chapter 11

IONIZATION AND EXCITATION OF ATOMIC LI BY FAST IONS

Analogies with Photons

J.A. Tanis
Department of Physics, Western Michigan University, Kalamazoo, MI 49008
tanis@wmich.edu

N. Stolterfoht
Hahn-Meitner-Institut Berlin GmbH, D14109 Berlin, Germany
stolterfoht@hmi.de

Abstract Ionization and excitation of atomic lithium caused by the impact of nearly relativistic ions are considered, and the analogies with corresponding processes induced by incident photons are examined. Specifically, the nature of the interactions leading to single-electron ejection, and the mechanisms for producing doubly vacant K-shell states, are investigated. The bombarding particle for the studies discussed here is 95 MeV/u Ar^{18+}, for which $v/c = 0.42$. First, single ionization (K-shell or L-shell) is found to consist of a part due to dipole transitions to the continuum (corresponding to photoionization) plus a part due to binary encounters between the fast projectile and a target electron (corresponding to Compton scattering). For the production of states with two K-shell vacancies, it is shown that the electron-electron interaction plays an important role in the formation of these so-called "hollow" states, in a manner similar to photon-induced processes. Atomic lithium provides a unique system in which to study ionization and excitation because of its tightly bound inner shell and weakly bound outer shell, and the fact that single- and double-K-shell vacancy production can be simultaneously investigated from a single Auger electron emission spectrum.

Keywords: ionization, excitation, electron emission, Auger emission, double–K–shell vacancies, electron–electron interactions, electron correlation, shake processes, dielectronic processes

F.J. Currell (ed.), The Physics of Multiply and Highly Charged Ions, Vol. 2, 339-368.

1. Introduction

Collisions of atomic systems with ions or photons can result in excitation, ionization, or charge changing. In addition to one-electron transitions, combinations of these transitions can occur, and the resulting two- (or multi-) electron processes often provide insight into the structure of the atomic system itself, as well as how this structure relates to the dynamics of the collision process. For nearly relativistic projectiles (on the order of half the speed of light), the projectile may be considered a source of virtual photons in its interaction with a target system. In this case, significant analogies are expected between fast ions and photons in their interactions with matter. Such similarities have, in fact, been recognized since the very beginnings of atomic collision physics [1, 2] and have subsequently been examined in detail [3, 4].

For ions traveling near relativistic speeds the interaction time with a target atom is about 10^{-18} s, while single ionization times ($\sim 10^{-16}$ s) and atomic relaxation and autoionization times ($> 10^{-15}$ s) are typically much longer than this. So, the projectile is already far removed from the collision region when the residual target system "relaxes", and the ion plays no further role in the subsequent relaxation processes. This could be likened to the purely classical situation of a high-speed bullet passing through a strawberry. By the time that the strawberry "explodes", the bullet is already far removed from the interaction region, and, hence, no longer has any influence on the "relaxing" strawberry [5]. This separation of the ionization and excitation phases of an interaction from the subsequent deexcitation is important because it plays a key role in the interpretation of high-velocity collision phenomena.

Because of the similarity to photon-induced interactions, dipole transitions ($\Delta L = 1$) are expected to be important in ionization and excitation by high-velocity projectiles [1]. Moreover, electron emission which results from P to S transitions is expected to show the $sin^2\,\theta$ (θ is the electron emission angle relative to the beam direction) angular dependence characteristic of dipole transitions. Thus, measurements of electron emission can provide a sensitive probe of the connections to photon-induced processes. Such comparisons, as well as deviations from photon- induced expectations, provide important insight into the dynamics of Coulombic interactions between atomic systems at high velocities.

In the interaction of an ion or photon with an atomic system, the strength of the perturbation is typically determined by the momentum transferred in the collision, i.e., $\Delta p = \Delta E/v$, where Δp is the momentum transferred, ΔE is the energy transferred, and v is the speed of the impacting ion or photon. For incident photons that can be considered as individual bombarding "particles", such as those from a synchrotron source, the photon can interact with only a single electron. Thus, the perturbation is weak because a photon cannot transfer large momentum. In ion-atom collisions, if the projectile velocity is significantly larger than the velocity of the active bound electron, then the collision interaction is weak so that perturbative methods can be used. For intermediate-to-high velocities ($v > 10$ a.u., where 1 a.u. of velocity is the Bohr velocity in hydrogen, namely, v_B = 2.19 x 10^8cm/s), the Born approximation is often utilized, for which the strength of the perturbation is governed by the parameter Z/v, where Z is the atomic number of the projectile and v is the collision velocity. For small values of this parameter, specifically, $Z/v \leq 0.5$ (in atomic units), the collision interaction is expected to closely resemble that of a photon interacting with an atom because the momentum transferred is also small [1, 2, 6].

Electron emission from a target atom by the impact of a heavy projectile is a fundamental manifestation of the ionization process [4, 7]. Electrons emitted with low energies are particularly important because these electrons have, by far, the largest probability for ejection. These low-energy electrons, referred to as *soft-collision* electrons, are produced mainly in large impact-parameter collisions. When the velocity of the projectile is much larger than the velocity of the active bound electron, the momentum transferred in a soft collision is small. Despite the small momentum transfer, an ejected electron may carry away a significant amount of momentum. When the final electron momentum is larger than the momentum transferred, a third body is required to balance the missing momentum. If no other particle is ejected in the collision reaction, then the target nucleus has to take part in the ionization process. Therefore, soft collisions at high projectile energies can be attributed to *three-body* collisions involving the projectile ion, the active electron, and the residual target ion.

In the case of electron ejection by photons, i.e., the photoelectric effect, the incident photon is annihilated. In order to conserve energy and momentum the annihilation of the photon cannot take place without the interaction of the electron with the residual ion. Therefore, the photoelectric effect also corresponds to a three-body process involving the incident photon, the active electron, and the residual target ion. It is well known that this interaction is mediated by a dipole transition involving the transfer of one unit of angular momentum ($\Delta L = 1$). Recently, considerable interest has been devoted to the analogy between fast ions and photons in the ionization process [8–11].

On the other hand, for close encounters between a fast projectile ion and a target atom, significant momentum can be transferred directly to a target electron. In this *binary-encounter* interaction [12] between the ion and the electron the target atom plays only a minor role. Thus, this interaction is essentially a *two-body* process involving only the projectile ion and a target electron, with the energy and momentum of the ejected electron being determined mainly by two-body kinematics. In such a process, the target nucleus serves only to give a momentum distribution, determined by the Compton profile, to the ejected target electron. This two-body approximation, and the corresponding neglect of the interaction with the residual ion, is adopted in the framework of the impulse approximation [13, 14]. This neglect of the target nucleus has been used extensively previously in the study of the recombination of ions interacting with bound quasi-free target electrons via a resonant inverse Auger type process [15].

As pointed out by Bethe [1], the two-body process in electron emission by incident ions is analogous to the Compton scattering of photons by "free" electrons. In this latter process, the scattered photon has a smaller momentum, with the recoiling electron accounting for the missing momentum. If the energy of the initially bound electron is small compared to the incident photon energy, then the target nucleus plays a minor role, as in the case of the binary-encounter process for fast ions, serving only to give a momentum distribution to the ejected electron.

In addition to single-electron transitions, ionization and excitation can be part of a two- (or multi-) electron process. If the interaction of an incident ion or photon with an atomic system causes a core (i.e., inner-shell) vacancy, the initial ionization or excitation can trigger a second core vacancy due to the *electron-electron* ($e-e$) interaction, commonly referred to as *electron correlation*. Studies of electron correlation leading to these so-called "hollow" (i.e., vacant inner shells) systems are of interest and importance because they delve into the interplay between atomic structure and collisional dynamics [10]. In this latter regard, the double-K-shell ionization of He by fast ions and photons has attracted much attention [16] for many years due to the insight it provides into electron correlation effects.

For a fully-stripped ion colliding with a target atom, two-electron transitions may occur independently through separate *nucleus-electron* ($n-e$) interactions with the perturbing partner, or the $e-e$ interaction may be important. In the latter case, double-core vacancy production in a target atom proceeds via an initial $n-e$ interaction followed by an $e-e$ interaction. This sequence is often referred to as TS1 (two-step with one projectile interaction), a process that implies *dynamic electron correlation* [17, 18]. Double-core vacancy production in ion-atom collisions may also come about by means of separate $n-e$ interactions, referred to as TS2 (two-step with two projectile interactions), but this process will not be considered here. These two-electron phenomena, and, in particular, $e-e$ effects, have been the subject of many recent studies [7, 10, 16].

In the case of photon-induced processes, the photon can similarly initiate a sequence of transitions [10, 19, 20] following the perturbation of a single electron, but only the $e-e$ interaction can cause a subsequent transition. These photon-induced processes in atoms and molecules have become a field of intense investigation [21]. For double-core vacancy production by photons, the initial excitation or ionization event is followed by a transition caused by the $e-e$

interaction to produce the hollow system. Thus, the correlation initiated by photons is similar to TS1 for fast ions.

In the following, we will examine the similarities between photon and fast-ion interactions with neutral lithium target atoms from two distinct standpoints: (1) single-electron ejection (ionization), and (2) two- (or multi-) electron transitions leading to doubly vacant K-shell states. Because of its characteristics, three-electron atomic lithium provides a unique system in which to examine the photon-like properties of fast ions, and, therefore provides a benchmark for the study of similar effects in other collision systems. The bombarding ion for the investigations discussed here is 95 MeV/u Ar^{18+}, corresponding to $v/c = 0.42$ (v is the ion velocity and c is the speed of light), and a perturbation strength of $Z/v = 0.31$ (in atomic units). This latter value is well within the region of validity of the Born approximation.

First, we will describe the experimental methods that are used, and show the kind of data that are obtained to reveal the aforementioned similarities between photons and ions. Then, we will discuss the two cases, namely, single ionization and double-K-shell vacancy production, separately to show how fast ions manifest themselves as photons in these processes. Despite the separate treatment of these two distinct aspects of vacancy production by fast ions, it is emphasized that the common thread connecting them is the analogy with corresponding photon-induced processes. These analogies will be noted frequently throughout the following discussion. Fuller reports of the single ionization [22, 23] and the double-core vacancy production processes discussed here [24, 25] have been previously published.

2. Experimental Methods

A schematic of the electron detection apparatus used to obtain the data presented here is shown in fig. 11.1. The essential components inside the scattering chamber are the electron spectrometer and the oven used to produce a lithium vapor target. The scattering chamber and the electron spectrometer are similar to ones used previously, and more details can be found in Ref. [26]. The experimental work described here was carried out at the GANIL facilty in Caen, France. An intense beam (1-2 μA) of 95 MeV/u Ar^{18+} ions made possible the measurement of highly accurate low-resolution single-electron emisson spectra, and high-resolution measurements for single- and double-K-shell vacancy production, resulting from collisions of these ions with atomic lithium. The beam was incident on a Li vapor target obtained by heating metallic Li in a temperature-controlled oven. The Li vapor emerging from the oven formed a jet of about 3-4 mm diameter, and was crossed by the beam that was collimated to a size of about 2 mm $\times$ 2 mm. Continuum electrons emitted from the Li were measured with a parallel-plate electron spectrometer as shown in the figure.

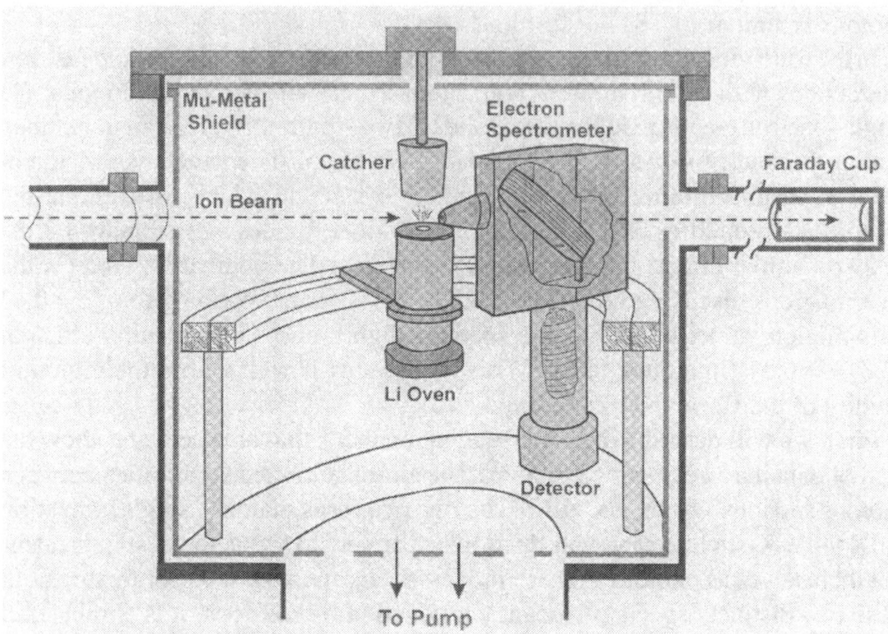

Figure 11.1. Schematic of the experimental apparatus. The scattering chamber contains a parallel-plate electron spectrometer and a high-temperature oven to produce a Li vapor target. The spectrometer is mounted on a movable ring to enable measurement of the angular distributions of electrons ejected in the collision. The vapor jet is directed into a cooled catcher to reduce Li contamination of the chamber.

In working with the lithium vapor target several instrumental difficulties had to be overcome. First, the metallic lithium was heated slowly to drive contaminants from the surface. Then, the lithium temperature was set just high enough above the melting point (180^0 C) to obtain a stable jet of Li atoms without producing significant amounts of molecular lithium, i.e., Li_2 which would complicate the analysis of the measured spectra. In this way, data could be recorded for several hours before it became necessary to refill the oven with lithium. Target thicknesses on the order of 1 x 10^{14} atoms/cm^2 were achieved. Moreover, for the reliable detection of low energy electrons ($<$ 100 eV), the possibility of perturbing effects due to the electric and magnetic fields associated with the relatively large current used to heat the metallic lithium, as well as effects due to lithium build-up on the spectrometer surfaces and insulators, had to be considered. In the former case, a bifilar heating wire was used and measurements were taken with the heating current on and off. No significant differences were noted, showing that the measured spectra were not altered by stray fields. In the latter case, an efficient baffle system was used to protect the sensitive parts of the spectrometer and to prevent electrical breakdowns. With the procedures used, yields could be measured reliably for electrons emitted with energies as low as ~ 3 eV.

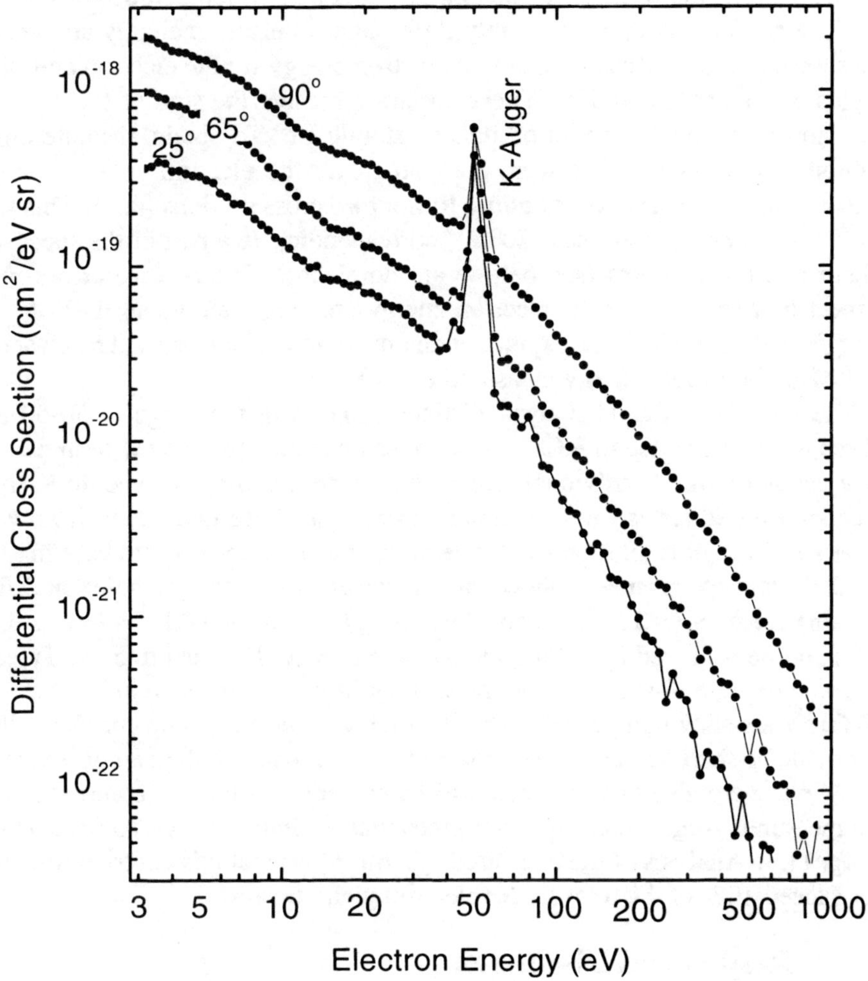

Figure 11.2. Energy distributions of electrons emitted in 95 MeV/u Ar^{18+} + Li collisions at observation angles of 25°, 65°, and 90°. The continuum part of each spectrum originates from single ionization of the Li target. The peak between ~ 50-60 eV is due to KLL Auger electron emission following excitation of a 1s electron to a higher bound state.

Continuum electrons emitted from the Li target were analyzed with the parallel-plate spectrometer in the electron energy range from about 3 to 1000 eV. The spectrometer had a solid angle of about 10^{-3} sr and a relative energy resolution ΔE/E of about 5.5%, where E is the passage energy of the electrons through the spectrometer. Since ΔE/E is constant, the absolute energy resolution ΔE is dependent on the electron energy. Ejected electrons could be observed in the energy range from 25^0 to 155^0, which could be varied in small steps. Typical spectra, doubly-differential in electron energy and angle, are shown in fig. 11.2 as a function of electron energy for the electron emission angles of 25^0, 65^0, and 90^0. These spectra represent the sum of the electron emission from the 1s and 2s orbitals in atomic Li. To obtain absolute cross sections the measured data were integrated over the electron emission angle and normalized to the corresponding Rutherford cross sections [4, 7]. This was done at one energy point near 200 eV corresponding to a part of the spectrum where two-body effects (see below) are dominant. The relative errors with respect to a variation of the electron energy and angle are about ± 25%. At energies below ~ 10 eV the experimental uncertainties increase and reach about ± 40% at the lowest energy measured (~3 eV).

The spectra of fig. 11.2 show distinct maxima in the range ~ 50-60 eV. These maxima are due to KLL Auger emission from core-excited lithium, and they contain several individual Auger lines corresponding to specific K-shell vacancy excited states. In a separate experiment, these individual lines were measured in high resolution by decelerating the electrons prior to entering the parallel-plate spectrometer. Since the relative electron energy resolution ΔE/E remains constant (~5.5%), an absolute energy resolution (ΔE) as low as 0.25 eV could be achieved by setting the passage energy (E) equal to 5 eV. Typical results for single-K-shell excitation resulting in Auger emission in the range ~ 50-60 eV are shown in fig. 11.3. Similar high-resolution spectra corresponding to double-K-shell vacancy production in Li, occurring in the energy range ~ 70-90 eV, were also obtained and will be presented below. A contribution to the measured Auger intensity from molecular lithium is observed near 52 eV in fig. 11.3. Analysis of the measured spectra shows that this contribution does not exceed 10% of the total Auger electron yield, however.

3. Results and Discussion

Spectra such as those shown in figs. 11.2 and 11.3 can be used to investigate important aspects of the collisional dynamics that relate to the structure of multielectron atomic systems. Furthermore, there can be an interplay between the dynamics of a collision and the structure of a system that depends sensitively on electron correlation effects. For high projectile velocities, such as those considered here, clear analogies with photon-induced processes are expected. In the following, we will examine these analogies in the collisional dynamics

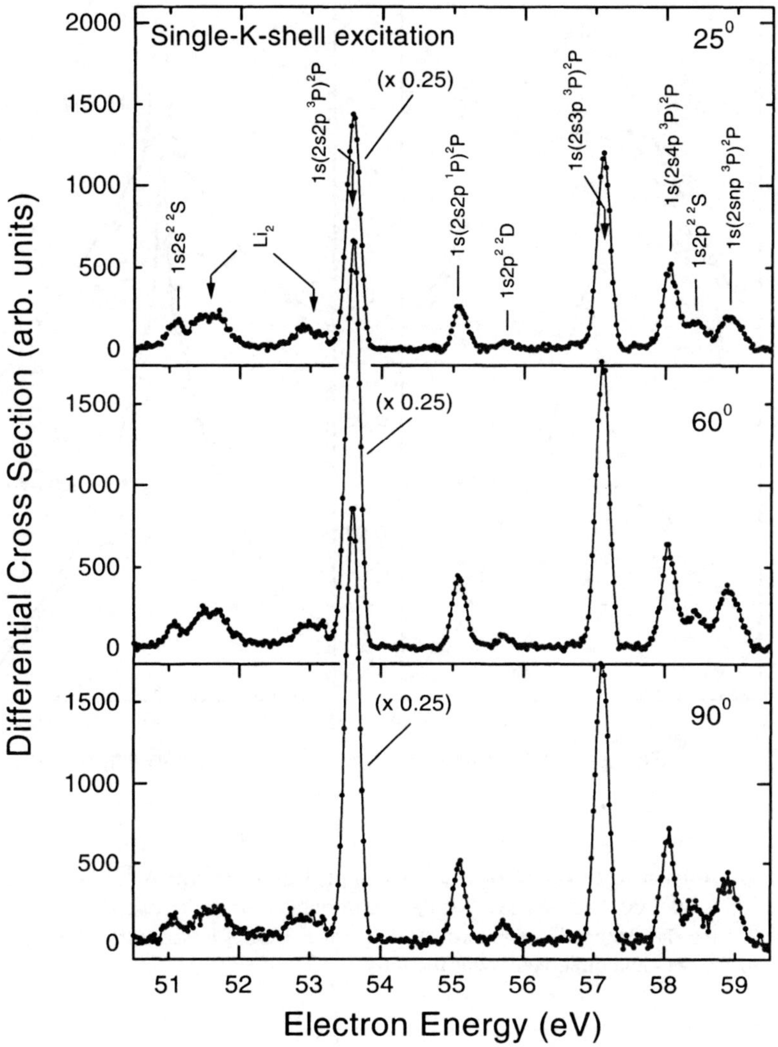

Figure 11.3. High-resolution single-K-shell-excitation Li Auger spectra for electron emission angles of 25^0, 60^0, and 90^0 induced by 95 MeV/u Ar^{18+} projectiles. The specific excited-state configurations are indicated. The strongly dominating $1s(2s2p\ ^3P)\ ^2P$ peak has been divided by a factor of four for display purposes. Note that the singly-K-shell excited $1s2p^2\ ^2D$ state is a doubly-excited state configuration ($1s \rightarrow 2p + 2s \rightarrow 2p$).

of single ionization and in the double-K-shell vacancy production of atomic lithium.

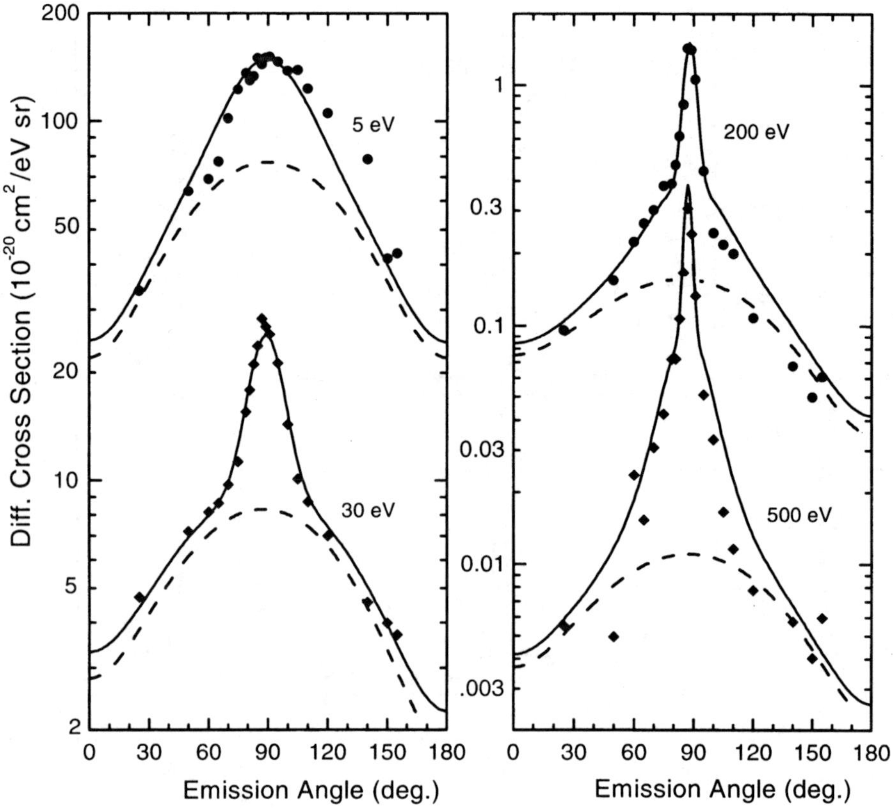

Figure 11.4. Angular distributions of electrons emitted in 95 MeV/u Ar^{18+} + Li collisions at energies of 5, 30, 200, and 500 eV. The dashed curve is a fit to the underlying three-body contribution using the extended dipole term $A + B \sin^2 \theta + C\cos \theta$ (see text). The solid curve is the sum of the two- and three-body contributions.

3.1 Single Ionization

The spectra of fig. 11.2 show emitted electron energy distributions measured for three observation angles. From these data and similar data for several other angles, the angular distributions for electron emission can be determined for specific ejected electron energies. The resulting angular distributions for emitted electron energies of 5, 30, 200, and 500 eV are shown in fig. 11.4. These spectra are characterized by a distinct peak near 90^0 that becomes narrower and relatively more intense as the electron energy increases, and an underlying part with a broad maximum centered at 90^0 that does not change its shape as the electron energy changes. It will be shown below that these narrow and broad contributions to electron emission are due to processes that correspond to Compton scattering and photoionization, respectively. Furthermore, there appears to be a 'kink' in the measured 30 eV and 200 eV spectra at angles just above and below the point where the sharp peak turns into the broad maximum. It will be shown that this kink is due to the node in the 2s wavefunction of atomic lithium. Thus, these spectra reveal important dynamical aspects of the collision process, as well as the effect of the atomic structure of Li on the collision dynamics.

3.1.1 General Considerations. To understand the spectra of fig. 11.4, we recall that photons can interact with electrons by means of the photoelectric effect or by Compton scattering. The photoelectric effect, which causes a dipole transition, is necessarily a *three-body* process involving the incident photon, the ejected electron, and the residual target atom because the photon is annihilated. On the other hand, Compton scattering is a *two-body* process involving a binary encounter between the incident photon and the ejected electron giving rise to a photon of lower energy and momentum in the final state.

Due to the uncertainty principle, the angular momentum transferred in a collision affects the angular distribution of the ejected electrons. Specifically, the angular momentum l and the emission angle θ are canonical quantities, subject to the uncertainty principle $\Delta l\, \Delta\theta = 1$ (analogous to linear momentum Δp and position Δx). Thus, low-order multipole transitions should produce a broad angular distribution, while high-order multipoles should produce a narrow angular distribution. Since dipole transitions involve the transfer of one unit of angular momentum ($\Delta l = 1$), these transitions should give rise to a broad angular distribution of ejected electrons. On the other hand, Compton scattering, because there is a large momentum transfer, involves high angular momentum transfer ($\Delta l >> 1$), and, consequently, should produce a narrow angular distribution of electrons.

As pointed out by Bethe [1], ionization by fast ions should be analogous to ionization by photons, possessing both a dipole-like part and a Compton-like part, corresponding to three- and two-body processes, respectively. Thus, for fast ions three-body effects in ionization are expected to give rise to a broad angular distribution of ejected electrons, while two-body effects should produce a sharp peak. These characteristic differences in the angular distributions of the binary- and soft-collision electrons provide an experimental method to separate the two- and three-body effects, respectively. Specifically, the distinct peak of narrow angular width observed near 90^0 in the spectra of fig. 11.4 corresponds to the transfer of large angular momentum ($\Delta l >> 1$), while the underlying broad angular distribution is due to dipole transitions ($\Delta l = 1$), in accordance with the uncertainty principle mentioned above. It is noted that an important criterion to observe a sharp two-body (binary-encounter) peak is the validity of the impulse approximation [13, 14], which, in turn, requires projectiles with high velocity.

For electron emission from an atom the principal quantity that affects two- and three-body aspects in the ionization process is the binding energy of the active electron. When the initial binding is large, the interaction of the ejected electron with the residual ion is also expected to be large during the collision, and vice versa.

In previous studies, He has often been used as a target. Since the binding energy of the outermost electron in He is the highest for any atom, three-body effects are expected to dominate electron emission from this atom. For example, the emission of soft-collision electrons in relativistic collisions of U^{92+} with He was uniquely attributed to three-body, dipole-type transitions [11]. However, this supposition may not be valid for the more general case of a target atom with less tightly bound electrons, in which case the two-body process is expected to gain importance. Furthermore, structures in the wave function, more complex than those of a 1s electron, may become important.

Here, we consider electron emission from Li, which has two orbitals, 1s and 2s, with largely different binding energies. The densities and wave functions [27] of these orbitals are shown in fig. 11.5. The 1s electron has a binding energy of about 59 eV and is localized close (< 1 a.u.) to the nucleus. The 2s electron has a much smaller binding energy of about 5.5 eV. Because of the node in the wave function, the 2s orbital has two parts, an inner part close to the nucleus (< 1 a.u.) and an outer part extending quite far from the nucleus (up to about 7 a.u.) as shown in the figure. Thus, atomic Li provides a unique system in which to test for two- and three-body effects.

With these general considerations in mind, we can now return to the spectra of fig. 11.4 and interpret the various characteristics of these data. It will be shown that specific features corresponding to two- and three-body effects of the ionization process can be identified, thereby demonstrating clear analogies with photoionization.

3.1.2 Analysis and Discussion. The doubly-differential, in energy and angle, cross sections for electrons emitted in ion-atom or photon-atom collisions can be written as a coherent sum of spherical harmonics $Y_l^m(\theta)$ as follows:

$$\frac{d\sigma}{d\varepsilon\, d\Omega} = \left| \sum_{lm} a_{lm}\, Y_l^m(\theta) \right|^2 \qquad (11.1)$$

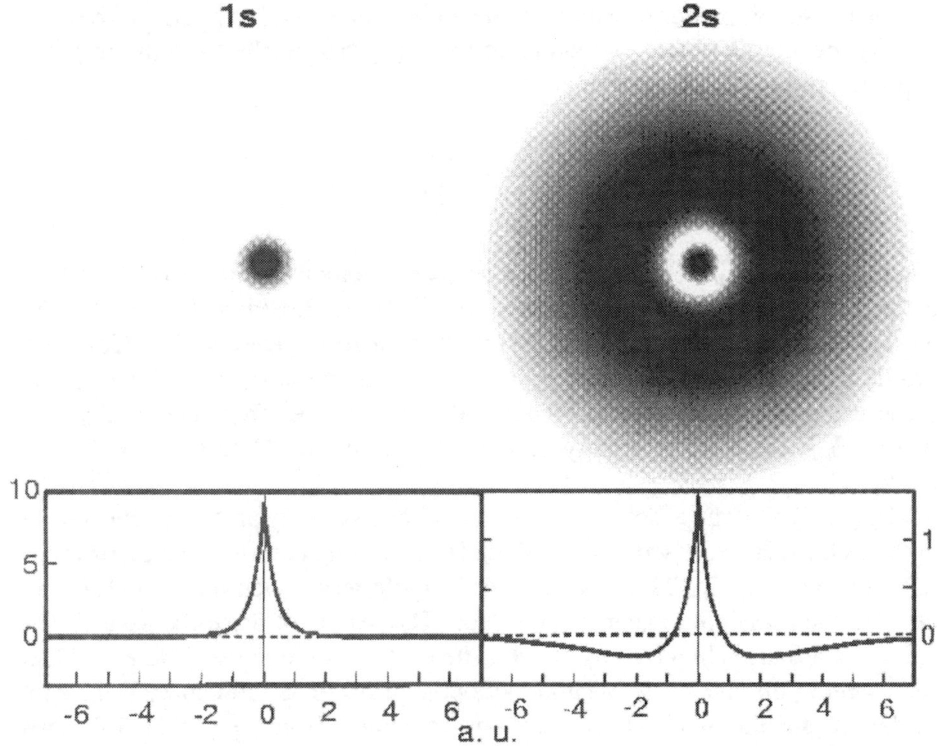

Figure 11.5. Electron densities (upper part) and wave functions (lower part) for the 1s and 2s atomic orbitals of Li. These were evaluated using the Cowan code [27].

where the a_{lm} are amplitudes for electron ejection with angular momentum l and magnetic quantum number m, and ε and θ are the energy and emission angle (relative to the beam direction) of the emitted electron. To treat the three-body part of the interaction, we use the dipole term for $l = 1$, which reduces the angular distribution of Eq. (11.1) to $A + B\sin^2\theta$, where A and B are the amplitudes a_{10} and $a_{1\pm1}$, respectively. To allow for asymmetries in this three-body part, we also keep the monopole term $l = 0$, which gives the cross term $C\cos\theta$ with $C = c_{00}$ due to an interference effect with the dipole term. However, for the high projectile velocities considered here, the monopole term is expected to be small. Then, the three-body part of the cross section is given by:

$$\frac{d\sigma_{three-body}}{d\varepsilon d\Omega} = A + B\sin^2\theta + C\cos\theta, \tag{11.2}$$

which will be referred to as the extended dipole term.

The two-body part of the interaction can be represented to good approximation by the simple binary-encounter formula [7] that results from the impulse approximation [13, 14]:

$$\frac{d\sigma_{two-body}}{d\varepsilon d\Omega} = \frac{4Z^2}{v^2 k_c^3} J(p_z), \tag{11.3}$$

where Z is the projectile charge, v the projectile velocity, $p_z = k \cos\theta - \Delta E/v$ is the initial momentum component along the beam direction, $k = (2\varepsilon)^{1/2}$ the ejected electron momentum, and ΔE is the transferred energy. The Compton profile $J(p_z)$ has a maximum at $p_z = 0$. The momentum $k_c = (2\Delta E)^{1/2}$ is the mean value of the momentum transferred and is derived from $\Delta E = E_b + \varepsilon$, where E_b is the binding energy of the active electron. Note that $\varepsilon \gg E_b \Rightarrow k_c \approx k$.

Eq. (11.3) gives an absolute calculation of the two-body part of the ionization cross section. By subtracting this calculated two-body part from the experimental data shown in fig. 11.4, the extended dipole term given by eq. (11.2) can be fit to the remaining experimental data. The results of these theoretical calculations are also shown in fig. 11.4. The dashed curve is the fit to eq. (11.2) representing the sum of dipole and monopole terms. The solid curve is the sum of this fit plus the two-body contribution calculated from eq. (11.3). It is seen that the data are extremely well represented by the sum of these components representing the two- and three-body contributions.

A more detailed analysis of this procedure applied to the data for electrons emitted at 30 eV is shown in fig. 11.6. Here, the two-body contribution from the 1s orbital (labeled 2[1s]), evaluated by means of eq. (11.3), is shown separately from the total two-body contribution (labeled 2) for the 1s and 2s orbitals. The fit to the extended dipole term given by eq. (11.2) is shown by the dashed curve (labeled 3). Again, the sum of all the theoretical components is given by the solid curve.

The results of the analysis in fig. 11.4 and 11.6 reveal that the main part of the distinct peak near 90^0 is due to two-body binary-encounter processes. The shape of this binary-encounter peak is determined by the Compton profile term for the target electrons that appears in eq. (11.3). An important consequence of using this equation to separate the contribution from the 1s orbital is that the dominant part of the binary-encounter peak can be attributed to the 2s orbital. The 1s electron, since it involves higher electron momenta, has a broader Compton profile as can be seen from the 2[1s] curve. Also, the inner part of the 2s wave function, since it is closer to the nucleus (see fig. 11.5), produces a broad component in the Compton profile as seen from the "wings" that occur for angles $\leq 60^0$ and $\geq 120^0$ in the curve labeled 2 in fig. 11.6. In fact, the calculations show that, apart from a constant factor, the Compton

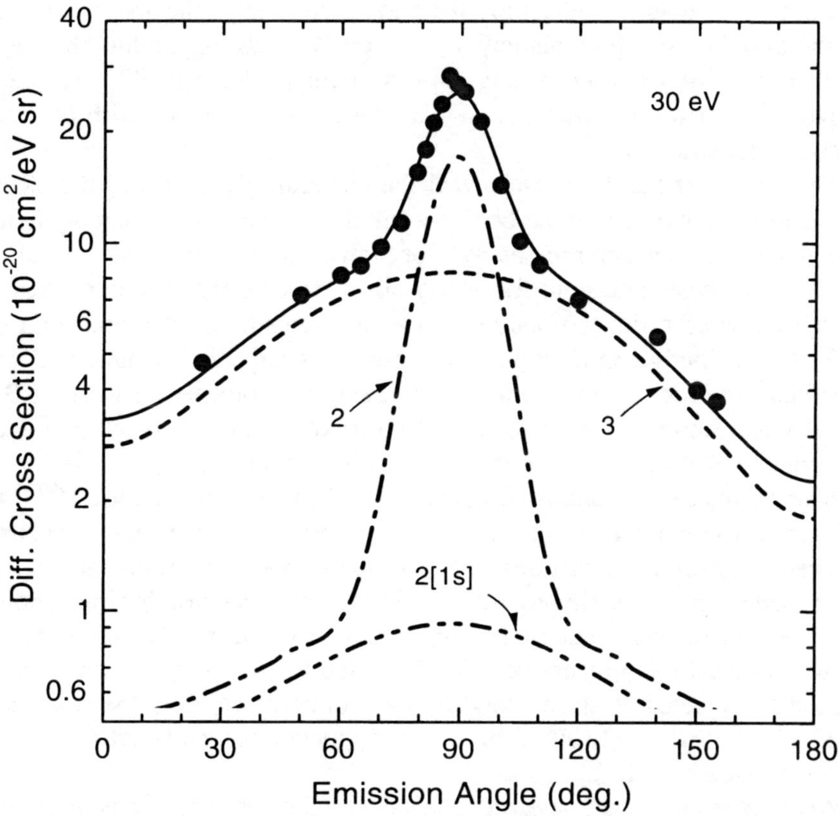

Figure 11.6. Angular distribution of 30 eV electrons emitted in collisions of 95 MeV/u Ar^{18+} + Li. The dot-dashed curve labeled 2 refers to calculations using the two-body theory given by eq. (11.3). The dot-dot-dashed curve labeled 2[1s] is the calculated two-body contribution from the 1s orbital only. The dashed curve labeled 3 is a fit to the underlying three-body part using the extended dipole term given by eq. (11.2). The solid curve is the sum of all contributions.

profiles for the 1s wave function and the inner part of the 2s wave function are essentially the same. Thus, the 1s wave function as well as the node in the 2s orbital give rise to a 'kink' in the Compton profile that is clearly seen in the data of fig. 11.6.

The spectra of fig. 11.4 show that the binary-encounter peak becomes narrower for higher electron energies. This increasing sharpness of the Compton profile may be readily understood from the derivative of the above expression for the initial momentum, namely, $p_z = k \cos\theta - \Delta E/v$, yielding the angular width $\Delta\theta = \Delta p_z/k$ when setting $\sin\theta \approx 1$ for angles near 90°. Hence, for a given Δp_z it follows that the angular width $\Delta\theta$ decreases with increasing electron momentum k.

The results of fig. 11.4 also reveal the changing dynamics of the electron emission as the two- and three-body contributions change considerably in magnitude with varying electron energy. For high electron energies ($\geq$ 500 eV) the two-body processes account for nearly all of the electron emission. At lower energies, however, the two-body part does not vanish. For electrons emitted at 5 eV the two-body contribution is still nearly as large as the three-body part. This latter finding is remarkable because, since the pioneering work of Bethe [1], it has become common practice to attribute the emission of low-energy electrons by fast projectiles to three-body dipole transitions [3, 6]. In fact, the present results can be contrasted with those of Ref. [11] for 1 GeV/u U^{92+} + He collisions where nearly all of the low-energy electron emission was attributed to three-body dipole transitions. This comparison illustrates convincingly the point made earlier that He presents a special nontypical case for ionization because of the relatively large binding energy of its electrons. The present results show, that for Li significant contributions due to two-body processes remain even at low ejected electron energies, and, furthermore, from the calculations of eq. (11.3), that nearly all of the two-body ionization can be attributed to the 2s orbital (see fig. 11.6).

Another result of the spectral analysis is that the three-body part of the ionization process can be separated from the dominating two-body part, even at high electron emission energies, as seen in fig. 11.4. To study further the properties of the three-body interaction, theoretical results have been evaluated [23] within the framework of the Born approximation using a modified version of the program by Gulyás *et al.* [28]. The measured double-differential cross sections represented by eq. (Taneq2) were integrated over the electron emission angle to give the single-differential cross sections for electron emission vs. emitted electron energy. In fig. 11.7 the experimental data are compared with theoretical results for dipole plus monopole, and monopole only, transitions. The curve representing the monopole term is seen to be relatively small. In fact, the sum of the dipole and monopole terms is about an order of magnitude larger than the monopole term alone. Only at the highest energies near 1000 eV does

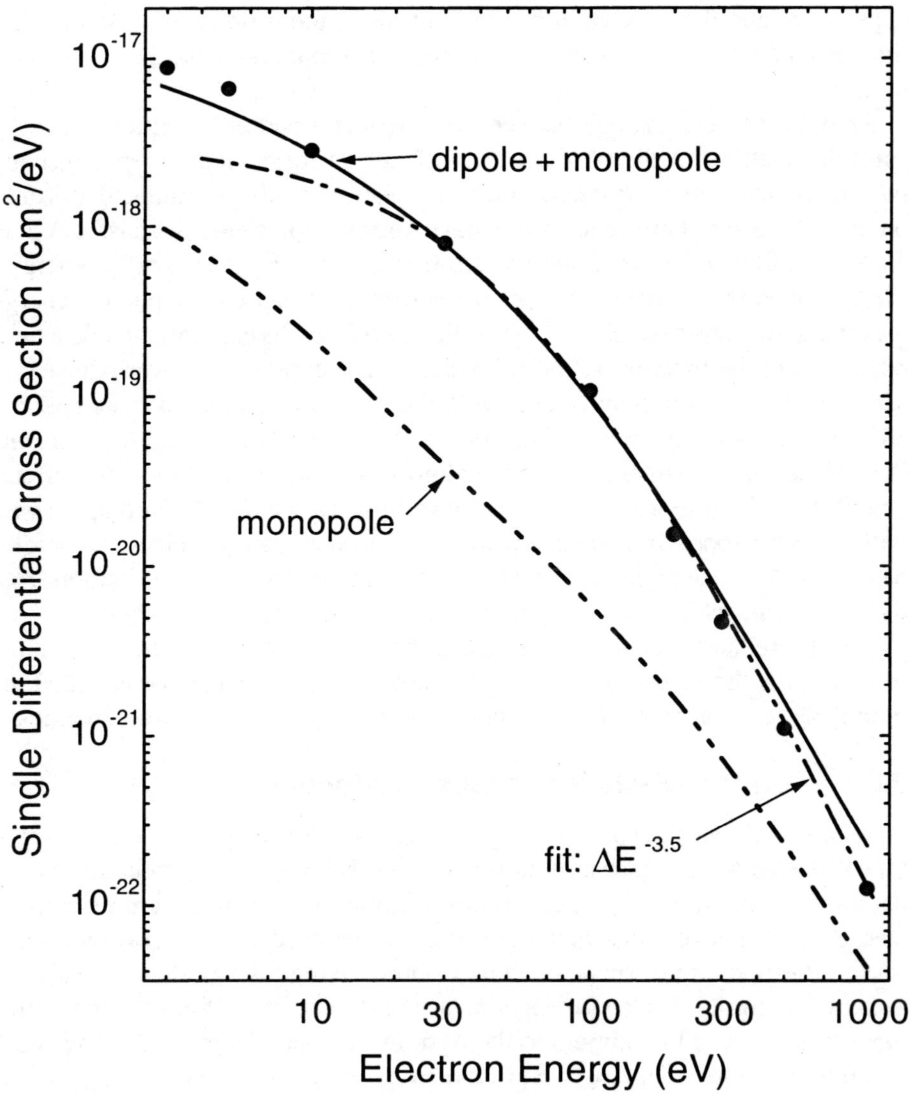

Figure 11.7. Angle integrated cross sections for electron emission due to three-body processes in 95 MeV/u Ar^{18+} + Li collisions. The data points originate from the fits of eq. (11.2) to the experimental results shown in fig. 11.4. The curves labeled "dipole+monopole" and "monopole" were evaluated using the Born approximation [28], and represent absolute calculations of these contributions. The "dipole+monopole" curve refers to dipole (l = 1) plus monopole (l = 0) transitions, while the "monopole" curve refers to monopole (l = 0) transitions only. The curve labeled "fit" represents the function $x\Delta E^{-3.5}$, where ΔE is the electronic energy transfer and x is a fitting parameter used to achieve the best overall agreement with the data (see text).

the monopole term become important. The good agreement obtained between experiment and theory seen in fig. 11.7 attests to the validity of attributing the fitted part of the experimental data to dipole transitions and, hence, to three-body effects.

Returning to the analogy between fast ions and photons, we note from the Einstein relation for the photoelectric effect that these electron spectroscopy measurements allow for determining the energy of the annihilated (virtual) photon, since this photon energy is equal to the electronic energy transfer $\Delta E = E_b + \varepsilon$ (recall that ε is the emitted electron energy and E_b is the binding energy). The probability for photoabsorption decreases strongly with photon energy following the power law $\Delta E^{-3.5}$ [8, 29]. Therefore, the experimental data have been fit with the function x$\Delta E^{-3.5}$, where x is a constant that was adjusted to achieve the best overall agreement with the experimental data between 30 and 300 eV. The binding energy of the 1s orbital (~59 eV) was used to determine ΔE, since the 1s orbital is governed almost exclusively by three-body effects (see Ref. [23] for more details). The fit results given in fig. 11.7 compare well with the Born approximation calculations. The discrepancy at electron energies below 30 eV is due to the fact that the fit does not include the contribution from the 2s orbital, which gains importance at lower energies. At energies $\geq$ 300 eV the discrepancy between the data and the fit originates from the fact that the fit represents dipole transitions only, whereas the Born results indicate that the monopole term increases in importance for higher emitted electron energies.

3.2 Double-K-shell Vacancy Production

As noted above, the spectra of fig. 11.2 show a distinct peak in the range 50-60 eV due to KLL Auger emission following single-K-shell excitation of the neutral lithium atom. By using high-resolution electron detection techniques (see Sec. II above), this maximum can be resolved into several individual lines corresponding to specific excited states having a single-K-shell vacancy as shown in fig. 11.3. Careful examination of the spectra of fig. 11.2 shows that there is also a small maximum to the right of the main Auger peak. This latter maximum results from Auger emission from excited states possessing two-K-shell vacancies. By using similar high-resolution techniques, this maximum can also be resolved into several constituent lines in the range ~ 70-90 eV as shown in fig. 11.8, where the specific states corresponding to the observed Auger emission lines are indicated. The energy resolution for these latter spectra was 0.55 eV compared to 0.25 eV for the spectra of fig. 11.3. It should be noted that the spectra of figs. 11.3 and 11.8 were measured for several additional angles and were verified to be symmetric with respect to 90^0 as expected. By using energies available in the literature [30, 31] for the main single- and double-K-shell vacancy lines in Li, the high-resolution spectra of figs. 11.3 and

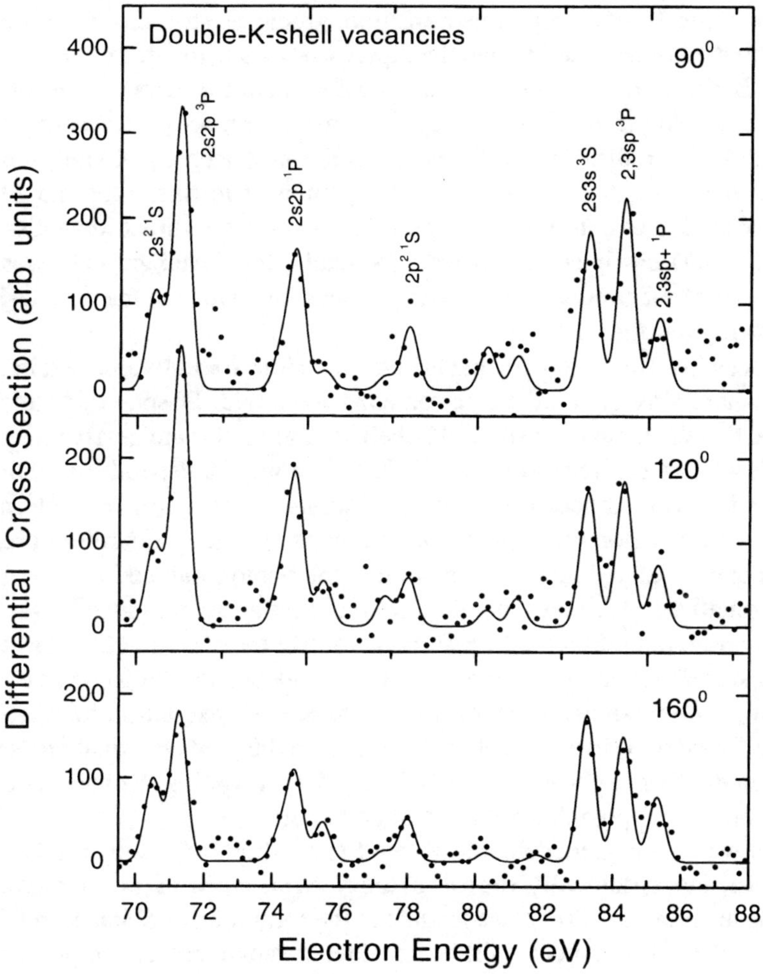

Figure 11.8. High-resolution double-K-shell vacancy Li Auger spectra for electron emission angles of 90^0, 120^0, and 160^0 induced by 95 MeV/u Ar^{18+} projectiles. The observed Auger lines correspond mainly to double-K-shell-vacancy configurations in excited Li^+ resulting from K-shell ionization plus K-shell excitation events. The large intensities observed for $2s^2\ ^1S$ and 2s3s ^3S are particularly significant, as is the fact that the ^3S intensity is greater than the ^1S intensity (see text).

11.8 were fit to give relative intensities for each of the observed lines. Then, by normalizing the peak intensities to the continuous electron background, absolute cross sections could be determined for each of the observed Auger lines in figs. 11.3 and 11.8 (see Ref. [25]).

The single K-shell excitation spectra of fig. 11.3 are seen to be dominated by the strong 1s(2s2p ^3P) ^2P line formed by the 1s→2p dipole transition. The corresponding 1s→2s monopole transition, giving rise to $1s2s^2\ ^2S$, is observed to be about 50 times smaller than the dipole transition. In fig. 11.8, the observed double-K-shell vacancy Auger lines, which amount to about 2-3% of the intensity of the single-K-shell vacancy lines, result primarily from configurations in excited Li^+, indicating that these lines are produced by K-shell ionization plus K-shell excitation. For the spectra displayed in figs. 11.3 and 11.8, the background due to continuous electron emission (from direct ionization of the lithium target) has been subtracted. Typically, this "background" was of the same order of magnitude as the observed peak intensities in the double-K-shell vacancy region (fig. 11.8).

It is seen from these figures that the cross sections for individual excited states depend quite strongly on the electron emission angle. Despite this variation, it is found that the double- to single-K-shell vacancy ratios are nearly independent of angle with an average ratio of 2.3% [25]. The angular dependence of the cross sections is expected because dipole excitations (leading to $L = 1$ intermediate states) dominate for the high velocity considered here. Furthermore, these predominantly dipole interactions point to the photon-induced nature of the K-shell excitations [1, 6]. An angular analysis similar to that given by Eq. (11.1) can be carried out [25] with the result that each of the P states ($L = 1$) is found to vary as $\sin^2\theta$, while the S states ($L = 0$) are independent of angle as expected. Moreover, the total cross section for single-K-shell excitation integrated over angle is found to be in excellent agreement with the predicted total cross section based on the Born approximation [25, 32], demonstrating the validity of using this theoretical approach for these fast collisions.

Here we will consider double-K-shell (i.e., "hollow" atom) production in atomic Li, and, specifically, the role of the *electron-electron* ($e-e$) interaction in forming these states. Processes leading to the formation of S states and P states with two K-shell vacancies will be considered separately, as the production of these states mainly involves different manifestations of the $e-e$ interaction. Also, the formation of these states will be related to the corresponding photon-induced processes, and compared with results obtained by other investigators for incident photons.

3.2.1 General Considerations. In the case of double-K-shell-vacancy production, we are primarily interested in the contribution of the $e-e$ interaction to the formation of these states. This interaction has two aspects corresponding to whether the first electron is emitted slowly or suddenly. For slow emission, subsequent excitation or ionization of a second electron involves the mutual scattering of two electrons, i.e., it is *dielectronic* in nature [18]. On the other hand, sudden emission can result in a subsequent electron transition due to the change in the potential seen by the second active electron as the excited system relaxes [33]. This latter type of transition is a "mean-field" effect referred to as a *shake* process.

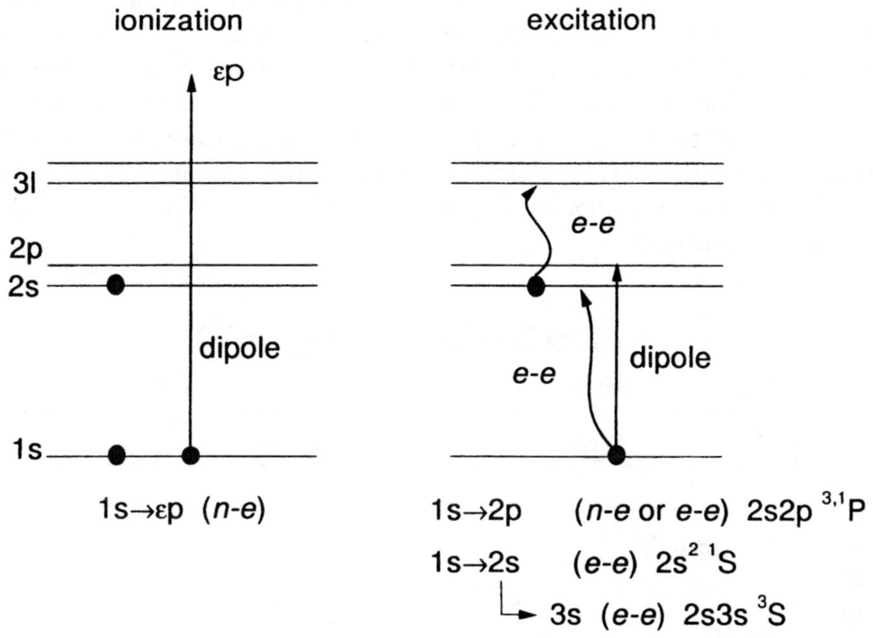

Figure 11.9. Schematic showing double-K-shell-vacancy production in Li via K-shell ionization followed by K-shell excitation. The most prominent transitions are shown, as well as the dominant interaction involved ($n-e$ or $e-e$) in producing the indicated doubly vacant K-shell intermediate excited states.

Double-K-shell vacancy production from ground-state Li^0 ($1s^2 2s$), resulting from K-shell ionization plus K-shell excitation, is shown schematically in fig. 11.9. As noted above, the double-K-shell vacancy spectra of fig. 11.8 are dominated by two-electron states in Li^+ corresponding to this ionization plus excitation process. The fact that the spectra of fig. 11.8 consist mainly of two-electron doubly vacant K-shell states is significant because it means that double-K-shell excitation is negligible. Thus, fig. 11.9 shows ionization via a $1s\rightarrow \varepsilon p$ dipole transition accompanied by excitation via a $1s\rightarrow 2l$ transition (for l=1, the excitation is dipole, while for l=0 it is monopole), thereby producing two K-shell vacancies in Li^+.

The figure shows the excitation possibilities that are expected to dominate. The most probable excitation, mediated by the $n-e$ or $e-e$ interaction, is the dipole transition $1s\rightarrow 2p$, giving rise to the $2s2p\,^{3,1}P$ states. Since the probability for monopole transitions ($1s\rightarrow 2s$, 3s) via an $n-e$ interaction is negligible, the

$2s^2\ ^1S$ and $2s3s\ ^3S$ states can only result from the $e-e$ interaction. In this latter regard, transitions mediated by shake processes (following sudden electron emission), can *only* give rise to Δl=0 (and ΔL=0) transitions because these processes cause internal rearrangement of the residual ion, with the consequence that the total orbital angular momentum of the system cannot change. On the other hand, dielectronic transitions (following slow electron emission) can give rise to Δl=0 or 1 (and ΔL=0 or 1) transitions. Moreover, we note that ionization must precede excitation for the formation of doubly vacant K-shell S states, since initial 1s→2p excitation would give rise to the $1s(2s2p\ ^{3,1}P)$ states (see fig. 11.3), from which the $2s^2\ ^1S$ or $2s3s\ ^3S$ states cannot be formed via subsequent ionization.

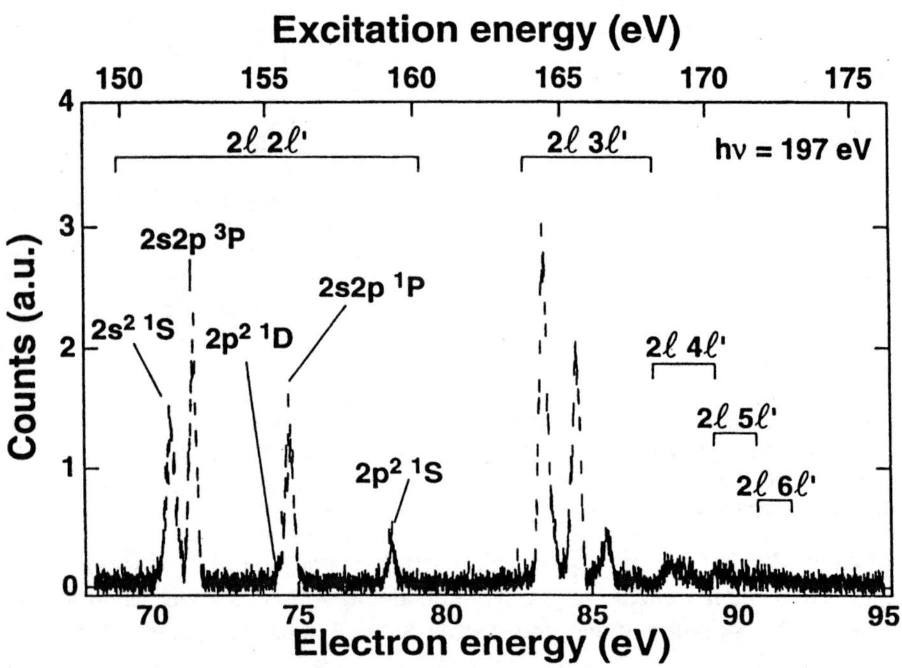

Figure 11.10. High-resolution double-K-shell vacancy spectrum resulting from incident 197 eV photons, as reported by Diehl *et al.* in Ref. [34]. The similarity of this spectrum to those shown in Fig. 8 should be noted.

A spectrum similar to those of fig. 11.8, shown here as fig. 11.10, was obtained by Diehl *et al.* [34] for the excitation of Li^0 $(1s^22s)$ by 197 eV photons, an energy value which lies slightly above the Li^0 double-K-shell vacancy threshold. It is seen that there are striking similarities between the spectra of figs. 11.8

and 11.10, indicating similar mechanisms leading to the formation of double-K-shell vacancy states by incident ions and photons. Since only $e - e$ interactions, following initial photoionization, can lead to the spectrum observed for incident photons (fig. 11.10), it is concluded that these same interactions must dominate for incident fast ions (fig. 11.8) as well. While the S states can be formed by either shake or dielectronic processes, the P states can only come about from dielectronic processes.

3.2.2 Analysis and Discussion. Noteworthy features of the double-K-shell vacancy spectra presented in fig. 11.8 (and fig. 11.10) are the large intensities of the $2s^2\ ^1S$ and $2s3s\ ^3S$ states, as well as the fact that the ^3S intensity is larger than the ^1S intensity. Furthermore, the total intensity involving 2l3l' configurations is seen to be as large, or even larger, than the total intensity due to 2l2l' configurations. These observations would not be expected based on independent $n - e$ interactions, since the probabilities for exciting $1s \rightarrow nl$ transitions, as calculated in the Born approximation, are expected to decrease strongly as n increases for the high velocity considered here [32].

To explain the relatively large intensities observed for the $2s^2\ ^1S$ and $2s3s\ ^3S$ states, we calculate the probabilities for the formation of these states via shake from initially ionized $1s2s\ ^{1,3}S$ states. We recognize that, in addition to shake, dielectronic interactions due to slow electron emission could contribute to the formation of these S states, but calculations of this contribution are beyond the scope of the present work.

Shake probabilities can be calculated within the sudden approximation from the overlap of the two-electron initial- and final-state wave functions. In the sudden approximation, the initial-state wave functions are those of "frozen" neutral Li, while the final-state wave functions are those of the relaxed Li^+ ion. These wave functions can be obtained using the Grant atomic structure code [35]. Then, the shake probabilities for forming the $2s^2\ ^1S$ and $2s3s\ ^3$ S states are calculated from the transition matrix elements $|\langle 2s^2\ ^1S|1s_0 2s_0\ ^1S\rangle|^2$ and $3|\langle 2s3s\ ^3S|1s_0 2s_0\ ^3S\rangle|^2$, respectively. The subscript "0" refers to the "frozen" initial-state neutral Li wave functions, and the factor of 3 for the triplet state accounts for the statistical weighting of this latter state. The $2s^2\ ^1S$ state can only result from the intermediate ("frozen") $1s_0 2s_0\ ^1S$ state, where the $1s_0$ electron goes to 2s, and the $2s_0$ electron remains in the 2s orbital of the Li^+ ion. Thus, there is only one pathway for this transition, which we denote as *direct*. Consequently, we use this transition strength as a benchmark for the relative strengths of the other double-K-shell vacancy transitions.

On the other hand, for the $2s3s\ ^3S$ state two possibilities must be considered (see fig. 11.9). This configuration can come about by means of a *direct* $1s_0 \rightarrow 3s$ transition (not shown in the figure), or by an *exchange-shake* mechanism where the $1s_0$ electron goes to 2s and *simultaneously* the $2s_0$ electron goes to the

3s level, as indicated in the figure. Since the 2s3s ^{3}S state is a triplet, it can result only from the "frozen" intermediate $1s_0 2s_0$ ^{3}S state because this internal rearrangement cannot change the spin angular momentum of the system. For the same reason, this latter state cannot produce the 2s^2 ^{1}S state. Because there are two pathways, direct and exchange, to the 2s3s ^{3}S state, there can be interference between the amplitudes for these processes.

From the calculated shake probabilities [25], it is found that the exchange mechanism dominates strongly over the direct mechanism for the formation of the 2s3s ^{3}S state. Thus, it is concluded that formation of the 2s3s ^{3}S state is principally a *three-step* process involving an initial $n - e$ interaction, or a photon-electron interaction in the case of incident photons, causing 1s$\rightarrow$ εp ionization, followed by two $e-e$ interactions ($1s_0 \rightarrow 2s + 2s_0 \rightarrow 3s$ excitations). Moreover, the interference between the direct and exchange mechanisms gives rise to constructive interference [25] further enhancing the strength of this three-step transition. Since a triplet state for the 2s^2 configuration cannot exist due to Pauli blocking, this constructive interference can be interpreted as the $1s_0$ electron "pushing" the initial $2s_0$ electron to 3s, while itself going to the 2s orbital. In other words, the Pauli blocking that prevents formation of the 2s^2 ^{3}S state gives rise to an enhancement, via constructive interference, of the 2s3s ^{3}S state to conserve the total transition probability.

Table 11.1. Measured and calculated relative intensities for the various Li double-K-shell vacancy states observed in this work. Results for incident Ar^{18+} ions are compared with those of ref. [34] for incident photons. In each case, the intensities have been normalized to that observed for the 2s^2 ^{1}S state. The measured relative intensities were calculated from experimentally determined angle integrated cross sections. The calculated relative intensity for the 2s3s ^{3}S state is the shake probability obtained from overlap integrals using the Grant code [35] as discussed in the text.

	Relative Intensities		
	95 MeV/u Ar^{18+} Ions [25]		197 eV Photons [34]
Configuration	Measured	Calculated	Measured
2s^2 ^{1}S	1	1	1
2s2p ^{3}P	2.6	—	1.1
2s2p ^{1}P	1.3	—	0.80
2p^2 ^{1}S	0.42	—	0.18
2s3s ^{3}S	1.7	1.7	1.6
2,3sp ^{3}P	1.5	—	1.2

The calculated relative strength of the 2s3s ^{3}S state compared to the 2s^2 ^{1}S state via shake is 1.7, a value which is in excellent agreement with the measured strength of this line in both figs. 11.8 and 11.10. The values of these relative intensities, as well as those of the P states, are listed in Table 11.1. The photon energy of 197 eV used in the work of Ref. [34] is significant because it is only slightly above the double-K-shell vacancy threshold, and so only slow electrons are ejected. This is similar to the present case of fast ion-induced electron emission where most of the electrons are emitted with relatively low velocities (corresponding to energies < 10 eV). Indeed, from Table I it is seen that the relative intensities obtained from the photon-induced spectrum of fig. 11.10 are comparable, in most cases, to those obtained from the ion-induced spectra of fig. 11.8.

It is significant that the observed intensity of the 2s3s ^{3}S state relative to the 2s^2 ^{1}S state is well represented by the shake calculation alone (see Table I), without including any contribution from dielectronic transitions. While the reason for this result is not entirely clear, it is noted that photoionization studies have shown that shake calculations could account for observed Auger satellite lines due to ionization plus excitation [20, 36] occurring only a few eV above threshold. Thus, it appears that mean-field effects may also be significant for rather slow ejected electrons. On the other hand, a possible explanation for the seeming absence of dielectronic processes may be that these latter $e-e$ interactions nearly always lead to an angular momentum exchange in the excitation process, thereby leaving the S states to be formed mainly by shake processes. This point needs further investigation, however.

Next, we consider the 2s2p ^{3}P configuration. While this state can be formed by separate $n-e$ interactions (1s$\rightarrow$ εp plus 1s$\rightarrow$2p) in the case of ion bombardment, it can also have a contribution from the $e-e$ interaction. This latter process takes place when the ionized εp electron, as it leaves the atom, interacts with the remaining 1s electron, exciting it to 2p while simultaneously giving up its l=1 unit of angular momentum to become a continuum εs electron, i.e., $(1s \rightarrow \varepsilon p) \Rightarrow (\varepsilon p + 1s' \rightarrow \varepsilon s + 2p)$, thereby producing the 2s2p ^{3}P state via a dielectronic (slow electron) process. Such dielectronic interactions involving the exchange of angular momentum have been previously observed in low-velocity highly-charged ion-atom collisions [37]. We note that this 2s2p ^{3}P state cannot be formed via shake since an internal rearrangement cannot change the angular momentum of the intermediate 1s2s ^{3}S residual ion. Thus, identification of an $e-e$ contribution to 2s2p ^{3}P makes it possible to distinguish experimentally the dielectronic process from the shake processes. Until the work of Refs. [24, 25], the distinction between these competing $e-e$ mechanisms was not achieved experimentally.

To determine the $e-e$ contribution to 2s2p ^{3}P, we consider again the photon-induced electron emission spectrum of fig. 11.10. In this spectrum, the 2s2p ^{3}P state is produced by the dielectronic $e-e$ interaction following the initial photoionization event. On the other hand, for incident ions, the 2s2p ^{3}P state observed in fig. 11.8 can be formed through a combination of $n-e$ and $e-e$ interactions. Since the 2s^2 ^{1}S line is attributed entirely to the $e-e$ interaction (following an initial ionizing $n-e$ or photon-electron interaction), the 2s2p ^{3}P line induced by fast ions (fig. 11.8) must be at least as large as this same line in the photon-induced spectrum (fig. 11.10). This is indeed the case as seen from Table I, where the relative intensity of the 2s2p ^{3}P state for fast ions is seen to be significantly larger than for photons. This excess for ions is attributed to the $n-e$ interaction. A similar situation occurs for the 2s2p ^{1}P state, but the difference is not as large. These comparisons show that the $e-e$ interaction plays a significant role in the formation of the double-K-shell vacancy ^{3}P and ^{1}P states by incident fast ions, and, furthermore, provides spectral identification for the dielectronic (slow electron) process. Unfortunately, no theoretical calculations of the dielectronic process exist to compare with the observations.

Finally, we consider the 2p^2 ^{1}S state observed at ~78 eV in figs. 11.8 and 11.10. From Table I it is seen that the relative intensity of this line is significantly larger for fast ions than for photons. The 2p^2 ^{1}S state can be formed by a combination of ionization plus excitation processes involving $n-e$ or $e-e$ interactions for fast ions, or involving $e-e$ interactions for photons. However, a primary consideration for forming this state is the configuration mixing with the 2s^2 ^{1}S state, which has the same total spin and orbital angular momentum. A calculation [38] of this mixing based on the Fischer code [39] predicts the 2p^2 ^{1}S intensity to be 33% of the 2s^2 ^{1}S intensity, a value which is somewhat larger than that obtained from the photon-induced spectrum of fig. 11.10 (see Table I). On the other hand, in the fast ion work the observed 2p^2 intensity is about 40% of the 2s^2 intensity. Consequently, there appears to be a contribution from the $n-e$ interaction to the intensity of the 2p^2 ^{1}S line in the ion-induced spectrum of fig. 11.8.

4. Concluding Remarks

We have considered two aspects of fast collisions, specifically, (1) mechanisms of single-electron emission (ionization) and (2) processes leading to double-K-shell-vacancy production. The principal features characterizing the single ionization discussed here are exhibited by the spectra of fig. 11.4, which were obtained from the continuous part of the spectra shown in fig. 11.2. On the other hand, the main features relevant to double-K-shell vacancy production are shown by the spectra of fig. 11.8, which were obtained from high-resolution measurements of the K-shell Auger emission between ~ 50-90 eV seen in fig. 11.2. Each of these aspects has been examined for collisions of 95 MeV/u Ar^{18+} ions (v/c ≅ 0.4, Z/v ≅ 0.3) with atomic lithium. At this velocity, significant connections with photon-induced processes are expected and observed. For single ionization, it is found that the ionization can be separated into *two-* and *three-body* phenomena corresponding to Compton scattering (binary encounters) and photoionization (dipole transitions), respectively. For double-K-shell vacancy production, the *electron-electron* ($e-e$) interaction is found to play a significant role, manifesting itself in the form of either *shake* or *dielectronic* processes that correspond to the identical processes initiated by incident photons.

For fast projectiles colliding with atomic Li, not only can the two- and three-body effects in single ionization be separated, these effects are also found to play essentially different roles in the ionization of the Li 1s and 2s orbitals. Analysis shows that three-body effects dominate the removal of the 1s electron, which still contributes significantly to the ionization at intermediate- to high-ejected electron energies ($\geq$ 500 eV). On the other hand, removal of the 2s electron occurs predominantly via two-body effects that contribute significantly to the ionization for electron energies even as low as 5 eV. Thus, in contrast to ionization of atomic He, the two-body process remains important for soft collisions in atomic Li. In the analysis the use of high-energy projectiles is essential because (a) perturbation theory can be used, thus justifying the Born approximation, (b) the minimum momentum transfer in soft collisions is small so that ionization by fast ions resembles photoionization, where the incident photon is annihilated, and (c) in binary-encounter two-body collisions, the initial binding of the ejected electron to the target nucleus can be neglected.

From these findings the question arises as to how far the conclusions of Ref. [11] for He or of Refs. [22, 23] for Li can be generalized to other target atoms. With its small binding energy and large spatial extension the Li 2s orbital is rather unique. On the other hand, the He 1s orbital is also unique but its properties are opposite to lithium because helium has the largest outer shell binding energy of any atom and a rather small spatial extension. Hence, it might be revealing to study the role of two-body effects for other target atoms having electrons in the 2s orbital or higher shells with binding energies between those of Li and He. From the results presented here for the Li 2s orbital significant contributions from two-body effects for multielectron atoms with several shells might be expected. However, to fully answer this question, further detailed work is required.

In the case of double-K-shell-vacancy production of Li by fast Ar^{18+} projectiles, the two K-shell vacancies were found to originate mainly from ionization plus excitation of the Li target. Additionally, the production of 2l3l' doubly-K-shell excited configurations was found to be as large, or larger, than the 2l2l' configurations. High-resolution measurements of the double-K-shell vacancy Auger lines permitted the spectral identification of contributions from the $e-e$ interaction, and, furthermore, separate identification of the *shake* and *dielectronic* contributions to the $e-e$ interaction.

Double-K-shell vacancy S states were attributed almost entirely to the $e-e$ interaction and analyzed in terms of two-electron state *shake* calculations, i.e., the overlap integrals for the initial ("frozen") and final states. These calculations gave results that generally agreed with the observed intensity for the 2s3s ^{3}S state relative to the $2s^2\ ^1S$ state. A significant finding was that the $2s3s\ ^3S$ state comes about mainly by means of a three-electron transition involving an *exchange-shake* process that is a dynamical manifestation of the Pauli exclusion principle. While the dielectronic manifestation of the $e-e$ interaction might be expected to contribute to the formation of S states as well, this does not appear to be the case. The reason for this result is not clear and must await further investigation for an explanation. Analysis of the $2s2p\ ^{3,1}P$ intensities indicated that these states are formed largely by the dielectronic process following slow electron emission, with an additional contribution from $n-e$ interactions (shake cannot give rise to this state). Finally, the $2p^2\ ^1S$ line was found to be almost entirely due to configuration mixing with the $2s^2\ ^1S$ state, although $n-e$ interactions could contribute to the formation of this state as well.

Acknowledgments

The authors are indebted to the several collaborators with whom they have worked closely and who have contributed substantially to various aspects of the work presented here. J.A.T. expresses special appreciation to the Berlin group for its kind hospitality during several visits to Berlin, without which much of the work presented here would not have been possible. Considerable financial support for this work was provided by the Chemical Sciences, Geosciences, and Biosciences Division, Office of Basic Energy Sciences, Office of Science, U.S. Department of Energy.

References

[1] H.A. Bethe, Ann. Physics (Leipzig) **5**, 325 (1930)

[2] E.J. Williams, Phys. Rev. **45**, 729 (1934)

[3] M. Inokuti, Rev. Mod. Phys. **43**, 297 (1971); M. Inokuti, Y. Itikawa, and J.E. Turner, Rev. Mod. Phys. **50**, 23 (1978)

[4] M.E. Rudd, Y.-K. Kim, D.H. Madison, and T.J. Gay, Rev. Mod. Phys. **64**, 441 (1992)

[5] J.F. Reading is acknowledged for this analogy.

[6] M. Inokuti and Y.-K. Kim, Phys. Rev. **186**, 100 (1969); Y.-K. Kim and M. Inokuti, Phys. Rev. A **1**, 1132 (1970)

[7] N. Stolterfoht, R.D. DuBois, and R.D. Rivarola, *Electron Emission in Heavy Ion-Atom Collisions*, Springer-Verlag, Berlin, (1997)

[8] J. Burgdörfer, L.R. Andersson, J.H. McGuire, and T. Ishihara, Phys. Rev. A **50**, 349 (1994)

[9] J. Wang, J.H. McGuire, J. Burgdörfer, and Y. Qiu., Phys. Rev. Lett. **77**, 1723 (1996)

[10] J.H. McGuire, *Introduction to Dynamic Correlation*, Cambridge University Press, Cambridge, England, (1997)

[11] R. Moshammer et al., Phys. Rev. Lett. **79**, 3621 (1997)

[12] M. Gryzinski, Phys. Rev. **138**, A305 (1965); **138**, A332 (1965); **138** A336 (1965)

[13] P. Eisenberger and P.M. Platzman, Phys. Rev. A **2**, 415 (1970)

[14] D. Brandt, Phys. Rev. A **27**, 1314 (1983)

[15] See J.A. Tanis p. 241-257 in *Recombination of Atomic Ions*, eds. W.G. Graham, W. Fritsch, Y. Hahn, and J.A. Tanis, Plenum Press, New York, (1992),

[16] See J.H. McGuire, N. Berrah, R.J. Bartlett, J.A.R. Samson, J.A. Tanis, C.L. Cocke, and A.S. Schlachter, J. Phys. B **28**, 913 (1995)

[17] J.F. Reading and A.L. Ford, p. 693-698 in *Electronic and Atomic Collisions, Invited Papers*, ed. by H.B. Gilbody *et al*, North-Holland, Amsterdam, (1988)

[18] N. Stolterfoht, Nucl. Instrum. Meth. Phys. Res. B **53**, 477 (1991)

[19] B. Crasemann, Jour. de Physique, **48**, C9-389 (1987)

[20] B. Crasemann, Comm. At. Mol. Phys. **22**, 163 (1989)

[21] See, for example, R.H. Pratt, p. 59-80 in *X-Ray and Inner-Shell Processes, 18th International Conference*, ed. by R.W. Dunford *et al*, AIP, New York, (2000)

[22] N. Stolterfoht, J.-Y. Chesnel, M. Grether, B. Skogvall, F. Frémont, D. Lecler, D. Hennecart, X. Husson, J.P. Grandin, B. Sulik, L. Gulyás, and J.A. Tanis, Phys. Rev. Lett. **80**, 4649 (1998)

[23] N. Stolterfoht, J.-Y. Chesnel, M. Grether, J.A. Tanis, B. Skogvall, F. Frémont, D. Lecler. D. Hennecart, X. Husson, J.P. Grandin, Cs. Koncz, L. Gulyás, and B. Sulik, Phys. Rev. A**59**, 1262 (1999)

[24] J.A. Tanis, J.-Y. Chesnel, F. Frémont, D. Hennecart, X. Husson, A. Cassimi, J.P. Grandin, B. Skogvall, B. Sulik, J.-H. Bremer, and N. Stolterfoht, Phys. Rev. Lett. **83**, 1131 (1999)

[25] J.A. Tanis, J.-Y. Chesnel, F. Frémont, D. Hennecart, X. Husson, D. Lecler, A. Cassimi, J.P. Grandin, J. Rangama, B. Skogvall, B. Sulik, J.-H. Bremer, and N. Stolterfoht, Phys. Rev. A**62**, 032715 (2000)

[26] N. Stolterfoht, Z. Phys. **248**, 81 (1971)

[27] R.D. Cowan, *The Theory of Atomic Structure and Spectra*, University of California Press, Berkeley, (1981)

[28] L. Gulyás, P.D. Fainstein, and A. Salin, J. Phys. B **28**, 245 (1995)

[29] T. Ishihara, K. Hino, and J. McGuire, Phys. Rev. A **44**, R6980 (1991)

[30] P. Ziem, R. Bruch, and N. Stolterfoht, J. Phys. B **8**, L480 (1976)

[31] M. Rødbro, R. Bruch, and P. Bisgaard, J. Phys. B **12**, 2413 (1979)

[32] Calculations of excitation and ionization were performed using Born approximation codes provided by A. Salin (unpublished)

[33] J.H. McGuire, Phys. Rev. Lett. **49**, 1153 (1982); Phys. Rev. A **36**, 1114 (1987)

[34] S. Diehl et al., J. Phys. B: At. Mol Opt. Phys. **30** L595 (1997)

[35] K.G. Dyall, I.P. Grant, C.T. Johnson, F.A. Parpia, and E.P. Plummer, Comput. Phys. Commun. **55**, 425 (1989)

[36] J. Stöhr, R. Jaeger, and J.J. Rehr, Phys. Rev. Lett., **51**, 821 (1983)

[37] F.W. Meyer, D.C. Griffin, C.C. Havener, M.S. Huq, R.A. Phaneuf, J.K. Swenson, and N. Stolterfoht, Phys. Rev. Lett. **60**, 1821 (1988)

[38] T. Gorczyca, private communication.

[39] C.F. Fischer, Comput. Phys. Commun. **64**, 369 (1991)

Chapter 12

ION–ION COLLISIONS

R. Trassl

Institut für Kernphysik, Universität Giessen, 35392 Giessen, Germany

Roland.H.Trassl@strz.uni-giessen.de

Abstract Ion–ion collisions belong to the fundamental processes in all types of plasmas in astrophysical objects (e.g. the earth's ionoshpere or the solar corona) as well as in laboratory discharges (for example in thermonuclear fusion). The knowledge of cross sections for ion–ion collision processes is therefore essential for the modeling of these plasmas.

In the field of ion–ion collisions, the most promising method for the investigation of cross sections is the use of two crossed ion beams. The main experimental difficulty arises from the fact that ion beams are very dilute targets. However, as it is described in this chapter, it is possible to investigate charge–exchange and ionisation processes by using coincidence and beam–pulsing techniques, respectively. Examples for the measurement of total cross sections for few–electron systems will be given.

Furthermore, the investigation of angular–differential cross sections allows a more detailed insight into the collision process and provides a test of existing theories on a fundamental level. In order to extract an angular–differential cross section from a measured scattering distribution, a transformation between the laboratory system and the center–of–mass system as well as a deconvolution with respect to the primary beam profile has to be performed. The results for such a measurement will be shown for an exemplary reaction.

Keywords: ion–ion collisions, crossed beams, charge–exchange, ionisation, angular–differential cross section

1. Introduction

Atomic collision processes play an important role in any kind of plasma. Dense gases or plasmas are in thermal equilibrium and can therefore be described by the laws of thermodynamics without a detailed knowledge about the elementary atomic processes. However, most astrophysical as well as laboratory plasmas (e.g. the earth's ionosphere or thermonuclear plasmas) are

F.J. Currell (ed.), The Physics of Multiply and Highly Charged Ions, Vol. 2, 369-395.

usually far from thermodynamic equilibrium [1]. The modeling of these plasmas requires precise data about the fundamental atomic collision processes, i.e. electron–ion, ion–atom and ion–ion collisions.

The systematic investigation of the electron–impact ionisation of atoms and ions already started at the beginning of the century with the experiments of Lenard [2, 3] and Bloch [4] looking at the ionisation efficiency of cathode beams in gases and the classical description of the process by Thomson [5]. However, it was not before 1961 that Dolder and co–workers [6] performed a milestone experiment in atomic collision physics. They measured the electron–impact ionisation of He^+–ions using a *crossed–beams* set–up.

Charge–exchange in collisions between ions and atoms was first observed in 1922 by Henderson [7] who did his experiments with α–particles from a radioactive sample. Since then and connected with the development of ion sources the field of ion–atom collisions boomed.

Compared to that and due to the experimental difficulties given later in the article, the field of charge–changing processes in collisions between two ions is relatively new. In 1969 the first experiments investigating the mutual neutralisation

$$A^+ + B^- \longrightarrow A^0 + B^0 \tag{12.1}$$

were performed by Ferguson and co–workers [8] using the so-called *Flowing Afterglow* technique. They pumped a gas through a pipe and ignited a discharge at the beginning of the tube. The ions produced here traveled through the pipe and were detected in a mass spectrometer. From the decreasing of the ion concentration ρ_i they could derive the decay constant α of the specific ion species:

$$\frac{d\rho_i}{dt} = -\alpha \cdot \rho_i^2 \tag{12.2}$$

In order to measure the rate coefficients for the recombination between positive and negative ions quantitatively, it is necessary to determine the absolute ion densities in the streaming plasma. In 1972 the *Flowing Afterglow/Langmuir Probe* technique was developed by Smith and Plumb [9]. With this method it was possible to measure the ion density in the plasma depending on the position of the Langmuir probe.

The investigation of ion–ion collisions in a plasma leads to high reaction rates due to the high densities which are possible because of the quasi–neutrality of the plasma. Here, as already mentioned before, the determination of the ion densities raises a problem and therefore quantitative measurements are very difficult to perform. Furthermore, different elementary processes can lead to the same final state and so it is almost impossible to identify the respective reaction channel. In order to perform an ion–ion collision experiment under well–

defined conditions, one has to prepare both reaction partners. One possibility is to store the ions in an ion trap, as for example a Penning [10] or a Paul trap [11]. By using an **E**lectron **B**eam **I**on **T**rap, for example the SuperEBIT [12] in Livermore, it is possible to store ions up to U^{92+}. The problem with using an ion trap is that, because of the Coulomb repulsion, it is not possible to store more than a few 1000 ions in the trap. Therefore, when the trap is used as a target for ion–ion collisions, experiments are only possible for reactions with a very large cross section due to the extremely low reaction rates.

The most promising method in this field of research is the intersection of two ion beams at an interaction angle α. When $\alpha = 0°$ it is called a *merged beams* experiment; when α is finite we talk about *crossed beams.*

2. Experimental Techniques

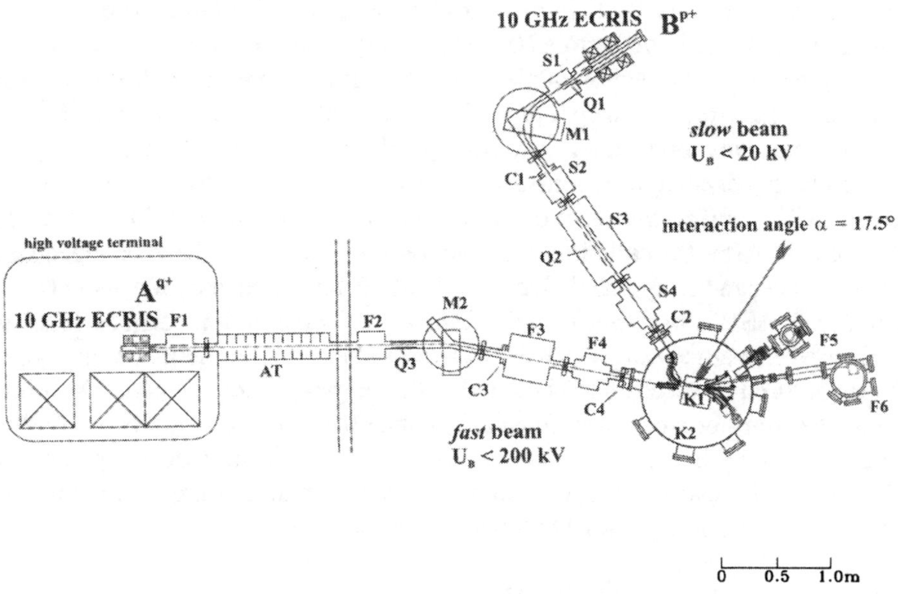

Figure 12.1. Schematic set–up of the Giessen ion–ion *crossed–beams* experiment.

The basic set–up of the Giessen ion–ion *crossed–beams* experiment is shown in Fig. 12.1. It consists of two separate beam lines which intersect in a reaction chamber under ultra–high vacuum conditions (p$\approx 10^{-11}$–10^{-10} mbar). In one beam line the ions are produced in a 10 GHz ECR ion source and are accelerated with voltages of up to 20 kV (in the following this beam line is called *slow beam*). After the extraction the ion beam is focused with an electrostatic quadrupole triplet (Q1). Since an ECR ion source produces a whole charge state distribution and not only ions in a single charge state, the analysing magnet M1 selects the desired charge–to–mass ratio of the ions. In the following, two pairs of slits (C1, C2) collimate the ion beam to typically $1.5 \cdot 1.5$ mm^2. These slits, together with the electrostatic quadrupole quadruplet (Q2), determine the ion beam properties at the interaction point. Here, one can choose between a *high–current* mode (high intensity with a focus at the interaction region, therefore a divergent beam) and a *high–brilliance* mode (parallel ion beam, hence low intensity).

In the second beam line (*fast beam*) the ions are generated in an identical 10 GHz ECR ion source on a high–voltage terminal. After the extraction from the ion source at typically 10 kV, the ions are further accelerated in an acceleration tube (AT) and focused with the electrostatic quadrupole triplet Q3. The charge–to–mass ratio is selected with the magnet M2 and the ion beam is then collimated with two pairs of slits (C3, C4). A blow–up of the interaction region is shown in Fig. 12.2. Shortly before the interaction both ion beams pass through electrostatic deflectors ES1, ES2 (*slow beam*) and ES3 (*fast beam*) and are guided to the interaction angle of 17.5°. Furthermore, the deflection cleans the ion beams from impurity ions which were produced in collisions with residual gas particles along their way through the beam lines. At the interaction point charge–exchange or ionisation takes place and due to the low momentum transfer during the reaction, the product ions remain within their parent ion beams. After the collision, the reaction products which have changed their charge state can be separated electrostatically from the primary ion beam (AFB: *fast beam*, ASB: *slow beam*). In the following, the parent ion beam currents are measured in Faraday cups (F1, F2). The product ions in the *fast beam* which have captured an electron are counted in the single–particle detector D1 (either a channeltron– or a position–sensitive channelplates detector). The ionised reaction products in the *slow beam* are deflected out of the reaction plane with the spherical capacitor SC, reducing the background, and are also counted in the single–particle detector D2 (channeltron detector).

3. Total Cross Sections

3.1 Definition of the Total Cross Section

The quantity that can be measured in an ion–ion *crossed–beams* experiment is the so–called cross section, which is related to the probability that a specific reaction takes place under certain circumstances. Let us assume we have an ion beam with N_i particles in the charge state i hitting a target of density n and a length l. The number of particles N_f in charge state f after passing the target is then given by:

$$N_f = N_i \cdot n \cdot l \cdot \sigma_t \tag{12.3}$$

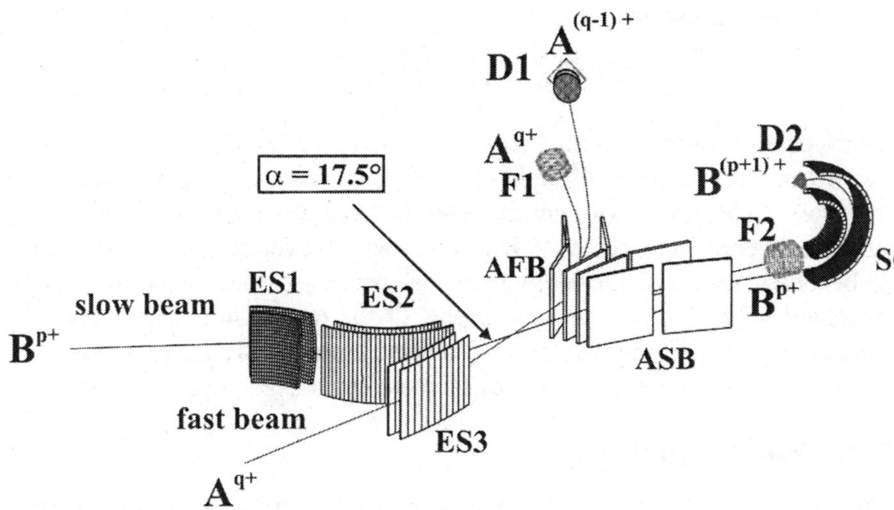

Figure 12.2. 3–dimensional view of the interaction region of the Giessen ion–ion *crossed beams* experiment.

Here σ_t denotes the total cross section of the specific reaction. In a classical picture σ_t is equivalent to an area centered around a target particle. An incoming particle has to travel through that area to undergo the reaction.

With well–collimated ion beams, a condition can be achieved where all incoming ions have the same parallel momentum vectors. Under this assumption the total cross section can be written as ([13]):

$$\sigma_t = \frac{v_1 v_2 \sin\alpha}{v_{rel}} \cdot \frac{q_1 q_2 F}{I_1 I_2} \cdot \frac{R}{\epsilon_1 \epsilon_2} \tag{12.4}$$

Here the first term is a kinematic factor which is determined by the absolute values of the ion velocities v_1 and v_2 in the laboratory system, the interaction angle α and the relative velocity v_{rel}:

$$v_{rel} = |\vec{v_1} - \vec{v_2}| = \sqrt{v_1^2 + v_2^2 - 2v_1 v_2 \cos\alpha} \tag{12.5}$$

Cross sections are usually given as a function of the relative velocity or of the center–of–mass energy which is given by:

$$E_{cm} = \frac{1}{2}\mu v_{rel} = \mu \left(\frac{E_1}{m_1} + \frac{E_2}{m_2} - 2\sqrt{\frac{E_1 E_2}{m_1 m_2}} \cos\alpha \right) \tag{12.6}$$

where μ is the reduced mass:

$$\mu = \frac{m_1 m_2}{m_1 + m_2} \tag{12.7}$$

The accessible center–of–mass energies calculated with equation (12.6) are shown in Fig. 12.3 for three different interaction angles α in the collision system He^+ + He^+. The energy of one ion beam is fixed at E_1=10 keV whereas E_2 is varied between 1 and 200 keV. It can be seen that for α=0 (*merged beams*) E_{cm} becomes zero and therefore very low center–of–mass energies can be investigated with high laboratory energies of the ion beams. These energies are necessary for an efficient beam transport through the beam lines of an experiment. Larger interaction angles increase the accessible E_{cm} energy.

3.2 The Form Factor

The second term in equation (12.4) consists of the ion charges q_1, q_2 and their ion currents I_1, I_2. The ion charges are separated by magnetic sector fields and the ion currents are measured in Faraday cups. F is the so–called form factor which describes the geometrical overlap of the two ion beams and is given by:

$$F = \frac{\int\limits_z J_1(z)dz \cdot \int\limits_z J_2(z)dz}{\int\limits_z J_1(z) \cdot J_2(z)dz} \tag{12.8}$$

where $J_1(z)$ and $J_2(z)$ are the one–dimensional current densities of the two ion beams in z–direction perpendicular to the collision plane. In order to determine the form factor, both ion beam profiles have to be measured simultaneously. This is done by moving a slotted plate through the interaction point of the two ion beams (slit width δz). At a fixed position of the slit z_k, the one–dimensional current densities $j_{1,2}(k)$ are measured in the Faraday cups. Assuming the slit width δz as well as the distance between two positions of the slit Δz to be small compared to the size of the ion beam, the one–dimensional current densities can be regarded as constant and the form factor can be written as:

$$F = \frac{\sum\limits_{k=1}^{n} j_1(z_k) \cdot \sum\limits_{k=1}^{n} j_2(z_k)}{\sum\limits_{k=1}^{n} j_1(z_k) \cdot j_2(z_k)} \cdot \Delta z \tag{12.9}$$

A typical form factor measurement is shown in Fig. 12.4. Because of the technique used to determine the beam overlap, the measurement of the form factor cannot be performed simultaneously with the detection of the reaction products but has to be done separately. Although the form factor is measured repeatedly during a cross section measurement, this parameter is the biggest

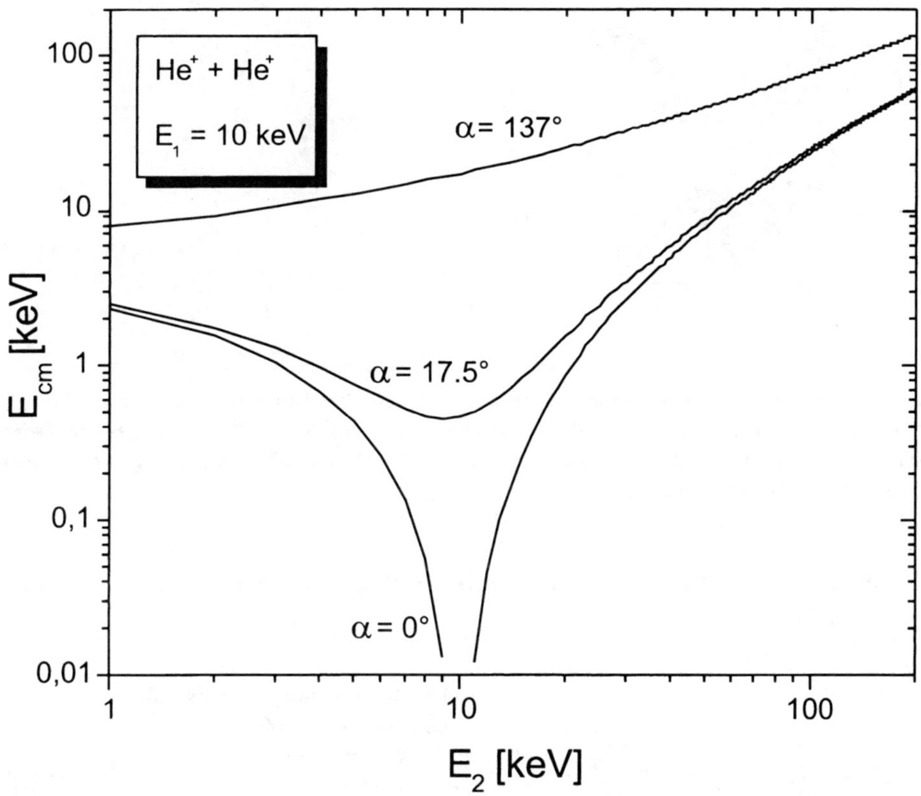

Figure 12.3. Accessible center–of–mass energies in an ion-ion collision experiment for different interaction angles α. The values were calculated for the He^+ + He^+ collision system. The energy of beam 1 is fixed at 10 keV whereas the energy of beam 2 is varied between 1 and 200 keV.

error source because it has to be assumed that it stays constant during the detection of the reaction products.

In the last term, R is the reaction rate of the collision process and ϵ_1, ϵ_2 denote the detection efficiencies of the product ions in the two detectors. The reaction rate is the most difficult quantity to be measured in an ion–ion experiment as will be shown in the following.

3.3 The Reaction Rate

In a mass–to–charge analysed ion beam all ions have the same charge state and therefore they repel each other causing the beam to widen up. A simple

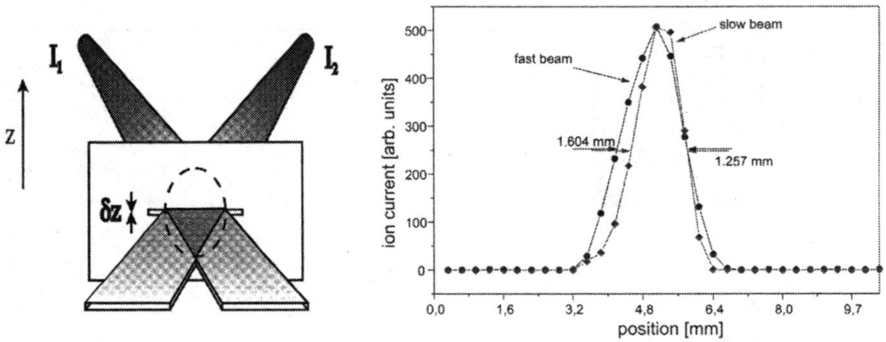

Figure 12.4. Measurement of the form factor. A slotted plate (slit width δz) is moved through the interaction point and the one–dimensional current densities of the two ion beams (with intensities I_1, I_2) are measured in Faraday cups. The right side of the picture shows a typical measurement of the form factor.

formula for the estimation of the ion beam diameter after a distance l can be found in [14]:

$$\frac{d}{d_0} \approx \sqrt{4.1 \cdot 10^{10} \frac{\sqrt{M/q} \cdot j \cdot l^2}{U^{3/2}}} \qquad (12.10)$$

l : distance along the axis [m]
d_0 : diameter at l=0
d : diameter at l
M : ion mass
q : ion charge
j : current density at l=0 [A/cm^2]
U : acceleration voltage [V]

Let us assume an ion beam of Ar^+–ions with an energy of 5 keV and a diameter d_0 of 5 mm. Allowing an angle of divergence of 1°, the beam diameter is 22 mm after a flight distance of 0.5 m and therefore $d/d_0 = 22mm/5mm = 4.4$. From equation (12.10) we can now calculate the maximum current density at l=0: $j \approx 100\mu A/cm^2$ which corresponds to a maximum ion current of 20 μA. The particle density in such an ion beam is $4 \cdot 10^7$ ions/cm^3 which is the same as a residual gas density at a pressure of 10^{-9} mbar at room temperature. In a real experiment the ion beams have to be even more collimated in order to reduce the background in the single–particle detectors which arises from scattered particles. Therefore the particle densities that can be used for an experiment are equal to the residual gas density of ultra–high vacuum at about 10^{-10} mbar and thus a lot of effort has to be put into the production of ultra–high vacuum conditions.

For this reason, ion beams are very dilute targets which is illustrated in Fig. 12.5. In this picture the light grey dots represent the residual gas particles

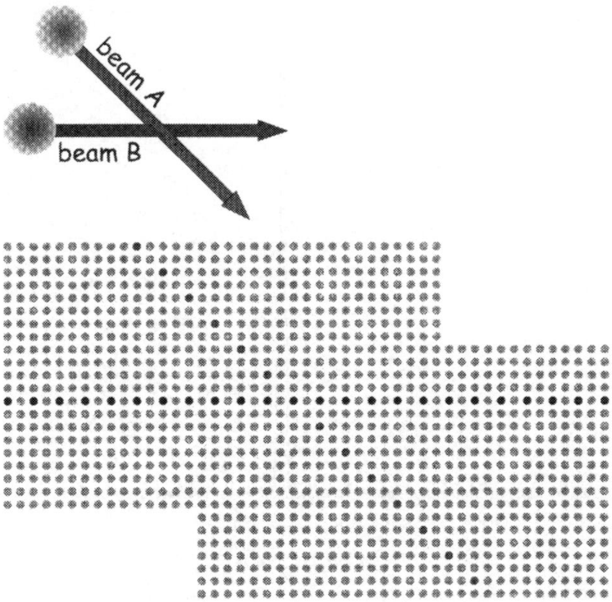

Figure 12.5. Illustration of the achievable particle density of collimated ion beams under ultra–high vacuum conditions. The light grey dots represent the residual gas particles at a pressure of 10^{-10} mbar whereas the two ion beams are shown as dark dots.

at a pressure of 10^{-10} mbar whereas the dark dots are the two ion beams. It can easily be seen that collisions of an ion with a residual gas molecule or atom are much more likely than the collision with another ion. In a typical experiment at a pressure of $5 \cdot 10^{-11}$ mbar and ion currents of about 50 nA, the background from residual gas collisions is about 3–4 orders of magnitude higher than the signal from the true ion–ion collisions. The identification of the reaction products will be explained in more detail in the following. Furthermore, the typical reaction rates that can be achieved are in the order of 1 per second (but they can also be much lower depending on the available ion currents from the ion sources and on the investigated cross section). These reaction products have to be separated very carefully from the incident ions which are about 10^{13} ions per second.

For the measurement of the reaction rate R (equation 12.4), two different methods can be used.

3.3.1 Reaction Rate for Charge–Exchange Reactions. In a charge–exchange reaction

$$A^{q+} + B^{p+} \rightarrow A^{(q-1)+} + B^{(p+1)+} \qquad (12.11)$$

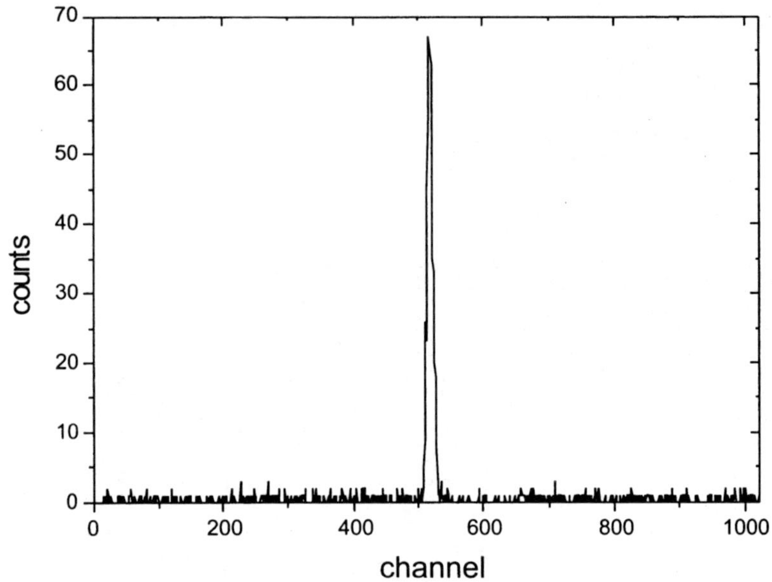

Figure 12.6. Coincidence spectrum for the measurement of the reaction rate for a charge–exchange reaction.

both collision partners alter their charge state. Here, the reaction products can be separated electrostatically from the parent ion beams and can be identified by using a coincidence method.

Reaction products from an ion-ion collision are formed simultaneously at the intersection point. After the interaction they travel to the single–particle detectors at fixed times t_{fb}, t_{sb}. Here, fb and sb denote the *fast* and *slow* beam. The reaction products in both beams reach the detectors at a fixed time difference $\Delta t = |t_{fb} - t_{sb}|$. On the other hand, there is no time correlation between product ions that were formed in collisions with residual gas particles. The reaction products from one ion beam are taken as a start signal for a time–to–amplitude converter which is stopped by the signals from the second beam. The true ion–ion signals can now be separated from the background by plotting the number of counts in certain time intervals Δt as a function of Δt.

Figure 12.6 shows a typical example of such a coincidence measurement where a sharp peak can be seen on a flat background. The reaction rate is then given as the area under the peak divided by the measuring time.

3.3.2 Reaction Rate for the Total Electron–Loss Process. This coincidence technique, however, cannot be applied for the measurement of an

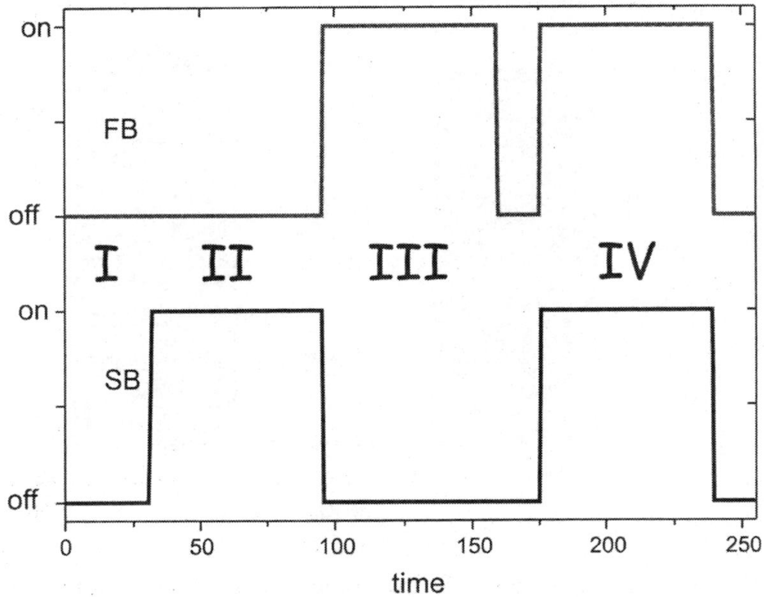

Figure 12.7. Beam–pulsing scheme for the measurement of the reaction rate for an electron–loss reaction.

ionisation process

$$A^{q+} + B^{p+} \rightarrow A^{q+} + B^{(p+1)+} + e^- \qquad (12.12)$$

where only one of the collision partners alters its charge state. Here, it is possible to measure the total electron–loss cross section which is the sum of reactions 12.11 and 12.12. This measurement is performed by using a beam–pulsing technique as shown in Fig. 12.7.

In the first phase (I), both beams are switched off. Here, only the electronic noise of the detectors is determined. In phase II, only the *slow beam* is switched on and the background of charge–changing collisions with residual gas particles is measured. In phase III, where only the *fast beam* is on, we count the background of this beam in the detector of the *slow beam*. In phase IV, eventually both beams are switched on and now the true ion–ion signals are additionally detected. The reaction rate R_{loss} is then given by

$$R_{loss} = R_{IV} - R_{III} - R_{II} + R_I \qquad (12.13)$$

Fig.12.8 shows a typical example for this kind of measurement. It can be seen that in phase I there is only a small electronic background but a high rate of

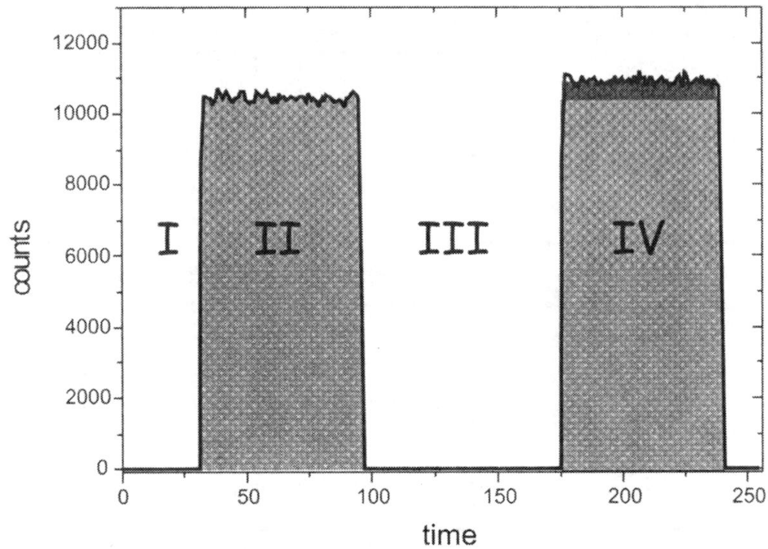

Figure 12.8. Beam–pulsing spectrum for the measurement of the reaction rate for a total electron–loss reaction.

ion–residual gas collisions in phase II. The background of the *fast beam* in the detector of the *slow beam* is again small because this detector is far away from the interaction region (detector D2 in Fig.12.2). In phase IV, where both beams are switched on, there is a small additional signal on top of the background (highlighted in dark grey) which is due to the true ion-ion collisions. This figure also shows the main experimental difficulty of this technique. Since the reaction rate for the total electron loss is given as the difference of two almost equal reaction rates (see eq.12.13), a long measuring time is necessary in order to achieve a reasonable statistical error.

3.4 Results for Total Cross Sections

A typical result of a total cross section measurement is shown in Fig. 12.9 for the reactions

$$B^{2+} + He^{2+} \longrightarrow B^{3+} + He^{+} (charge\, exchange) \qquad (12.14)$$

$$B^{2+} + He^{2+} \longrightarrow B^{3+} + He^{2+} + e^{-} (ionisation) \qquad (12.15)$$

obtained by Skiera et al. [15]. This collision system is a quasi–one–electron system where the active 2s–electron is transferred. The 2 K–shell electrons are bound too strongly in order to take part in the collision. Their influence is a

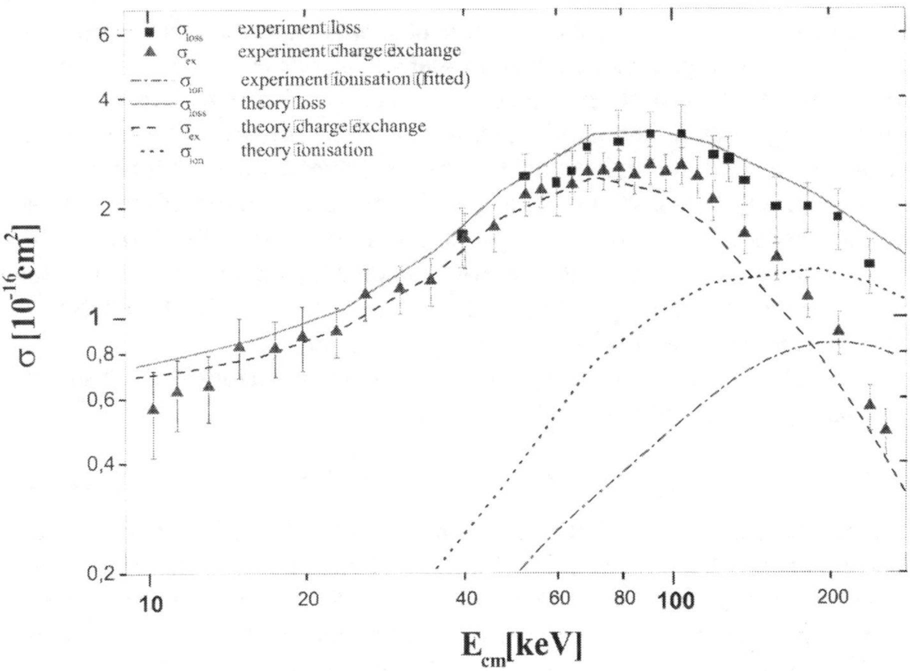

Figure 12.9. Measured and theoretical cross sections for charge exchange and electron loss in collisions between B^{2+} and He^{2+} ions as a function of the center–of–mass energy [15]. The experimental ionisation cross section was obtained by subtraction of two fits through the measured values.

shielding of the nuclear Coulomb potential. This system is therefore an ideal testing ground for existing theories.

In the figure, the black squares are the result of the total electron–loss measurement ($\sigma_{ion} + \sigma_{exch}$) obtained with the beam–pulsing technique which has been described in section 12.3.3. The experimental charge–exchange cross section is plotted as the triangles. The dash–dotted line is the experimental result for the ionisation cross section which has been obtained by fitting the loss and charge–exchange cross sections and subtracting them. In this picture one can see the typical behaviour of cross sections in ion–ion collisions. At low center–of–mass energies, the charge–exchange cross section dominates, whereas the ionisation cross section becomes important at higher energies.

Also shown in Fig. 12.9 are theoretical calculations using the Basis–Generator–Method (BGM) which is a special method to solve the time–dependent Schrödinger equation in the form of coupled–channel equations [16] [17]. The solid line represents the calculation for the total electron–loss cross section and shows an excellent agreement with the experimental data over the whole range of center–of–mass energies on an absolute scale. The dashed curve is the theoretical result for the charge–exchange cross section. Whereas at low center–of–mass energies there is a good agreement between theory and experiment, the calculation seems to underestimate the experimental values at higher energies. This deviation is due to problems with the representation of the moving projectile orbitals in the used BGM basis at high energies [18]. Since in the theory the ionisation cross section is also obtained by subtraction of the total loss and the charge–exchange cross sections, there is also a deviation between the theoretical and experimental ionisation cross sections.

Another example for total cross sections in ion–ion collisions shall be shown for the charge–exchange reaction

$$He^{2+} + He^{+} \longrightarrow He^{+} + He^{2+} \qquad (12.16)$$

which differs essentially from the above–mentioned $B^{2+} + He^{2+}$ system. The collision system 12.16 is symmetric (identical nuclei) and the electron is transferred resonantly (energy defect Q=0, see section 12.4.2). Furthermore it is the most fundamental ion–ion collision system (only one electron and two nuclei) in which the collision partners repulse each other by the Coulomb interaction in the entry as well as in the exit channel.

The results of various experiments and theories are shown in Fig. 12.10 for the total charge–exchange cross section.

Comparing the experimental data of Peart et al. (open triangles) [22] and Jognaux et al. (solid triangles) [23] with the theories of Forster et al. (open squares) [19], Bardsley et al. (solid diamonds) [20] and Dickinson et al. (dashed line) [21], a substantial discrepancy can be observed. Whereas in the high–energy region experiments and theory merge, they differ at lower center–of–mass energies. This discrepancy lead to the assumption that large scattering angles might play an important role in the collision process. If the scattering angle becomes larger than the acceptance of the used detectors, a part of the reaction products will not reach the detector and therefore an apparently smaller cross section will be measured. The particle loss due to large scattering angles also would explain the fact that the discrepancy between experiment and theory becomes larger at lower energies. Here, due to the lower velocity, the interaction time between the ions becomes larger and thus leads to larger scattering angles.

In order to clarify this situation, a measurement of the scattering distribution of the product ions has been performed. As it is shown in section 12.4.4, there are large scattering angles involved in the collision process. An analysis of the acceptances used in the above–mentioned experiments showed that indeed only a fraction of the particles has been counted. Taking the losses into account, Melchert et al. [24] obtained the experimental data also shown as solid circles in Fig. 12.10. These data confirm the existing theories.

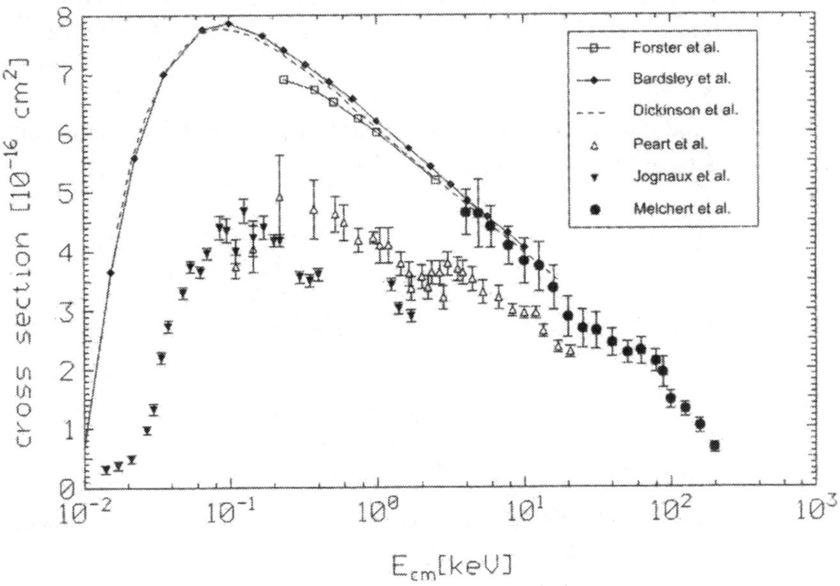

Figure 12.10. Charge–exchange cross sections for the reaction $He^{2+} + He^{+} \longrightarrow He^{+} + He^{2+}$ (open squares: theory by Forster et al. [19]; solid diamonds: theory by Bardsley et al. [20]; dashed line: theory by Dickinson et al. [21]; open triangles: experiment by Peart et al. [22]; solid triangles: experiment by Jognaux et al. [23]; solid circles: experiment by Melchert et al. [24].)

Apart from the experimental point of view, the measurement of scattering distributions allows the determination of *angular–differential* cross sections. These give a more detailed insight into the collision processes and also allow the test of existing theories on a more sophisticated level.

4. Angular–Differential Cross Sections

4.1 Definition of the Angular–Differential Cross Section

When a collision process is described with respect to the solid–angle element $d\Omega$ into which the reaction products are scattered, the concept of the *angular-differential* cross section is used. The *angular-differential* cross section $\left(\frac{d\sigma}{d\Omega}\right)_\theta$ is given as the ratio between the number of projectiles which have been scattered into a solid–angle element $d\Omega$ and the current density j_0 of the incoming projectiles referring to one target particle (see for example [25]):

$$\left(\frac{d\sigma}{d\Omega}\right)_\theta = \frac{number\,of\,projectiles\,scattered\,into\,d\Omega\,per\,time\,interval}{(current\,density\,j_0\,of\,incoming\,projectiles)\cdot(number\,of\,target\,particles)} \tag{12.17}$$

In Fig.12.11 the collision process of a parallel particle beam with current density j_0 and a target is shown schematically.

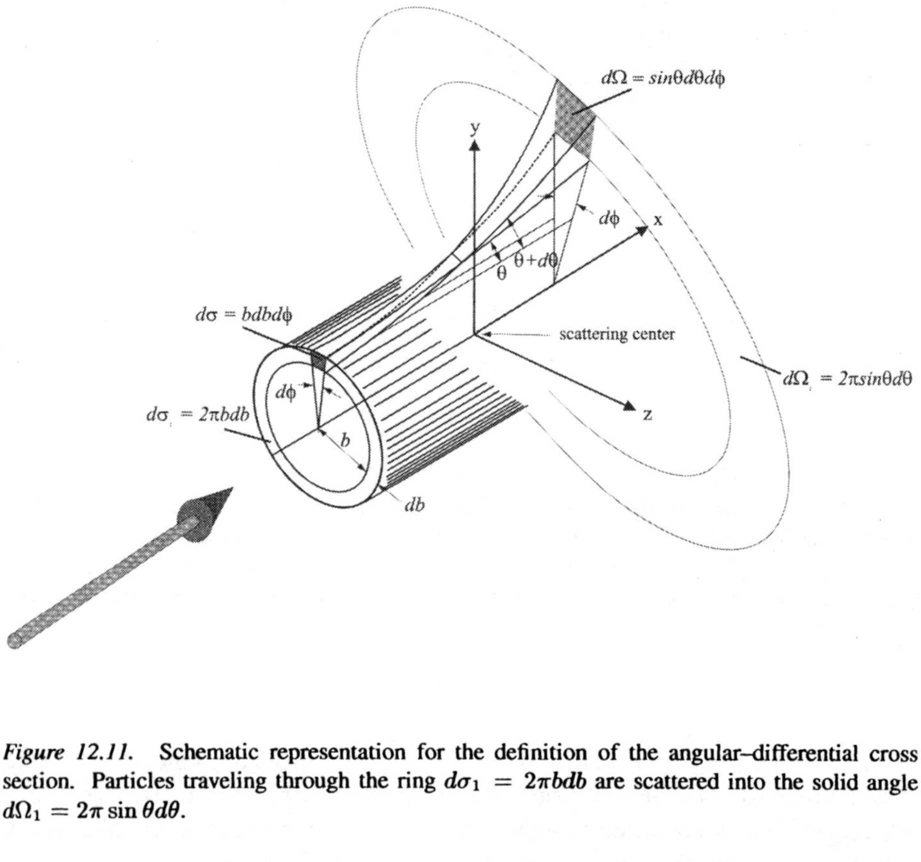

Figure 12.11. Schematic representation for the definition of the angular–differential cross section. Particles traveling through the ring $d\sigma_1 = 2\pi bdb$ are scattered into the solid angle $d\Omega_1 = 2\pi\sin\theta d\theta$.

The target is at rest on the axis of the incoming beam and therefore — unlike in ion–ion collisions with *crossed beams*— the laboratory and center–of–mass velocities are parallel. For unpolarized projectiles and target particles, this leads to a scattering distribution with rotational symmetry referring to ϕ. Considering the conservation of the number of particles, the relation between the impact parameter b and the scattering angle θ can be deduced. The $number\,of\,projectiles\,scattered\,into\,d\Omega\,per\,time\,interval$ is therefore equal to the number of incoming projectiles per time interval traveling through the ring $d\sigma_1 = 2\pi bdb$. Together with eq. 12.17 one obtains for the θ–dependence of the angular–differential cross section:

$$j_0 \cdot 2\pi \cdot bdb = j_0 \cdot 2\pi \cdot \sin\theta \cdot d\theta \cdot \left(\frac{d\sigma}{d\Omega}\right)_\theta \tag{12.18}$$

$$\left(\frac{d\sigma}{d\Omega}\right)_\theta = \frac{b}{\sin\theta} \cdot \left|\frac{db}{d\theta}\right| \tag{12.19}$$

The absolute value in eq. 12.19 is due to the fact that the cross section, by definition, can only have positive values.

The total cross section of a reaction σ_t is then given by the integration of the angular–differential cross section

$$\sigma_t = \int_{\phi=0}^{2\pi} \int_{\theta=0}^{\pi} \left(\frac{d\sigma}{d\Omega}\right)_\theta \sin\theta d\theta d\phi \tag{12.20}$$

and for collision processes with rotational symmetry, integration over the azimuth angle ϕ leads to:

$$\sigma_t = 2\pi \cdot \int_0^{\pi} \left(\frac{d\sigma}{d\Omega}\right)_\theta \sin\theta d\theta \tag{12.21}$$

Unlike the total cross section, the *angular–differential* cross section depends on the choice of the coordinate system. However, in order to obtain results that are independent of the used experimental set–up and which can be compared with theories or other experiments, it is necessary to look at the cross sections in the center–of–mass system.

4.2 Transformation Laboratory – Center–of–Mass System

From the conservation of the number of particles and from definition 12.17 of the *angular–differential* cross section follows that the number of particles scattered into the solid angle $d\Omega_{CM}$ with the coordinates (θ_{CM}, ϕ_{CM}) in the center–of–mass system, must be equal to the number of particles scattered into the solid angle $d\Omega_L$ with the coordinates (θ_L, ϕ_L) in the laboratory system. Therefore, we have the following relation between *angular–differential* cross sections in the laboratory and center–of–mass system:

$$\left(\frac{d\sigma}{d\Omega}\right)_L (\theta_L, \phi_L) \cdot d\Omega_L = \left(\frac{d\sigma}{d\Omega}\right)_{CM} (\theta_{CM}, \phi_{CM}) \cdot d\Omega_{CM} \tag{12.22}$$

or

$$\left(\frac{d\sigma}{d\Omega}\right)_{CM} = \left(\frac{d\sigma}{d\Omega}\right)_L \frac{d\Omega_L}{d\Omega_{CM}} \tag{12.23}$$

Transforming the angular–differential cross sections, which are measured in the laboratory system, into the center–of–mass system, this relation 12.23 has to be taken into account.

The dynamics and the microscopic behaviour of an ion–ion collision process has to be treated quantum–mechanically. However, the classical conservation of energy and momentum has to be fulfilled asymptotically and therefore the reaction kinematics in ion–ion collisions can be described under the assumption of an infinitesimal small collision time.

The collision process in a *crossed–beams* experiment is illustrated schematically in Fig. 12.12. Two ions with masses $m1$ and $m2$ and laboratory velocities $\vec{v}_1$ and $\vec{v}_2$ prior to the collision are made to intersect at an interaction angle α. The whole system moves with a constant center–of–mass velocity $\vec{v}_{CM}$; before the collision the ion velocities in the center–of–mass system are $\vec{s}_1$ and $\vec{s}_2$, respectively.

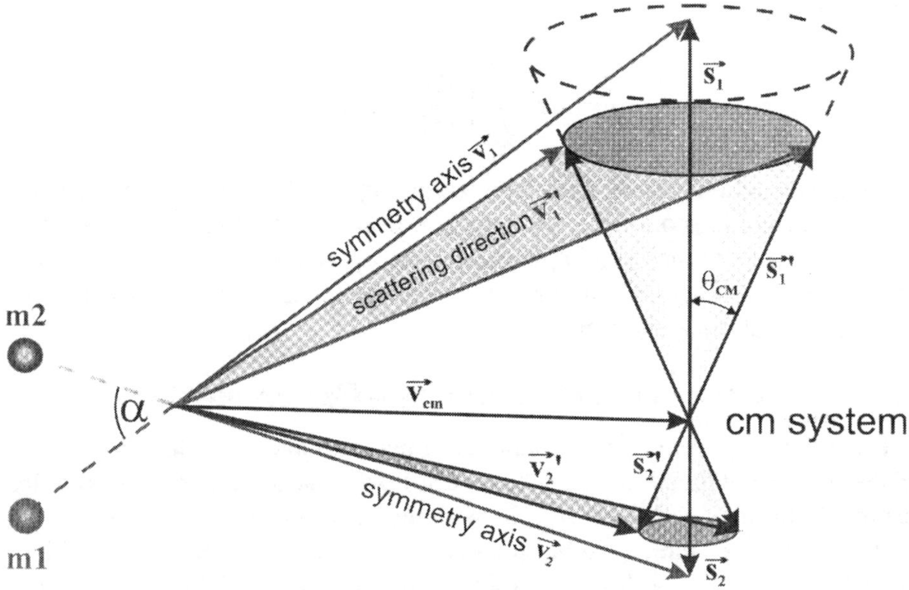

Figure 12.12. Schematic illustration of the collision process in a *crossed–beams* experiment.

During the collision the two ions are scattered with the scattering angle θ_{CM} in the center–of–mass system and their center–of–mass velocities $\vec{s}_1\,'$ and $\vec{s}_2\,'$ open a cone of reaction products which has rotational symmetry.

Having a purely *elastic* scattering process, the absolute values of the center–of–mass velocities after the collision $\vec{s}_1\,'$ and $\vec{s}_2\,'$ are equal to $\vec{s}_1$ and $\vec{s}_2$. In this case we obtain the dashed cone of reaction products for beam 1 in Fig. 12.12. The same is true for *resonant* charge–exchange reactions, where the binding energy of the transferred electron before and after the collision is the same. However, if the binding energy of the electron is different after the collision, the potential energy of the system has changed during the collision and we thus speak of *inelastic* or *non–resonant* processes. The change in the potential energy is described by the energy defect ΔE (or Q–value) of the reaction. Due to the conservation of energy, a change of the potential energy causes also a change of the kinetic energy of the reaction products. Therefore, when an *inelastic* collision occurs ($\Delta E \neq 0$), the absolute values of the center–of–mass velocities $\vec{s}_1{}'$ and $\vec{s}_2{}'$ change as well after the collision. This is shown in Fig. 12.12 for the case of an endothermic reaction ($\Delta E = E'_{kin} - E_{kin} = E_{pot} - E'_{pot} < 0 \implies s'_i < s_i$). Here, the vectors $\vec{s}_1{}'$ and $\vec{s}_2{}'$ are shorter than $\vec{s}_1$ and $\vec{s}_2$.

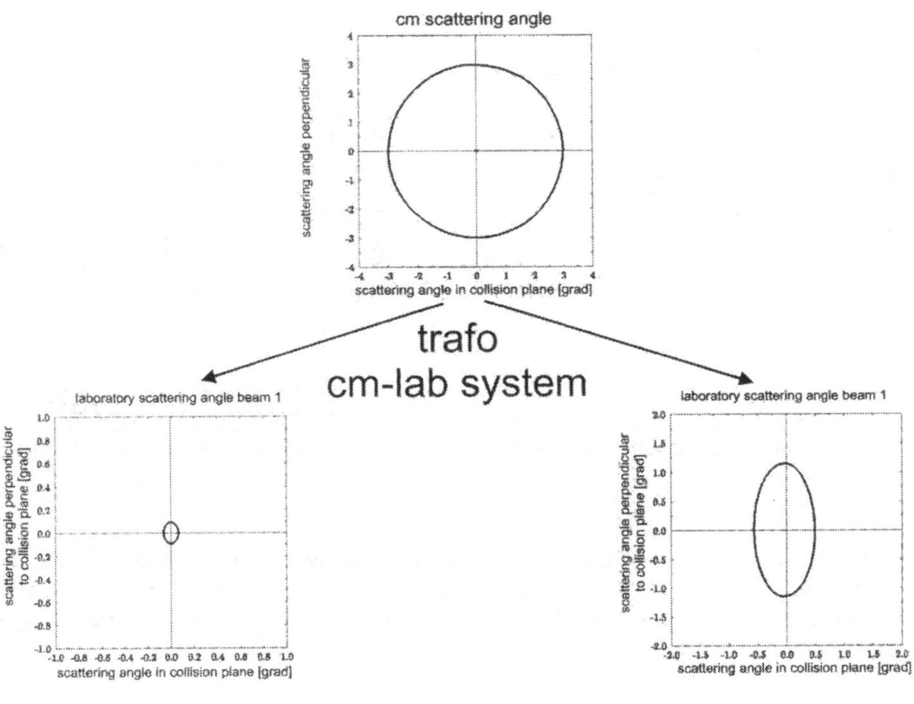

Figure 12.13. Example of the transformation of a scattering angle $\theta_{CM} = 3°$ from the center–of–mass system into the laboratory system of the two ion beams. The collision system is $Ar^{6+} + He^{+}$ at a center–of–mass energy of 1.30 keV with laboratory energies of $E_1 = 125\ keV$ and $E_2 = 8\ keV$, respectively.

The detection of the ion–ion reaction products happens in the laboratory system. This means that the cones of the reaction products in the center–of–mass system are observed from the respective axes of symmetry, which are the laboratory velocities $\vec{v}_1$ and $\vec{v}_2$ prior to the collision. Therefore, the scattering distributions observed in the laboratory system result from conic sections through the rotationally symmetric reaction cones in the center–of–mass system. This destroys the rotational symmetry in the laboratory system and the distributions of the scattered reaction products appear as elliptical structures. The size and the ratio of the semiaxes of the ellipses vary with different combinations of laboratory energies and masses.

Fig. 12.13 shows an example of such a transformation for the collision system $Ar^{6+} + He^{+}$ at a center–of–mass energy of 1.30 keV with laboratory energies of $E_1 = 125\ keV$ and $E_2 = 8\ keV$, respectively. The scattering angle in the center–of–mass system was chosen to be $\theta_{CM} = 3°$. This transformation was performed for an energy defect of $\Delta E = 0$.

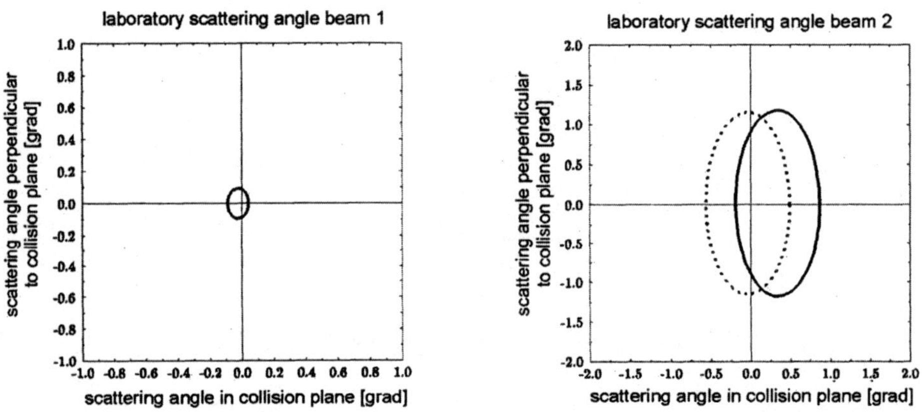

Figure 12.14. Comparison of scattering distributions in the laboratory system for $\Delta E = 0$ (dashed ellipse) and $\Delta E = +50\ eV$ (solid ellipse). The shift of the ellipse which is due to the energy defect is stronger in beam 2 and practically not visible in beam 1 because of its higher mass and energy.

Choosing an arbitrary energy defect of $\Delta E = +50\ eV$ (exothermic reaction) and performing the same transformation leads to a horizontal shift of the elliptical scattering distribution as shown in Fig. 12.14. This shift is stronger for beam 1 (He^{+}) and, in this representation, practically not visible for beam 2 (Ar^{6+}) due to the higher mass and energy.

A precise measurement of the shift of the reaction products from the axes of symmetry (laboratory velocities prior to the collision) therefore allows the determination of the energy defect of a specific reaction. This has been achieved for the first time by Pfeiffer et al. [26] for the reaction

$$Ar^{6+} + He^{+}(1s) \longrightarrow Ar^{5+}(nl) + He^{2+} + Q \tag{12.24}$$

In this collision system the $1s$–electron of the He^{+}–ion can be captured into different states (nl) of the Ar^{5+}–ion, leading to different Q–values. The determination of these Q–values then allows the performance of a state–selective measurement of the electron–capture process.

4.3 Measurement of the Angular–Differential Cross Section

In 1997, the first *angular–differential* cross sections in ion–ion collisions have been measured [27]. The measurement of *angular–differential* cross sections requires not only the time information as obtained from the coincidence technique described in section 12.3.3 but also the spatial distribution of the reaction products. This is achieved by the use of position–sensitive detectors. The principle method of the measurement of a scattering distribution is shown in Fig. 12.15. Again, as described above, the product ions in one ion beam start a time–to–amplitude converter which is stopped by the products from the second beam (one of the detectors in now position–sensitive). This leads to a coincidence spectrum shown on top of the figure. The dark grey area under the coincidence peak contains signals from both ion–ion collisions and ion–residual gas collisions. These events are sorted into a matrix (left side of the figure). The light grey regions have no time correlation and contain only events from collisions with residual gas particles. These are sorted into another matrix (right side of the figure). The subtraction of these two matrices, after suitable normalization, gives the true ion–ion scattering distribution.

In order to extract an *angular–differential* cross section from such a scattering distribution, first of all a transformation into the center–of–mass system has to be carried out (section 12.4.2). Furthermore, one has to take into account that a real ion beam is not a mathematical line but has a finite beam profile. This broadens the observed scattering distribution and smears out possible structures. However, the experimental data can be deconvoluted with respect to the primary beam profile using e.g. the Bayesian deconvolution [28] or Fourier methods [29] .

4.4 Results for Angular–Differential Cross Sections

A typical example of the measurement of a scattering distribution is shown in Fig. 12.16 as it was obtained for the resonant charge–exchange reaction

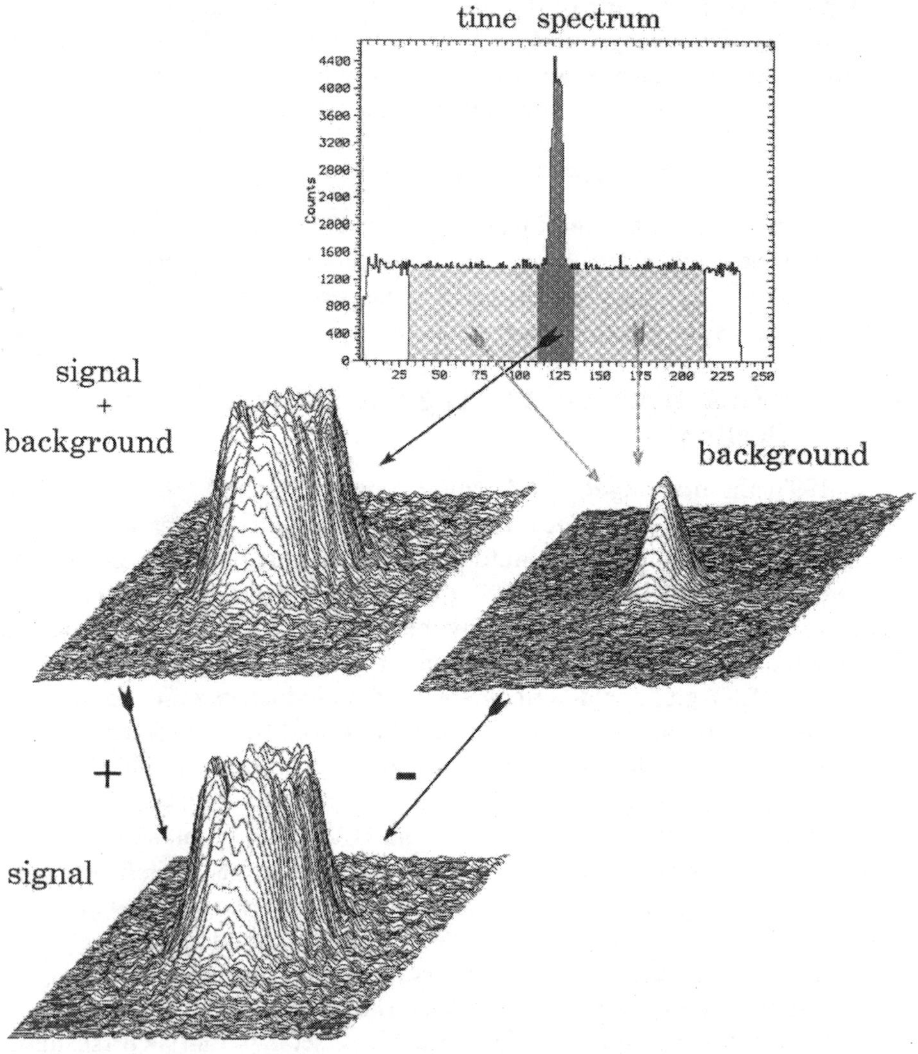

Figure 12.15. Determination of the scattering distribution of the reaction products. The top of the figure shows the time spectrum as already described in section 12.3.3. The dark grey area contains signals resulting from both ion–ion and ion–residual gas collisions; these are sorted into one matrix. Signals in the light grey regions are only due to ion–residual gas collisions and are sorted into a background matrix. The subtraction of both matrices leads to the scattering distribution of the reaction products.

$$He^{2+} + He^{+} \longrightarrow He^{+} + He^{2+} \tag{12.25}$$

measured by Krüdener [27].

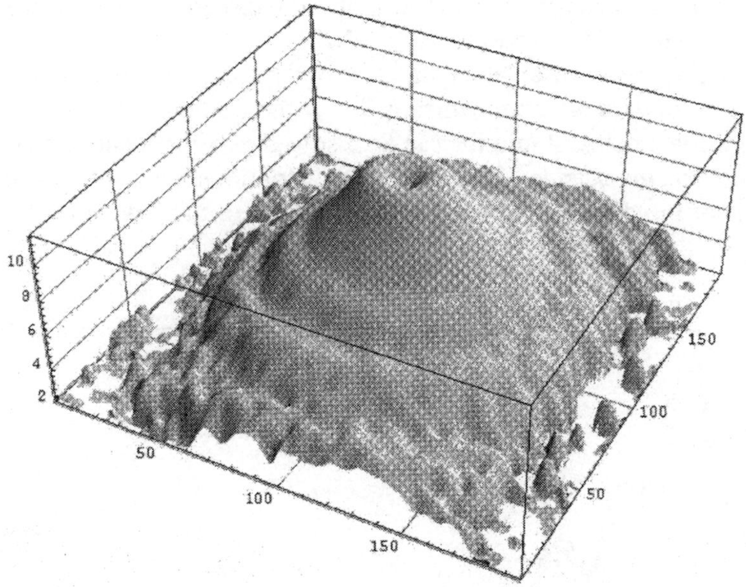

Figure 12.16. Scattering distribution of the He^+ reaction products measured for the charge-exchange reaction $He^{2+} + He^+ \longrightarrow He^+ + He^{2+}$ measured in [27].

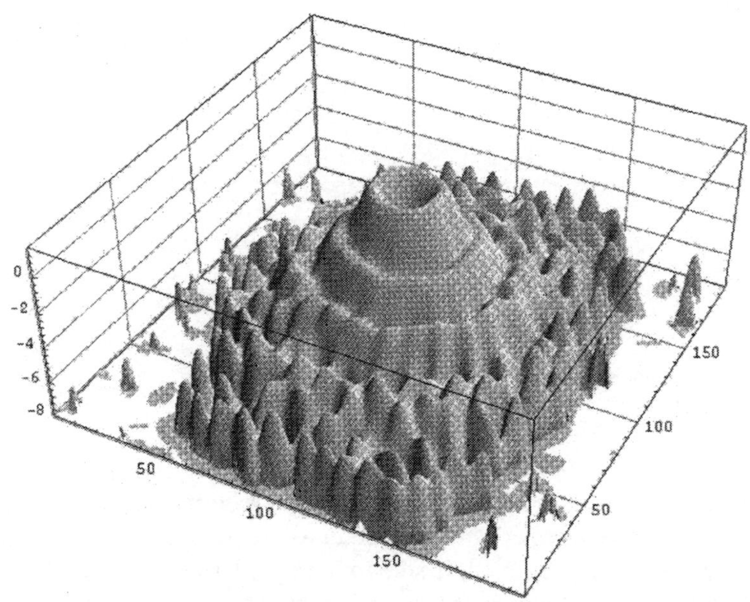

Figure 12.17. Deconvoluted scattering distribution of the He^+ reaction products measured for the charge–exchange reaction $He^{2+} + He^+ \longrightarrow He^+ + He^{2+}$ measured in [27].

The result of the deconvolution of this scattering distribution with respect to the primary beam profile is shown in Fig. 12.17.

One can clearly see an oscillatory structure of the scattering distribution with a minimum in forward direction which is due to the Coulomb repulsion of the ions. In addition, another 4 minima can be resolved experimentally (the peak–like structures on the outside of the scattering distribution are due to numerical fluctuations of the deconvolution algorithm because of the low count rates in this region).

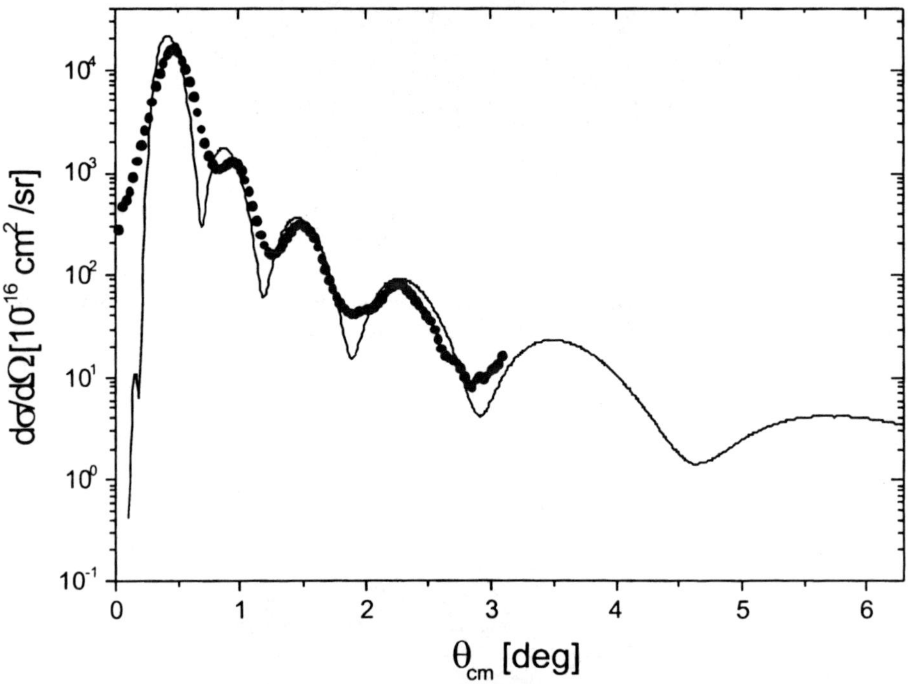

Figure 12.18. Angular–differential cross section for the reaction $He^{2+} + He^{+} \longrightarrow He^{+} + He^{2+}$ at a center–of–mass energy of 2.5 keV [27].

In order to obtain an *angular–differential* cross section, this scattering distribution has to be transformed into the center–of–mass system and integrated over the azimuth angle. The result is shown in Fig. 12.18 for a center–of–mass energy of 2.5 keV. Also shown in the figure is a quantum–mechanical theory by Uskov and Presnyakov [27] which is based on a partial–wave analysis using a molecular basis. Theory and experiment are compared on an absolute scale and the astonishing agreement allows the conclusion that one–electron systems are fairly well understood. It has to be mentioned, however, that this is absolutely not true for heavier systems with more electrons. Here, theory and experiment sometimes differ by orders of magnitude.

The oscillations in the *angular–differential* cross section can be interpreted as a quantum–mechanical interference between *gerade* and *ungerade* electronic states of the ionic quasi–molecule which is formed during the collision. This behaviour is illustrated in Fig. 12.19 which shows schematically the potential energy curves as a function of the internuclear distance R.

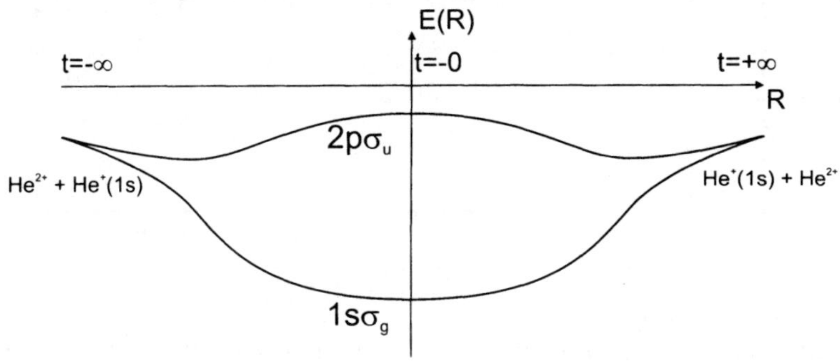

Figure 12.19. Schematic potential energy curves of the σ–orbitals during the collision as a function of the internuclear distance R.

At t=-∞, the collision system consists of two separate ions (He^{2+} and He^{+} in the 1s–state). During the collision process these two ions approach each other forming a quasi–molecular ion (He_2^{3+}) at R=0. In this quasi–molecule there are *gerade* ($1s\sigma_g$) and *ungerade* ($2p\sigma_u$) orbitals. After the electron is transferred from the He^{+} to the He^{2+}–ion, the quasi–molecule separates again and at t=+∞ we have again two ions $He^{+}(1s)$ and He^{2+}. During the collision, the electron can travel on each of the potential curves, both leading to the same final state. Since, at the end, we do not know which way the electron took, quantum–mechanical interference is observed. Quantitatively the charge–exchange reaction has been described by Lichten [30] using an impact parameter approximation with molecular orbitals. According to this theory the electron capture probability P_0 is given by

$$P_0 = \sin^2\left(\frac{1}{2v}\cdot\int\limits_{-\infty}^{+\infty}\left(E_{2p\sigma_u} - E_{1s\sigma_g}\right) ds\right) \tag{12.26}$$

The observed oscillations are due to the $\sin^2$ term in eq. 12.26.

In Fig. 12.20 another *angular–differential* cross section is shown for the same reaction at a lower center–of–mass energy (E_{CM}=0.5 keV). A smaller collision

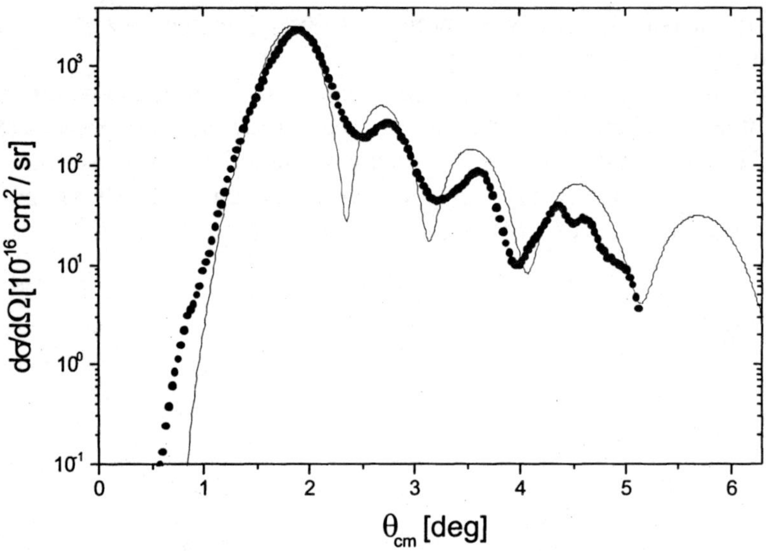

Figure 12.20. Angular–differential cross section for the reaction $He^{2+} + He^{+} \longrightarrow He^{+} + He^{2+}$ at a center–of–mass energy of 0.5 keV [27].

energy also means smaller velocity and therefore the interaction time of the ions in the strong Coulomb field increases. This leads to larger scattering angles and explains the discrepancy between theories and experiments which was shown in Fig. 12.10.

References

[1] A. Dalgarno, In: *Physics of Ion–Ion and Electron–Ion Collisions*, F. Brouillard and J. W. McGowan, eds., 1. Plenum Press, New York (1983)

[2] P. Lenard, Wied. Ann. **51**, 225 (1894)

[3] P. Lenard, Ann. d. Phys. **15**, 485 (1904)

[4] S. Bloch, Ann. d. Phys. **38**, 559 (1912)

[5] J. J. Thomson, Phil. Mag. **23**, 449 (1912)

[6] K. T. Dolder, M. F. A. Harrison and P. C. Thoneman, Proc. Roy. Soc. **A264**, 367 (1961)

[7] G. H. Henderson, Proc. Roy. Soc. **A102**, 496 (1922)

[8] E. E. Ferguson, F. C. Fehsenfeld and D. L. Albritton, In: *Gas Phase Ion Chemistry Vol.1*, M. T. Bowers, ed., 45. Academic Press, New York (1975)

[9] D. Smith and I. C. Plumb, J. Phys. D **5**, 1226 (1972)

[10] F. M. Penning, Physica **4**, 71 (1937)

[11] W. Paul, H. P. Reinhardt and U. von Zahn, Zeitschr. f. Physik **152**, 143 (1958)

[12] D. A. Knapp, R. E. Marrs, S. R. Elliot, E. W. Magee and R. Zasadzinski, Nucl. Inst. & Meth. **A334**, 305 (1993)

[13] F. Brouillard and W. Claeys, In: *Physics of Ion–Ion and Electron–Ion Collisions*, F. Brouillard and J. W. McGowan, eds., 415. Plenum Press, New York (1983)

[14] M. von Ardenne, *Tabellen zur angewandten Physik, 1. Band.* Deutscher Verlag der Wissenschaften, Berlin (1962)

[15] D. Skiera, R. Trassl, K. Huber, H. Bräuning, E. Salzborn, M. Keim, A. Achenbach, T. Kirchner, H. J. Lüdde and R. M. Dreizler, Physica Scripta **T92**, 423 (2001)

[16] H. J. Lüdde, A. Henne, T. Kirchner and R. M. Dreizler, J. Phys. B **29**, 4423 (1996)

[17] O. J. Kroneisen, H. J. Lüdde, T. Kirchner and R. M. Dreizler, J. Phys. A **32**, 2141 (1999)

[18] M. Keim, private communication (2000)

[19] C. Forster, R. Shingal, D. R. Flower, B. H. Brandsen and A. S. Dickinson, J. Phys. B **21**, 3941 (1988)

[20] J. N. Bardsley, P. Gangopadhyay and B. M. Penetrante, Phys. Rev. A **40**, 2742 (1989)

[21] A. S. Dickinson and D. J. W. Hardie, J. Phys. B **12**, 4147 (1979)

[22] B. Peart and K. Dolder, J. Phys. B **12**, 4155 (1979)

[23] A. Jognaux, F. Brouillard and S. Szücs, J. Phys. B **11**, L669 (1978)

[24] F. Melchert, S. Krüdener, R. Schulze, S. Pfaff and E. Salzborn, J. Phys. B **28**, L355 (1995)

[25] D. J. Blochinzew, *Grundlagen der Quantenmechanik.* Verlag Harri Deutsch, Frankfurt a.M. (1977)

[26] A. Pfeiffer, F. Melchert, K. Diemar, K. Huber and E. Salzborn, Physica Scripta **T80**, 95 (1999)

[27] S. Krüdener, F. Melchert, K. v. Diemar, A. Pfeiffer, K. Huber, E. Salzborn, D. B. Uskov and L. P. Presnyakov, Phys. Rev. Lett. **79**, 1002 (1997)

[28] W. H. Richardson, J. Opt. Soc. Am. **62**, 55 (1972)

[29] S. Krüdener, In: *Proceedings of the XIXth International Conference on the Physics of Electronic and Atomic Collisions*, L. J. Dube, J. B. A. Mitchell, J. W. McConkey and C. E. Brion, eds., 677. AIP Conf. Proc. 360 (1995)

[30] W. Lichten, Phys. Rev. **131**, 229 (1963)

Index

Lightning Source UK Ltd.
Milton Keynes UK
UKOW04f0005151113

221082UK00003B/16/P

9 789048 164097